Texts in Applied Mathematics 54

Texts in Applied Mathematics

(continued after index)

Jan S. Hesthaven   Tim Warburton

Nodal Discontinuous Galerkin Methods

Algorithms, Analysis, and Applications

 Springer

Jan S. Hesthaven
Division of Applied Mathematics
Brown University
Providence, RI 02912
USA
jan.hesthaven@brown.edu

Tim Warburton
Department of Computational
 and Applied Mathematics
Rice University
Houston, TX 77005
USA
Tim.Warburton@Rice.edu

Series Editors
J.E. Marsden
Control and Dynamic Systems, 107-81
California Institute of Technology
Pasadena, CA 91125
USA

L. Sirovich
Division of Applied Mathematics
Brown University
Providence, RI 02912
USA

S.S. Antman
Department of Mathematics
and
Institute for Physical Science
 and Technology
University of Maryland
College Park, MD 20742-4015
USA
ssa@math.umd.edu

ISBN: 978-1-4419-2463-6   e-ISBN: 978-0-387-72067-8
DOI: 10.1007/978-0-387-72067-8

Mathematics Subject Classification (2000): 65M60, 65M70, 65N35

Printed on acid-free paper

9 8 7 6 5 4 3 2 1

springer.com

To Aminia and Sarah

Series Preface

Mathematics is playing an ever more important role in the physical and biological sciences, provoking a blurring of boundaries between scientific disciplines and a resurgence of interest in the modern as well as the classical techniques of applied mathematics. This renewal of interest, both in research and teaching, has led to the establishment of the series Texts in Applied Mathematics (TAM).

The development of new courses is a natural consequence of a high level of excitement on the research frontier as newer techniques, such as numerical and symbolic computer systems, dynamical systems, and chaos, mix with and reinforce the traditional methods of applied mathematics. Thus, the purpose of this textbook series is to meet the current and future needs of these advances and to encourage the teaching of new courses.

TAM will publish textbooks suitable for use in advanced undergraduate and beginning graduate courses, and will complement the Applied Mathematical Sciences (AMS) series, which will focus on advanced textbooks and research-level monographs.

Pasadena, California   J.E. Marsden
Providence, Rhode Island   L. Sirovich
College Park, Maryland   S.S. Antman

Preface

The algorithms, methods, and Matlab implementations described in this text have been developed during almost a decade of collaboration. During this time we have worked to simplify the basic methods and make the ideas more accessible to a broad audience. Many people, both students and colleagues, have helped during the development of this project and we are grateful for all their suggestions and input.

A special thank goes to A.P. Engsig-Karup for help with DistMesh and for agreeing to contribute Appendix B. A. Schiemenz and A. Narayan have spent long hours reading drafts and offering feedback. Helpful suggestions and numerous corrections have also been given by C. Chauviere, D. Leykekhman, N. Nunn, A. Potschka, J. Rodriguez, and T. Vdovina. What remains of misprints and mistakes does in no way reflect on their efforts and the responsibility for these rests entirely with us.

The first author (JSH) would like to thank his colleagues in the Division of Applied Mathematics at Brown University for years of support and encouragement. In particular, however, he wishes to expressed his deep appreciation for the many years of friendship and support of his closest colleague David Gottlieb. A substantial part of this text was written while JSH spent a sabbatical year during 2005-2006 at the Institute of Mathematical Modeling at the Technical University of Denmark. The relaxed yet inspiring atmosphere in the Numerical Analysis group offered an ideal opportunity for this project to materialize and JSH wishes to express his appreciation for the hospitality during this year.

The second author (TW) wishes to express sincere gratitude to his former and current colleagues of the Division of Applied Mathematics at Brown University, the Department of Mathematics and Statistics at the University of New Mexico, and the Department of Computational and Applied Mathematics at Rice University for their support and advice during this time. TW would like to specifically single out Thomas Hagstrom for all his contributions to this work, in particular helping to improve the Courant-Friedrichs-Lewy condition of discontinuous Galerkin methods.

We would also like to thank Springer-Verlag, New York, for making the publication of this manuscript possible. We would in particular like to thank senior editor for applied mathematics Achi Dosanjh for her support and encouragement during the whole process. We would like to thank Frank Ganz for his valuable technical assistance and Barbara Tompkins for her immaculate attention to detail during copy-editing. They have all helped to dramatically improve the look and feel of the manuscript.

Naturally, an effort like this has relied on financial support from numerous sources. In particular, we wish to thank the National Science Foundation (NSF, USA), the US Airforce Office of Scientific Research (AFOSR, USA), the US Army Research Office (ARO, USA), the Sandia University Research Program (SURP, USA), the Defense Advanced Research Projects Agency (DARPA, USA), the Alfred P. Sloan Foundation (USA), and the Otto Mønsted Foundation (Denmark) for their generous support during the last years. Without this support at different stages of the work, this text would not have been possible.

Brown University, Providence, Rhode Island   *Jan S Hesthaven*
Rice University, Houston, Texas   *Tim Warburton*
October 2007

Contents

1

Introduction

When faced with the task of solving a partial differential equation computationally, one quickly realizes that there is quite a number of different methods for doing so. Among these are the widely used finite difference, finite element, and finite volume methods, which are all techniques used to derive discrete representations of the spatial derivative operators. If one also needs to advance the equations in time, there is likewise a wide variety of methods for the integration of systems of ordinary differential equations available to choose among. With such a variety of successful and well tested methods, one is tempted to ask why there is a need to consider yet another method.

To appreciate this, let us begin by attempting to understand the strengths and weaknesses of the standard techniques. We consider the one-dimensional scalar conservation law for the solution $u(x,t)$

$$\frac{\partial u}{\partial t} + \frac{\partial f}{\partial x} = g, \ x \in \Omega, \tag{1.1}$$

subject to an appropriate set of initial conditions and boundary conditions on the boundary, $\partial\Omega$. Here $f(u)$ is the flux, and $g(x,t)$ is some prescribed forcing function.

The construction of any numerical method for solving a partial differential equation requires one to consider the two choices:

- How does one represent the solution $u(x,t)$ by an approximate solution $u_h(x,t)$?
- In which sense will the approximate solution $u_h(x,t)$ satisfy the partial differential equation?

These two choices separate the different methods and define the properties of the methods. It is instructive to seek a detailed understanding of these choices and how they impact the schemes to appreciate how to address problems and limitations associated with the classic schemes.

Let us begin with the simplest and historically oldest method, known as the finite difference method. In this approach, a grid, $x^k, k = 1 \ldots K$, is laid down in space and spatial derivatives are approximated by difference methods; that is, the conservation law is approximated as

$$\frac{du_h(x^k, t)}{dt} + \frac{f_h(x^{k+1}, t) - f_h(x^{k-1}, t)}{h^k + h^{k-1}} = g(x^k, t), \tag{1.2}$$

where u_h and f_h are the numerical approximations to the solution and the flux, respectively, and $h^k = x^{k+1} - x^k$ is the local grid size. The construction of a finite difference method requires that, in the neighborhood of each grid point x^k, the solution and the flux are assumed to be well approximated by local polynomials

$$x \in [x^{k-1}, x^{k+1}] : \quad u_h(x, t) = \sum_{i=0}^{2} a_i(t)(x - x^k)^i, \quad f_h(x, t) = \sum_{i=0}^{2} b_i(t)(x - x^k)^i,$$

where the coefficients $a_i(t)$ and $b_i(t)$ are found by requiring that the approximate function interpolates at the grid points, x^k. Inserting these local approximations into Eq. (1.1), results in the residual

$$x \in [x^{k-1}, x^{k+1}] : \quad \mathcal{R}_h(x, t) = \frac{\partial u_h}{\partial t} + \frac{\partial f_h}{\partial x} - g(x, t).$$

Clearly, $\mathcal{R}_h(x, t)$ is not zero, as in that case, $u_h(x, t)$ would satisfy Eq. (1.1) exactly and would be the solution $u(x, t)$. Thus, we need to specify in which way u_h must satisfy the equation, which amounts to a statement about the residual, $\mathcal{R}_h(x, t)$. If we have a total of K grid points and, thus, K unknown grid point values, $u_h(x^k, t)$, a natural choice is to require that the residual vanishes exactly at these grid points. This results in exactly K finite difference equations of the type in Eq. (1.2) for the K unknowns, completing the scheme.

One of the most appealing aspects of this method is its simplicity; that is the discretization of general problems and operators is often intuitive and, for many problems, leads to very efficient schemes. Furthermore, the explicit semidiscrete form gives flexibility in the choice of timestepping methods if needed. Finally, these methods are supported by an extensive body of theory (see, e.g., [142]), they are sufficiently robust and efficient to be used for a variety of problems, and extensions to higher order approximations by using a local polynomial approximation of higher degree is relatively straightforward.

It is also, however, the reliance on the local one-dimensional polynomial approximation that is the Achilles' heel of the method, as that enforces a simple dimension-by-dimension structure in higher dimensions. Additional complications caused by the simple underlying structure are introduced around boundaries and discontinuous internal layers (e.g., discontinuous material coefficients). This makes the native finite difference method ill-suited to deal with complex geometries, both in terms of general computational domains and internal discontinuities as well as for local order and grid size changes to reflect local features of the solution.

The above discussion highlights that to ensure geometric flexibility one needs to abandon the simple one-dimensional approximation in favor of something more general. The most natural approach is to introduce an element-based discretization. Hence, we assume that Ω is represented by a collection of elements, D^k, typically simplexes or cubes, organized in an unstructured manner to fill the physical domain.

A method closely related to the finite difference method, but with added geometric flexibility, is the finite volume method. In its simplest form, the solution $u(x,t)$ is approximated on the element by a constant, $\overline{u}^k(t)$, at the center, x^k, of the element. This is introduced into Eq. (1.1) to recover the cellwise residual

$$x \in \mathsf{D}^k: \ \mathcal{R}_h(x,t) = \frac{\partial \overline{u}^k}{\partial t} + \frac{\partial f(\overline{u}^k)}{\partial x} - g(x,t),$$

where the element is defined as $\mathsf{D}^k = [x^{k-1/2}, x^{k+1/2}]$ with $x^{k+1/2} = \frac{1}{2}(x^k + x^{k+1})$. In the finite volume method we require that the cell average of the residual vanishes identically, leading to the scheme

$$h^k \frac{d\overline{u}^k}{dt} + f^{k+1/2} - f^{k-1/2} = h^k \overline{g}^k, \tag{1.3}$$

for each cell. Note that the approximation and the scheme is purely local and, thus, imposes no conditions on the grid structure. In particular, all cells can have different sizes, h^k. The flux term reduces to a pure surface term by the use of the divergence theorem, also known as Gauss' theorem. This step introduces the need to evaluate the fluxes at the boundaries. However, since our unknowns are the cell averages of the numerical solution u_h, the evaluation of these fluxes is not straightforward.

This reconstruction problem and the subsequent evaluation of the fluxes at the interfaces can be addressed in many different ways and the details of this lead to different finite volume methods. A simple solution to the reconstruction problem is to use

$$u^{k+1/2} = \frac{\overline{u}^{k+1} + \overline{u}^k}{2}, \ \ f^{k+1/2} = f(u^{k+1/2}),$$

and likewise for $f^{k-1/2}$. Alternatively, one could be tempted to simply take

$$f^{k+1/2} = \frac{f(\overline{u}^k) + f(\overline{u}^{k+1})}{2},$$

although this turns out to not be a good idea for general nonlinear problems. For linear problems and equidistant grids these methods all reduce to the finite difference method. However, one easily realizes that the formulation is less restrictive in terms of the grid structure; that is, the reconstruction of solution values at the interfaces is a local procedure and generalizes straightforwardly to unstructured grids in high dimensions, thus ensuring the desired geometric flexibility. Furthermore, the construction of the interface fluxes can be done in various ways that are closely related to the particular equations (see, e.g., [218, 303]). This is particularly powerful when one considers nonlinear conservation laws.

If, however, we need to increase the order of accuracy of the method, a fundamental problem emerges. Consider again the problem in one dimension. We wish to reconstruct the solution, u_h, at the interface and we seek a local polynomial, $u_h(x)$ of the form

$$x \in [x^{k-1/2}, x^{k+3/2}]: \quad u_h(x) = a + bx.$$

We then require

$$\int_{x^{k-1/2}}^{x^{k+1/2}} u_h(x)\,dx = h^k \overline{u}^k, \quad \int_{x^{k+1/2}}^{x^{k+3/2}} u_h(x)\,dx = h^{k+1} \overline{u}^{k+1}$$

to recover the two coefficients. The reconstructed value of the solution, u_h, and therefore also $f(u_h(x^{k+1/2}))$ can then be evaluated.

To reconstruct the interface values at a higher accuracy we can continue as above and seek a local solution of the form

$$u_h(x) = \sum_{j=0}^{N} a_j (x - x^k)^j.$$

However, to find the $N+1$ unknown coefficients, we will need information from at least $N+1$ cells. In the simple one-dimensional case, this can be done straightforwardly, as for the finite difference scheme (i.e., by extending the size of the stencil). However, the need for a high-order reconstruction reintroduces the need for a particular grid structure and thus destroys the geometric flexibility of the finite volume method in higher dimensions. This defeats our initial motivation for considering the finite volume method. On unstructured grids this approach requires a reconstruction based on genuinely multivariate polynomials with general cell center locations which is both complex and prone to stability problems. The main limitation of finite volume methods is found in its inability to extend to higher-order accuracy on general unstructured grids.

Looking back at the finite volume method, we realize that the problem with the high-order reconstruction is that it must span multiple elements as the numerical approximation, $u_h(x,t)$, is represented by cell averages only. One could be tempted to take a different approach and introduce more degrees of freedom on the element. To pursue this idea, let us first redefine the element D^k as the interval bounded by the grid points $[x^k, x^{k+1}]$ and with a total of K elements and $K+1$ grid points. Note that this is slightly different from the finite volume scheme where the element was defined by staggered grid points as $[x^{k-1/2}, x^{k+1/2}]$. Inside this element, we assume that the local solution is expressed in the form

$$x \in \mathsf{D}^k : \quad u_h(x) = \sum_{n=1}^{N_p} b_n \psi_n(x),$$

where we have introduced the use of a locally defined basis function, $\psi_n(x)$. In the simplest case, we can take these basis functions be linear; that is,

$$x \in \mathsf{D}^k : \quad u_h(x) = u(x^k)\frac{x - x^{k+1}}{x^k - x^{k+1}} + u(x^{k+1})\frac{x - x^k}{x^{k+1} - x^k} = \sum_{i=0}^{1} u(x^{k+i})\ell_i^k(x),$$

where the linear Lagrange polynomial, $\ell_i^k(x)$, is given as

$$\ell_i^k(x) = \frac{x - x^{k+1-i}}{x^{k+i} - x^{k+1-i}}.$$

With this local element-based model, each element shares the nodes with one other element (e.g., D^{k-1} and D^k share x^k). We have a global representation of u_h as

$$u_h(x) = \sum_{k=1}^{K} u(x^k) N^k(x) = \sum_{k=1}^{K} u^k N^k(x),$$

where the piecewise linear shape function, $N^i(x^j) = \delta_{ij}$ is the basis function and $u^k = u(x^k)$ remain as the unknowns.

To recover the scheme to solve Eq. (1.1), we define a space of test functions, V_h, and require that the residual is orthogonal to all test functions in this space as

$$\int_\Omega \left(\frac{\partial u_h}{\partial t} + \frac{\partial f_h}{\partial x} - g_h \right) \phi_h(x)\, dx = 0, \quad \forall \phi_h \in \mathsf{V}_h.$$

The details of the scheme is determined by how this space of test functions is defined. A classic choice, leading to a Galerkin scheme, is to require the that spaces spanned by the basis functions and test functions are the same. In this particular case we thus assume that

$$\phi_h(x) = \sum_{k=1}^{K} v(x^k) N^k(x).$$

Since the residual has to vanish for all $\phi_h \in \mathsf{V}_h$, this amounts to

$$\int_\Omega \left(\frac{\partial u_h}{\partial t} + \frac{\partial f_h}{\partial x} - g_h \right) N^j(x)\, dx = 0,$$

for $j = 1 \ldots K$. Straightforward manipulations yield the scheme

$$\mathcal{M} \frac{d\boldsymbol{u}_h}{dt} + \mathcal{S} \boldsymbol{f}_h = \mathcal{M} \boldsymbol{g}_h, \qquad (1.4)$$

where

$$\mathcal{M}_{ij} = \int_{\Omega} N^i(x) N^j(x)\, dx, \quad \mathcal{S}_{ij} = \int_{\Omega} N^i(x) \frac{dN^j}{dx}\, dx,$$

reflects the globally defined mass matrix and stiffness matrix, respectively. We also have the vectors of unknowns, $\boldsymbol{u}_h = [u^1, \ldots, u^{K+1}]^T$, of fluxes, $\boldsymbol{f}_h = [f^1, \ldots, f^{K+1}]^T$, and the forcing, $\boldsymbol{g}_h = [g^1, \ldots, g^{K+1}]^T$, given on the N_p nodes.

This approach, which reflects the essence of the classic finite element method [182, 293, 340, 341], clearly allows different element sizes. Furthermore, we recall that a main motivation for considering methods beyond the finite volume approach was the interest in higher-order approximations. Such extensions are relatively simple in the finite element setting and can be achieved by adding additional degrees of freedom to the element while maintaining shared nodes along the faces of the elements [197]. In particular, one can have different orders of approximation in each element, thereby enabling local changes in both size and order, known as hp-adaptivity (see, e.g., [93, 94]).

However, the above discussion also highlights disadvantages of the classic continuous finite element formulation. First, we see that the globally defined basis functions and the requirement that the residual be orthogonal to the same set of globally defined test functions implies that the semidiscrete scheme becomes implicit and $\mathcal{M}$ must be inverted. For time dependent problems, this is a clear disadvantage compared to finite difference and finite volume methods. On the other hand, for problems with no explicit time dependence, this is less of a concern.

There is an additional subtle issue that relates to the structure of the basis. If we recall the discussion above, we recognize that the basis functions are symmetric in space. For many types of problems (e.g., a heat equation), this is a natural choice. However, for problems such as wave problems and conservation laws, in which information flows in specific directions, this is less natural and can causes stability problems if left unchanged (see, e.g., [182, 339]). In finite difference and finite volume methods, this problem is addressed by the use of upwinding, either through the stencil choice or through the design of the reconstruction approach.

In Table 1.1 we summarize some of the issues discussed so far. Looking at it, one should keep in mind that this comparison reflects the most basic methods and that many of the problems and restrictions can be addressed and overcome in a variety of ways. Nevertheless, the comparison does highlight which shortcoming one should strive to resolve when attempting to formulate a new method.

Reflecting on the previous discussion, one realizes that to ensure geometric flexibility and support for locally adapted resolution, we must strive for an element-based method where high-order accuracy is enabled through the local approximation, as in the finite element method. However, the global Galerkin statement, introduced by the globally defined basis and test functions, on the

Table 1.1. We summarize generic properties of the most widely used methods for discretizing partial differential equations [i.e., finite difference methods (FDM), finite volume methods (FVM), and finite element methods (FEM), as compared with the discontinuous Galerkin finite element method (DG-FEM)]. A ✓ represents success, while ✗ indicates a short-coming in the method. Finally, a (✓) reflects that the method, with modifications, is capable of solving such problems but remains a less natural choice.

	Complex geometries	High-order accuracy and hp-adaptivity	Explicit semi- discrete form	Conservation laws	Elliptic problems
FDM	✗	✓	✓	✓	✓
FVM	✓	✗	✓	✓	(✓)
FEM	✓	✓	✗	(✓)	✓
DG-FEM	✓	✓	✓	✓	(✓)

residual destroys the locality of the scheme and introduces potential problems with the stability for wave-dominated problems. On the other hand, this is precisely the regime where the finite volume method has several attractive features.

An intelligent combination of the finite element and the finite volume methods, utilizing a space of basis and test functions that mimics the finite element method but satisfying the equation in a sense closer to the finite volume method, appears to offer many of the desired properties. This combination is exactly what leads to the discontinuous Galerkin finite element method (DG-FEM).

To achieve this, we maintain the definition of elements as in the finite element scheme such that $D^k = [x^k, x^{k+1}]$. However, to ensure the locality of the scheme, we duplicate the variables located at the nodes x^k. Hence the vector of unknowns is defined as

$$u_h = [u^1, u^2, u^2, u^3, \ldots, u^{K-1}, u^K, u^K, u^{K+1}]^T,$$

and is now $2K$ long rather than $K + 1$ as in the finite element method. In each of these elements we assume that the local solution can be expressed as

$$x \in D^k : \quad u_h^k(x) = u^k \frac{x - x^{k+1}}{x^k - x^{k+1}} + u^{k+1} \frac{x - x^k}{x^{k+1} - x^k} = \sum_{i=0}^{1} u^{k+i} \ell_i^k(x) \in V_h,$$

and likewise for the flux, f_h^k. The space of basis functions is defined as $V_h = \oplus_{k=1}^{K} \{\ell_i^k\}_{i=0}^{1}$, i.e., as the space of piecewise polynomial functions. Note in particular that there is no restrictions on the smoothness of the basis functions between elements.

As in the finite element case, we now assume that the local solution can be well represented by a linear approximation $u_h \in V_h$ and form the local residual

$$x \in \mathsf{D}^k : \ \mathcal{R}_h(x,t) = \frac{\partial u_h^k}{\partial t} + \frac{\partial f_h^k}{\partial x} - g(x,t),$$

for each element. Going back to the finite element scheme, we recall that the global conditions on this residual are the source of the global nature of the operators $\mathcal{M}$ and $\mathcal{S}$ in Eq. (1.4). To overcome this, we require that the residual is orthogonal to all test functions $\phi_h \in \mathsf{V}_h$, leading to

$$\int_{\mathsf{D}^k} \mathcal{R}_h(x,t) \ell_j^k(x) \, dx = 0,$$

for all the test functions, $\ell_j^k(x)$. The strictly local statement is a direct consequence of V_h being a broken space and the fact that we have duplicated solutions at all interface nodes.

At first, the locality also appears problematic as this statement does not allow one to recover a meaningful global solution. Furthermore, the points at the ends of the elements are shared by two elements so how does one ensure uniqueness of the solution at these points?

These problems are overcome by observing that the above local statement is very similar to that recovered in the finite volume method. Following this line of thinking, let us use Gauss' theorem to obtain the local statement

$$\int_{\mathsf{D}^k} \frac{\partial u_h^k}{\partial t} \ell_j^k - f_h^k \frac{d\ell_j^k}{dx} - g\ell_j^k \, dx = - \left[f_h^k \ell_j^k \right]_{x^k}^{x^{k+1}}.$$

What remains now is to understand what the right-hand side means. This is, however, easily understood by considering the simplest case where $\ell_j^k(x)$ is a constant, in which case we recover the finite volume scheme in Eq. (1.3). Hence, the main purpose of the right-hand side is to connect the elements. This is further made clear by observing that both element D^k and element D^{k+1} depends on the flux evaluation at the point, x^{k+1}, shared among the two elements. This situation is identical to the reconstruction problem discussed previously for the finite volume method where the interface flux is recovered by combining the information of the two cell averages appropriately.

At this point it suffices to introduce the numerical flux, f^*, as the unique value to be used at the interface and obtained by combining information from both elements. With this we recover the scheme

$$\int_{\mathsf{D}^k} \frac{\partial u_h^k}{\partial t} \ell_j^k - f_h^k \frac{d\ell_j^k}{dx} - g\ell_j^k \, dx = - \left[f^* \ell_j^k \right]_{x^k}^{x^{k+1}},$$

or, by applying Gauss' theorem once again,

$$\int_{\mathsf{D}^k} \mathcal{R}_h(x,t) \ell_j^k(x) \, dx = \left[(f_h^k - f^*) \ell_j^k \right]_{x^k}^{x^{k+1}}.$$

These two formulations are the discontinuous Galerkin finite element (DG-FEM) schemes for the scalar conservation law in weak and strong form, respectively. Note that the choice of the numerical flux, f^*, is a central element of the scheme and is also where one can introduce knowledge of the dynamics of the problem [e.g., by upwinding through the flux as in a finite volume method (FVM)].

To mimic the terminology of the finite element scheme, we write the two local elementwise schemes as

$$\mathcal{M}^k \frac{d\boldsymbol{u}_h^k}{dt} - (\mathcal{S}^k)^T \boldsymbol{f}_h^k - \mathcal{M}^k \boldsymbol{g}_h^k = -f^*(x^{k+1})\boldsymbol{\ell}^k(x^{k+1}) + f^*(x^k)\boldsymbol{\ell}^k(x^k)$$

and

$$\mathcal{M}^k \frac{d\boldsymbol{u}_h^k}{dt} + \mathcal{S}^k \boldsymbol{f}_h^k - \mathcal{M}^k \boldsymbol{g}_h^k = (f_h^k(x^{k+1}) - f^*(x^{k+1}))\boldsymbol{\ell}^k(x^{k+1})$$
$$-(f_h^k(x^k) - f^*(x^k))\boldsymbol{\ell}^k(x^k). \qquad (1.5)$$

Here we have the vectors of local unknowns, $\boldsymbol{u}_h^k$, of fluxes, $\boldsymbol{f}_h^k$, and the forcing, $\boldsymbol{g}_h^k$, all given on the nodes in each element. Given the duplication of unknowns at the element interfaces, each vector is $2K$ long. Furthermore, we have $\boldsymbol{\ell}^k(x) = [\ell_1^k(x), \ldots, \ell_{N_p}^k(x)]^T$ and the local matrices

$$\mathcal{M}_{ij}^k = \int_{\mathsf{D}^k} \ell_i^k(x)\ell_j^k(x)\,dx, \quad \mathcal{S}_{ij}^k = \int_{\mathsf{D}^k} \ell_i^k(x)\frac{d\ell_j^k}{dx}\,dx.$$

While the structure of the DG-FEM is very similar to that of the finite element method (FEM), there are several fundamental differences. In particular, the mass matrix is local rather than global and thus can be inverted at very little cost, yielding a semidiscrete scheme that is explicit. Furthermore, by carefully designing the numerical flux to reflect the underlying dynamics, one has more flexibility than in the classic FEM to ensure stability for wave-dominated problems. Compared with the FVM, the DG-FEM overcomes the key limitation on achieving high-order accuracy on general grids by enabling this through the local element-based basis. This is all achieved while maintaining benefits such as local conservation and flexibility in the choice of the numerical flux.

All of this, however, comes at a price – most notably through an increase in the total degrees of freedom as a direct result of the decoupling of the elements. For linear elements, this yields a doubling in the total number of degrees of freedom compared to the continuous FEM as discussed above. For certain problems, this clearly becomes an issue of significant importance; that is, if we seek a steady solution and need to invert the discrete operator $\mathcal{S}$ in Eq. (1.5) then the associated computational work scales directly with the size of the matrix. Furthermore, for problems where the flexibility in the flux choices and the locality of scheme is of less importance (e.g., for elliptic problems), the DG-FEM is not as efficient as a better suited method like the FEM.

On the other hand, for the DG-FEM, the operator tends to be more sparse than for a FEM of similar order, potentially leading to a faster solution procedure. This overhead, caused by the doubling of degrees of freedom along element interfaces, decreases as the order of elements increase, thus reducing the penalty of the doubling of the interface nodes somewhat. In Table 1.1 we also summarize the strengths and weaknesses of the DG-FEM. This simple – and simpleminded – comparative discussion also highlights that time dependent wave-dominated problems emerge as the main candidates for problems where DG-FEM will be advantageous. As we will see shortly, this is also where they have achieved most prominence and widespread use during the last decade.

1.1 A brief account of history

The discontinuous Galerkin finite element method (DG-FEM) appears to have been proposed first in [269] as a way of solving the steady-state neutron transport equation

$$\sigma u + \nabla \cdot (au) = f,$$

where σ is a real constant, $a(x)$ is piecewise constant, and u is the unknown. The first analysis of this method was presented in [217], showing $\mathcal{O}(h^N)$-convergence on a general triangulated grid and optimal convergence rate, $\mathcal{O}(h^{N+1})$, on a Cartesian grid of cell size h and with a local polynomial approximation of order N. This result was later improved in [190] to $\mathcal{O}(h^{N+1/2})$-convergence on general grids. The optimality of this convergence rate was subsequently confirmed in [257] by a special example. These results assume smooth solutions, whereas the linear problems with nonsmooth solutions were analyzed in [63, 225]. Techniques for postprocessing on Cartesian grids to enhance the accuracy to $\mathcal{O}(h^{2N+1})$ for linear problems have been developed in [72, 86, 274]. A general discussion of the hyperbolic system and its discretization is given in [33, 112], and attempts to design numerical fluxes to minimize the dissipation are discussed in [39, 61].

The extension to the nonlinear scalar conservation law was first attempted in [53] by using linear elements in space and a forward Euler method to integrate in time, resulting in an unstable method unless a very restrictive timestep is used. In [54] this was resolved by introducing a slope limiter, resulting in a convergent scheme, albeit of low order in both space and time. This was overcome in [77] by combining the nonlinearly stable Runge-Kutta methods, introduced in [286], with an improved slope limiter [285], hence ensuring formal accuracy in smooth regions, sharp shock resolution, and convergence to the entropy solution. By extending this approach through the introduction of a generalized slope limited and high-order nonlinearly stable Runge-Kutta methods, these results were extended to high-order accuracy in [75] and to one-dimensional systems in [71]. The extension to the multidimensional case is complicated by the complexities of multidimensional slope-limiting, but this development was completed for scalar problems [64] and extended to systems

[76] by the introduction of slope limiters enforcing a local maximum principle only. A new family of limiters based on ideas from essentially nonoscillatory finite difference schemes have recently been proposed and shows some promise [264, 265, 266, 267], although mostly limited to simple grids. An extension to unstructured grids is discussed in [232]. DG-FEM using nonpolynomial basis functions has been discussed in [334] and a novel scheme using a patch-like approach is discussed in [58] for the wave equation.

These recent developments have lead to a dramatic expansion in the use of DG-FEM methods for a variety of applications based on hyperbolic systems, e.g., Maxwell's equations [70, 80, 81, 129, 130, 164, 205, 270, 318], equations of acoustics [14, 306], elasticity [92, 106, 107, 199, 200], shallow water equations and free surface wave descriptions [109, 110, 113, 114, 127], plasma physics [184, 224, 233], and gas dynamics [21, 75, 76].

A closely related development of DG-FEM for Hamilton-Jacobi equations with applications to control and level set methods has been pursued in [178, 216] utilizing the close connection between conservation laws and Hamilton-Jacobi equations. Alternatives to this most straightforward approach have been developed in [56, 221] and the development of limiters is discussed in [268].

The extension of the classic DG-FEM approach to problems with second order operators was initiated in [22] with the rewriting of higher-order operators into systems of first-order equations. This subsequently lead to the introduction of the local discontinuous Galerkin (LDG) method in [78] which has $\mathcal{O}(h^{N+1})$ optimal convergence. An alternative formulation is suggested in [24, 249] with forms of this formulation, known as internal penalty methods, having been developed much earlier [12, 102]. Comparative discussions are presented in [288, 335]. Generalizations to higher-order operators for the LDG and central schemes has been pursued in a number of papers [219, 332, 333].

These developments have opened up for a new range of applications, most notably viscous fluid flows, both compressible [22, 148, 149] and incompressible [66, 68, 226, 242], general advection-diffusion problems [51, 174, 254, 342] viscous magnetohydrodynamics [322], and semi conductor modeling [227]. Many examples in heat conduction and fluid dynamics applications can be found in [220].

The development of DG-FEM for elliptic problems goes far back in time with the introduction of the interior penalty methods [12, 102], although it was not associated with DG-FEM at that time. These methods have been further developed recently in the context of DG-FEM in [32, 48, 126, 281] with a particular emphasis on the understanding of the correct choice of the penalty terms and stabilization methods in general. Local error analysis for internal penalty methods are discussed in [194]. Both central fluxes, often termed Bassi-Rebay fluxes [22], and local discontinuous Galerkin fluxes (LDG) [78], are also widely studied and used [48, 50, 283]. An excellent overview of these and other methods, their properties, and differences is given in [13]. Extensions to biharmonic problems are discussed in [241, 243] and recent results for nonlinear elliptic problems are discussed in [250]. The more complex question of the behavior of the eigenvalue spectrum of the discrete Laplacian is analyzed in [11] for the interior penalty discretizations, highlighting the subtle interplay with the stabilization parameters. Closely related efforts for saddlepoint and noncoersive problems include the Stokes problem [66, 67, 278, 279] and the equations of linearized elasticity [73, 172], the second order wave equation [130], and Maxwell's equations [129].

As the prototype noncoercive problem, there has been significant developments towards the understanding of the DG-FEM approximations to Maxwell's equations. These developments include both the low-frequency case [255, 256] and the high-frequency, indefinite case [166, 167, 168]. A nonstabilized scheme is discussed in [169]. Understanding the properties of the spectrum of the discrete Maxwell operator is complex due to a large nullspace, the potential for spurious eigenvalues, and the delicate interplay with the stabilization. This line of research was initiated in [165] and further studied in [34, 82, 319], with a thorough analysis just recently presented in [35].

The last decade has seen a rapid growth in the development of rigorous approaches for the error controlled adaptive solution using DG-FEM. In [28] the first large-scale adaptive solution of conservation laws was demonstrated, using local ad hoc error estimators. A rigorous approach is very complicated and for the nonlinear problem, only a few results are available [62]. A posteriori error norm estimates have recently been developed for some classes of linear problems [170, 171, 173] and for quasilinear elliptic problems [177]. The use of adjoint methods for error estimation for conservation laws has been discussed extensively in [146, 147, 175, 176], highlighting the potential of such techniques for time dependent conservation laws with shocks (e.g., the Euler equations of compressible gas dynamics). An alternative approach to develop local error estimators is based on superconvergence. This phenomenon was shown for both elliptic [65] and hyperbolic problems [6, 120] and subsequently extended to multidimensional problems [5, 7, 211] and nonlinear problems [8].

Specialized techniques for time integration has received some attention, starting with the strongly stable or TVD Runge-Kutta methods, enabling nonlinear stability [139, 140, 286]. Lax-Wendroff type methods are discussed in [262] and the use of more complicated Runge-Kutta type methods are discussed in [191]. The local nature of the discontinuous Galerkin method suggests that local timestepping is an interesting approach and a robust and systematic approach to this, known as ADER-DG, has been developed in [107] with the generalization to general nonlinear system discussed in [299, 302].

The development of efficient solvers for the discrete systems originating from the DG-FEM approach remains less advanced than in many other areas. However, there has been a recent surge of activity in this very important area, with the first steps in the development of Schwarz preconditioners [117, 133, 193, 213], focusing on advection-diffusion problems. The development of multigrid methods has received a renewed interest through the encouraging results of local p multigrid [119, 150, 245, 246] as well as by combining classic

h-type multigrid with a p-type multigrid [245]. These developments focus on problems derived from compressible fluid dynamics.

A wealth of further theoretical developments and numerous applications can be found by consulting the conference proceeding [69] as well as one of the recent special volumes of leading journals dedicated to DG-FEM's and applications [3, 74, 91, 195].

It is noteworthy that in parallel with the development of DG-FEM's during the last two decades, there have been closely related developments in various communities. The focus of many of these recent activities has been based on realizing the advantages of using weakly imposed interface and boundary conditions. In high-order accurate FDM's and compact schemes, the development of simultaneous approximation term (SAT) schemes [44, 47, 248] for imposing boundary conditions weakly is a close parallel to DG-FEM's. Extensions of this approach to deal with complex boundaries are discussed in [1]. Within the community of spectral methods, the first ideas along such lines were introduced in [121] for elliptic problems and in [122, 123] for linear hyperbolic problems. This approach, now known as spectral penalty methods, was extensively developed in [152, 153, 157] for advection-diffusion problems, compressible gas dynamics, and complex geometries. Further generalizations to unstructured grids were pursued in [158, 163]. A review of other developments and applications is given in [154]. Most of these methods have been developed with the provable stability of the schemes as a prime objective and with analysis based on semidiscrete approximations of linear or linearized problems. A related, but different, approach is the spectral volume method [316, 317], which combines a classic DG-FEM formulation with an element-based finite volume reconstruction. For linear problems, this becomes essentially equivalent to a nodal DG-FEM whereas for nonlinear problems, there are minor differences, as discussed in [295, 336].

1.2 Summary of the chapters

While the discussion in the preceding section highlights the basic elements of the DG-FEM schemes and their unique qualities when compared to other widely used methods, it is also clear that a deeper discussion is needed. In particular, we need to develop a better understanding of the choices involved (e.g., local basis functions and numerical fluxes), as well as an understanding of more basic questions regarding accuracy, stability, and convergence of the scheme. Beyond this, we also need to address the often overlooked or neglected issue of how to implement the methods in a robust and efficient manner.

We begin on this ambitious agenda in Chapter 2, where we return to the basics of the DG-FEM and discuss in more detail the formulation of these methods for linear wave problems. The emphasis is on gaining a better understanding of the key steps to be taken in the formulation, what role the flux plays, and which criteria should be applied when choosing these. We will also consider a few examples to gain an appreciation of what can be expected in terms of accuracy of the schemes.

In Chapter 3, we continue the discussion of the one-dimensional wave problem but focus on how to implement the DG-FEM's in a general, flexible, and robust manner. In particular, we discuss in detail several different choices of the local basis, leading to both modal and nodal bases. We also discuss how to, in practice, assemble the computational grid and compute all the entities required to enable the implementation of the scheme. The emphasis remains on the one-dimensional linear case and we conclude with several examples.

This sets the stage for Chapter 4, where we return to the fundamentals and discuss basic properties of the schemes, such as stability and accuracy. This requires insight into polynomial approximation properties, linear stability for systems, and error estimates for the semidiscrete schemes, confirming the observations we made in Chapters 2 and 3. We also discuss fully discrete stability in a methods-of-lines setting to offer guidelines on how to choose the timestep in practical situations. We will conclude with a discussion on how to improve on the timestep restrictions through various techniques as well as a brief discussion of local timestepping. While some of the results are of a general nature, the focus remains on results for the one-dimensional case.

In Chapter 5 we finally leave the world of linear problems with piecewise constant coefficients and engage in the more complex analysis of problems with variable coefficients and nonlinear conservation laws. We begin by discussing, briefly, well-known properties of conservation laws and techniques to analyze them. This enables a parallel discussion of the basic properties of the DG-FEM's, highlighting some fundamental and very important properties that render these schemes particularly well suited for solving conservation laws. A particularly important issue is that to mimic the analysis exactly, one is required to perform exact integrations of volume and surface integrals. This, however, often becomes prohibitively expensive for general nonlinear problems and relaxing this requirement is an essential feature of the nodal DG-FEM. This introduces some new complications that we shed some light on and show how to overcome by borrowing techniques known from spectral collocation methods. This leads to a discussion of the second generic difficulty associated with the solution of nonlinear conservation laws caused by the potential for the development of discontinuous solutions. We discuss the Gibbs phenomenon and ways to control its impact through filtering before introducing the use of limiters to ensure a fully nonoscillatory solution, albeit possibly at the expense of the high-order accuracy. A final element of the classic Runge-Kutta based DG-FEM's is the strong stability preserving Runge-Kutta methods which ensure nonlinear stability, hence enabling rigorous convergence proofs, even in the nonlinear case. We conclude this lengthy but central chapter with a detailed example involving the solution of the Euler equations for compressible gas dynamics.

With the fundamentals in place for the one-dimensional situation, Chapter 6 addresses the extensions needed to solve problems in two spatial dimensions. In many ways, the chapter mimics Chapter 3, discussing two-dimensional nodal and modal basis on triangles to enable geometric flexibility. This again sets the stage for the development of all the components needed to enable a simple and efficient implementation on general grids. We discuss how to choose the timestep in practical situations before offering some examples of two-dimensional hyperbolic problems; in this case, the Maxwell equations and the two-dimensional Euler equations. For the latter case we return to some of the issues discussed in Chapter 5 (e.g., limiting techniques on general grids). We conclude this chapter with an overview of theoretical results for the multidimensional case.

Whereas all discussions in the first 6 chapters have been for first-order problems, we discuss in Chapter 7 the development of DG-FEM's for higher-order and mixed problems (e.g, advection-diffusion problems). We first discuss issues related to time dependent problems, proper formulations, and the definitions of the numerical fluxes leading to stable schemes. This leads into a discussion of DG-FEM's for elliptic problems with an emphasis on numerical fluxes and stabilization techniques and their impact. Examples illustrate the accuracy of the different formulations. Following this, we highlight the key theoretical results for the different formulations, before concluding this chapter with a brief discussion and illustrations of these methods in two dimensions. As final nontrivial examples, combining many of the elements discussed thus far, we develop schemes for the solution of the two-dimensional Navier-Stokes equations for both a compressible and an incompressible fluid.

The following two chapters, Chapters 8 and 9, are slightly different in nature from the previous parts of the text. In these two chapters, we discuss more advanced topics which may, nevertheless, prove useful for certain applications.

In Chapter 8 we return to the two-dimensional second order problem but consider the eigenvalues of the discrete operators as a much more sensitive measure of how well the continuous operators are represented. We identify possible problems with nonphysical spurious eigenvalues and discuss in some detail the impact of the stabilization on these spurious eigenvalues. These issues are discussed both for the Helmholtz operators and the Maxwell curl-curl operator, using different choices of fluxes and local basis functions. In particular, the delicate interplay between the flux, the stabilization, and the local basis is highlighted.

Chapter 9 is devoted to a number of more advanced aspects of DG-FEM when solving problems of realistic complexity. In particular, we discuss how to use elements with general curved edges/faces, suitable for high-order body-conforming discretizations. Furthermore, we discuss nonconforming discretizations as a first step toward adaptive solution of partial differential equations. Both techniques will be demonstrated for the solution of two-dimensional problems with curvilinear grids and/or nonconforming discretizations.

Chapter 10 contains a condensed discussion of the extension of most of the basis material in Chapters 3 and 6 to the full three-dimensional case with discretizations based on tetrahedra. The emphasis here is entirely on the algorithmic aspect and is included to offer the reader a platform for simple experimentation and algorithm development in three spatial dimensions. However, there are no essential theoretical differences between the two-dimensional case, discussed in detail in Chapter 6, and the general three-dimensional scheme. A couple of examples, including the Maxwell equations and the Poisson equation, are included for illustration.

The three appendices contain various useful information. In particular, in Appendix A we discuss the computation of Jacobi polynomials and their extension to simplexes. The methods and routines discussed there are used throughout the text. Appendix B provides a brief introduction to a simple Matlab-based grid generator, DistMesh. This offers a flexible and simple way of generating grids for algorithm development and simple applications. In Appendix C we give an overview of the software used throughout the text, including a list of important variable names, as well as a brief description of some useful utility scripts.

1.3 On the use and abuse of the Matlab codes

Throughout the text, Matlab codes are used to illustrate the many concepts discussed in the text as well as to provide a platform for experimentation and further algorithm development. It is, however, imperative that one appreciates that the Matlab codes are not suitable for large-scale computations, as they are neither fast not robust enough for that.

If one has, after having developed and tested algorithms using the Matlab codes, a wish to go beyond this platform, we refer to the book webpage http://www.nudg.org. There one will find C/C++ packages, based on the same principles as the Matlab codes used here, but developed with large-scale computing in mind and with support for parallel computation and fully nonconforming discretizations in two and three spatial dimensions.

1.4 Scope of text and audience

The primary goal of this text is to serve as an extensive introduction to discontinuous Galerkin methods for the solution of partial differential equations. The flexibility of these methods allows them to be formulated both to high-order accuracy, for complex geometries, and for unsteady and nonlinear problems. However, the development of these methods relies on the combination of a number of apparently disconnected advances made during the last three decades. This makes the learning curve for the analysis, implementation, and application of these methods somewhat steep, often holding back researchers and students from using these methods.

The current text seeks to help make this effort a bit less daunting by collecting, in one place, both the required basic mathematical tools and ideas and a number of examples on how to implement these methods for a variety of problems. In particular, this latter aspect is too often neglected although is essential to enable the development of robust and efficient computational tools. The combination of theoretical insight and practical implementations should also ensure that the text appeals broadly to researchers and students in the sciences and engineering.

The level of the text is primarily geared to graduate students and researchers in the applied sciences and engineering. In particular, a first course in partial differential equations and their numerical solution is a good background.

However, the combined emphasis of analysis, algorithms, and application examples allows the text to be used in a variety of different ways. For practitioners and researchers wanting a quick introduction into the basic ideas, Chapters 2, 3, 6, 7, and 10 will provide just that, with an emphasis mostly on linear problems. For nonlinear problems, one can supplement with parts of Chapter 5. For researchers needing a more thorough understanding of the methods, a careful study of Chapters 1–5 and 7 will give insight into both analytic and practical aspects of the methods as well as enable one to seek more advanced results through the extensive bibliography.

As a textbook, the manuscript can likewise be used in various ways. While the general level of the text is best suited for a graduate-level course, one can choose to focus on basic ideas and algorithms by considering Chapters 2–4, and 7. This may also be done as part of a larger course in an advanced undergraduate-level course. Likewise, by selecting only parts of the text, one can use the text as part of a larger graduate-level course. By going through the whole text, and in particular Chapters 1 to 7, one will have material enough for a full-semester graduate course in the analysis and applications of discontinuous Galerkin methods. If needed, one can supplement this with more specialized papers taken from the extensive bibliography. The exercises at the end of most chapters should be helpful in the development of an understanding the basic concepts and encourage broad experimentation with the embedded software.

2

The key ideas

2.1 Briefly on notation

While we initially strive to stay away from complex notation a few fundamentals are needed. We consider problems posed on the physical domain Ω with boundary $\partial\Omega$ and assume that this domain is well approximated by the computational domain Ω_h. This is a space filling triangulation composed of a collection of K geometry-conforming nonoverlapping elements, D^k. The shape of these elements can be arbitrary although we will mostly consider cases where they are d-dimensional simplexes.

We define the local inner product and $L^2(\mathsf{D}^k)$ norm

$$(u, v)_{\mathsf{D}^k} = \int_{\mathsf{D}^k} uv \, dx, \ \ \|u\|^2_{\mathsf{D}^k} = (u, u)_{\mathsf{D}^k},$$

as well as the global broken inner product and norm

$$(u, v)_{\Omega, h} = \sum_{k=1}^{K} (u, v)_{\mathsf{D}^k}, \ \ \|u\|^2_{\Omega, h} = (u, u)_{\Omega, h}.$$

Here, (Ω, h) reflects that Ω is only approximated by the union of D^k, that is

$$\Omega \simeq \Omega_h = \bigcup_{k=1}^{K} \mathsf{D}^k,$$

although we will not distinguish the two domains unless needed.

Generally, one has local information as well as information from the neighboring element along an intersection between two elements. Often we will refer to the union of these intersections in an element as the trace of the element. For the methods we discuss here, we will have two or more solutions or boundary conditions at the same physical location along the trace of the element.

To fix the notation, we refer to the interior information of the element by a superscript "−" and to the exterior information by a superscript "+". Using this notation, it is useful to define the average

$$\{\{u\}\} = \frac{u^- + u^+}{2},$$

where u can be both a scalar and a vector. In a similar fashion we also define the jumps along a normal, $\hat{n}$, as

$$[\![u]\!] = \hat{n}^- u^- + \hat{n}^+ u^+, \quad [\![u]\!] = \hat{n}^- \cdot u^- + \hat{n}^+ \cdot u^+.$$

Note that it is defined differently depending on whether u is a scalar or a vector, u.

2.2 Basic elements of the schemes

In the following, we introduce the key ideas behind the family of discontinuous element methods that are the main topic of this text. Before getting into generalizations and abstract ideas, let us develop a basic understanding through a simple example.

2.2.1 The first schemes

Consider the linear scalar wave equation

$$\frac{\partial u}{\partial t} + \frac{\partial f(u)}{\partial x} = 0, \quad x \in [L, R] = \Omega, \tag{2.1}$$

where the linear flux is given as $f(u) = au$. This is subject to the appropriate initial conditions

$$u(x, 0) = u_0(x).$$

Boundary conditions are given when the boundary is an inflow boundary, that is

$$u(L, t) = g(t) \quad \text{if} \quad a \geq 0,$$
$$u(R, t) = g(t) \quad \text{if} \quad a \leq 0.$$

We approximate Ω by K nonoverlapping elements, $x \in [x_l^k, x_r^k] = \mathsf{D}^k$, as illustrated in Fig. 2.1. On each of these elements we express the local solution as a polynomial of order $N = N_p - 1$

$$x \in \mathsf{D}^k : \quad u_h^k(x, t) = \sum_{n=1}^{N_p} \hat{u}_n^k(t) \psi_n(x) = \sum_{i=1}^{N_p} u_h^k(x_i^k, t) \ell_i^k(x).$$

Here, we have introduced two complementary expressions for the local solution. In the first one, known as the modal form, we use a local polynomial

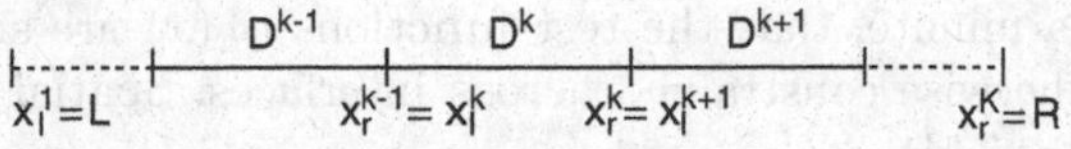

Fig. 2.1. Sketch of the geometry for simple one-dimensional example.

basis, $\psi_n(x)$. A simple example of this could be $\psi_n(x) = x^{n-1}$. In the alternative form, known as the nodal representation, we introduce $N_p = N + 1$ local grid points, $x_i^k \in D^k$, and express the polynomial through the associated interpolating Lagrange polynomial, $\ell_i^k(x)$. The connection between these two forms is through the definition of the expansion coefficients, $\hat{u}_n^k$. We return to a discussion of these choices in much more detail later; for now it suffices to assume that we have chosen one of these representations.

The global solution $u(x, t)$ is then assumed to be approximated by the piecewise N-th order polynomial approximation $u_h(x, t)$,

$$u(x, t) \simeq u_h(x, t) = \bigoplus_{k=1}^{K} u_h^k(x, t),$$

defined as the direct sum of the K local polynomial solutions $u_h^k(x, t)$.

We continue, as discussed briefly in Chapter 1, and form the residual

$$\mathcal{R}_h(x, t) = \frac{\partial u_h}{\partial t} + \frac{\partial a u_h}{\partial x},$$

and we must decide in which sense this should vanish.

Let us introduce a globally defined space V_h of test functions, ϕ_h, as $V_h = \oplus_{k=1}^{K} V_h^k$ where the locally defined spaces are defined as $V_h^k = \text{span}\{\psi_n(D^k)\}_{n=1}^{N_p}$. In the limit of $(N_p, K) \to \infty$ we assume that ϕ_h converges uniformly to ϕ.

We recognize V_h as the space of piecewise smooth functions defined on Ω_h. A general locally defined member $\phi_h^k \in V_h^k$ of this space is given as

$$x \in D^k: \quad \phi_h^k(x) = \sum_{n=1}^{N_p} \hat{\phi}_n^k \psi_n(x).$$

We now require that the residual is orthogonal to all test functions in V_h, resulting in the local statement

$$\int_{D^k} \mathcal{R}_h(x, t) \psi_n(x) \, dx = 0, \quad 1 \le n \le N_p, \tag{2.2}$$

on all K elements. This yields exactly N_p equations for the N_p local unknowns on each element. However, we have not imposed any particular constraints on the basis or test functions, and, thus, neglected the issue of how to impose boundary conditions and how to recover the global solution from the K local solutions. In other words, Eq. (2.2) is not yet a method suitable for solving the global problem.

Assume, for a minute, that the test functions $\phi_h(x)$ are smooth but not continuous or otherwise constrained across interfaces. Spatial integration by parts of Eq. (2.2) yields

$$\int_{\mathsf{D}^k} \left(\frac{\partial u_h^k}{\partial t} \psi_n - a u_h^k \frac{d\psi_n}{dx} \right) dx = - \left[a u_h^k \psi_n \right]_{x_l^k}^{x_r^k}$$

$$= - \int_{\partial \mathsf{D}^k} \hat{\boldsymbol{n}} \cdot a u_h^k \psi_n \, dx, \;\; 1 \le n \le N_p,$$

where $\hat{\boldsymbol{n}}$ represents the local outward pointing normal. The use of a surface integral may seem a bit artificial in this simple case but, as we will see later, it makes generalizations very natural. In this one-dimensional case, $\hat{\boldsymbol{n}}$ is simply a scalar and takes the value of $+1$ and -1 at the right and the left interface, respectively.

Revisiting Fig. 2.1, we see that as a consequence of the lack of conditions on the local solution and the test functions, the solution at the interfaces between elements is multiply defined and we will need to choose which solution, or combination of solutions, is correct. Delaying the details of this choice for a bit, we refer to this solution as $(au_h)^*$, known as the numerical flux. This leads to the semidiscrete scheme

$$\int_{\mathsf{D}^k} \left(\frac{\partial u_h^k}{\partial t} \psi_n - a u_h^k \frac{d\psi_n}{dx} \right) dx = - \int_{\partial \mathsf{D}^k} \hat{\boldsymbol{n}} \cdot (au_h)^* \psi_n \, dx, \;\; 1 \le n \le N_p. \quad (2.3)$$

We recover a total of $K \times N_p$ equations for the same number of unknowns; that is, we have defined a method for obtaining the globally defined solution. We refer to this form, obtained after integration by parts once, as the weak form.

This should be contrasted by the strong form, recovered by doing integration by parts once again, as

$$\int_{\mathsf{D}^k} \mathcal{R}_h(x,t) \psi_n \, dx = \int_{\partial \mathsf{D}^k} \hat{\boldsymbol{n}} \cdot \left(a u_h^k - (au_h)^* \right) \psi_n \, dx, \;\; 1 \le n \le N_p. \quad (2.4)$$

As in the equivalent weak form, it is the right-hand side that is responsible for recovering the global solution from the local solutions and imposing the boundary conditions. This emphasizes the key role played by the numerical flux, $(au_h)^*$. Note also that in this strong form, we do not have to require any smoothness of the test function, ϕ_h. In particular, it could be a delta-function. Let us further illustrate a few options for the choice of the test function through some examples.

Example 2.1. Assume that we express the local solution using a modal expansion,

$$x \in \mathsf{D}^k : \; u_h^k(x,t) = \sum_{n=1}^{N_p} \hat{u}_n^k(t) \psi_n(x),$$

in which case the N_p expansion coefficients are the unknowns.

If we choose the space of test functions to be same as the solution space (i.e., a Galerkin approach as discussed above) we recover the local semidiscrete weak statement

$$\hat{\mathcal{M}}^k \frac{d}{dt}\hat{\boldsymbol{u}}_h^k - \left(\hat{\mathcal{S}}^k\right)^T a\hat{\boldsymbol{u}}_h^k = -(au_h)^*\boldsymbol{\psi}(x_r^k) + (au_h)^*\boldsymbol{\psi}(x_l^k), \qquad (2.5)$$

where

$$\hat{\mathcal{M}}_{ij}^k = (\psi_i, \psi_j)_{\mathsf{D}^k}, \quad \hat{\mathcal{S}}_{ij}^k = \left(\psi_i, \frac{d\psi_j}{dx}\right)_{\mathsf{D}^k},$$

are the local mass and stiffness matrices, respectively. Furthermore, we have

$$\hat{\boldsymbol{u}}_h^k = [\hat{u}_1^k, \dots, \hat{u}_{N_p}^k]^T, \quad \boldsymbol{\psi} = [\psi_1(x), \dots, \psi_{N_p}(x)]^T,$$

as the vectors of the local solution and the local test functions, respectively. The scheme in Eq. (2.5) is the classical discontinuous Galerkin finite element method (DG-FEM) in weak form.

If we proceed in exactly the same way but consider the strong form, we recover

$$\hat{\mathcal{M}}^k \frac{d}{dt}\hat{u}_h^k + \hat{\mathcal{S}}^k a\hat{u}_h^k = \left(au_h^k - (au_h)^*\right)\boldsymbol{\psi}(x_r^k) - \left(au_h^k - (au_h)^*\right)\boldsymbol{\psi}(x_l^k). \quad (2.6)$$

The two formulations are mathematically equivalent but computationally different.

While the discontinuous Galerkin methods introduced in the above are the most widely used, we stress that they are not unique, as illustrated in the following example.

Example 2.2. We express the local solution using a nodal expression,

$$x \in \mathsf{D}^k: \ u_h^k(x,t) = \sum_{i=1}^{N_p} u_h^k(x_i^k, t)\ell_i^k(x);$$

that is, the N_p nodal values, $u_h^k(x_i^k, t)$, are the unknowns.

Now, choose N_p nodes, y_i^k, in each element and associate with these the test functions based on the shifted delta-function, $\delta(x - y_i^k)$. Note that there is no connection between x_i^k, on which the local approximation is based, and y_i^k, used to define the test functions (see [159]).

Since the weak form does not allow a space of nonsmooth test function, we consider the strong form, Eq. (2.4), as

$$\mathcal{M}^k \frac{d}{dt}\boldsymbol{u}_h^k + \mathcal{S}^k a\boldsymbol{u}_h^k = \left(au_h^k - (au_h)^*\right)\boldsymbol{\ell}^k(x_r^k) - \left(au_h^k - (au_h)^*\right)\boldsymbol{\ell}^k(x_l^k), \quad (2.7)$$

where

$$\mathcal{M}_{ij}^k = \ell_j^k(y_i^k), \ \ \mathcal{S}_{ij}^k = \left.\frac{d\ell_j^k}{dx}\right|_{y_i^k}$$

are the local mass and stiffness matrices, respectively. Furthermore, we have

$$\boldsymbol{u}_h^k = [u_1^k, \dots, u_{N_p}^k]^T, \ \ \boldsymbol{\ell}^k(x) = [\ell_1^k(x), \dots, \ell_{N_p}^k(x)]^T$$

as the vectors of the local nodal solution and the vector of Lagrange polynomials, respectively.

Schemes of this kind, using delta-functions as test functions, are known as collocation penalty methods in the literature. In the special case of $x_i^k = y_i^k$, the scheme simplifies further, as the mass matrix becomes the identity matrix.

The specification of the numerical flux, $(au_h)^*$, is most naturally related to the dynamics of the partial differential equation being solved. At the left end of the local domain, D^k, this numerical flux should be a function of $[au_h^{k-1}(x_r^{k-1}), au_h^k(x_l^k)]$, while the right end depends on $[au_h^k(x_r^k), au_h^{k+1}(x_l^{k+1})]$. A simple interpretation is that $(au_h)^*$ is the flux one would wish to know at the interface. This also suggests that at a physical boundary, a reasonable, but not unique, choice is that $(au_h)^* = ag(t)$ whenever a boundary condition as $u(x,t) = g(t)$ needs to be specified.

To further appreciate the role of the numerical flux, let us continue Example 2.1.

Example 2.3. A simple way to establish boundedness for the solution to the wave equation on an interval is to recognize that

$$\frac{d}{dt}\|u\|_\Omega^2 = -a\left(u^2(R) - u^2(L)\right),$$

which is derived by multiplication of Eq. (2.1) by $u(x,t)$, integration over the domain, D^k, followed by integration by parts. This approach, known as the energy method, yields information about where boundary conditions are needed. Clearly, it is reasonable that the numerical approximation to the wave equation displays similar behavior.

Assume that we use a nodal approach, i.e.,

$$x \in \mathsf{D}^k : \ u_h^k(x,t) = \sum_{i=1}^{N_p} u_h^k(x_i^k, t)\ell_i^k(x),$$

and consider the strong form given as

$$\mathcal{M}^k \frac{d}{dt}\boldsymbol{u}_h^k + \mathcal{S}^k\left(a\boldsymbol{u}_h^k\right) = \left[\boldsymbol{\ell}^k(x)(au_h^k - (au_h)^*)\right]_{x_l^k}^{x_r^k}. \tag{2.8}$$

Here, we have introduced the nodal versions of the local operators

$$\mathcal{M}_{ij}^k = \left(\ell_i^k, \ell_j^k\right)_{\mathsf{D}^k}, \quad \mathcal{S}_{ij}^k = \left(\ell_i^k, \frac{d\ell_j^k}{dx}\right)_{\mathsf{D}^k}.$$

First, realize that

$$\boldsymbol{u}_h^T \mathcal{M}^k \boldsymbol{u}_h = \int_{\mathsf{D}^k} \sum_{i=1}^{N_p} u_h^k(x_i^k)\ell_i^k(x) \sum_{j=1}^{N_p} u_h^k(x_j^k)\ell_j^k(x)\, dx = \left\| u_h^k \right\|_{\mathsf{D}^k}^2;$$

that is, it recovers the local energy. Furthermore, consider

$$\boldsymbol{u}_h^T \mathcal{S}^k \boldsymbol{u}_h = \int_{\mathsf{D}^k} \sum_{i=1}^{N_p} u_h^k(x_i^k)\ell_i^k(x) \sum_{j=1}^{N_p} u_h^k(x_j^k)\frac{d\ell_j^k}{dx}\, dx = \int_{\mathsf{D}^k} u_h^k(x)(u_h^k(x))'\, dx$$

$$= \frac{1}{2}[(u_h^k)^2]_{x_l^k}^{x_r^k}.$$

Thus, it mimics an integration-by-parts form.

By multiplying Eq. (2.8) with $\boldsymbol{u}_h^T$ we directly recover

$$\frac{d}{dt}\left\| u_h^k \right\|_{\mathsf{D}^k}^2 = -a[(u_h^k)^2]_{x_l^k}^{x_r^k} + 2\left[u_h^k(au_h^k - (au)^*) \right]_{x_l^k}^{x_r^k}. \tag{2.9}$$

For stability, we must, as for the original equation, require that

$$\sum_{k=1}^K \frac{d}{dt}\left\| u_h^k \right\|_{\mathsf{D}^k}^2 = \frac{d}{dt}\|u_h\|_{\Omega,h}^2 \le 0.$$

It suffices to control the terms associated with the coupling of the interfaces, each of which looks like

$$\hat{\boldsymbol{n}}^- \cdot \left(au_h^2(x^-) - 2u_h(x^-)(au)^*(x^-) \right) + \hat{\boldsymbol{n}}^+ \cdot \left(au_h^2(x^+) - 2u_h(x^+)(au)^*(x^+) \right) \le 0$$

at every interface. Here (x^-, x^+) refers to the left (e.g., x_r^k), and right side (e.g., x_l^{k+1}) of the interface, respectively. We also have that $\hat{\boldsymbol{n}}^- = -\hat{\boldsymbol{n}}^+$.

Let us consider a numerical flux like

$$f^* = (au)^* = \{\!\{au\}\!\} + |a|\frac{1-\alpha}{2}[\![u]\!].$$

At any internal interface, this gives a contribution to the energy integral as

$$-|a|(1-\alpha)[\![u_h]\!]^2,$$

which is nonpositive, provided $0 \le \alpha \le 1$.

Assume first that $a > 0$. Then taking $f_L = 0$ and $f_R = au_h(x_r^K)$ as numerical boundary fluxes, reflect a Dirichlet boundary condition of $u(x_l^1) = 0$

and a purely internal choice at the outflow, as is reasonable. In this case, we recover the global energy condition from Eq. (2.9) by summing over all elements

$$\frac{d}{dt}\|u_h\|_{\Omega,h}^2 = -|a|(1-\alpha)\sum_{k=1}^{K-1}[\![u_h^k(x_r^k)]\!]^2 - (1-\alpha)a(u_h^1(x_l^1))^2 - a(u_h^K(x_r^K))^2.$$

For $0 \leq \alpha \leq 1$ we have global stability.

An alternative choice of the numerical boundary fluxes is $f_L = -au_h^1(x_l^1)$ and $f_R = au_h(x_r^K)$, which likewise reflects a Dirichlet boundary condition of $u(x_l^1) = 0$, imposed by a mirror principle, and a purely internal choice at the outflow. This yields a global energy estimate as

$$\frac{d}{dt}\|u_h\|_{\Omega,h}^2 = -|a|(1-\alpha)\sum_{k=1}^{K-1}[\![u_h^k(x_r^k)]\!]^2 - a(u_h^1(x_l^1))^2 - a(u_h^K(x_r^K))^2.$$

Again, we recover stability for $0 \leq \alpha \leq 1$. A particularly interesting case is that of $\alpha = 1$, corresponding to purely central fluxes. In this case, we do not get any contributions from the internal boundaries; that is, for a periodic case, the energy is constant as for the original equation.

As highlighted in the above, the role of the flux is to guarantee stability of the formulation by mimicking the flow of information in the underlying partial differential equation. In the particularly simple case we are studying here (i.e., the wave equation) a natural choice for the flux is

$$(au)^* = \{\!\{au\}\!\} + |a|\frac{1-\alpha}{2}[\![u]\!], \quad 0 \leq \alpha \leq 1, \tag{2.10}$$

as also illustrated in the example above. If $\alpha = 1$, the numerical flux is simply the average of the two solutions, known as a central flux. For $\alpha = 0$, we recover a flux which always takes information from where it is coming; that is, it is an upwind flux.

We have not said anything about whether the above methods are good – or even convergent, although the example does highlight the role of the numerical flux. However, discontinuous Galerkin methods and penalty methods are examples of specific members of a much wider class of methods, all with a number of important properties:

- The solutions are piecewise smooth, often polynomial, but discontinuous between elements.
- Boundary conditions and interface continuity are enforced only weakly.
- All operators are local.
- The schemes are well suited to variable order and element sizes, as all information exchange is across the interfaces only.

Before we continue with the extension of these schemes to more general situations, let us illustrate their computational performance. Discussions on how these methods are implemented are postponed to Chapter 3.

Example 2.4. Consider Eq. (2.1) as

$$\frac{\partial u}{\partial t} - 2\pi \frac{\partial u}{\partial x} = 0, \quad x \in [0, 2\pi],$$

with periodic boundary conditions and initial condition as

$$u(x, 0) = \sin(lx), \quad l = \frac{2\pi}{\lambda},$$

where λ is the wavelength. We use the strong form, Eq. (2.8), although for this simple example, the weak form yields identical results. The nodes are chosen as the Legendre-Gauss-Lobatto nodes as we shall discuss in detail in Chapter 3. An upwind flux is used and a fourth-order explicit Runge-Kutta method is employed to integrate the equations in time with the timestep chosen small enough to ensure that timestep errors can be neglected (See Chapter 3 for details on the implementation).

In Table 2.1 we list a number of results, showing the global L^2-error at final time $T = \pi$ as a function of the number of elements, K, and the order of the local approximation, N. Inspecting these results, we observe several things. First, the scheme is clearly convergent and there are two roads to a converged result; one can increase the local order of approximation, N, and/or one can increase the number of elements, K.

The rate by which the results converge are not, however, the same when changing N and K. If we define $h = 2\pi/K$ as a measure of the size of the local element, we observe that

$$\|u - u_h\|_{\Omega,h} \leq Ch^{N+1}.$$

Thus, it is the order of the local approximation that gives the fast convergence rate. The constant, C, does not depend on h, but it may depend on the final time, T, of the solution. To highlight this, we consider in Table 2.2 the same problem but solved at different final times, T.

This indicates a linear scaling in time as

$$\|u - u_h\|_{\Omega,h} \leq C(T)h^{N+1} \simeq (c_1 + c_2 T)h^{N+1},$$

Table 2.1. Global L^2-errors when solving the wave equation using K elements each with a local order of approximation, N. Note that for $N = 8$, the finite precision dominates and destroys the expected convergence rate.

N\K	2	4	8	16	32	64	Convergence rate
1	–	4.0E-01	9.1E-02	2.3E-02	**5.7E-03**	1.4E-03	2.0
2	2.0E-01	4.3E-02	**6.3E-03**	8.0E-04	1.0E-04	1.3E-05	3.0
4	**3.3E-03**	3.1E-04	9.9E-06	3.2E-07	1.0E-08	3.3E-10	5.0
8	2.1E-07	2.5E-09	4.8E-12	2.2E-13	5.0E-13	6.6E-13	$\simeq 9.0$

Table 2.2. Global L^2-errors as a function of the final time T when solving the wave equation using K elements each with a local order of approximation, N.

Final time (T)	π	10π	100π	1000π	2000π
(N,K)=(2,4)	4.3E-02	7.8E-02	5.6E-01	>1	>1
(N,K)=(4,2)	3.3E-03	4.4E-03	2.8E-02	2.6E-01	4.8E-01
(N,K)=(4,4)	3.1E-04	3.3E-04	3.4E-04	7.7E-04	1.4E-03

Table 2.3. Scaled execution times when solving the wave equation using K elements each with a local order of approximation, N.

N\K	2	4	8	16	32	64
1	1.00	2.19	3.50	8.13	**19.6**	54.3
2	2.00	3.75	**7.31**	15.3	38.4	110.
4	**4.88**	8.94	20.0	45.0	115.	327.
8	15.1	32.0	68.3	163.	665.	1271.
16	57.8	121.	279.	664.	1958.	5256.

Clearly, c_1 and c_2 are problem-dependent constants.

The high accuracy reflected in Table 2.1, however, comes at a price. In Table 2.3 we show the approximate execution times, scaled to one for $(N, K) = (1, 2)$, for all the examples in Table 2.1. The execution time and, thus, the computational effort scales approximately like

$$\text{Time} \simeq C(T)K(N + 1)^2,$$

where C, again, scales with the final time in a linear fashion. We recall that the timestep is taken to be very small and constant for all cases; that is, the constant C depends both on T and the size of the timestep. As we will discuss in more detail in Chapter 4, this timestep also depends on N and K, but here we neglect this effect.

At first it could look as if high order accuracy (i.e., high values of N) is not really worth considering. A closer look at the results above, however, speaks to the contrary. Clearly, if one wishes very high accuracy, then only the use of high values of N offers this. However, even for a moderate accuracy, a high-order method can be superior in terms of execution time.

Consider, as an example, an error of $\mathcal{O}(5 \times 10^{-3})$. From the results in Table 2.1 we see that this can be achieved through different combinations of (N, K) [e.g., $(1, 32)$, $(2, 8)$, and $(4, 2)$]. Comparing with the result in Table 2.3, however, the latter combination, having the highest order of approximation, is clearly also the fastest. Additionally, based on the results listed in Table 2.2, one furthermore benefits from using high values of N if long time integration is required.

To further emphasize some important aspects of the scheme, we show in Fig. 2.2 the computed solution obtained in different ways. Using $N = 1$

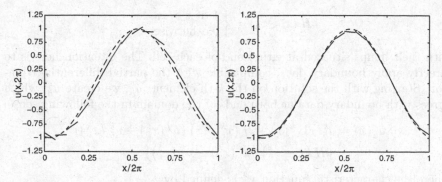

Fig. 2.2. On the left we compare the solution to the wave equation, computed using an $N = 1$, $K = 12$ discontinuous Galerkin method with a central flux. On the right the same computation employs an upwind flux. In both cases, the dashed line represents the exact solution.

and $K = 12$ elements, we show the solution at $T = 2\pi$, computed using a central flux. We note in particular the discontinuous nature of the solution, which is a characteristic of the family of methods discussed here. To contrast this, we show on the right of Fig. 2.2 the same solution computed using a pure upwind flux. This leads to a solution with smaller jumps between the elements. However, the dissipative nature of the upwind flux is also apparent. An important lesson to learn from this is that a visually smoother solution is not necessarily a more accurate solution, although in the case considered here, the global errors are comparable.

Many of the observations made in the above example regarding high-order methods can be put on firmer ground through an analysis of the phase errors. Although we will return to some of this again in Chapter 4, other insightful details can be found in [155, 159, 208, 307].

The example illustrates that to solve a given problem to a specific accuracy, one is most likely best off by having the ability to choose the element size, h, as well as the order of the scheme, N, independently, and preferably in a local manner. The ability to do this is one of the main advantages of the family of schemes discussed in this text.

2.2.2 An alternative viewpoint

Before we continue, it is instructive to consider an alternative derivation. The main dilemma posed by the choice of a piecewise polynomial representation of u_h with no a priori assumptions about continuity is how to evaluate a gradient. Starting with a solution defined on each of the elements of the grid, u_h^k, we can imagine continuing the function from the boundaries and beyond the element. In the one-dimensional case, this is achieved by adding two scaled Heaviside functions, defined as

$$H\left(x - x_0\right) = \begin{cases} 1 & \text{if } x > x_0 \\ 0 & \text{otherwise,} \end{cases}$$

with their jumps situated at either end of each cell. The intention here is to directly apply boundary data, compatible with the partial differential equation. Starting with the solution on the k-th element, u_h^k, we create $\tilde{u}_h^k$, which agrees with boundary data at both ends of the domain in the following sense:

$$\tilde{u}_h^k(x) = u_h^k(x)\chi^k(x) + H\left(x_l^k - x\right)\left(u^*\left(x_l^k\right) - u_h^k\left(x_l^k\right)\right)$$
$$+ H\left(x - x_r^k\right)\left(u^*\left(x_r^k\right) - u_h^k\left(x_r^k\right)\right),$$

where the characteristic function χ^k is defined by

$$\chi^k\left(x\right) = \begin{cases} 1 & \text{if } x \in D^k \\ 0 & \text{otherwise.} \end{cases}$$

The boundary conditions desired for u_h^k are introduced by u^*, which may depend both on the internal and external traces of u_h; that is, it is the numerical solution defined along the boundary of the element.

This continued function, $\tilde{u}_h^k$, is unlikely to have a well-defined strong gradient because of the addition of the discontinuous Heaviside functions; thus, we resort to differentiating in the sense of distributions. If we introduce the smooth compactly supported test function ϕ, the weak derivative of $\tilde{u}_h^k$, call it $d\tilde{u}_h^k$, is given as

$$\left(\phi, d\tilde{u}_h^k\right)_{D^k} := \left(\phi, \frac{d\tilde{u}_h^k}{dx}\right)_{D^k} \tag{2.11}$$

$$= \left(\phi, \frac{du_h^k}{dx}\right)_{D^k} + \left(\phi, \frac{dH\left(x_l^k - x\right)}{dx}\left(u^*\left(x_l^k\right) - u_h^k\left(x_l^k\right)\right)\right)_{D^k}$$

$$+ \left(\phi, \frac{dH\left(x - x_r^k\right)}{dx}\left(u^*\left(x_r^k\right) - u_h^k\left(x_r^k\right)\right)\right)_{D^k}$$

$$= \left(\phi, \frac{du_h^k}{dx}\right)_{D^k} - \phi\left(x_l^k\right)\left(u^*\left(x_l^k\right) - u_h^k\left(x_l^k\right)\right)$$

$$+ \phi\left(x_r^k\right)\left(u^*\left(x_r^k\right) - u_h^k\left(x_r^k\right)\right).$$

The last step follows immediately from the selection principle of the distributional derivative of the Heaviside function (see, e.g., [271]). If we choose $u^*\left(x_l^k\right) = u_h^{k-1}\left(x_r^{k-1}\right)$ and $u^*\left(x_r^k\right) = u_h^k\left(x_r^k\right)$, the terms on the right hand side reduce exactly to those appearing in the spatial terms of the upwind scheme discussed previously. By defining u^* slightly differently, it is easily seen that more general schemes can be recovered.

We can succinctly represent the upwind scheme as

$$\left(\phi_h, \frac{\partial u_h^k}{\partial t}\right)_{D^k} + \left(\phi_h, du_h^k\right)_{D^k} = 0.$$

In this formulation, we formally replace the gradients of the Heaviside functions with Dirac delta-functions whose selecting properties yield exactly the kind of statement proposed above. In the specific case of the advection equation, the boundary condition states are derived using an upwind principle. This interlude on the derivation of the basic DG scheme highlights the fact that despite its apparent complexity, the DG operator in strong form is just the result of encoding a distributional derivative to account for boundary conditions introduced at the end points of each element as a consequence of the discontinuous representation of solution.

2.3 Toward more general formulations

While the formulation of the discontinuous Galerkin schemes for the simple wave equation is illustrative and highlights many key properties, we need to consider more general problems (e.g., systems of equations, nonlinear problems, and multidimensional problems).

There are, however, only a few things that need to be addressed to achieve all of this. Let us first consider the nonlinear, scalar, conservation law

$$\frac{\partial u}{\partial t} + \frac{\partial f(u)}{\partial x} = 0, \quad x \in [L, R], \tag{2.12}$$

subject to appropriate initial conditions

$$u(x, 0) = u_0(x).$$

The boundary conditions are provided when the boundary is an inflow boundary; that is,

$$u(L, t) = g_1(t) \quad \text{when} \quad f_u(u(L, t)) \geq 0,$$
$$u(R, t) = g_2(t) \quad \text{when} \quad f_u(u(R, t)) \leq 0.$$

Proceeding exactly as for the linear case discussed previously, we assume that the global solution can be well approximated by a space of piecewise polynomial functions, defined on the union of D^k, and require the residual to be orthogonal to space of test functions, $\phi_h = \bigoplus_{k=1}^{K} \phi_h^k = \mathsf{V}_h$, to recover the locally defined weak formulation

$$\int_{\mathsf{D}^k} \left(\frac{\partial u_h^k}{\partial t} \phi_h^k - f_h^k(u_h^k) \frac{d\phi_h^k}{dx} \right) dx = -\int_{\partial \mathsf{D}^k} \hat{\boldsymbol{n}} \cdot f^* \phi_h^k \, dx,$$

and the strong form

$$\int_{\mathsf{D}^k} \left(\frac{\partial u_h^k}{\partial t} + \frac{\partial f_h^k(u_h^k)}{\partial x} \right) \phi_h^k \, dx = \int_{\partial \mathsf{D}^k} \hat{\boldsymbol{n}} \cdot \left(f_h^k(u_h^k) - f^* \right) \phi_h^k \, dx,$$

for all test functions $\phi_h^k \in V_h^k$. Assume that all local test functions can be represented as

$$x \in \mathsf{D}^k : \quad \phi_h^k(x) = \sum_{n=1}^{N_p} \hat{\phi}_n^k \psi_n(x), \tag{2.13}$$

this leads to N_p equations as

$$\int_{\mathsf{D}^k} \left(\frac{\partial u_h^k}{\partial t} \psi_n - f_h^k(u_h^k) \frac{d\psi_n}{dx} \right) dx = -\int_{\partial \mathsf{D}^k} \hat{\boldsymbol{n}} \cdot f^* \psi_n \, dx, \tag{2.14}$$

and

$$\int_{\mathsf{D}^k} \left(\frac{\partial u_h^k}{\partial t} + \frac{\partial f_h^k(u_h^k)}{\partial x} \right) \psi_n \, dx = \int_{\partial \mathsf{D}^k} \hat{\boldsymbol{n}} \cdot \left(f_h^k(u_h^k) - f^* \right) \psi_n \, dx, \tag{2.15}$$

for the weak and strong semi-discrete form, respectively. In both cases $1 \leq n \leq N_p$.

We have introduced the N-th-order approximation to the numerical flux:

$$x \in \mathsf{D}^k : \quad f_h^k(u_h^k) = \sum_{n=1}^{N_p} \hat{f}_n^k \psi_n(x) = \sum_{i=1}^{N_p} f_h(x_i^k) \ell_i^k(x).$$

As for the simple linear problem, there are several options for the test function, ϕ_h, although the Galerkin approach is the most popular.

The choice of the numerical flux, $f^* = f^*(u_h^-, u_h^+)$, is again at the very heart of the formulation. We must, of course, make sure that it is consistent [i.e., single valued as $f(u_h) = f^*(u_h, u_h)$] when specifying it. Another guideline, originating in the hugely successful development of monotone finite volume methods through the 1980s, is to require that the flux is chosen so that the scheme reduces to a monotone scheme in the low order/finite volume limit. This is ensured by requiring that $f^*(a, b)$ is nondecreasing in the first argument and nonincreasing in the second argument [218].

Another definition of a monotone flux is as an E-flux [251], satisfying

$$\forall v \in [a, b] : \ (f^*(a, b) - f(v))(b - a) \leq 0,$$

where (a, b) represent the internal and external value, respectively. The classic finite volume literature [79, 218] is filled with numerical fluxes with the above properties and we will not attempt to give an exhaustive list.

To keep things simple, we will primarily consider the Lax-Friedrichs flux along the normal, $\hat{\boldsymbol{n}}$,

$$f^{LF}(a, b) = \frac{f(a) + f(b)}{2} + \frac{C}{2}\hat{\boldsymbol{n}} \cdot (a - b),$$

as the simplest choice. The definition of C allows for some variation; that is, the classic definition

$$C \geq \max_{\inf u_h(x) \leq s \leq \sup u_h(x)} |f_u(s)|$$

is global in nature through the global inf / sup limits. Here and in the following, $f_u = \frac{\partial f}{\partial u}$ is the flux Jacobian. The global Lax-Friedrichs flux is generally more dissipative than the local Lax-Friedrichs flux given as

$$C \geq \max_{\min(a,b) \leq s \leq \max(a,b)} |f_u(s)|.$$

In both cases it is straightforward to see that the flux is monotone.

The Lax-Friedrichs flux is perhaps the simplest numerical flux and often the most efficient flux, but, generally, not leading to the most accurate scheme. Clearly, it contains as special cases the central flux and the upwind flux for the scalar linear problem discussed previously. Note that for $f(u) = au$, we recover Eq. (2.10), emphasizing the role in the dissipation played by the constant C.

It should be emphasized that many other choices for the numerical flux are available and some will likely yield superior results for specific problems. In Chapter 6.6 we shall return to this topic in some more depth and we also refer to a detailed comparison of different numerical fluxes for the Euler equations of gas dynamics in [263]. A comprehensive discussion of fluxes and approximate Riemann solvers can be found in [218]. Fluxes designed to minimize the dissipation are discussed in [39, 61].

If we consider systems of conservation laws

$$\frac{\partial \boldsymbol{u}}{\partial t} + \frac{\partial \boldsymbol{f}(\boldsymbol{u})}{\partial x} = 0, \quad x \in [L, R], \tag{2.16}$$

with initial conditions

$$\boldsymbol{u}(x, 0) = \boldsymbol{u}_0(x),$$

no essentially new component is introduced to obtain the semi-discrete schemes. In this case, $\boldsymbol{u} = [u_1(x, t), \dots, u_m(x, t)]^T$ is an m-vector and the vector-valued flux $\boldsymbol{f} : \mathrm{R}^m \to \mathrm{R}^m$.

The question of boundary conditions, and, ultimately, the choice of the numerical flux, is more complicated for the system than for the scalar problem. In general terms, the boundary conditions are

$$\mathcal{B}_L \boldsymbol{u}(L, t) = \boldsymbol{g}_1(t) \quad \text{at } x = L,$$
$$\mathcal{B}_R \boldsymbol{u}(R, t) = \boldsymbol{g}_2(t) \quad \text{at } x = R,$$

where the sum of the ranks of the boundary operators, $\mathcal{B}_L$ and $\mathcal{B}_R$, equals the required number of inflow conditions.

To construct the numerical fluxes, one can use the Lax-Friedrichs fluxes as above. The only difference from the scalar case to the system case is that the constant, C, is determined by the maximum eigenvalue of the flux Jacobian

$$\boldsymbol{f}_u = \frac{\partial \boldsymbol{f}}{\partial \boldsymbol{u}}.$$

The final extension of the formulation to multidimensional systems is entirely straightforward; that is, we assume that the solution, $\boldsymbol{u}(\boldsymbol{x}, t)$, is approximated by a multidimensional piecewise polynomial, $\boldsymbol{u}_h$. Proceeding as above, we recover the weak formulation

$$\int_{\mathsf{D}^k} \left(\frac{\partial \boldsymbol{u}_h^k}{\partial t} \phi_h^k - \boldsymbol{f}_h^k(\boldsymbol{u}_h^k) \cdot \nabla \phi_h^k \right) d\boldsymbol{x} = -\int_{\partial \mathsf{D}^k} \hat{\boldsymbol{n}} \cdot \boldsymbol{f}^* \phi_h^k \, d\boldsymbol{x}, \qquad (2.17)$$

and the strong form

$$\int_{\mathsf{D}^k} \left(\frac{\partial \boldsymbol{u}_h^k}{\partial t} + \nabla \cdot \boldsymbol{f}_h^k(\boldsymbol{u}_h^k) \right) \phi_h^k \, d\boldsymbol{x} = \int_{\partial \mathsf{D}^k} \hat{\boldsymbol{n}} \cdot \left(\boldsymbol{f}_h^k(\boldsymbol{u}_h^k) - \boldsymbol{f}^* \right) \phi_h^k \, d\boldsymbol{x}. \quad (2.18)$$

for all locally defined test functions, $\phi_h^k \in \mathsf{V}_h^k$. Naturally, $\boldsymbol{u}_h^k$ and the test functions, ϕ_h^k, are now multidimensional functions of $\boldsymbol{x} \in \mathsf{R}^d$. The semi-discrete formulation then follows immediately by expressing the local test functions as in Eq. (2.13).

The definition of the numerical fluxes follows the path discussed in the above, e.g., the Lax-Friedrichs flux along the normal, $\hat{\boldsymbol{n}}$, is

$$\boldsymbol{f}^* = \{\{\boldsymbol{f}_h(\boldsymbol{u}_h)\}\} + \frac{C}{2} [\![\boldsymbol{u}_h]\!].$$

Alternatives are possible, but this flux generally leads to both efficient, accurate, and robust methods. The constant in the Lax-Friedrichs flux is given as

$$C = \max_u \left| \lambda \left(\hat{\boldsymbol{n}} \cdot \frac{\partial \boldsymbol{f}}{\partial \boldsymbol{u}} \right) \right|,$$

where $\lambda(\cdot)$ indicates the eigenvalue of the matrix.

2.4 Interlude on linear hyperbolic problems

For linear systems, the construction of the upwind numerical flux is particularly simple and we will discuss this in a bit more detail. Important application areas include Maxwell's equations and the equations of acoustics and elasticity.

To illustrate the basic approach, let us consider the two-dimensional system

$$\mathcal{Q}(\boldsymbol{x}) \frac{\partial \boldsymbol{u}}{\partial t} + \nabla \cdot \mathcal{F} = \mathcal{Q}(\boldsymbol{x}) \frac{\partial \boldsymbol{u}}{\partial t} + \frac{\partial \boldsymbol{F}_1}{\partial x} + \frac{\partial \boldsymbol{F}_2}{\partial y} = 0, \qquad (2.19)$$

where the flux is assumed to be given as

$$\mathcal{F} = [\boldsymbol{F}_1, \boldsymbol{F}_2] = [\mathcal{A}_1(\boldsymbol{x})\boldsymbol{u}, \mathcal{A}_2(\boldsymbol{x})\boldsymbol{u}].$$

Furthermore, we will make the natural assumption that $\mathcal{Q}(\boldsymbol{x})$ is invertible and symmetric for all $\boldsymbol{x} \in \Omega$. To formulate the numerical flux, we will need an approximation of $\hat{n} \cdot \mathcal{F}$ utilizing information from both sides of the interface. Exactly how this information is combined should follow the dynamics of the equations.

Let us first assume that $\mathcal{Q}(\boldsymbol{x})$ and $\mathcal{A}_i(\boldsymbol{x})$ vary smoothly throughout Ω. In this case, we can rewrite Eq. (2.19) as

$$\mathcal{Q}(\boldsymbol{x}) \frac{\partial \boldsymbol{u}}{\partial t} + \mathcal{A}_1(\boldsymbol{x}) \frac{\partial \boldsymbol{u}}{\partial x} + \mathcal{A}_2(\boldsymbol{x}) \frac{\partial \boldsymbol{u}}{\partial y} + \mathcal{B}(\boldsymbol{x}) \boldsymbol{u} = 0,$$

where $\mathcal{B}$ collects all low-order terms; for example, it vanishes if $\mathcal{A}_i$ is constant. Since we are interested in the formulation of a flux along the normal, $\hat{n}$, we will consider the operator

$$\Pi = (\hat{n}_x \mathcal{A}_1(\boldsymbol{x}) + \hat{n}_y \mathcal{A}_2(\boldsymbol{x})).$$

Note in particular that

$$\hat{n} \cdot \mathcal{F} = \Pi \boldsymbol{u}.$$

The dynamics of the linear system can be understood by considering $\mathcal{Q}^{-1}\Pi$. Let us assume that $\mathcal{Q}^{-1}\Pi$ can be diagonalized as

$$\mathcal{Q}^{-1}\Pi = \mathcal{S} \Lambda \mathcal{S}^{-1},$$

where the diagonal matrix, Λ, has purely real entries; that is, Eq. (2.19) is a strongly hyperbolic system [142]. We express this as

$$\Lambda = \Lambda^+ + \Lambda^-,$$

corresponding to the elements of Λ that have positive and negative signs, respectively. Thus, the nonzero elements of Λ^- correspond to those elements of the characteristic vector $\mathcal{S}^{-1}\boldsymbol{u}$ where the direction of propagation is opposite to the normal (i.e., they are incoming components). In contrast, those elements corresponding to Λ^+ reflect components propagating along $\hat{n}$ (i.e., they are leaving through the boundary). The basic picture is illustrated in Fig. 2.3, where we, for completeness, also illustrate a $\lambda_2 = 0$ eigenvalue (i.e., a nonpropagating mode).

With this basic understanding, it is clear that a numerical upwind flux can be obtained as

$$(\hat{n} \cdot \mathcal{F})^* = \mathcal{Q}\mathcal{S} \left(\Lambda^+ \mathcal{S}^{-1}\boldsymbol{u}^- + \Lambda^- \mathcal{S}^{-1}\boldsymbol{u}^+ \right),$$

by simply combining the information from the two sides of the shared edge in the appropriate manner.

As intuitive as this approach is, it hinges on the assumption that $\mathcal{Q}^{-1}$ and Π vary smoothly with $\boldsymbol{x}$ throughout Ω. Unfortunately, this is not the case for many types of application (e.g., electromagnetic or acoustic problems with piecewise smooth materials).

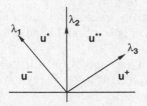

Fig. 2.3. Sketch of the characteristic wave speeds of a three-wave system at a boundary between two states, u^- and u^+. The two intermediate states, u^* and u^{**}, are used to derive the upwind flux. In this case, $\lambda_1 < 0$, $\lambda_2 = 0$, and $\lambda_3 > 0$.

To derive the proper numerical upwind flux for such cases, we need to be more careful. One should keep in mind that the much simpler local Lax-Friedrichs flux also works in this case, albeit most likely leading to more dissipation than if the upwind flux is used.

For simplicity, we assume that we have only three entries in Λ, given as

$$\lambda_1 = -\lambda, \quad \lambda_2 = 0, \quad \lambda_3 = \lambda,$$

with $\lambda > 0$; that is, the wave corresponding to λ_1 is entering the domain, the wave corresponding to λ_3 is leaving, and λ_2 corresponds to a stationary wave as illustrated in Fig. 2.3.

Following the well-developed theory or Riemann solvers [218, 303], we know that

$$\forall i : \quad -\lambda_i \mathcal{Q}[u^- - u^+] + [(\Pi u)^- - (\Pi u)^+] = 0, \tag{2.20}$$

must hold across each wave. This is also known as the Rankine-Hugoniot condition and is a simple consequence of conservation of u across the point of discontinuity. To appreciate this, consider the scalar wave equation

$$\frac{\partial u}{\partial t} + \lambda \frac{\partial u}{\partial x} = 0, \quad x \in [a, b].$$

Integrating over the interval, we have

$$\frac{d}{dt} \int_a^b u \, dx = -\lambda \left(u(b, t) - u(a, t) \right) = f(a, t) - f(b, t),$$

since $f = \lambda u$. On the other hand, since the wave is propagating at a constant speed, λ, we also have

$$\frac{d}{dt} \int_a^b u \, dx = \frac{d}{dt} \left((\lambda t - a) u^- + (b - \lambda t) u^+ \right) = \lambda (u^- - u^+).$$

Taking $a \to x^-$ and $b \to x^+$, we recover the jump conditions

$$-\lambda (u^- - u^+) + (f^- - f^+) = 0.$$

The generalization to Eq. (2.20) is now straightforward.

Returning to the problem in Fig. 2.3, we have the system of equations

$$\lambda Q^-(u^* - u^-) + \left[(\Pi u)^* - (\Pi u)^-\right] = 0,$$
$$\left[(\Pi u)^* - (\Pi u)^{**}\right] = 0,$$
$$-\lambda Q^+(u^{**} - u^+) + \left[(\Pi u)^{**} - (\Pi u)^+\right] = 0,$$

where (u^*, u^{**}) represents the intermediate states.

The numerical flux can then be obtained by realizing that

$$(\hat{n} \cdot \mathcal{F})^* = (\Pi u)^* = (\Pi u)^{**},$$

which one can attempt to express using (u^-, u^+) through the jump conditions above. This leads to an upwind flux for the general discontinuous case.

To appreciate this approach, we consider a few examples.

Example 2.5. Consider first the linear hyperbolic problem

$$\frac{\partial q}{\partial t} + \mathcal{A}\frac{\partial q}{\partial x} = \frac{\partial}{\partial t}\begin{bmatrix} u \\ v \end{bmatrix} + \begin{bmatrix} a(x) & 0 \\ 0 & -a(x) \end{bmatrix}\frac{\partial}{\partial x}\begin{bmatrix} u \\ v \end{bmatrix} = 0,$$

with $a(x)$ being piecewise constant. For this simple equation, it is clear that $u(x,t)$ propagates right while $v(x,t)$ propagates left and we could use this to form a simple upwind flux.

However, let us proceed using the Riemann jump conditions. If we introduce $a^{\pm}$ as the values of $a(x)$ on two sides of the interface, we recover the conditions

$$a^-(q^* - q^-) + (\Pi q)^* - (\Pi q)^- = 0,$$
$$-a^+(q^* - q^+) + (\Pi q)^* - (\Pi q)^+ = 0,$$

where q^* refers to the intermediate state, $(\Pi q)^*$ is the numerical flux along $\hat{n}$, and

$$(\Pi q)^{\pm} = \hat{n} \cdot (\mathcal{A}q)^{\pm} = \hat{n} \cdot \begin{bmatrix} a^{\pm} & 0 \\ 0 & -a^{\pm} \end{bmatrix}\begin{bmatrix} u^{\pm} \\ v^{\pm} \end{bmatrix} = \hat{n} \cdot \begin{bmatrix} a^{\pm}u^{\pm} \\ -a^{\pm}v^{\pm} \end{bmatrix}.$$

A bit of manipulation yields

$$(\Pi q)^* = \frac{1}{a^+ + a^-}\left(a^+(\Pi q)^- + a^-(\Pi q)^+ + a^+a^-(q^- - q^+)\right),$$

which simplifies as

$$(\Pi q)^* = \frac{2a^+a^-}{a^+ + a^-}\hat{n} \cdot \left(\begin{bmatrix} \{\{u\}\} \\ -\{\{v\}\} \end{bmatrix} + \frac{1}{2}\begin{bmatrix} [\![u]\!] \\ [\![v]\!] \end{bmatrix}\right),$$

and the numerical flux $(\mathcal{A}q)^*$ follows directly from the definition of $(\Pi q)^*$. We observe that if $a(x)$ is smooth (i.e., $a^- = a^+$) then the numerical flux is simply upwinding. Furthermore, for the general case, the above is equivalent to defining an intermediate wave speed, a^*, as

$$a^* = \frac{2a^- a^+}{a^+ + a^-},$$

which is the harmonic average.

Let us also consider a slightly more complicated problem, originating in electromagnetics.

Example 2.6. Consider the one-dimensional Maxwell's equations

$$\begin{bmatrix} \varepsilon(x) & 0 \\ 0 & \mu(x) \end{bmatrix} \frac{\partial}{\partial t} \begin{bmatrix} E \\ H \end{bmatrix} + \begin{bmatrix} 0 & 1 \\ 1 & 0 \end{bmatrix} \frac{\partial}{\partial x} \begin{bmatrix} E \\ H \end{bmatrix} = 0. \tag{2.21}$$

Here $(E, H) = (E(x,t), H(x,t))$ represent the electric and magnetic field, respectively, while ε and μ are the electric and magnetic material properties, known as permittivity and permeability, respectively.

To simplify the notation, let us write this as

$$\mathcal{Q} \frac{\partial q}{\partial t} + \mathcal{A} \frac{\partial q}{\partial x} = 0,$$

where

$$\mathcal{Q} = \begin{bmatrix} \varepsilon(x) & 0 \\ 0 & \mu(x) \end{bmatrix}, \quad \mathcal{A} = \begin{bmatrix} 0 & 1 \\ 1 & 0 \end{bmatrix}, \quad q = \begin{bmatrix} E \\ H \end{bmatrix},$$

reflect the spatially varying material coefficients, the one-dimensional rotation operator, and the vector of state variables, respectively.

The flux is given as $\mathcal{A}q$ and the eigenvalues of $\mathcal{Q}^{-1}\mathcal{A}$ are $\pm(\varepsilon\mu)^{-1/2}$, reflecting the two counter-propagating light waves propagating at the local speed of light, $c = (\varepsilon\mu)^{-1/2}$.

Proceeding by using the Riemann conditions, we obtain

$$c^- \mathcal{Q}^-(q^* - q^-) + (\Pi q)^* - (\Pi q)^- = 0,$$
$$-c^+ \mathcal{Q}^+(q^* - q^+) + (\Pi q)^* - (\Pi q)^+ = 0,$$

where q^* refers to the intermediate state, $(\mathcal{A}q)^*$ is the numerical flux, and

$$(\Pi q)^\pm = \hat{n} \cdot (\mathcal{A}q)^\pm = \hat{n} \cdot \begin{bmatrix} H^\pm \\ E^\pm \end{bmatrix}.$$

Simple manipulations yield

$$(c^+ \mathcal{Q}^+ + c^- \mathcal{Q}^-)(\Pi q)^* = c^+ \mathcal{Q}^+ (\Pi q)^- + c^- \mathcal{Q}^- (\Pi q)^+ + c^- c^+ \mathcal{Q}^- \mathcal{Q}^+ (q^- - q^+),$$

or

$$H^* = \frac{1}{\{\{Z\}\}}\left(\{\{ZH\}\} + \frac{1}{2}[E]\right), \quad E^* = \frac{1}{\{\{Y\}\}}\left(\{\{YE\}\} + \frac{1}{2}[H]\right),$$

where

$$Z^{\pm} = \sqrt{\frac{\mu^{\pm}}{\varepsilon^{\pm}}} = (Y^{\pm})^{-1},$$

represents the impedance of the medium.

If we again consider the simplest case of a continuous medium, things simplify considerably as

$$H^* = \{\{H\}\} + \frac{Y}{2}[E], \quad E^* = \{\{E\}\} + \frac{Z}{2}[H],$$

which we recognize as the Lax-Friedrichs flux since

$$\frac{Y}{\varepsilon} = \frac{Z}{\mu} = \frac{1}{\sqrt{\varepsilon\mu}} = c$$

is the speed of light (i.e., the fastest wave speed in the system).

2.5 Exercises

1. Consider the scalar problem

$$\frac{\partial u}{\partial t} = \nu \frac{\partial^2 u}{\partial x^2}, \quad x \in [-1, 1].$$

   a) Use an energy method to determine how many boundary conditions are needed and suggest different combinations.

   b) Does the problem preserve energy in a periodic domain?

2. Consider the scalar problem

$$\frac{\partial u}{\partial t} = \frac{\partial^3 u}{\partial x^3}, \quad x \in [-1, 1].$$

   a) Use an energy method to determine how many boundary conditions are needed and suggest different combinations.

   b) Does the problem preserve energy in a periodic domain?

3. Show that for the linear scalar problem, the local Lax-Friedrichs, the global Lax-Friedrichs, and the upwind flux are all the same.

4. Consider the scalar problem

$$\frac{\partial u}{\partial t} + a\frac{\partial u}{\partial x} = bu, \quad x \in [-1, 1],$$

with proper initial conditions and a and b being real constants.

a) Propose a discontinuous Galerkin method for solving this problem.

b) Prove that the semidiscrete scheme is stable.

5. Consider the scalar problem

$$\frac{\partial u}{\partial t} + a\frac{\partial u}{\partial x} = f(x, t), \quad x \in [-1, 1],$$

with proper initial conditions and a is a real constant.

a) Propose a discontinuous Galerkin method for solving this problem.

b) Prove that the scheme is stable in a semidiscrete sense. (Hint: To prove stability, consider the problem for the error between the computed and the exact solution – known as the error equation).

6. Consider the system

$$\frac{\partial u}{\partial t} + \frac{\partial v}{\partial x} = 0, \quad \frac{\partial v}{\partial t} + \frac{\partial u}{\partial x} = 0, \quad x \in [-1, 1],$$

subject to proper initial conditions.

a) Discuss how many boundary conditions are needed and which type they must be.

b) Propose a discontinuous Galerkin method for solving this problem.

c) Prove that the semidiscrete scheme is stable.

d) Can one use the same scheme to solve the almost identical problem

$$\frac{\partial u}{\partial t} + \frac{\partial v}{\partial x} = 0, \quad \frac{\partial v}{\partial t} - \frac{\partial u}{\partial x} = 0, \quad x \in [-1, 1],$$

subject to proper initial conditions?

7. Consider the one-dimensional model for acoustics in a mean flow, given as

$$\frac{\partial u}{\partial t} + M(x)\frac{\partial u}{\partial x} + M_x u = -\frac{\partial p}{\partial x},$$

$$\frac{\partial p}{\partial t} + M(x)\frac{\partial p}{\partial x} + M_x p = -\frac{\partial u}{\partial x},$$

where the unknowns are the velocity, $u(x, t)$, and the pressure, $p(x, t)$. The Mach number, $M(x) = u_0(x)/c_0(x)$, reflects the velocity of the steady mean flow and is assumed to be piecewise constant.

a) Using an energy technique, discuss how many boundary conditions are needed in a finite domain at each end – note that this depends on $M(x)$.

b) In the case where the problem is periodic, does the system conserve energy?

c) Assume that $M(x)$ is smooth and derive an upwind flux for it.

d) Assume now that $M(x)$ can be discontinuous (e.g., modeling sound propagating through a stationary shock), and derive the upwind flux at a discontinuous point using the Rankine-Hugoniot conditions.

3

Making it work in one dimension

As simple as the formulations in the last chapter appear, there is often a leap between mathematical formulations and an actual implementation of the algorithms. This is particularly true when one considers important issues such as efficiency, flexibility, and robustness of the resulting methods.

In this chapter we address these issues by first discussing details such as the form of the local basis and, subsequently, how one implements the nodal DG-FEMs in a flexible way. To keep things simple, we continue the emphasis on one-dimensional linear problems, although this results in a few apparently unnecessarily complex constructions. We ask the reader to bear with us, as this slightly more general approach will pay off when we begin to consider more complex nonlinear and/or higher-dimensional problems.

3.1 Legendre polynomials and nodal elements

In Chapter 2 we started out by assuming that one can represent the global solution as a the direct sum of local piecewise polynomial solution as

$$u(x,t) \simeq u_h(x,t) = \bigoplus_{k=1}^{K} u_h^k(x^k,t).$$

The careful reader will note that this notation is a bit careless, as we do not address what exactly happens at the overlapping interfaces. However, a more careful definition does not add anything essential at this point and we will use this notation to reflect that the global solution is obtained by combining the K local solutions as defined by the scheme.

The local solutions are assumed to be of the form

$$x \in \mathsf{D}^k = [x_l^k, x_r^k] : \ u_h^k(x,t) = \sum_{n=1}^{N_p} \hat{u}_n^k(t)\psi_n(x) = \sum_{i=1}^{N_p} u_h^k(x_i^k,t)\ell_i^k(x).$$

We did not, however, discuss the specifics of this representation, as this is less important from a theoretical point of view. The results in Example 2.4 clearly illustrate, however, that the accuracy of the method is closely linked to the order of the local polynomial representation and some care is warranted when choosing this.

Let us begin by introducing the affine mapping

$$x \in \mathsf{D}^k : \ x(r) = x_l^k + \frac{1+r}{2} h^k, \ \ h^k = x_r^k - x_l^k, \tag{3.1}$$

with the reference variable $r \in \mathsf{I} = [-1, 1]$. We consider local polynomial representations of the form

$$x \in \mathsf{D}^k : \ u_h^k(x(r), t) = \sum_{n=1}^{N_p} \hat{u}_n^k(t) \psi_n(r) = \sum_{i=1}^{N_p} u_h^k(x_i^k, t) \ell_i^k(r).$$

Let us first discuss the local modal expansion,

$$u_h(r) = \sum_{n=1}^{N_p} \hat{u}_n \psi_n(r).$$

where we have dropped the superscripts for element k and the explicit time dependence, t, for clarity of notation.

As a first choice, one could consider $\psi_n(r) = r^{n-1}$ (i.e., the simple monomial basis). This leaves only the question of how to recover $\hat{u}_n$. A natural way is by an L^2-projection; that is, by requiring that

$$(u(r), \psi_m(r))_\mathsf{I} = \sum_{n=1}^{N_p} \hat{u}_n \left(\psi_n(r), \psi_m(r) \right)_\mathsf{I},$$

for each of the N_p basis functions ψ_n. We have introduced the inner product on the interval I as

$$(u, v)_\mathsf{I} = \int_{-1}^1 uv \, dx.$$

This yields

$$\mathcal{M}\hat{u} = u,$$

where

$$\mathcal{M}_{ij} = (\psi_i, \psi_j)_\mathsf{I}, \ \ \hat{u} = [\hat{u}_1, \dots, \hat{u}_{N_p}]^T, \ \ u_i = (u, \psi_i)_\mathsf{I},$$

leading to N_p equations for the N_p unknown expansion coefficients, $\hat{u}_i$. However, note that

$$\mathcal{M}_{ij} = \frac{1}{i+j-1} \left[1 + (-1)^{i+j} \right], \tag{3.2}$$

which resembles a Hilbert matrix, known to be very poorly conditioned. If we compute the condition number, $\kappa(\mathcal{M})$, for $\mathcal{M}$ for increasing order of approximation, N, we observe in Table 3.1 the very rapidly deteriorating conditioning of $\mathcal{M}$. The reason for this is evident in Eq. (3.2), where the coefficient

Table 3.1. Condition number of $\mathcal{M}$ based on the monomial basis for increasing value of N.

N	2	4	8	16
$\kappa(\mathcal{M})$	1.4e+1	3.6e+2	3.1e+5	3.0e+11

$(i + j - 1)^{-1}$ implies an increasing close to linear dependence of the basis as (i, j) increases, resulting in the ill-conditioning. The impact of this is that, even at moderately high order (e.g., $N > 10$), one cannot accurately compute $\hat{u}$ and therefore not a good polynomial representation, u_h.

A natural solution to this problem is to seek an orthonormal basis as a more suitable and computationally stable approach. We can simply take the monomial basis, r^n, and recover an orthonormal basis from this through an L^2-based Gram-Schmidt orthogonalization approach. This results in the orthonormal basis [296]

$$\psi_n(r) = \tilde{P}_{n-1}(r) = \frac{P_{n-1}(r)}{\sqrt{\gamma_{n-1}}},$$

where $P_n(r)$ are the classic Legendre polynomials of order n and

$$\gamma_n = \frac{2}{2n + 1}$$

is the normalization. An easy way to compute this new basis is through the recurrence

$$r\tilde{P}_n(r) = a_n \tilde{P}_{n-1}(r) + a_{n+1}\tilde{P}_{n+1}(r),$$

$$a_n = \sqrt{\frac{n^2}{(2n + 1)(2n - 1)}},$$

with

$$\psi_1(r) = \tilde{P}_0(r) = \frac{1}{\sqrt{2}}, \quad \psi_2(r) = \tilde{P}_1(r) = \sqrt{\frac{3}{2}}r.$$

Appendix A discusses how to implement these recurrences and a number of other useful manipulations of $\tilde{P}_n(r)$, which is a special example of the much larger family of normalized Jacobi polynomials [159, 296].

To evaluate $\tilde{P}_n(r)$ we use the library routine JacobiP.m as

```
>> P =JacobiP(r,0,0,n);
```

Using this basis, the mass matrix, $\mathcal{M}$, is the identity and the problem of conditioning is resolved. It leaves open the question of how to compute $\hat{u}_n$, which is given as

$$\hat{u}_n = (u, \psi_n)_\mathrm{I} = (u, \tilde{P}_{n-1})_\mathrm{I}.$$

For a general function u, the evaluation of this inner product is nontrivial. The only practical way is to approximate the integral with a sum. In particular, one can use a Gaussian quadrature of the form

$$\hat{u}_n \simeq \sum_{i=1}^{N_p} u(r_i) \tilde{P}_{n-1}(r_i) w_i,$$

where r_i are the quadrature points and w_i are the quadrature weights. An important property of such quadratures is that they are exact, provided u is a polynomial of order $2N_p - 1$ [90]. The weights and nodes for the Gaussian quadrature can be computed using JacobiGQ.m, as discussed in Appendix A. To compute the $N_p = N+1$ nodes and weights $(\boldsymbol{r}, \boldsymbol{w})$ for an $(2N+1)$-th-order accurate quadrature, one executes the command

$$>> [\boldsymbol{r},\boldsymbol{w}] = \text{JacobiGQ}(0,0,N);$$

For the purpose of later generalizations, we consider a slightly different approach and define $\hat{u}_n$ such that the approximation is interpolatory; that is

$$u(\xi_i) = \sum_{n=1}^{N_p} \hat{u}_n \tilde{P}_{n-1}(\xi_i),$$

where ξ_i represents a set of N_p distinct grid points. Note that these need not be associated with any quadrature points. We can now write

$$\mathcal{V}\hat{\boldsymbol{u}} = \boldsymbol{u},$$

where

$$\mathcal{V}_{ij} = \tilde{P}_{j-1}(\xi_i), \quad \hat{\boldsymbol{u}}_i = \hat{u}_i, \quad \boldsymbol{u}_i = u(\xi_i).$$

This matrix, $\mathcal{V}$, is recognized as a generalized Vandermonde matrix and will play a pivotal role in all of the subsequent developments of the scheme. This matrix establishes the connection between the modes, $\hat{\boldsymbol{u}}$, and the nodal values, $\boldsymbol{u}$, and we will benefit from making sure that it is well conditioned.

Since we have already determined that $\tilde{P}_n$ is the optimal basis, we are left with the need to choose the grid points ξ_i to define the Vandermonde matrix. There is significant freedom in this choice, so let us try to develop a bit of intuition for a reasonable criteria.

We first recognize that if

$$u(r) \simeq u_h(r) = \sum_{n=1}^{N_p} \hat{u}_n \tilde{P}_{n-1}(r),$$

is an interpolant (i.e., $u(\xi_i) = u_h(\xi_i)$ at the grid points, ξ_i), then we can write it as

$$u(r) \simeq u_h(r) = \sum_{i=1}^{N_p} u(\xi_i) \ell_i(r).$$

Here,

$$\ell_i(r) = \prod_{\substack{j=1 \\ j \neq i}}^{N_p} \frac{r - \xi_j}{\xi_i - \xi_j},$$

is the interpolating Lagrange polynomial with the property $\ell_i(r_j) = \delta_{ij}$. It is well known that $\ell_i(r)$ exists and is unique as long as the ξ_i's are distinct.

If we define the Lebesque constant

$$\Lambda = \max_r \sum_{i=1}^{N_p} |\ell_i(r)|,$$

we realize that

$$\|u - u_h\|_\infty = \|u - u^* + u^* - u_h\|_\infty \leq \|u - u^*\|_\infty + \|u^* - u_h\|_\infty \leq (1 + \Lambda)\|u - u^*\|_\infty,$$

where $\|\cdot\|_\infty$ is the usual maximum norm and u^* represents the best approximating polynomial of order N. Hence, the Lebesque constant indicates how far away the interpolation may be from the best possible polynomial representation u^*. Note that Λ is determined solely by the grid points, ξ_i. To get an optimal approximation, we should therefore aim to identify those points, ξ_i, that minimize the Lebesque constant.

To appreciate how this relates to the conditioning of $\mathcal{V}$, recognize that as a consequence of uniqueness of the polynomial interpolation, we have

$$\mathcal{V}^T \boldsymbol{\ell}(r) = \tilde{\boldsymbol{P}}(r),$$

where $\boldsymbol{\ell} = [\ell_1(r), \ldots, \ell_{N_p}(r)]^T$ and $\tilde{\boldsymbol{P}}(r) = [\tilde{P}_0(r), \ldots, \tilde{P}_N(r)]^T$. We are interested in the particular solution, $\boldsymbol{\ell}$, which minimizes the Lebesque constant. If we recall Cramer's rule for solving linear systems of equations

$$\ell_i(r) = \frac{\mathrm{Det}[\mathcal{V}^T(:,1), \mathcal{V}^T(:,2), \ldots, \tilde{\boldsymbol{P}}(r), \mathcal{V}^T(:,i+1), \ldots, \mathcal{V}^T(:,N_p)]}{\mathrm{Det}(\mathcal{V}^T)}.$$

It suggests that it is reasonable to seek ξ_i such that the denominator (i.e., the determinant of $\mathcal{V}$), is maximized.

For this one dimensional case, the solution to this problem is known in a relatively simple form as the N_p zeros of [151, 159]

$$f(r) = (1 - r^2)\tilde{P}_N'(r).$$

These are closely related to the normalized Legendre polynomials and are known as the Legendre-Gauss-Lobatto (LGL) quadrature points. Using the library routine JacobiGL.m in Appendix A, these nodes can be computed as

```
>> [r] = JacobiGL(0,0,N);
```

To illustrate the importance of choosing these points correctly, consider the following example.

Example 3.1. We first compare the growth of the determinant of $\mathcal{V}$ as a function of N, using either the LGL points or the equidistant points

$$\xi_i = -1 + \frac{2(i-1)}{N}, \quad i \in [1, N_p].$$

In Fig. 3.1 we show the entries of $\mathcal{V}$ for different choices of the basis, $\psi_n(r)$, and the interpolation points, ξ_i. For Fig. 3.1a and Fig. 3.1b, both relying on the monomial basis, we see indications that the last columns are almost linearly dependent and that the situation is worse in the case of the equidistant points. In Fig. 3.1b the situation is slightly improved through the use of the LGL grid points.

In Fig. 3.2 we further illustrate this by computing the determinant of the Vandermonde matrix $\mathcal{V}$, highlighting a significant qualitative difference in the behavior for the two sets of nodes as N increases. For the LGL nodes, the determinant continues to increase in size with N, reflecting a well-behaved interpolation where $\ell_i(r)$ takes small values in between the nodes. For the equidistant nodes, the situation is entirely different.

While the determinant has comparable values for small values of N, the determinant for the equidistant nodes begins to decay exponentially fast once N exceeds 16, caused by the close to linear dependence of the last columns

a)
$$\begin{bmatrix}
1.00 & -1.00 & 1.00 & -1.00 & 1.00 & -1.00 & 1.00 \\
1.00 & -0.67 & 0.44 & -0.30 & 0.20 & -0.13 & 0.09 \\
1.00 & -0.33 & 0.11 & -0.04 & 0.01 & -0.00 & 0.00 \\
1.00 & 0.00 & 0.00 & 0.00 & 0.00 & 0.00 & 0.00 \\
1.00 & 0.33 & 0.11 & 0.04 & 0.01 & 0.00 & 0.00 \\
1.00 & 0.67 & 0.44 & 0.30 & 0.20 & 0.13 & 0.09 \\
1.00 & 1.00 & 1.00 & 1.00 & 1.00 & 1.00 & 1.00
\end{bmatrix}$$

b)
$$\begin{bmatrix}
1.00 & -1.00 & 1.00 & -1.00 & 1.00 & -1.00 & 1.00 \\
1.00 & -0.83 & 0.69 & -0.57 & 0.48 & -0.39 & 0.33 \\
1.00 & -0.47 & 0.22 & -0.10 & 0.05 & -0.02 & 0.01 \\
1.00 & 0.00 & 0.00 & 0.00 & 0.00 & 0.00 & 0.00 \\
1.00 & 0.47 & 0.22 & 0.10 & 0.05 & 0.02 & 0.01 \\
1.00 & 0.83 & 0.69 & 0.57 & 0.48 & 0.39 & 0.33 \\
1.00 & 1.00 & 1.00 & 1.00 & 1.00 & 1.00 & 1.00
\end{bmatrix}$$

c)
$$\begin{bmatrix}
0.71 & -1.22 & 1.58 & -1.87 & 2.12 & -2.35 & 2.55 \\
0.71 & -0.82 & 0.26 & 0.49 & -0.91 & 0.72 & -0.04 \\
0.71 & -0.41 & -0.53 & 0.76 & 0.03 & -0.78 & 0.49 \\
0.71 & 0.00 & -0.79 & 0.00 & 0.80 & 0.00 & -0.80 \\
0.71 & 0.41 & -0.53 & -0.76 & 0.03 & 0.78 & 0.49 \\
0.71 & 0.82 & 0.26 & -0.49 & -0.91 & -0.72 & -0.04 \\
0.71 & 1.22 & 1.58 & 1.87 & 2.12 & 2.35 & 2.55
\end{bmatrix}$$

d)
$$\begin{bmatrix}
0.71 & -1.22 & 1.58 & -1.87 & 2.12 & -2.35 & 2.55 \\
0.71 & -1.02 & 0.84 & -0.35 & -0.28 & 0.81 & -1.06 \\
0.71 & -0.57 & -0.27 & 0.83 & -0.50 & -0.37 & 0.85 \\
0.71 & 0.00 & -0.79 & 0.00 & 0.80 & 0.00 & -0.80 \\
0.71 & 0.57 & -0.27 & -0.83 & -0.50 & 0.37 & 0.85 \\
0.71 & 1.02 & 0.84 & 0.35 & -0.29 & -0.81 & -1.06 \\
0.71 & 1.22 & 1.58 & 1.87 & 2.12 & 2.35 & 2.55
\end{bmatrix}$$

Fig. 3.1. Entries of $\mathcal{V}$ for $N = 6$ and different choices of the basis, $\psi_n(r)$, and evaluation points, ξ_i. For (a) and (c) we use equidistant points and (b) and (d) are based on LGL points. Furthermore, (a) and (b) are based on $\mathcal{V}$ computed using a simple monomial basis, $\psi_n(r) = r^{n-1}$, while (c) and (d) are based on the orthonormal basis, $\psi_n(r) = \tilde{P}_{n-1}(r)$.

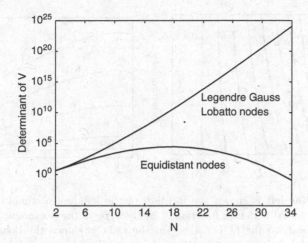

Fig. 3.2. Growth of the determinant of the Vandermonde matrix, $\mathcal{V}$, when computed based on LGL nodes and the simple equidistant nodes.

in $\mathcal{V}$. For the interpolation, this manifests itself as the well-known Runge phenomenon [90] for interpolation at equidistant grids. The rapid decay of the determinant enables an exponentially fast growth of $\ell_i(r)$ between the grid points, resulting in a very poorly behaved interpolation. This is also reflected in the corresponding Lebesque constant, which grows like 2^{N_p} for the equidistant nodes while it grows like $\log N_p$ in the optimal case of the LGL nodes (see [151, 307] and references therein).

The main lesson here is that it is generally not a good idea to use equidistant grid points when representing the local interpolating polynomial solution. This point is particularly important when the interest is in the development of methods performing robustly at higher values of N.

One can ask, however, whether there is anything special about the LGL points, beyond the fact that they maximize the Vandermonde determinant. In particular, can one find other families of points with the similar property that the Vandermonde determinant continues to grow with the order of approximation.

To investigate this, we compute the determinant of $\mathcal{V}$ for fixed values of N but with points which are Gauss-Lobatto points for the symmetric Jacobi polynomials $P_n^{(\alpha,\alpha)}(r)$ (see Appendix A). A special case of these are the Legendre polynomials obtained with $\alpha = 0$.

In Fig. 3.3 we show the spatial variation of the nodes for $N = 16$ as well as the variation of the Vandermonde determinant as a function of α for different values of N. While this confirms that $\alpha = 0$ maximizes the determinant for all N, we also see that there is a broad range of values of α where the quantitative behavior is identical to that obtained with the LGL nodes. Only when α becomes significantly larger than zero do we begin to see a significant

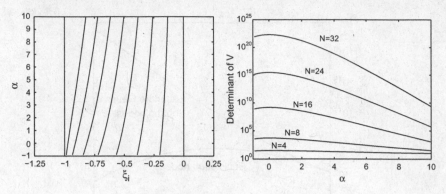

Fig. 3.3. On the left is shown the location of the left half of the Jacobi-Gauss-Lobatto nodes for $N = 16$ as a function of α, the type of the polynomial. Note that $\alpha = 0$ corresponds to the LGL nodes. On the right is shown the behavior of the Vandermonde determinant as a function of α for different values of N, confirming the optimality of the LGL nodes but also showing robustness of the interpolation to this choice.

decay in the value of the determinant, indicating possible problems in the interpolation. Considering the corresponding nodal distribution in Fig. 3.3 for these values of α, this is perhaps not surprising.

This highlights that it is the overall structure of the nodes rather than the details of the individual node position that is important; for example, one could optimize these nodal sets for various applications.

To summarize matters, we have local approximations of the form

$$u(r) \simeq u_h(r) = \sum_{n=1}^{N_p} \hat{u}_n \tilde{P}_{n-1}(r) = \sum_{i=1}^{N_p} u(r_i)\ell_i(r), \qquad (3.3)$$

where $\xi_i = r_i$ are the Legendre-Gauss-Lobatto quadrature points. A central component of this construction is the Vandermonde matrix, $\mathcal{V}$, which establishes the connections

$$\boldsymbol{u} = \mathcal{V}\hat{\boldsymbol{u}}, \quad \mathcal{V}^T \boldsymbol{\ell}(r) = \tilde{\boldsymbol{P}}(r), \quad \mathcal{V}_{ij} = \tilde{P}_j(r_i).$$

By carefully choosing the orthonormal Legendre basis, $\tilde{P}_n(r)$, and the nodal points, r_i, we have ensured that $\mathcal{V}$ is a well-conditioned object and that the resulting interpolation is well behaved. A script for initializing $\mathcal{V}$ is given in Vandermonde1D.m.

————————————————— | Vandermonde1D.m | —————————————————

```
function [V1D] = Vandermonde1D(N,r)

% function [V1D] = Vandermonde1D(N,r)
% Purpose : Initialize the 1D Vandermonde Matrix, V_{ij} = phi_j(r_i);

V1D = zeros(length(r),N+1);
for j=1:N+1
    V1D(:,j) = JacobiP(r(:), 0, 0, j-1);
end;
return
```

Using $\mathcal{V}$, we can transform directly between modal representations, using $\hat{u}_n$ as the unknowns, and nodal representations, using $u(r_i)$ as the unknowns. The two representations are mathematically equivalent but computationally different and they each have certain advantages and disadvantages, depending on the application at hand.

In what remains, we will focus on the nodal representations, which has advantages for more complex problems as we will see later. However, it is important to keep the alternative in mind as an optional representation.

3.2 Elementwise operations

With the development of a suitable local representation of the solution, we are now prepared to discuss the various elementwise operators needed to implement the algorithms from Chapter 2.

We begin with the two operators $\mathcal{M}^k$ and $\mathcal{S}^k$. For the former, the entries are given as

$$\mathcal{M}_{ij}^k = \int_{x_l^k}^{x_r^k} \ell_i^k(x)\ell_j^k(x)\,dx = \frac{h^k}{2}\int_{-1}^{1} \ell_i(r)\ell_j(r)\,dr = \frac{h^k}{2}(\ell_i,\ell_j)_I = \frac{h^k}{2}\mathcal{M}_{ij},$$

where the coefficient in front is the Jacobian coming from Eq. (3.1) and $\mathcal{M}$ is the mass matrix defined on the reference element I.

Recall that

$$\ell_i(r) = \sum_{n=1}^{N_p}(\mathcal{V}^T)_{in}^{-1}\tilde{P}_{n-1}(r),$$

from which we recover

$$\begin{aligned}
\mathcal{M}_{ij} &= \int_{-1}^{1}\sum_{n=1}^{N_p}(\mathcal{V}^T)_{in}^{-1}\tilde{P}_{n-1}(r)\sum_{m=1}^{N_p}(\mathcal{V}^T)_{jm}^{-1}\tilde{P}_{m-1}(r)\,dr \\
&= \sum_{n=1}^{N_p}\sum_{m=1}^{N_p}(\mathcal{V}^T)_{in}^{-1}(\mathcal{V}^T)_{jm}^{-1}(\tilde{P}_{n-1},\tilde{P}_{m-1})_I = \sum_{n=1}^{N_p}(\mathcal{V}^T)_{in}^{-1}(\mathcal{V}^T)_{jn}^{-1},
\end{aligned}$$

where the latter follows from orthonormality of $\tilde{P}_n(r)$. Thus, we recover

$$\mathcal{M}^k = \frac{h^k}{2}\mathcal{M} = \frac{h^k}{2}\left(\mathcal{V}\mathcal{V}^T\right)^{-1}.$$

Let us now consider $\mathcal{S}^k$, given as

$$\mathcal{S}^k_{ij} = \int_{x^k_l}^{x^k_r} \ell^k_i(x)\frac{d\ell^k_j(x)}{dx}\,dx = \int_{-1}^{1} \ell_i(r)\frac{d\ell_j(r)}{dr}\,dr = \mathcal{S}_{ij}.$$

Note in particular that no metric constant is introduced by the transformation. To realize a simple way to compute this, we define the differentiation matrix, $\mathcal{D}_r$, with the entries

$$\mathcal{D}_{r,(i,j)} = \left.\frac{d\ell_j}{dr}\right|_{r_i}.$$

The motivation for this is found in Eq. (3.3), where we observe that $\mathcal{D}_r$ is the operator that transforms point values, $u(r_i)$, to derivatives at these same points (e.g., $\boldsymbol{u}'_h = \mathcal{D}_r\boldsymbol{u}_h$).

Consider now the product $\mathcal{M}\mathcal{D}_r$, with entries

$$(\mathcal{M}\mathcal{D}_r)_{(i,j)} = \sum_{n=1}^{N_p} \mathcal{M}_{in}\mathcal{D}_{r,(n,j)} = \sum_{n=1}^{N_p} \int_{-1}^{1} \ell_i(r)\ell_n(r)\left.\frac{d\ell_j}{dr}\right|_{r_n}\,dr$$

$$= \int_{-1}^{1} \ell_i(r)\sum_{n=1}^{N_p}\left.\frac{d\ell_j}{dr}\right|_{r_n}\ell_n(r)\,dr = \int_{-1}^{1} \ell_i(r)\frac{d\ell_j(r)}{dr}\,dr = \mathcal{S}_{ij}.$$

In other words, we have the identity

$$\mathcal{M}\mathcal{D}_r = \mathcal{S}.$$

The entries of the differentiation matrix, $\mathcal{D}_r$, can be found directly,

$$\mathcal{V}^T\boldsymbol{\ell}(r) = \tilde{\boldsymbol{P}}(r) \quad \Rightarrow \quad \mathcal{V}^T\frac{d}{dr}\boldsymbol{\ell}(r) = \frac{d}{dr}\tilde{\boldsymbol{P}}(r),$$

leading to

$$\mathcal{V}^T\mathcal{D}_r^T = (\mathcal{V}_r)^T, \quad \mathcal{V}_{r,(i,j)} = \left.\frac{d\tilde{P}_j}{dr}\right|_{r_i}.$$

To compute the entries of $\mathcal{V}_r$ we use the identity

$$\frac{d\tilde{P}_n}{dr} = \sqrt{n(n+1)}\tilde{P}^{(1,1)}_{n-1}(r),$$

where $\tilde{P}^{(1,1)}_{n-1}$ is the Jacobi polynomial (see Appendix A). This is implemented in GradJacobiP.m to initialize the gradient of $\mathcal{V}$ evaluated at the grid points, $\boldsymbol{r}$, as in GradVandermonde1D.m.

─────────────────── GradJacobiP.m ───────────────────

```
function [dP] = GradJacobiP(r, alpha, beta, N);

% function [dP] = GradJacobiP(r, alpha, beta, N);
% Purpose: Evaluate the derivative of the Jacobi polynomial of type
%                (alpha,beta)>-1, at points r for order N and returns
%                dP[1:length(r))]

dP = zeros(length(r), 1);
if(N == 0)
  dP(:,:) = 0.0;
else
  dP = sqrt(N*(N+alpha+beta+1))*JacobiP(r(:),alpha+1,beta+1, N-1);
end;
return
```

─────────────────── GradVandermonde1D.m ───────────────────

```
function [DVr] = GradVandermonde1D(N,r)

% function [DVr] = GradVandermonde1D(N,r)
% Purpose : Initialize the gradient of the modal basis (i) at (r)
%                at order N

DVr = zeros(length(r),(N+1));

% Initialize matrix
for i=0:N
   [DVr(:,i+1)] = GradJacobiP(r(:),0,0,i);
end
return
```

The entries of $\mathcal{D}_r$ now follow directly from

$$\mathcal{D}_r = \mathcal{V}_r \mathcal{V}^{-1},$$

as in Dmatrix1D.m. This allows us to recover $\mathcal{S}$ directly, although as we will see later, we will frequently use $\mathcal{D}_r$ rather than $\mathcal{S}$.

─────────────────── Dmatrix1D.m ───────────────────

```
function [Dr] = Dmatrix1D(N,r,V)

% function [Dr] = Dmatrix1D(N,r,V)
% Purpose : Initialize the (r) differentiation matrices
%                on the interval, evaluated at (r) at order N

Vr = GradVandermonde1D(N, r);
Dr = Vr/V;
return
```

$$
\begin{bmatrix} -0.50 & 0.50 \\ -0.50 & 0.50 \end{bmatrix}, \quad
\begin{bmatrix} -1.50 & 2.00 & -0.50 \\ -0.50 & 0.00 & 0.50 \\ 0.50 & -2.00 & 1.50 \end{bmatrix}, \quad
\begin{bmatrix}
-5.00 & 6.76 & -2.67 & 1.41 & -0.50 \\
-1.24 & 0.00 & 1.75 & -0.76 & 0.26 \\
0.38 & -1.34 & 0.00 & 1.34 & -0.38 \\
-0.26 & 0.76 & -1.75 & 0.00 & 1.24 \\
0.50 & -1.41 & 2.67 & -6.76 & 5.00
\end{bmatrix},
$$

$$
\begin{bmatrix}
-18.00 & 24.35 & -9.75 & 5.54 & -3.66 & 2.59 & -1.87 & 1.28 & -0.50 \\
-4.09 & 0.00 & 5.79 & -2.70 & 1.67 & -1.15 & 0.82 & -0.56 & 0.22 \\
0.99 & -3.49 & 0.00 & 3.58 & -1.72 & 1.08 & -0.74 & 0.49 & -0.19 \\
-0.44 & 1.29 & -2.83 & 0.00 & 2.85 & -1.38 & 0.86 & -0.55 & 0.21 \\
0.27 & -0.74 & 1.27 & -2.66 & 0.00 & 2.66 & -1.27 & 0.74 & -0.27 \\
-0.21 & 0.55 & -0.86 & 1.38 & -2.85 & 0.00 & 2.83 & -1.29 & 0.44 \\
0.19 & -0.49 & 0.74 & -1.08 & 1.72 & -3.58 & 0.00 & 3.49 & -0.99 \\
-0.22 & 0.56 & -0.82 & 1.15 & -1.67 & 2.70 & -5.79 & 0.00 & 4.09 \\
0.50 & -1.28 & 1.87 & -2.59 & 3.66 & -5.54 & 9.75 & -24.35 & 18.00
\end{bmatrix}
$$

Fig. 3.4. Examples of differentiation matrices, $\mathcal{D}_r$, for orders $N = 1, 2, 4$ in the top row and $N = 8$ in the bottom row.

Examples of differentiation matrices for different orders of approximation are shown in Fig. 3.4. A few observations are worth making regarding these operators. Inspection of Fig. 3.4 shows that

$$
\mathcal{D}_{r,(i,j)} = -\mathcal{D}_{r,(N-i,N-j)},
$$

known as skew-antisymmetric. This is a general property as long as $r_i = r_{N-i}$ (i.e., the interpolation points are symmetric around $r = 0$) and this can be explored to compute fast matrix-vector products [291]. Furthermore, we notice that each row-sum is exactly zero, reflecting that the derivative of a constant is zero. This also implies that $\mathcal{D}_r$ has at least one zero eigenvalue. A more careful analysis will in fact show that $\mathcal{D}_r$ is nilpotent; that is, it has $N + 1$ zero eigenvalues and only one eigenvector. Thus, $\mathcal{D}_r$ is not diagonizable but is similar only to a rank $N + 1$ Jordan block. These and other details are discussed in depth in [159].

Example 3.2. To illustrate the accuracy of derivatives computed using $\mathcal{D}_r$, let us consider a few examples.

We first consider the following analytic function

$$
u(x) = \exp(\sin(\pi x)), \quad x \in [-1, 1],
$$

and compute the errors of the computed derivatives as a function of the order of the approximation, N. In Fig. 3.5 we show the L^2-error as a function of N and observe an exponentially fast decay of the error, known as spectral convergence [159]. This is a clear indication of the strength of very high-order methods for the approximation of smooth functions.

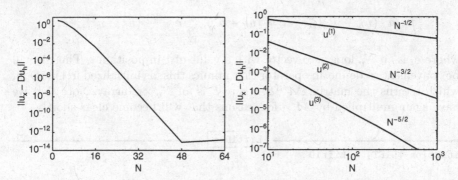

Fig. 3.5. On the left, we show the global L^2-error for the derivative of the smooth function in Example 3.2, highlighting exponential convergence. On the right, we show the global L^2-error for the derivative of the functions in Example 3.2, illustrating algebraic convergence closely connected to the regularity of the function.

To offer a more complete picture of the accuracy of the discrete derivative, let us also consider the sequence of functions

$$u^{(0)}(x) = \begin{cases} -\cos(\pi x), & -1 \le x < 0 \\ \cos(\pi x), & 0 \le x \le 1, \end{cases} \qquad \frac{du^{(i+1)}}{dx} = u^{(i)}, \quad i = 0, 1, 2, 3 \ldots.$$

Note that $u^{(i+1)} \in C^i$; that is, $u^{(1)}$ is continuous but its derivative is discontinuous. In Fig. 3.5 we show the L^2-error for the discrete derivative of $u^{(i)}$ for $i = 1, 2, 3$. We now observe a more moderate convergence rate as

$$\left\| \frac{du^{(i)}}{dx} - \mathcal{D}_r u_h^{(i)} \right\|_\Omega \propto N^{1/2-i}.$$

If we recall that

$$\left(\hat{u}_x^{(i)} \right)_n = \hat{u}_n^{(i-1)} \propto \frac{1}{n^i}$$

then a rough estimate follows directly since

$$\left\| \frac{du^{(i)}}{dx} - \mathcal{D}_r u_h^{(i)} \right\|_\Omega^2 \le \sum_{n=N+1}^\infty \frac{1}{n^{2i}} \le \frac{1}{N^{2i-1}},$$

confirming the close connection between accuracy and smoothness indicated in Fig. 3.5. We shall return to a more detailed discussion of these aspects in Chapter 4 where we shall also realize that things are a bit more complex than the above result indicates.

To complete the discussion of the local operators, we consider the remaining local operator that is responsible for extracting the surface terms of the form

$$\oint_{-1}^{1} \hat{n} \cdot (u_h - u^*)\ell_i(r)\, dr = (u_h - u^*)|_{r_{N_p}}\, e_{N_p} - (u_h - u^*)|_{r_1}\, e_1,$$

where e_i is an N_p long zero vector with a value of 1 in position i. Thus, it will be convenient to define an operator that mimics this, as initialized in Lift1D.m, which returns the matrix, $\mathcal{M}^{-1}\mathcal{E}$, where $\mathcal{E}$ is an $N_p \times 2$ array. Note that we have again multiplied by $\mathcal{M}^{-1}$ for reasons that will become clear shortly.

──────────────── Lift1D.m ────────────────

```
function [LIFT] = Lift1D

% function [LIFT] = Lift1D
% Purpose  : Compute surface integral term in DG formulation

Globals1D;
Emat = zeros(Np,Nfaces*Nfp);

% Define Emat
Emat(1,1) = 1.0; Emat(Np,2) = 1.0;

% inv(mass matrix)*\s_n (L_i,L_j)_{edge_n}
LIFT = V*(V'*Emat);
return
```

3.3 Getting the grid together and computing the metric

With the local operators in place, we now discuss how to compute the metric associated with the grid and how to assemble the global grid. We begin by assuming that we provide two sets of data:

- A row vector, VX, of N_v $(= K + 1)$ vertex coordinates. These represent the end points of the intervals or elements. They do not need to be ordered in any particular way although they must be distinct. These vertices are numbered consecutively from 1 to N_v.
- An integer matrix, EToV, of size $K \times 2$, containing in each row k, the numbers of the two vertices forming element k. It is the responsibility of the user/grid generator to ensure that the elements are meaningful; that is, the grid should consist of one connected set of nonoverlapping elements. It is furthermore assumed that the nodes are ordered so elements are defined counter-clockwise (in one dimension this reflects a left-to-right ordering of the elements).

These two pieces of information are provided by any grid generator, or, alternatively, a user should provide them to specify the grid for a problem (see Appendix B for a further discussion of this).

If we specify the order, N, of the local approximations, we can compute the corresponding LGL nodes, r, in the reference element, I, and map them to each of the physical elements using the affine mapping, Eq. (3.1), as

```
>> va = EToV(:,1)'; vb = EToV(:,2)';
>> x = ones(Np,1)*VX(va) + 0.5*(r+1)*(VX(vb)-VX(va));
```

The array, x, is $N_p \times K$ and contains the physical coordinates of the grid points. The metric of the mapping is

$$r_x = \frac{1}{x_r} = \frac{1}{J} = \frac{1}{\mathcal{D}_r x},$$

as implemented in GeometricFactors1D.m. Here, J represents the local transformation Jacobian.

GeometricFactors1D.m

```
function [rx,J] = GeometricFactors1D(x,Dr)

% function [rx,J] = GeometricFactors1D(x,Dr)
% Purpose  : Compute the metric elements for the local mappings
%            of the 1D elements

xr  = Dr*x; J = xr; rx = 1./J;
return
```

To readily access the elements along the faces (i.e., the two end points of each element), we form a small array with these two indices:

```
>> fmask1 = find( abs(r+1) < NODETOL)';
>> fmask2 = find( abs(r-1) < NODETOL)';
>> Fmask = [fmask1;fmask2]';
>> Fx = x(Fmask(:), :);
```

where the latter, Fx, is a $2 \times K$ array containing the physical coordinates of the edge nodes.

We now build the final piece of local information: the local outward pointing normals and the inverse Jacobian at the interfaces, as done in Normals1D.m.

Normals1D.m

```
function [nx] = Normals1D

% function [nx] = Normals1D
% Purpose : Compute outward pointing normals at elements faces

Globals1D;
nx = zeros(Nfp*Nfaces, K);
```

```
% Define outward normals
nx(1, :) = -1.0; nx(2, :) = 1.0;
return
```

>> Fscale = 1./(J(Fmask,:));

So far, all the elements are disconnected and we need to obtain information about which vertex on each element connects to which vertex on neighboring elements to form the global grid. If a vertex is not connected, it will be assumed to be a boundary with the possible need to impose a boundary condition.

We form an integer matrix, FToV, which is $(2K) \times N_v$. The $2K$ reflects the total number of faces, each of which have been assigned a global number. This supplies a map between this global face numbering, running from 1 to $2K$, and the global vertex numbering, running from 1 to N_v.

To recover information about which global face connects to which global face, we compute the companion matrix

$$\text{FToF} = (\text{FToV})(\text{FToV})^T,$$

which is a $(2K) \times (2K)$ integer matrix with ones on the diagonal, reflecting that each face connects to itself. Thus, a pure connection matrix is found by removing the self-reference

$$\text{FToF} = (\text{FToV})(\text{FToV})^T - \text{I}.$$

In each row, corresponding to a particular face, one now finds either a value of 1 in one column corresponding to the connecting face or only zeros, reflecting that the particular face is an unconnected face (i.e., a boundary point).

Let us consider a simple example to make this approach clearer.

Example 3.3. Assume that we have read in the two arrays

$$\boldsymbol{VX} = \begin{bmatrix} 0.0, 0.5, 1.5, 3.0, 2.5 \end{bmatrix}, \quad \text{EToV} = \begin{bmatrix} 1\,2 \\ 2\,3 \\ 3\,5 \\ 5\,4 \end{bmatrix},$$

representing a grid with $K = 4$ elements and $N_v = 5$ vertices. We form

$$\text{FToV} = \begin{bmatrix} 1\,0\,0\,0\,0 \\ 0\,1\,0\,0\,0 \\ 0\,1\,0\,0\,0 \\ 0\,0\,1\,0\,0 \\ 0\,0\,1\,0\,0 \\ 0\,0\,0\,0\,1 \\ 0\,0\,0\,1\,0 \\ 0\,0\,0\,0\,1 \end{bmatrix}.$$

to connect the $2K$ local faces to the N_v globally numbered faces.

The connectivity can then be recovered from the operator

$$(\text{FToV})(\text{FToV})^T = \begin{bmatrix} 1 & 0 & 0 & 0 & 0 & 0 & 0 & 0 \\ 0 & 1 & 1 & 0 & 0 & 0 & 0 & 0 \\ 0 & 1 & 1 & 0 & 0 & 0 & 0 & 0 \\ 0 & 0 & 0 & 1 & 1 & 0 & 0 & 0 \\ 0 & 0 & 0 & 1 & 1 & 0 & 0 & 0 \\ 0 & 0 & 0 & 0 & 0 & 1 & 0 & 1 \\ 0 & 0 & 0 & 0 & 0 & 0 & 1 & 0 \\ 0 & 0 & 0 & 0 & 0 & 1 & 0 & 1 \end{bmatrix}.$$

The identity diagonal reflects the self-reference. Furthermore, we see that beyond the diagonal, rows 1 and 7 indicate that these vertices have no connection, and are assumed to reflect the outer boundaries of the computational domain. Finally, we see that face 2 connects to face 3 and face 3 to face 2, in agreement with the basic list of input data.

From the FToF-matrix we can immediately recover the element-to-element, EToE, and element-to-face, EToF, information as each segment of two rows or columns corresponds to one element and its two faces. The whole procedure is implemented in Connect1D.m.

--- Connect1D.m ---

```
function [EToE, EToF] = Connect1D(EToV)

% function [EToE, EToF] = Connect1D(EToV)
% Purpose  : Build global connectivity arrays for 1D grid based
%                 on standard EToV input array from grid generator

Nfaces = 2;
% Find number of elements and vertices
K = size(EToV,1); TotalFaces = Nfaces*K; Nv = K+1;

% List of local face to local vertex connections
vn = [1,2];

% Build global face to node sparse array
SpFToV = spalloc(TotalFaces, Nv, 2*TotalFaces);
sk = 1;
for k=1:K
  for face=1:Nfaces
    SpFToV( sk, EToV(k, vn(face))) = 1;
    sk = sk+1;
  end
end
```

```
% Build global face to global face sparse array
SpFToF = SpFToV*SpFToV' - speye(TotalFaces);

% Find complete face to face connections
[faces1, faces2] = find(SpFToF==1);

% Convert face global number to element and face numbers
element1 = floor( (faces1-1)/Nfaces )  + 1;
face1    =   mod( (faces1-1), Nfaces ) + 1;
element2 = floor( (faces2-1)/Nfaces )  + 1;
face2    =   mod( (faces2-1), Nfaces ) + 1;

% Rearrange into Nelements x Nfaces sized arrays
ind = sub2ind([K, Nfaces], element1, face1);
EToE       = (1:K)'*ones(1,Nfaces);
EToF       = ones(K,1)*(1:Nfaces);
EToE(ind) = element2; EToF(ind) = face2;
return
```

The final step is to extract the indices of the connecting vertices. For this we use a global numbering of all face-based degrees of freedom in the grid (i.e., $K \times 2$). Using the connectivity information built in Connect1D.m, these are found by simply running through all element faces and constructing the vertex maps, **vmapM** and **vmapP**, corresponding to u^- and u^+ in the usual terminology; that is, **vmapM** refers to interior information and **vmapP** contains the node numbers of the exterior data. Furthermore, if we initialize **vmapM = vmapP** and if it remains so after visiting all elements, these edges represent boundaries. For simplicity, we assume the first node of element 1 and the last node of element K to be boundaries. The procedure is illustrated in BuildMaps1D.m.

```
                              ┌─────────────┐
─────────────────────────────│ BuildMaps1D.m │─────────────────────
                              └─────────────┘
function [vmapM, vmapP, vmapB, mapB] = BuildMaps1D

% function [vmapM, vmapP, vmapB, mapB] = BuildMaps1D
% Purpose: Connectivity and boundary tables for nodes given in the K #
%          of elements, each with N+1 degrees of freedom.

Globals1D;

% number volume nodes consecutively
nodeids = reshape(1:K*Np, Np, K);
vmapM   = zeros(Nfp, Nfaces, K);
vmapP   = zeros(Nfp, Nfaces, K);

for k1=1:K
  for f1=1:Nfaces
```

```
    % find index of face nodes with respect to volume node ordering
    vmapM(:,f1,k1) = nodeids(Fmask(:,f1), k1);
  end
end

for k1=1:K
  for f1=1:Nfaces
    % find neighbor
    k2 = EToE(k1,f1); f2 = EToF(k1,f1);

    % find volume node numbers of left and right nodes
    vidM = vmapM(:,f1,k1); vidP = vmapM(:,f2,k2);

    x1  = x(vidM); x2  = x(vidP);

    % Compute distance matrix
    D = (x1 -x2 ).^2;
    if (D<NODETOL) vmapP(:,f1,k1) = vidP; end;
  end
end

vmapP = vmapP(:); vmapM = vmapM(:);

% Create list of boundary nodes
mapB = find(vmapP==vmapM); vmapB = vmapM(mapB);

% Create specific left (inflow) and right (outflow) maps
mapI = 1; mapO = K*Nfaces; vmapI = 1; vmapO = K*Np;
return
```

Everything discussed above must be done for any problem solved with a one-dimensional discontinuous Galerkin finite element method. We have assembled all these components in the startup script StartUp1D.m

―――――――――――――――――――― | StartUp1D.m | ――――――――――――――――――――

```
% Purpose : Setup script, building operators, grid, metric and
%                connectivity for 1D solver.

% Definition of constants

Globals1D; NODETOL = 1e-10;
Np = N+1; Nfp = 1; Nfaces=2;

% Compute basic Legendre Gauss Lobatto grid
r = JacobiGL(0,0,N);

% Build reference element matrices
V  = Vandermonde1D(N, r); invV = inv(V);
```

```
Dr = Dmatrix1D(N, r, V);

% Create surface integral terms
LIFT = Lift1D();

% build coordinates of all the nodes
va = EToV(:,1)'; vb = EToV(:,2)';
x = ones(N+1,1)*VX(va) + 0.5*(r+1)*(VX(vb)-VX(va));

% calculate geometric factors
[rx,J] = GeometricFactors1D(x,Dr);

% Compute masks for edge nodes
fmask1 = find( abs(r+1) < NODETOL)';
fmask2 = find( abs(r-1) < NODETOL)';
Fmask  = [fmask1;fmask2]';
Fx = x(Fmask(:), :);

% Build surface normals and inverse metric at surface
[nx] = Normals1D();
Fscale = 1./(J(Fmask,:));

% Build connectivity matrix
[EToE, EToF] = Connect1D(EToV);

% Build connectivity maps
[vmapM, vmapP, vmapB, mapB] = BuildMaps1D;
```

To transfer information between these routines and the problem-specific routines discussed shortly, we use Globals1D.m to declare variables as part of the global Matlab environment.

————————————————— Globals1D.m —————————————————

```
% Purpose: declare global variables

global N Nfp Np K
global r x  VX
global Dr LIFT
global nx Fx Fscale
global vmapM vmapP vmapB mapB Fmask
global vmapI vmapO mapI mapO
global rx J
global rk4a rk4b rk4c
global Nfaces EToE EToF
global V invV
global NODETOL

% Low storage Runge-Kutta coefficients
```

```
rk4a = [                    0.0 ...
            -567301805773.0/1357537059087.0 ...
            -2404267990393.0/2016746695238.0 ...
            -3550918686646.0/2091501179385.0  ...
            -1275806237668.0/842570457699.0];
rk4b = [ 1432997174477.0/9575080441755.0 ...
            5161836677717.0/13612068292357.0 ...
            1720146321549.0/2090206949498.0  ...
            3134564353537.0/4481467310338.0  ...
            2277821191437.0/14882151754819.0];
rk4c = [                    0.0 ...
            1432997174477.0/9575080441755.0 ...
            2526269341429.0/6820363962896.0 ...
            2006345519317.0/3224310063776.0 ...
            2802321613138.0/2924317926251.0];
```

3.4 Dealing with time

The emphasis so far has been on the spatial dimension and a discrete representation of this. This reflects a method-of-lines approach where we discretize space and time separately, and use some standard technique to solve the ordinary differential equations for the latter.

We follow this approach here and, furthermore, focus on the use of explicit Runge-Kutta (RK) methods for integration in the temporal dimension. To discretize the semidiscrete problem

$$\frac{du_h}{dt} = \mathcal{L}_h(u_h, t),$$

where u_h is the vector of unknowns, we can use the standard fourth-order four stage explicit RK method (ERK)

$$k^{(1)} = \mathcal{L}_h(u_h^n, t^n),$$

$$k^{(2)} = \mathcal{L}_h\left(u_h^n + \frac{1}{2}\Delta t k^{(1)}, t^n + \frac{1}{2}\Delta t\right),$$

$$k^{(3)} = \mathcal{L}_h\left(u_h^n + \frac{1}{2}\Delta t k^{(2)}, t^n + \frac{1}{2}\Delta t\right),$$

$$k^{(4)} = \mathcal{L}_h\left(u_h^n + \Delta t k^{(3)}, t^n + \Delta t\right),$$

$$u_h^{n+1} = u_h^n + \frac{1}{6}\Delta t\left(k^{(1)} + 2k^{(2)} + 2k^{(3)} + k^{(4)}\right), \tag{3.4}$$

to advance from u_h^n to u_h^{n+1}, separated by the timestep, Δt.

As simple and widely used as this classic ERK approach is, it has the disadvantage that it requires four extra storage arrays, $k^{(i)}$. An attractive

Table 3.2. Coefficients for the low-storage five-stage fourth-order ERK method (LSERK).

i	a_i	b_i	c_i
1	0	$\dfrac{1432997174477}{9575080441755}$	0
2	$-\dfrac{567301805773}{1357537059087}$	$\dfrac{5161836677717}{13612068292357}$	$\dfrac{1432997174477}{9575080441755}$
3	$-\dfrac{2404267990393}{2016746695238}$	$\dfrac{1720146321549}{2090206949498}$	$\dfrac{2526269341429}{6820363962896}$
4	$-\dfrac{3550918686646}{2091501179385}$	$\dfrac{3134564353537}{4481467310338}$	$\dfrac{2006345519317}{3224310063776}$
5	$-\dfrac{1275806237668}{842570457699}$	$\dfrac{2277821191437}{14882151754819}$	$\dfrac{2802321613138}{2924317926251}$

alternative to this is a low-storage version [46] of the fourth-order method (LSERK) of the form

$$\boldsymbol{p}^{(0)} = \boldsymbol{u}^n,$$

$$i \in [1, \ldots, 5] : \begin{cases} \boldsymbol{k}^{(i)} = a_i \boldsymbol{k}^{(i-1)} + \Delta t \mathcal{L}_h \left(\boldsymbol{p}^{(i-1)}, t^n + c_i \Delta t \right), \\ \boldsymbol{p}^{(i)} = \boldsymbol{p}^{(i-1)} + b_i \boldsymbol{k}^{(i)}, \end{cases}$$

$$\boldsymbol{u}_h^{n+1} = \boldsymbol{p}^{(5)}. \tag{3.5}$$

The coefficients needed in the LSERK are given in Table 3.2. The main difference here is that only one additional storage level is required, thus reducing the memory usage significantly. On the other hand, this comes at the price of an additional function evaluation, as the low-storage version has five stages.

At first, it would seem that the additional stage makes the low-storage approach less interesting due to the added cost. However, as we will discuss in more detail in Chapter 4, the added cost in the low-storage RK is offset by allowing a larger stable timestep, Δt.

It should be emphasized that these methods are not exclusive and many alternatives exist [40, 143, 144]. In particular, for strongly nonlinear problems, the added nonlinear stability of strong stability-preserving methods may be advantageous, as we will discuss further in Chapter 5.

3.5 Putting it all together

Let us now return to the simple linear wave equation

$$\frac{\partial u}{\partial t} + 2\pi \frac{\partial u}{\partial x} = 0, \quad x \in [0, 2\pi],$$
$$u(x, 0) = \sin(x),$$
$$u(0, t) = -\sin(2\pi t).$$

It is easy to see that the exact solution is

$$u(x, t) = \sin(x - 2\pi t).$$

Following the developments in Chapter 2, we assume the local solution is well approximated as

$$u_h^k(x, t) = \sum_{i=1}^{N_p} u_h^k(x_i^k, t)\ell_i^k(x),$$

and the combination of these local solutions as the approximation to the global solution. This yields the local semidiscrete scheme

$$\mathcal{M}^k \frac{d\boldsymbol{u}_h^k}{dt} + 2\pi \mathcal{S} \boldsymbol{u}_h^k = \left[\boldsymbol{\ell}^k(x)(2\pi u_h^k - (2\pi u)^*) \right]_{x_l^k}^{x_r^k},$$

or

$$\frac{d\boldsymbol{u}_h^k}{dt} + 2\pi (\mathcal{M}^k)^{-1} \mathcal{S} \boldsymbol{u}_h^k = (\mathcal{M}^k)^{-1} \left[\boldsymbol{\ell}^k(x)(2\pi u_h^k - (2\pi u)^*) \right]_{x_l^k}^{x_r^k},$$

The analysis in Example 2.3 shows that choosing the flux as

$$(2\pi u)^* = \{\{2\pi u\}\} + 2\pi \frac{1 - \alpha}{2} [\![u]\!], \ \ 0 \le \alpha \le 1$$

results in a stable scheme.

An implementation of this is shown in AdvecRHS1D.m. The parameter α can be used to adjust the flux; for example $\alpha = 1$ is a central flux and $\alpha = 0$ reflects an upwind flux.

──────────────── AdvecRHS1D.m ────────────────

```
function [rhsu] = AdvecRHS1D(u,time, a)

% function [rhsu] = AdvecRHS1D(u,time)
% Purpose  : Evaluate RHS flux in 1D advection

Globals1D;

% form field differences at faces
alpha=1;
du = zeros(Nfp*Nfaces,K);
du(:) = (u(vmapM)-u(vmapP)).*(a*nx(:)-(1-alpha)*abs(a*nx(:)))/2;

% impose boundary condition at x=0
uin = -sin(a*time);
du (mapI) = (u(vmapI)- uin ).*(a*nx(mapI)-(1-alpha)*abs(a*nx(mapI)))/2;
du (mapO) = 0;

% compute right hand sides of the semi-discrete PDE
rhsu = -a*rx.*(Dr*u) + LIFT*(Fscale.*(du));
return
```

To complete the scheme for solving the advection equation, we need only integrate the system in time. As discussed in Section 3.4, we do this using a low-storage RK method. This is implemented in Advec1D.m.

```
———————————————————————— Advec1D.m ————————————————————
function [u] = Advec1D(u, FinalTime)

% function [u] = Advec1D(u, FinalTime)
% Purpose  : Integrate 1D advection until FinalTime starting with
%            initial the condition, u

Globals1D;
time = 0;

% Runge-Kutta residual storage
resu = zeros(Np,K);

% compute time step size
xmin = min(abs(x(1,:)-x(2,:)));
CFL=0.75; dt   = CFL/(2*pi)*xmin; dt = .5*dt;
Nsteps = ceil(FinalTime/dt); dt = FinalTime/Nsteps;

% advection speed
a = 2*pi;

% outer time step loop
for tstep=1:Nsteps
    for INTRK = 1:5
        timelocal = time + rk4c(INTRK)*dt;
        [rhsu] = AdvecRHS1D(u, timelocal, a);
        resu = rk4a(INTRK)*resu + dt*rhsu;
        u = u+rk4b(INTRK)*resu;
    end;
    % Increment time
    time = time+dt;
end;
return
```

One issue we have not discussed yet is how to choose the timestep to ensure a discretely stable scheme. In Advec1D.m we use the guideline

$$\Delta t \leq \frac{C}{2\pi} \min_{k,i} \Delta x_i^k,$$

where $\Delta x_i^k = x_{i+1}^k - x_i^k$ and 2π is the maximum wave speed. The constant, C, is a Courant-Friedrichs-Levy (CFL)-like constant that can be expected to be $\mathcal{O}(1)$. We will discuss this in more detail in Chapter 4 but the above expression is an excellent guideline.

Finally, we need a driver routine where the grid information is read in, the global grid and the metric are assembled as discussed in Section 3.3, and the initial conditions are set. An example of this is shown in AdvecDriver1D.m. All results presented in Chapter 2 have been obtained using this simple code.

AdvecDriver1D.m

```
% Driver script for solving the 1D advection equations
Globals1D;

% Order of polymomials used for approximation
N = 8;

% Generate simple mesh
[Nv, VX, K, EToV] = MeshGen1D(0.0,2.0,10);

% Initialize solver and construct grid and metric
StartUp1D;

% Set initial conditions
u = sin(x);

% Solve Problem
FinalTime = 10;
[u] = Advec1D(u,FinalTime);
```

It is worth emphasizing that the above three routines, AdvecRHS1D.m, Advec1D.m, and AdvecDriver1D.m, are all that is needed to solve the problem at any order and using any grid. In fact, only AdvecRHS1D.m requires substantial changes to solve another problem as we will see next, where we discuss the routines for a more complex problem. The grid information is generated in MeshGen1D.m, which is described in Appendix B.

3.6 Maxwell's equations

As a slightly more complicated problem, let us consider the one-dimensional Maxwell's equations

$$\varepsilon(x)\frac{\partial E}{\partial t} = -\frac{\partial H}{\partial x}, \quad \mu(x)\frac{\partial H}{\partial t} = -\frac{\partial E}{\partial x}, \tag{3.6}$$

where (E, H) represent the electric and magnetic fields, respectively, and the material parameters, $\varepsilon(x)$ and $\mu(x)$, reflect the electric permittivity and magnetic permeability, respectively.

We solve these equations in a fixed domain $x \in [-2, 2]$ with both $\varepsilon(x)$ and $\mu(x)$ changing discontinuously at $x = 0$. For simplicity, we assume that the materials are otherwise piecewise constant. Relaxing this would only require minor modifications. Furthermore, we assume that $E(-2,0) = E(2,0) = 0$, corresponding to a physical situation where the computational domain is enclosed in a metallic cavity.

To develop the scheme, we take the usual path and seek an approximation $(E, H) \simeq (E_h, H_h)$ being composed as the direct sum of K local polynomials (E_h^k, H_h^k) on the form

$$\begin{bmatrix} E_h^k(x,t) \\ H_h^k(x,t) \end{bmatrix} = \sum_{i=1}^{N_p} \begin{bmatrix} E_h^k(x_i^k,t) \\ H_h^k(x_i^k,t) \end{bmatrix} \ell_i^k(x).$$

Introducing this into Eq. (3.6) and requiring Maxwell's equations to be satisfied locally on the strong discontinuous Galerkin form yields the semidiscrete scheme

$$\frac{dE_h^k}{dt} + \frac{1}{J^k \varepsilon^k} \mathcal{D}_r H_h^k = \frac{1}{J^k \varepsilon^k} \mathcal{M}^{-1} \left[\ell^k(x)(H_h^k - H^*) \right]_{x_l^k}^{x_r^k}$$

$$= \frac{1}{J^k \varepsilon^k} \mathcal{M}^{-1} \oint_{x_l^k}^{x_r^k} \hat{n} \cdot (H_h^k - H^*) \ell^k(x)\, dx,$$

$$\frac{dH_h^k}{dt} + \frac{1}{J^k \mu^k} \mathcal{D}_r E_h^k = \frac{1}{J^k \mu^k} \mathcal{M}^{-1} \left[\ell^k(x)(E_h^k - E^*) \right]_{x_l^k}^{x_r^k}$$

$$= \frac{1}{J^k \varepsilon^k} \mathcal{M}^{-1} \oint_{x_l^k}^{x_r^k} \hat{n} \cdot (E_h^k - E^*) \ell^k(x)\, dx.$$

As the fluxes, we will use those derived in Example 2.6 on the form

$$H^* = \frac{1}{\{\{Z\}\}} \left(\{\{ZH\}\} + \frac{1}{2}[\![E]\!] \right), \quad E^* = \frac{1}{\{\{Y\}\}} \left(\{\{YE\}\} + \frac{1}{2}[\![H]\!] \right),$$

where

$$Z^{\pm} = \sqrt{\frac{\mu^{\pm}}{\varepsilon^{\pm}}} = (Y^{\pm})^{-1}.$$

This yields the terms

$$H^- - H^* = \frac{1}{2\{\{Z\}\}} \left(Z^+ [\![H]\!] - [\![E]\!] \right),$$

$$E^- - E^* = \frac{1}{2\{\{Y\}\}} \left(Y^+ [\![E]\!] - [\![H]\!] \right),$$

as the penalty terms entering on the right-hand side of the semidiscrete scheme. This is all implemented in MaxwellRHS1D.m.

```
───────────────────────── MaxwellRHS1D.m ─────────────────────────
function [rhsE, rhsH] = MaxwellRHS1D(E,H,eps,mu)

% function [rhsE, rhsH] = MaxwellRHS1D(E,H,eps,mu)
% Purpose  : Evaluate RHS flux in 1D Maxwell

Globals1D;

% Compute impedance
Zimp = sqrt(mu./eps);

% Define field differences at faces
dE = zeros(Nfp*Nfaces,K); dE(:) = E(vmapM)-E(vmapP);
dH = zeros(Nfp*Nfaces,K); dH(:) = H(vmapM)-H(vmapP);
Zimpm = zeros(Nfp*Nfaces,K); Zimpm(:) = Zimp(vmapM);
Zimpp = zeros(Nfp*Nfaces,K); Zimpp(:) = Zimp(vmapP);
Yimpm = zeros(Nfp*Nfaces,K); Yimpm(:) = 1./Zimpm(:);
Yimpp = zeros(Nfp*Nfaces,K); Yimpp(:) = 1./Zimpp(:);

% Homogeneous boundary conditions, Ez=0
Ebc = -E(vmapB); dE (mapB) = E(vmapB) - Ebc;
Hbc =  H(vmapB); dH (mapB) = H(vmapB) - Hbc;

% evaluate upwind fluxes
fluxE = 1./(Zimpm + Zimpp).*(nx.*Zimpp.*dH - dE);
fluxH = 1./(Yimpm + Yimpp).*(nx.*Yimpp.*dE - dH);

% compute right hand sides of the PDE's
rhsE = (-rx.*(Dr*H) + LIFT*(Fscale.*fluxE))./eps;
rhsH = (-rx.*(Dr*E) + LIFT*(Fscale.*fluxH))./mu;
return
```

The temporal integration is done using Maxwell1D.m which is essentially unchanged from Advec1D.m in the previous section. The only change is that we are now integrating a system and, thus, need to integrate all components of the system at each stage.

```
───────────────────────── Maxwell1D.m ─────────────────────────
function [E,H] = Maxwell1D(E,H,eps,mu,FinalTime);

% function [E,H] = Maxwell1D(E,H,eps,mu,FinalTime)
% Purpose  : Integrate 1D Maxwell's until FinalTime starting with
%            conditions (E(t=0),H(t=0)) and materials (eps,mu).

Globals1D;
time = 0;

% Runge-Kutta residual storage
```

```
resE = zeros(Np,K); resH = zeros(Np,K);

% compute time step size
xmin = min(abs(x(1,:)-x(2,:)));
CFL=1.0;  dt = CFL*xmin;
Nsteps = ceil(FinalTime/dt); dt = FinalTime/Nsteps;

% outer time step loop
for tstep=1:Nsteps
  for INTRK = 1:5
    [rhsE, rhsH] = MaxwellRHS1D(E,H,eps,mu);

    resE = rk4a(INTRK)*resE + dt*rhsE;
    resH = rk4a(INTRK)*resH + dt*rhsH;

    E = E+rk4b(INTRK)*resE;
    H = H+rk4b(INTRK)*resH;
  end
  % Increment time
  time = time+dt;
end
return
```

In the final routine, MaxwellDriver1D.m, the only significant difference from AdvecDriver1D.m is the need to also specify the spatial distribution of ε and μ. In this particular case, we have assumed that these are constant in each element but can jump between elements. Furthermore, we assume here that the jump is between the two middle elements. Clearly, this specification is problem dependent and can be changed as needed. Also, if ε and/or μ vary smoothly within the elements, this can be specified here with no further changes elsewhere.

──────────────────────── MaxwellDriver1D.m ────────────────────────
```
% Driver script for solving the 1D Maxwell's equations
Globals1D;

% Polynomial order used for approximation
N = 6;

% Generate simple mesh
[Nv, VX, K, EToV] = MeshGen1D(-2.0,2.0,80);

% Initialize solver and construct grid and metric
StartUp1D;

% Set up material parameters
eps1 = [ones(1,K/2), 2*ones(1,K/2)];
mu1 = ones(1,K);
```

```
epsilon = ones(Np,1)*eps1; mu = ones(Np,1)*mu1;

% Set initial conditions
E = sin(pi*x).*(x<0); H = zeros(Np,K);

% Solve Problem
FinalTime = 10;
[E,H] = Maxwell1D(E,H,epsilon,mu,FinalTime);
```

To illustrate the performance of the developed algorithm, we consider a problem consisting of a cavity, $x \in [-1, 1]$, with metallic end walls [i.e., $E(\pm 1, t) = 0$]. Furthermore, we assume that $x = 0$ represents an interface between the two halves of the cavity and that each half is a different homogeneous material. For simplicity we assume the materials are nonmagnetic (i.e., $\mu^{(m)} = 1, m = 1, 2$).

This cavity supports a number of resonant modes of the form

$$E^{(m)}(x,t) = - \left[A^{(m)} \exp(i\omega n^{(m)} x) - B^{(m)} \exp(-i\omega n^{(m)} x) \right] \exp(i\omega t),$$

$$H^{(m)}(x,t) = \left[A^{(m)} \exp(i\omega n^{(m)} x) + B^{(m)} \exp(-i\omega n^{(m)} x) \right] \exp(i\omega t),$$

where

$$B^{(1)} = \exp(-i2n^{(1)}\omega)A^{(1)}, \;\; B^{(2)} = -\exp(i2n^{(2)}\omega)A^{(2)},$$

due to the boundary conditions,

$$A^{(1)} = \frac{n^{(2)} \cos(n^{(2)}\omega)}{n^{(1)} \cos(n^{(1)}\omega)}, \;\; A^{(2)} = \exp(i\omega(n^{(1)} + n^{(2)})),$$

as a convenient normalization and ω is a solution to the equation

$$-n^{(2)} \tan(\omega n^{(1)}) = n^{(1)} \tan(\omega n^{(2)}),$$

found by requiring continuity of the field components across the material interface. Here and elsewhere above, the index of refraction is defined as $n^{(m)} = \sqrt{\varepsilon^{(m)}}$.

For $n^{(1)} = n^{(2)}$, this is a simple sinusoidal solution, whereas for $n^{(1)} \neq n^{(2)}$ the solution is more complex and loses smoothness across the material interface where the solution is only continuous.

We first solve the above problem with $n^{(1)} = 1.0$ (i.e., vacuum), and $n^{(2)} = 1.5$, similar to glass, using different numbers of elements, K, and order of approximation, N, in each element. All elements have the same size, $h = 2/K$, and in the first case, we assume that K is even which guarantees that an element interface is located at $x = 0$. On the left in Fig. 3.6 we show the results, illustrating optimal convergence, $\mathcal{O}(h^{N+1})$, once the fields are reasonably resolved.

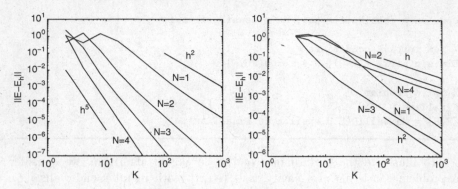

Fig. 3.6. We show the global L^2-error as a function of order of approximation, N, and number of elements, $K \propto h^{-1}$, for the computed E-field solution to Maxwell's equations in a metallic cavity with a material interface located at $x = 0$. On the left, the grid is conforming to the geometry; that is, there is an element boundary located at $x = 0$. On the right, this is violated and the material interface is located in the middle of an element. The loss of optimal convergence highlights the importance of using a body-conforming discretization.

Also shown in Fig. 3.6 are the results for the case where K is odd, implying that $x = 0$ falls in the middle of an element. This shows a dramatic reduction in the accuracy of the scheme, which now is second order accurate. We also observe that, for N even, the order is further reduced to first order. This can be attributed to the fact that for N even, there is a grid point exactly at $x = 0$, whereas for N odd, $x = 0$ falls between two grid points. For N being even, one can redefine the material constant at $x = 0$ to be the average in order to improve the convergence rate. A closer inspection of the result in Fig. 3.6 for $N = 3$ indicates a less than second order convergence and a detailed analysis reveals that the asymptotic rate should be $\mathcal{O}(h^{3/2})$ [234].

This latter example emphasizes the importance of using a geometry-conforming grid in which the elements align with features where the solution loses smoothness in order to maintain the global high-order accuracy. However, even if this is not possible, the results in Fig. 3.6 also show that although the rate of convergence is reduced, the absolute error is lower for high-order approximations; that is, the constant in front of the error term decreases with increasing order.

3.7 Exercises

1. Modify the solver for the scalar advection equation in Section 3.5 to solve

$$\frac{\partial u}{\partial t} + a \frac{\partial u}{\partial x} = 0, \ x \in [0, 2\pi],$$

with $a \geq 0$ and
$$u(0,t) = g(t), \quad u(x,0) = f(x).$$

Choose appropriate functions for the initial conditions, $f(x)$, and time-dependent boundary conditions, $g(t)$, and verify the convergence rate of the scheme, depending on both K and N.

2. Modify the solver for the scalar advection equation in Section 3.5 to solve
$$\frac{\partial u}{\partial t} + a\frac{\partial u}{\partial x} = h(x,t), \quad x \in [0, 2\pi],$$

with $a \geq 0$ and
$$u(0,t) = g(t), \quad u(x,0) = f(x).$$

Choose appropriate functions for the initial conditions, $f(x)$, the time-dependent boundary conditions, $g(t)$, and the forcing function, $h(x,t)$, and verify the convergence rate of the scheme, depending on both K and N.

3. Modify the solver for Maxwell's equations in Section 3.6 to solve the following problem
$$E(x,0) = (1 - x^2)x^2, \quad H(x,0) = 0, \quad x \in [0,1],$$

with the ends of the domain being assumed to be purely metallic (i.e., $E(0,t) = E(1,t) = 0$) and $\varepsilon = \mu = 1$. The initial condition reassembles at even integer times (i.e., $T = 2, 4, 6, \ldots$).
Verify the convergence rate of the scheme, depending on both K and N, and attempt to explain your observations.

4. Consider the material loaded cavity in Section 3.6 in the case where the material interface is placed exactly in the center of an element.

   a) Write a function that evaluates the exact solution and show that the first order convergence, observed for N even, can be improved by defining the permittivity at this center grid point as

   $$\bar{\varepsilon}(x = 0) = \frac{\varepsilon^{(1)} + \varepsilon^{(2)}}{2}.$$

   b) Show computationally that this recovers a convergence rate like $\mathcal{O}(h^{3/2})$ or better, with this rate being reached at very fine grids.

5. Consider the one-dimensional model for acoustics in a mean flow, given as
$$\frac{\partial u}{\partial t} + M(x)\frac{\partial u}{\partial x} = -\frac{\partial p}{\partial x},$$
$$\frac{\partial p}{\partial t} + M(x)\frac{\partial p}{\partial x} = -\frac{\partial u}{\partial x},$$

where the unknowns are the velocity, $u(x,t)$, and the pressure, $p(x,t)$. The Mach number, $M(x) = u_0(x)/c_0(x)$, reflects the velocity of the steady mean flow and is assumed to be piecewise constant.

a) Modify the solver for Maxwell's equations in Section 3.5 to solve this problem using a central flux or a local Lax-Friedrichs flux.

b) Test the accuracy of the code; that is, show that the error is proportional to Ch^s. What is s and can you determine how C depends on time and N?

Note: To test the accuracy of the code, you can use the old trick of a constructed solution: simply choose a solution (u, p), insert it into the equation and find the remainder. Adding this to the equation then guarantees that you know the exact solution.

4

Insight through theory

While the last chapters have focused on basic ideas and their implementation as a computational method, further insight into the performance of the method can be gained by revisiting some issues in more detail. To keep things relatively simple, we focus on the properties of the scheme for one-dimensional linear problems. However, as we will see later, many of the results obtained here carry over to multidimensional problems and even nonlinear problems with just a few modifications.

4.1 A bit more notation

Before we embark on a more rigorous discussion, we need to introduce some additional notation. In particular, both global norms, defined on Ω, and broken norms, defined over Ω_h as sums of K elements D^k, need to be introduced.

For the solution, $u(\boldsymbol{x}, t)$, we define the global L^2-norm over the domain $\Omega \in \mathsf{R}^d$ as

$$\|u\|_\Omega^2 = \int_\Omega u^2 \, d\boldsymbol{x}$$

as well as the broken norms

$$\|u\|_{\Omega,h}^2 = \sum_{k=1}^K \|u\|_{\mathsf{D}^{k'}}^2, \quad \|u\|_{\mathsf{D}^k}^2 = \int_{\mathsf{D}^k} u^2 \, d\boldsymbol{x}.$$

In a similar way, we define the associated Sobolev norms

$$\|u\|_{\Omega,q}^2 = \sum_{|\alpha|=0}^q \|u^{(\alpha)}\|_\Omega^2, \quad \|u\|_{\Omega,q,h}^2 = \sum_{k=1}^K \|u\|_{\mathsf{D}^k,q}^2, \quad \|u\|_{\mathsf{D}^k,q}^2 = \sum_{|\alpha|=0}^q \|u^{(\alpha)}\|_{\mathsf{D}^k}^2,$$

where α is a multi-index of length d. For the one-dimensional case, most often considered in this chapter, this norm is simply the L^2-norm of the q-th derivative. We will also define the space of functions, $u \in H^q(\Omega)$, as those functions for which $\|u\|_{\Omega,q}$ or $\|u\|_{\Omega,q,h}$ is bounded.

Finally, we will need the semi-norms

$$|u|^2_{\Omega,q,h} = \sum_{k=1}^{K} |u|^2_{\mathsf{D}^k,q}, \quad |u|^2_{\mathsf{D}^k,q} = \sum_{|\alpha|=q} \|u^{(\alpha)}\|^2_{\mathsf{D}^k}.$$

4.2 Briefly on convergence

Let us consider the one-dimensional hyperbolic system

$$\frac{\partial \boldsymbol{u}}{\partial t} + \mathcal{A}\frac{\partial \boldsymbol{u}}{\partial x} = 0,$$

where $\mathcal{A}$ is diagonalizable and we assume that appropriate boundary conditions are available to ensure well-posedness; that is, there exists constants C and α such that

$$\|\boldsymbol{u}(t)\|_\Omega \le C\exp(\alpha t)\|\boldsymbol{u}(0)\|_\Omega.$$

Let us assume that the solution is being approximated by an N-th-order piecewise polynomial, $\boldsymbol{u}_h$, which satisfies the semidiscrete scheme

$$\frac{d\boldsymbol{u}_h}{dt} + \mathcal{L}_h\boldsymbol{u}_h = 0.$$

Here, $\mathcal{L}_h$ represents the discrete approximation of $\mathcal{A}\partial_x$. Inserting the exact solution, $\boldsymbol{u}$, into the semidiscrete form yields

$$\frac{d\boldsymbol{u}}{dt} + \mathcal{L}_h\boldsymbol{u} = \mathcal{T}(\boldsymbol{u}(x,t)),$$

where $\mathcal{T}(\boldsymbol{u})$ is the truncation error, or the error by which the exact solution fails to satisfy the discrete approximation.

If we now introduce the error

$$\boldsymbol{\varepsilon}(\boldsymbol{x},t) = \boldsymbol{u}(\boldsymbol{x},t) - \boldsymbol{u}_h(\boldsymbol{x},t),$$

it is natural to seek convergence in the sense that

$$\forall t \in [0,T]: \lim_{\mathrm{dof}\to\infty} \|\boldsymbol{\varepsilon}(t)\|_{\Omega,h} \to 0.$$

We have introduced the notion of degrees of freedom (dof) to reflect that convergence can be achieved either by decreasing the cell size, h, by increasing the order of the approximation, N, or by doing both simultaneously, known as hp-convergence.

Proving convergence directly is, however, complicated due to the need to prove it for all time. Fortunately, there is shortcut. Consider the error equation

$$\frac{d}{dt}\varepsilon + \mathcal{L}_h\varepsilon = \mathcal{T}(\boldsymbol{u}(\boldsymbol{x},t)),$$

with the exact solution

$$\varepsilon(t) - \exp\left(-\mathcal{L}_h t\right)\varepsilon(0) = \int_0^t \exp\left(\mathcal{L}_h(s-t)\right)\mathcal{T}(\boldsymbol{u}(s))\,ds,$$

where we have suppressed the explicit dependence of $\boldsymbol{x}$ for simplicity. Integrating over the elements and summing up, we obtain

$$\|\varepsilon(t)\|_{\Omega,h} \le \|\exp\left(-\mathcal{L}_h t\right)\varepsilon(0)\|_{\Omega,h} + \left\|\int_0^t \exp\left(\mathcal{L}_h(s-t)\right)\mathcal{T}(\boldsymbol{u}(s))\,ds\right\|_{\Omega,h}.$$

Furthermore, since

$$\left\|\int_0^t \exp\left(\mathcal{L}_h(s-t)\right)\mathcal{T}(\boldsymbol{u}(s))\,ds\right\|_{\Omega,h} \le \int_0^t \|\exp\left(\mathcal{L}_h(s-t)\right)\|_{\Omega,h}\|\mathcal{T}(\boldsymbol{u}(s))\|_{\Omega,h}\,ds,$$

it suffices to ensure consistency in the sense that

$$\begin{cases} \lim_{\text{dof}\to\infty} \|\varepsilon(0)\|_{\Omega,h} = 0, \\ \lim_{\text{dof}\to\infty} \|\mathcal{T}(\boldsymbol{u}(t))\|_{\Omega,h} = 0 \end{cases}$$

and stability as

$$\lim_{\text{dof}\to\infty} \|\exp\left(-\mathcal{L}_h t\right)\|_{\Omega,h} \le C_h \exp(\alpha_h t), \quad t \ge 0, \tag{4.1}$$

to guarantee convergence. Note the close resemblance between stability and well-posedness.

This result, one half of the celebrated equivalence theorem by Lax and Richtmyer [215], suggests a natural approach to establish convergence for linear problems. First, we can focus on understanding consistency (e.g., the approximation of functions and operators) and, second, on the question of stability or boundedness of the operators.

4.3 Approximations by orthogonal polynomials and consistency

We will first discuss how well one can approximate functions and derivatives of functions in order to understand the consistency of the schemes.

Let us consider what happens in the local element, D^k, and how well one can approximate the local solution. In other words, if we approximate the global function, $u(x)$, as a piecewise N-th order polynomial function as

$$u(x) \simeq u_h(x) = \bigoplus_{k=1}^{K} u_h^k(x),$$

where

$$x \in \mathsf{D}^k : \quad u_h^k(x) = \sum_{n=1}^{N_p} \hat{u}_n^k \psi_n(x) = \sum_{i=1}^{N_p} u_h^k(x_i) \ell_i^k(x),$$

how well and in what sense can we expect u_h to approximate u?

As observed experimentally, we have two ways to improve the quality of u_h as an approximation of u: One can keep the order, N, of the local approximation fixed and increase the number of elements, K, known as h-refinement. Alternatively, one can keep K fixed and increase N, known as order or p-refinement. While both lead to a better approximation, they achieve this in different ways and it is important to understand when to use which path to achieve convergence.

Let us first estimate what can be expected under order refinement. For simplicity, we assume that all elements have length h (i.e., $\mathsf{D}^k = \overline{x}^k + \frac{r}{2}h$ where $\overline{x}^k = \frac{1}{2}(x_r^k + x_l^k)$ represents the cell center and $r \in [-1, 1]$ is the reference coordinate).

We begin by considering the standard interval and introduce the new variable

$$v(r) = u(hr) = u(x);$$

that is, v is defined on the standard interval, $\mathsf{I} = [-1, 1]$, and $\overline{x}^k = 0$, $x \in [-h, h]$. We discussed in Chapter 3 the advantage of using a local orthonormal basis – in this case, the normalized Legendre polynomials

$$\tilde{P}_n(r) = \frac{P_n(r)}{\sqrt{\gamma_n}}, \quad \gamma_n = \frac{2}{2n+1}.$$

Here, $P_n(r)$ are the classic Legendre polynomials of order n. A key property of these polynomials is that they satisfy a singular Sturm-Liouville problem

$$\frac{d}{dr}(1 - r^2)\frac{d}{dr}\tilde{P}_n + n(n+1)\tilde{P}_n = 0. \tag{4.2}$$

Let us consider the basic question of how well

$$v_h(r) = \sum_{n=0}^{N} \hat{v}_n \tilde{P}_n(r),$$

represents $v \in L^2(\mathsf{I})$. Note, that for simplicity of the notation and to conform with standard notation, we now have the sum running from 0 to N rather than from 1 to $N + 1 = N_p$, as used previously.

Prior to discussing interpolation, used throughout this text, we consider the properties of the projection where we utilize the orthonomality of $\tilde{P}_n(r)$ to find $\tilde{v}_n$ as

$$\tilde{v}_n = \int_{\mathsf{I}} v(r)\tilde{P}_n(r)\,dr.$$

An immediate consequence of the orthonormality of the basis is that

$$\|v - v_h\|_I^2 = \sum_{n=N+1}^{\infty} |\tilde{v}_n|^2,$$

recognized as Parseval's identity. A basic result for this approximation error follows directly.

Theorem 4.1. *Assume that $v \in H^p(I)$ and that v_h represents a polynomial projection of order N. Then*

$$\|v - v_h\|_{I,q} \leq N^{\rho-p}|v|_{I,p},$$

where

$$\rho = \begin{cases} \frac{3}{2}q, & 0 \leq q \leq 1 \\ 2q - \frac{1}{2}, & q \geq 1 \end{cases}$$

and $0 \leq q \leq p$.

Proof. We will just sketch the proof; the details can be found in [43]. Combining the definition of $\tilde{v}_n$ and Eq. (4.2), we recover

$$|\tilde{v}_n| \leq \left(\frac{1}{n(n+1)}\right)^p \int_I v^{(2p)}(r)\tilde{P}_n(r)\, dr,$$

by integration by parts $2p$ times. Combining this with Parseval's identity yields the required estimate in L^2:

$$\|v - v_h\|_{I,0} \leq N^{-p}|v|_{I,p},$$

for $v \in H^p(I)$, $p \geq 0$.

An intuitive understanding of the bound in the higher norms can be obtained by recalling the classic inverse inequality [43, 296]

$$\left\|\frac{d\tilde{P}_n(r)}{dr}\right\|_{I,0} \leq n^2 \left\|\tilde{P}_n(r)\right\|_{I,0};$$

that is, one should generally expect to lose two orders of convergence for each derivative, as reflected in the theorem. The more delicate estimate in the theorem relies on careful estimates of the behavior of the polynomials and properties of Sobolev spaces. We will not repeat the arguments here but refer to [43] for the details.   □

Note in particular that a consequence of this result is that if the local solution is smooth (i.e., $v \in H^p(I)$ for p large), convergence is very fast and exponential for an analytic function [297]. Thus, we recover the trademark of a classic spectral method [159] where the error decays exponentially fast with increasing N.

We will need a slightly refined estimate to get an improved insight into the convergence. For this, we need the following lemma [280].

Lemma 4.2. *Assume that for $v \in H^p(\mathsf{I})$, $p \geq 0$,*

$$v(r) = \sum_{n=0}^{N} \tilde{v}_n \tilde{P}_n(r).$$

Then

$$\int_{-1}^{1} |v^{(q)}|^2 (1 - r^2)^q dr = \sum_{n \geq q} |\tilde{v}_n|^2 \frac{(n+q)!}{(n-q)!} \leq |v|_{\mathsf{I},q}^2,$$

for $0 \leq q \leq p$.

The proof relies on properties of the orthogonal polynomials and the details can be found in [280]. With this one can establish the result:

Lemma 4.3. *If $v \in H^p(\mathsf{I})$, $p \geq 1$, then*

$$\|v - v_h\|_{\mathsf{I},0} \leq \left[\frac{(N+1-\sigma)!}{(N+1+\sigma)!} \right]^{1/2} |v|_{\mathsf{I},\sigma},$$

where $\sigma = \min(N+1, p)$.

Proof. Through Parseval's identity we have

$$\|v - v_h\|_{\mathsf{I},0}^2 = \sum_{n=N+1}^{\infty} |\tilde{v}_n|^2 = \sum_{n=N+1}^{\infty} |\tilde{v}_n|^2 \frac{(n-\sigma)!}{(n+\sigma)!} \frac{(n+\sigma)!}{(n-\sigma)!}.$$

Provided $\sigma \leq N + 1$, we have

$$\|v - v_h\|_{\mathsf{I},0}^2 \leq \frac{(N+1-\sigma)!}{(N+1+\sigma)!} \sum_{n=N+1}^{\infty} |\tilde{v}_n|^2 \frac{(n+\sigma)!}{(n-\sigma)!},$$

which, combined with Lemma 4.2 and $\sigma = \min(N+1, p)$, gives the result. $\square$

A generalization of this is the following result.

Lemma 4.4. *If $v \in H^p(\mathsf{I})$, $p \geq 1$ then*

$$\|v^{(q)} - v_h^{(q)}\|_{\mathsf{I},0} \leq \left[\frac{(N+1-\sigma)!}{(N+1+\sigma-4q)!} \right]^{1/2} |v|_{\mathsf{I},\sigma},$$

where $\sigma = \min(N+1, p)$ and $q \leq p$.

The proof follows the one above and is omitted. Note that in the limit of $N \gg p$, the Stirling formula gives

$$\|v^{(q)} - v_h^{(q)}\|_{\mathsf{I},0} \leq N^{2q-p} |v|_{\mathsf{I},p},$$

in agreement with Theorem 4.1.

The above estimates are all related to projections. However, as we discussed in Chapter 3, we are often concerned with the interpolations of v where

$$\boldsymbol{v} = \mathcal{V}\hat{\boldsymbol{v}},$$

is based on $N + 1$ points. The difference may be minor, but it is essential to appreciate it. Consider

$$v_h(r) = \sum_{n=0}^{N} \hat{v}_n \tilde{P}_n(r), \quad \tilde{v}_h(r) = \sum_{n=0}^{N} \tilde{v}_n \tilde{P}_n(r),$$

where $v_h(x)$ is the usual approximation based on interpolation and $\tilde{v}_h(r)$ refers to the approximation based on projection.

By the interpolation property we have

$$(\mathcal{V}\hat{\boldsymbol{v}})_i = v_h(r_i) = \sum_{n=0}^{\infty} \tilde{v}_n \tilde{P}_n(r_i) = \sum_{n=0}^{N} \tilde{v}_n \tilde{P}_n(r_i) + \sum_{n=N+1}^{\infty} \tilde{v}_n \tilde{P}_n(r_i),$$

from which we recover

$$\mathcal{V}\hat{\boldsymbol{v}} = \mathcal{V}\tilde{\boldsymbol{v}} + \sum_{n=N+1}^{\infty} \tilde{v}_n \tilde{P}_n(\boldsymbol{r}), \quad \boldsymbol{r} = (r_0, \ldots, r_N)^T.$$

This implies

$$v_h(r) = \tilde{v}_h(r) + \tilde{\boldsymbol{P}}^T(r)\mathcal{V}^{-1} \sum_{n=N+1}^{\infty} \tilde{v}_n \tilde{P}_n(\boldsymbol{r}).$$

Now, consider the additional term on the right-hand side,

$$\tilde{\boldsymbol{P}}^T(r)\mathcal{V}^{-1} \sum_{n=N+1}^{\infty} \tilde{v}_n \tilde{P}_n(\boldsymbol{r}) = \sum_{n=N+1}^{\infty} \tilde{v}_n \left(\tilde{\boldsymbol{P}}^T(r)\mathcal{V}^{-1} \tilde{P}_n(\boldsymbol{r}) \right),$$

which is allowed provided $v \in H^p(\mathsf{I})$, $p > 1/2$ [280]. Since

$$\tilde{\boldsymbol{P}}^T(r)\mathcal{V}^{-1}\tilde{P}_n(\boldsymbol{r}) = \sum_{l=0}^{N} \tilde{p}_l \tilde{P}_l(r), \quad \mathcal{V}\tilde{\boldsymbol{p}} = \tilde{P}_n(\boldsymbol{r}),$$

we can interpret the additional term as those high-order modes (recall $n > N$) that look like lower order modes on the grid. It is exactly this phenomenon that is known as aliasing and which is the fundamental difference between an interpolation and a projection.

The final statement, the proof of which is technical and given in [27], yields the following

Theorem 4.5. *Assume that* $v \in H^p(\mathsf{I})$, $p > \frac{1}{2}$, *and that* v_h *represents a polynomial interpolation of order* N. *Then*

$$\|v - v_h\|_{\mathsf{I},q} \le N^{2q-p+1/2} |v|_{\mathsf{I},p},$$

where $0 \le q \le p$.

Note in particular that the order of convergence can be up to one order lower due to the aliasing errors, but not worse than that.

Example 4.6. Let us revisit Example 3.2 where we considered the error when computing the discrete derivative. Using Theorem 4.5, we get a crude estimate as

$$\|v' - \mathcal{D}_r v_h\|_{\mathsf{I},0} \le \|v - v_h\|_{\mathsf{I},1} \le N^{5/2-p} |v|_{\mathsf{I},p}.$$

For the first example, that is,

$$v(r) = \exp(\sin(\pi r)),$$

this confirms the exponential convergence since $v(r)$ is analytic.

However, for the second problem,

$$v^{(0)}(r) = \begin{cases} -\cos(\pi r), & -1 \le r < 0 \\ \cos(\pi r), & 0 \le r \le 1, \end{cases} \quad \frac{dv^{(i+1)}}{dr} = v^{(i)}, \quad i = 0, 1, 2, 3 \ldots,$$

the situation is different. Note that $v^{(i)} \in H^i(\mathsf{I})$. In this case, the computations indicate

$$\|(v^{(i)})_r - \mathcal{D}_r v_h^{(i)}\|_{\mathsf{I},0} \le C N^{1/2-i},$$

which is considerably better than indicated by the general result. An improved estimate, valid only for interpolation of Gauss-Lobatto grid points, is [27]

$$\|v' - \mathcal{D}_r v_h\|_{\mathsf{I},0} \le N^{1-p} |v|_{\mathsf{I},p},$$

and is much closer to the observed behavior but suboptimal. This is, however, a sharp result for the general case, reflecting that special cases may show better results.

We return to the behavior for the general element of length h to recover the estimates for $u(x)$ rather than $v(r)$. We have the following bound:

Theorem 4.7. *Assume that* $u \in H^p(\mathsf{D}^k)$ *and that* u_h *represents a piecewise polynomial approximation of order* N. *Then*

$$\|u - u_h\|_{\Omega,q,h} \le C h^{\sigma-q} |u|_{\Omega,\sigma,h},$$

for $0 \le q \le \sigma$, *and* $\sigma = \min(N+1, p)$.

Proof. Introduce the new variable

$$v(r) = u(hr) = u(x).$$

Then

$$|v|_{\mathsf{I},q}^2 = \int_{\mathsf{I}} \left(v^{(q)}\right)^2 dr = \int_{\mathsf{D}^k} h^{2q-1} \left(u^{(q)}\right)^2 dx = h^{2q-1}|u|_{\mathsf{D}^k,q}^2.$$

Similarly fashion, we have

$$\|u\|_{\mathsf{D}^k,q}^2 = \sum_{p=0}^{q} |u|_{\mathsf{D}^k,p}^2 = \sum_{p=0}^{q} h^{1-2p}|v|_{\mathsf{I},p}^2 \le h^{1-2q}\|v\|_{\mathsf{I},q}^2.$$

We combine these estimates to obtain

$$\|u - u_h\|_{\mathsf{D}^k,q}^2 \le h^{1-2q}\|v - v_h\|_{\mathsf{I},q}^2 \le h^{1-2q}|v|_{\mathsf{I},\sigma}^2 = h^{2\sigma-2q}|u|_{\mathsf{D}^k,\sigma}^2,$$

where we have used Lemma 4.4 and defined $\sigma = \min(N+1, p)$. Summing over all elements yields the result.   □

For a more general grid where the element length is variable, it is natural to use $h = \max_k h^k$ (i.e., the maximum interval length). Combining this with Theorem 4.7 yields the main approximation result:

Theorem 4.8. *Assume that $u \in H^p(\mathsf{D}^k)$, $p > 1/2$, and that u_h represents a piecewise polynomial interpolation of order N. Then*

$$\|u - u_h\|_{\Omega,q,h} \le C \frac{h^{\sigma-q}}{N^{p-2q-1/2}} |u|_{\Omega,\sigma,h},$$

for $0 \le q \le \sigma$, and $\sigma = \min(N+1, p)$.

This result gives the essence of the approximation properties; that is, it shows clearly under which conditions on u we can expect consistency and what convergence rate to expect depending on the regularity of u and the norm in which the error is measured.

4.4 Stability

Stability is, in many ways, harder to deal with than consistency. In Chapter 2 we established stability for the simple scalar problem

$$\frac{\partial u}{\partial t} + a \frac{\partial u}{\partial x} = 0,$$

by using an energy method to obtain the result

$$\frac{1}{2}\frac{d}{dt}\|u_h\|^2_{\Omega,h} \leq c\|u_h\|^2_{\Omega,h},$$

which suffices to guarantee stability. In fact, we managed to prove that $c \leq 0$, ensuring that $\alpha_h \leq 0$ in the stability estimate, Eq. (4.1).

Stability of the linear system follows directly from this result with slight differences emerging only in the choice of the flux. Consider the general case

$$\frac{\partial \boldsymbol{u}}{\partial t} + \mathcal{A}\frac{\partial \boldsymbol{u}}{\partial x} = 0,$$

where $\mathcal{A}$ is an $m \times m$ diagonalizable matrix with purely real eigenvalues (i.e., a hyperbolic problem). We also assume that appropriate boundary conditions and initial conditions are provided. A simple scheme for this could be based on central fluxes as

$$\mathcal{M}^k\frac{d\boldsymbol{u}_i}{dt} + \sum_{j=1}^m \mathcal{A}_{ij}\mathcal{S}\boldsymbol{u}_j = \frac{1}{2}\int_{\partial\mathsf{D}^k} \sum_{j=1}^m \mathcal{A}_{ij}\hat{\boldsymbol{n}}\cdot[\![\boldsymbol{u}_j]\!]\,dx,$$

where $\boldsymbol{u}_i$ represents the i-th component of $\boldsymbol{u}$.

Since $\mathcal{A}$ is assumed to be uniformly diagonalizable, we have

$$\mathcal{A} = \mathcal{R}\Lambda\mathcal{R}^{-1},$$

and the above scheme transforms as

$$\mathcal{M}^k\frac{d\boldsymbol{v}_i}{dt} + \Lambda_{ii}\mathcal{S}\boldsymbol{v}_i = \frac{1}{2}\int_{\partial\mathsf{D}^k} \Lambda_{ii}\hat{\boldsymbol{n}}\cdot[\![\boldsymbol{v}_j]\!]\,dx,$$

where $\boldsymbol{v}_i = \mathcal{R}^{-1}\boldsymbol{u}_i$. This is the scheme for the scalar wave equation for which we have already established stability. Stability for the system follows then directly since

$$\begin{aligned}
\|\boldsymbol{u}_h(t)\|^2_{\Omega,h} &= \|\mathcal{R}\boldsymbol{v}_h(t)\|^2_{\Omega,h} \leq \|\mathcal{R}\|^2_{\Omega,h}\|\boldsymbol{v}_h(t)\|^2_{\Omega,h}\\
&\leq \|\mathcal{R}\|^2_{\Omega,h}\exp(\alpha_h t)\|\boldsymbol{v}_h(0)\|^2_{\Omega,h}\\
&\leq \|\mathcal{R}\|^2_{\Omega,h}\exp(\alpha_h t)\|\mathcal{R}^{-1}\boldsymbol{u}_h(0)\|^2_{\Omega,h}\\
&\leq \|\mathcal{R}\|^2_{\Omega,h}\|\mathcal{R}^{-1}\|^2_{\Omega,h}\exp(\alpha_h t)\|\boldsymbol{u}_h(0)\|^2_{\Omega,h}.
\end{aligned}$$

If $\|\mathcal{R}\|^2_{\Omega,h}\|\mathcal{R}^{-1}\|^2_{\Omega,h} \leq C_h$ stability follows. This is, however, ensured if $\mathcal{A}$ is uniformly diagonalizable.

In an entirely similar way, one can establish stability of the system with upwind fluxes provided upwinding is done on the characteristic variables (i.e., on $\boldsymbol{v}$ in the above example).

4.5 Error estimates and error boundedness

With stability and consistency established we can straightforwardly claim convergence for a number of different linear problems; for example

$$\frac{\partial u}{\partial t} + a\frac{\partial u}{\partial x} = 0. \tag{4.3}$$

Theorem 4.5 establishes consistency for the approximation of the spatial operator and we established stability in Chapter 2. From the equivalence theorem, convergence follows, albeit with time-dependent constants that may grow exponentially fast in time. The direct use of the consistency results would indicate an accuracy of the form

$$\|u - u_h\|_{\Omega,h} \leq \frac{h^N}{N^{p-5/2}}|u|_{\Omega,p,h},$$

for a smooth u. However, if we recall the results in Chapter 2, this appears to be suboptimal since we observed a behavior like

$$\|u(T) - u_h(T)\|_{\Omega,h} \leq h^{N+1}(C_1 + TC_2).$$

To recover these results, let us consider the problem in a bit more detail and establish convergence directly rather than through the equivalence theorem.

We define the bilinear form, $\mathcal{B}(u,\phi)$, as

$$\mathcal{B}(u,\phi) = (u_t,\phi)_\Omega + a(u_x,v)_\Omega = 0$$

for all smooth test functions, $\phi(x,t)$, and $u(x,t)$ being a solution to Eq. (4.3). For simplicity we assume that the problem is periodic, but this is not essential.

An immediate consequence of this is

$$\mathcal{B}(u,u) = 0 = \frac{1}{2}\frac{d}{dt}\|u\|_\Omega^2;$$

that is, the problem conserves energy, as we have already discussed. This also implies that we if solve Eq. (4.3) with two different initial conditions, $u_1(0)$ and $u_2(0)$, we have the error equation

$$\frac{1}{2}\frac{d}{dt}\|\varepsilon\|_\Omega^2 = 0,$$

or

$$\|\varepsilon(T)\|_\Omega = \|u_1(0) - u_2(0)\|_\Omega,$$

where we have defined the error $\varepsilon(t) = u_1(t) - u_2(t)$.

We will now mimic this for the discrete case as

$$\mathcal{B}_h(u_h,\phi_h) = ((u_h)_t,\phi_h)_{\Omega,h} + a((u_h)_x,\phi_h)_{\Omega,h} - (\hat{\boldsymbol{n}}\cdot(au_h - (au)^*),\phi_h)_{\partial\Omega,h} = 0,$$

for all test functions, $\phi_h \in \mathsf{V}_h$, and $u_h \in \mathsf{V}_h$ is the numerical solution, where

$$\mathsf{V}_h = \bigoplus_{k=1}^{K} \operatorname{span} \left\{ \tilde{P}_n(\mathsf{D}^k) \right\}_{n=0}^{N},$$

is the space of piecewise polynomials of order no larger than N defined on $\Omega_h (\simeq \Omega)$.

We again assume that a simple central flux of the type

$$(au)^* = \{\{au\}\},$$

is used and obtain

$$\mathcal{B}_h(u_h, \phi_h) = ((u_h)_t, \phi_h)_{\Omega,h} + a((u_h)_x, \phi_h)_{\Omega,h} - \frac{1}{2}(\llbracket au_h \rrbracket, \phi_h)_{\partial\Omega,h} = 0.$$

Since the jump term vanishes for the exact (smooth) solution, we have

$$\mathcal{B}_h(u, \phi_h) = 0,$$

which results in the error equation

$$\mathcal{B}_h(\varepsilon, \phi_h) = 0, \quad \varepsilon = u - u_h.$$

Observe that ε is not in the space of piecewise polynomials, V_h. Let us therefore write this as

$$\mathcal{B}_h(\varepsilon, \phi_h) = \mathcal{B}_h(\varepsilon_h, \phi_h) + \mathcal{B}_h(\varepsilon - \varepsilon_h, \phi_h) = 0, \qquad (4.4)$$

where we define

$$\varepsilon_h(r,t) = \sum_{n=0}^{N} \tilde{\varepsilon}_n(t) \tilde{P}_n(r) = \mathcal{P}_h \varepsilon(r,t),$$

as the N-th-order projection of ε onto the basis spanned by the orthonormal basis, $\tilde{P}_n(r)$. We recall that a consequence of this is the Galerkin orthogonality

$$n = 0, \dots, N : \int_{-1}^{1} (\varepsilon - \varepsilon_h) \tilde{P}_n(r) dr = 0.$$

If we take $\phi_h = \varepsilon_h$ to mimic the continuous case, we recover

$$\mathcal{B}_h(\varepsilon_h, \varepsilon_h) = ((\varepsilon_h)_t, \varepsilon_h)_{\Omega,h} + a((\varepsilon_h)_x, \varepsilon_h)_{\Omega,h} - \frac{1}{2}(\llbracket a\varepsilon_h \rrbracket, \varepsilon_h)_{\partial\Omega,h}.$$

After local integration by parts in space and using the periodicity, it is easily seen that the boundary term vanishes when both sides of an element are added and we recover

$$\mathcal{B}_h(\varepsilon_h, \varepsilon_h) = \frac{1}{2} \frac{d}{dt} \|\varepsilon_h\|_{\Omega,h}^2.$$

Thus, if we can estimate the left-hand side, we can estimate the behavior of the error. Recall Eq. (4.4) to recover

$$\frac{1}{2}\frac{d}{dt}\|\varepsilon_N\|_{\Omega,h}^2 = \mathcal{B}_h(\mathcal{P}_N u - u, \varepsilon_h),$$

since

$$\mathcal{P}_N\varepsilon - \varepsilon = \mathcal{P}_N(u - u_h) - (u - u_h) = \mathcal{P}_N u - u.$$

Here, $\mathcal{P}_N u$ represents the projection of the exact solution onto the space of piecewise N-th-order polynomials. Take $q = \mathcal{P}_N u - u$ and consider

$$\begin{aligned}
\mathcal{B}_h(q, \varepsilon_h)_{\Omega,h} &= (q_t, \varepsilon_h)_{\Omega,h} + a(q_x, \varepsilon_h)_{\Omega,h} - \frac{1}{2}([\![aq]\!], \varepsilon_h)_{\partial\Omega,h} \\
&= (q_t, \varepsilon_h)_{\Omega,h} - a(q, (\varepsilon_h)_x)_{\Omega,h} + (\hat{n}\cdot\{\!\{aq\}\!\}, \varepsilon_h)_{\partial\Omega,h} \\
&= (\hat{n}\cdot\{\!\{aq\}\!\}, \varepsilon_h)_{\partial\Omega,h},
\end{aligned}$$

due to the nature of the projection. To bound this, we need the following result:

Lemma 4.9. *If $u \in H^{p+1}(\mathrm{D}^k \cup \mathrm{D}^{k+1})$, then*

$$|\{\!\{aq\}\!\}|_{x_r^k} = |\{\!\{a(u - \mathcal{P}_N u)\}\!\}|_{x_r^k} \le C^k h^{\sigma-1/2}\frac{|a|}{2}|u|_{\mathrm{D}^k,\sigma},$$

where the constant, C^k, depends only on k, and $\sigma = \min(N+1, p)$.

The proof follows from the application of the Bramble-Hilbert lemma [59], Theorem 4.8, and the standard trace inequality [59]

$$\|u_h\|_{\partial\mathrm{D}} \le \frac{C(N)}{\sqrt{h}}\|u_h\|_{\mathrm{D}}.$$

We then have

$$\begin{aligned}
|\mathcal{B}_h(u - \mathcal{P}_N u, \varepsilon_h)| &\le \frac{1}{2}\left((\{\!\{aq\}\!\}, \{\!\{aq\}\!\})_{\partial\Omega,h} + (\varepsilon_h, \varepsilon_h)_{\partial\Omega,h}\right) \\
&\le C|a|h^{2\sigma-1}\|u\|_{\Omega,h,\sigma+1}^2,
\end{aligned}$$

by using Lemma 4.9 to bound the first part and Theorem 4.8 in combination with the trace inequality to bound the second term.

The final result is

$$\frac{d}{dt}\|\varepsilon_h\|_{\Omega,h}^2 \le C|a|h^{2\sigma-1}\|u\|_{\Omega,h,\sigma+1}^2,$$

from which we recover an improved result of the type

$$\|\varepsilon_h(T)\| \le (C_1 + C_2 T)h^{N+1/2},$$

in the limit where $p \gg N+1$ for $u \in H^p(\Omega)$. This very general result was first proven in [217]. Changing the flux to a more general flux does not change the result [60].

The recovery of the optimal $\mathcal{O}(h^{N+1})$ result relies on the use of a supercon-vergence property of the Legendre-Gauss-Radau points and is a special case that only applies to linear problem with strict upwinding. The main idea is to replace the orthogonal projection with the downwinded Gauss-Radau points. In this case, the Galerkin orthogonality no longer holds, but the boundary terms in the fluxes, estimated in Lemma 4.9 for the previous case, vanish identically and the main error contribution comes from the interior parts of the scheme that can be estimated by the results in Theorem 4.8. This results in the optimal estimate

$$\|\varepsilon_h(T)\|_{\Omega,h} \leq C(N)h^{N+1}(1 + C_1(N)T),$$

and confirms exactly what we observed previously. It also shows that we can-not hope for slower than linear growth in time of the error. The optimal convergence rate has been proven in [217] for the linear case and a further discussion of these techniques can be found in [60, 190].

4.6 Dispersive properties

Further insight into the accuracy of the scheme can be gained by discussing the dissipative and dispersive properties of the scheme.

For this, let us consider the simple wave equation

$$\frac{\partial u}{\partial t} + a\frac{\partial u}{\partial x} = 0, \tag{4.5}$$
$$u(x,0) = \exp(ilx),$$

where $x \in \mathsf{R}$ and l is the wavenumber of the initial condition. If we seek spatially periodic solutions of the form

$$u(x,t) = \exp(i(lx - \omega t)),$$

we easily recover that $\omega = al$, known as the dispersion relation with a being the exact phase velocity. The purpose of the discussion in this chapter is to understand how well the discontinuous Galerkin scheme reproduces this behavior.

We now assume that the computational domain is split into equidistant elements, D^k, all of length h. Following the approach in Section 2.2 we recover the basic semidiscrete local scheme

$$\frac{h}{2}\mathcal{M}\frac{d\boldsymbol{u}_h^k}{dt} + a\mathcal{S}\boldsymbol{u}^k = \boldsymbol{e}_N \left[(au_h^k) - (au_h^k)^*\right]_{x_r^k} - \boldsymbol{e}_0 \left[(au_h^k) - (au_h^k)^*\right]_{x_l^k},$$

where $x \in [x_l^k, x_r^k]$, $h = x_r^k - x_l^k$, and e_i is an $N+1$ long zero vector with 1 in entry i.

We consider the general flux

$$(au)^* = \{\{au\}\} + |a|\frac{1-\alpha}{2}[\![u]\!].$$

To further simplify matters, we assume that $a \geq 0$ and look for local solutions of the form

$$u_h^k(x^k, t) = U_h^k \exp[i(lx^k - \omega t)],$$

where U_h^k is a vector of coefficients. We seek to understand the relationship between l and ω for the discrete approach, known as the discrete or numerical dispersion relation. Inserting this into the scheme yields

$$\frac{h}{2}\mathcal{M}\frac{du_h^k}{dt} + a\mathcal{S}u^k = \frac{a\alpha}{2}e_N\left(u_h^k(x_r^k) - u_h^{k+1}(x_l^{k+1})\right)$$
$$-\frac{a(2-\alpha)}{2}e_0\left(u_h^k(x_l^k) - u_h^{k-1}(x_r^{k-1})\right).$$

Assume periodicity of the solution as

$$u_h^{k+1}(x_l^{k+1}) = \exp(ilh)u_h^k(x_l^k), \quad u_h^{k-1}(x_r^{k-1}) = \exp(-ilh)u_h^k(x_r^k)$$

to recover the expression

$$\left[\frac{i\omega h}{2}\mathcal{M} + a\mathcal{S} - \frac{a\alpha}{2}e_N\left(e_N^T - \exp(ilh)e_0^T\right)\right.$$
$$\left. + \frac{a(2-\alpha)}{2}e_0\left(e_0^T - \exp(-ilh)e_N^T\right)\right]U_h^k = 0.$$

This is recognized as a generalized eigenvalue problem

$$\left[2\mathcal{S} - \alpha e_N\left(e_N^T - \exp(iL(N+1))e_0^T\right)\right.$$
$$\left. + (2-\alpha)e_0\left(e_0^T - \exp(-iL(N+1))e_N^T\right)\right]U_h^k = i\Omega\mathcal{M}U_h^k.$$

We have normalized things as

$$L = \frac{lh}{N+1} = \frac{2\pi}{\lambda}\frac{h}{N+1} = 2\pi p^{-1}, \quad \Omega = \frac{\omega h}{a},$$

where

$$p = \frac{\lambda}{h/(N+1)}$$

is a measure of the number of degrees of freedom per wavelength. The minimum meaningful value for this is clearly 2 to uniquely identify a wave. We recognize that $L = \Omega/(N+1)$ is now the numerical dispersion relation, and by solving the eigenvalue problem for $\Omega = \Omega_r + i\Omega_i$, one recovers the dispersion relation of the numerical scheme, with Ω_r representing an approximation

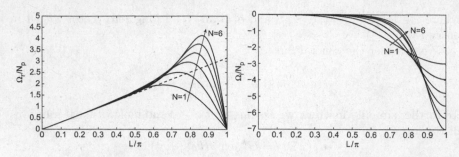

Fig. 4.1. On the left, we show the numerical dispersion relations for the linear advection operator with an upwind flux. The dashed line represents the exact case and the solid lines the numerical dispersion relation for the physical mode at different orders. On the right, we illustrate the dissipation associated with the different orders.

to the frequency ω, and Ω_i is the dissipation associated with the scheme. In particular, we note that

$$\frac{\Omega_r}{(N+1)L} = a_h(L)$$

provides a measure of the phase velocity of the numerical scheme and its dependence on the wavelength of the wave, measured in terms of degrees of freedom per wavelength, p.

In Fig. 4.1, we show the dispersion relation for the pure upwind scheme, $\alpha = 0$, for a range of orders of approximation, for the physical mode only. As expected, we see that for $L \ll 1$, the numerical phase velocity is very close to the physical wave speed and this agreement improves for a broader range of L as the order of the approximation increases, confirming the benefits of using high-order schemes for wave propagation. Furthermore, we see that there is a range of marginally resolved wavenumbers for which the numerical wave speed is faster than the physical speed. For the highest wavenumbers, one observes backward propagating spurious waves, as is also seen in most other methods. Also shown in Fig. 4.1 is Ω_i for the different orders of approximation, reflecting a significant dissipation of the high-frequency components (i.e., the wavenumbers that are only marginally resolved).

As we already know, the central flux yields a conservative scheme, reflected through $\Omega_i = 0$ for this case. The dispersion curves for this case are, however, more complex than for the upwind case, as shown in Fig. 4.2, where they are shown for $N = 2$. For $N = 1$ (not shown), the dispersion relation is similar to that of the upwinded scheme whereas the situation for $N = 2$ is very different, in particular for $L \simeq \pi/4$, indicating that the scheme could exhibit unphysical behavior even for well-resolved waves.

In contrast to this, we show in Fig. 4.3 the real and imaginary parts of all three modes for the upwinded $N = 2$ scheme. This clearly illustrates the physical mode but also that the unphysical modes are severely damped.

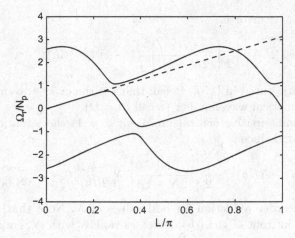

Fig. 4.2. Numerical dispersion relations for the linear advection operator with a purely central flux. The dashed line represents the exact case and the solid lines reflect the dispersion characteristics for the three numerical modes at $N = 2$.

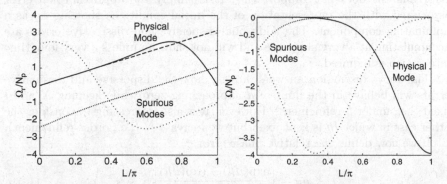

Fig. 4.3. On the left, we show the numerical dispersion relations for the linear advection operator with an upwind flux. The dashed line represents the exact case and the dotted curves are the spurious modes. On the right, we illustrate the dissipation associated with the three modes.

Computational studies of the dispersive and dissipative properties of DG-FEM were initiated in [181, 284], including extensions to two spatial dimensions. Steps toward a more rigorous analysis were taken in [179, 180], including extensions to two-dimensional problems and the impact of boundary conditions. A number of concluding conjectures made in that work have subsequently been proven in [9]. In the following, we will outline the key results of this analysis.

In the highly resolved region $lh \ll 1$, we have, provided $\alpha \neq 1$, the dispersive error [9]

$$\left| \mathcal{R}(\tilde{l}h) - \mathcal{R}(lh) \right| \simeq \frac{1}{2} \left[\frac{N!}{(2N+1)!} \right]^2 (lh)^{2N+3},$$

and the dissipative error is

$$\left| \mathcal{I}(\tilde{l}h) \right| \simeq \frac{1}{2} \left[\frac{N!}{(2N+1)!} \right]^2 (1-\alpha)^{(-1)^N} (lh)^{2N+2},$$

as was also conjectured in [179]. Recall that l is the exact wavenumber and $\tilde{l}$ reflects the numerical wavenumber (recall $\tilde{l}h = \Omega$).

For the nondissipative central flux (i.e., $\alpha = 1$) the situation is a little more complicated since

$$\left| \mathcal{R}(\tilde{l}h) - \mathcal{R}(lh) \right| \simeq \frac{1}{2} \left[\frac{N!}{(2N+1)!} \right]^2 \begin{cases} -\frac{1}{2}(lh)^{2N+3}, & N \text{ even} \\ 2(\tilde{l}h)^{2N+1}, & N \text{ odd}, \end{cases}$$

illustrating an order reduction for odd values of N. Note that, in all cases, the coefficient in front of the (lh) decreases rapidly with N, emphasizing the benefits of using large values of N.

These excellent dispersive properties help to understand why one very often observes $\mathcal{O}(h^{N+1})$ error in the computational experiments even if the error analysis does not support this. It is simply the approximation error associated with the representation of the initial conditions showing up as a dominating component. The additional dispersive and dissipative errors are accumulating at a very slow rate and will not show up unless a very long time integration is required.

The above discussion answers directly how the dispersive and dissipative errors will behave in the limit of h approaching zero while keeping N and l fixed (i.e., under h-refinement). However, it is also interesting to consider the other case in which lh is kept fixed but we increase N (i.e., order refinement).

If we now define the relative phase error

$$\rho_N = \left| \frac{\exp(ilh) - \exp(i\tilde{l}h)}{\exp(ilh)} \right|,$$

the convergence of ρ_N falls into three separate regions with distinct behavior [9]:

$$\rho_N \simeq \begin{cases} 2N+1 < lh - C(lh)^{1/3}, & \text{no convergence} \\ lh-o(lh)^{1/3} < 2N+1 < lh+o(lh)^{1/3}, & \mathcal{O}(N^{-1/3}) \text{ convergence} \\ 2N+1 \gg lh, & \mathcal{O}(hl/(2N+1))^{2N+2} \text{ convergence} \end{cases}$$

It is interesting to note that the threshold for no convergence is

$$2 \simeq \frac{lh}{N+1} = 2\pi p^{-1};$$

that is, one needs $p \geq \pi$ to achieve convergence, in agreement with classic results from spectral methods [136].

A final theorem provides a useful guideline to determine the resolution requirements for a general wave problem [9].

Theorem 4.10. *Let $\kappa > 1$ be fixed. If $N, lh \to \infty$ such that $(2N+1)/lh \to \kappa$, then*

$$\rho_N \simeq C \exp(-\beta(N+1/2)),$$

where $\beta > 0$ and C are constants, depending on κ but not on N.

Thus, as long as $\kappa > 1$ (i.e., $p > \pi$), there is exponential decay of the phase error.

In [10] the analysis of dispersion and dissipation for wave problems is extended to the second-order wave equation, using some of the schemes discussed in Chapter 7 for the spatial approximation. We will not discuss these results here and refer to the original work for the details. Dispersion and dissipation characteristics of DG schemes specialized to Maxwell's equations are discussed in [240, 276].

4.7 Discrete stability and timestep choices

Most of the above analysis pertains to stability and convergence of the semidiscrete form; that is, we have discretized space but kept time continuous, to obtain a system of ordinary differential equations as

$$\frac{\partial u}{\partial t} + a\frac{\partial u}{\partial x} = 0 \quad \Rightarrow \quad \frac{du_h}{dt} + \mathcal{L}_h u_h = 0.$$

Here, u_h is the vector of unknowns (nodes or modes) and $\mathcal{L}_h$ represents the discrete approximation to the periodic operator, $a\partial_x$.

To obtain a fully discrete scheme, we then use a method for the integration of systems of ordinary differential equations. As we already discussed in Chapter 3, the most popular choice is a Runge-Kutta method, but other choices are certainly also possible [40, 143, 144].

Regardless of the choice of the time integration method, however, we need to choose the time-step, Δt, used to advance the discrete form from t^n to t^{n+1}. This choice has to take into account the balance between accuracy and stability of the temporal scheme.

A standard technique, resulting in necessary but not sufficient conditions on Δt, is to consider the scalar test problem,

$$u_t = \lambda u, \quad \text{Real}(\lambda) \le 0,$$

and define the stability region, C, associated with the timestepping scheme as that region in the complex plane of $\lambda\Delta t$ for which the scheme is stable; examples for two explicit Runge-Kutta methods are shown in Fig. 4.4.

The connection between this simple case and the more general case is that the eigenvalues of $\mathcal{L}_h$ play the role of λ, and to ensure stability, we must choose Δt small enough so that the full eigenvalue spectrum fits inside the stability region of the time integration scheme.

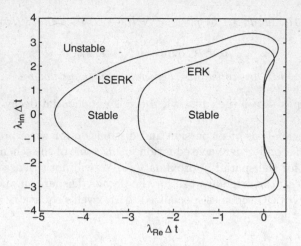

Fig. 4.4. Regions of absolute stability the explicit classic fourth-order Runge-Kutta (ERK) methods and the low-storage fourth-order, five-stage method (LSERK).

Before we continue, it is worthwhile trying to understand the scaling of $\mathcal{L}_h$ as a function of N and h. Recall first that

$$\mathcal{L}_h = \frac{2a}{h}\mathcal{M}^{-1}\left[\mathcal{S} - \mathcal{E}\right],$$

where $\mathcal{E}$ is a zero matrix with unity entries in the two diagonal corners. Consider now

$$\frac{h^2}{4a^2}\|\mathcal{L}_h\|_{\mathsf{I}}^2 = \frac{h^2}{4a^2}\sup_{\|u_h\|=1}\|\mathcal{L}_h u_h\|_{\mathsf{I}}^2$$

$$\leq \|\mathcal{D}_r\|_{\mathsf{I}}^2 + \|\mathcal{M}^{-1}\mathcal{E}\|_{\mathsf{I}}^2 + 2\sup_{\|u_h\|=1}\left(\mathcal{D}_r u_h, \mathcal{M}^{-1}\mathcal{E}u_h\right)_{\mathsf{I}}$$

$$\leq C_1 N^4 + C_2 N^2 + C_3 N^3 \leq C N^4,$$

where we have used the well-known inequality [280]

$$\left\|\frac{du_h}{dx}\right\|_{\mathsf{I}} \leq \sqrt{3}N^2\|u_h\|_{\mathsf{I}},$$

and the inverse trace inequality [320]

$$\|u_h\|_{\partial\mathsf{I}} \leq \frac{N+1}{\sqrt{2}}\|u_h\|_{\mathsf{I}}.$$

This results in a scaling as

$$\|\mathcal{L}_h\|_{\mathsf{D}^k} \leq C\frac{a}{h^k}N^2$$

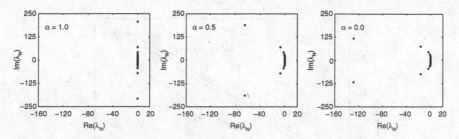

Fig. 4.5. Structure of the eigenvalue spectrum, λ_N, for $N = 24$ and periodic boundary conditions. On the left, for a purely central flux; in the middle for partially upwinded flux; and on the right, for a fully upwinded flux.

for element D^k of size h^k. Note, however, that this is a crude estimate as it does not account for differences introduced through the fluxes and other choices; that is, it is only an upper bound and the situation may be better than this.

Let us consider the wave equation again and assume that we have only one domain, $r \in [-1, 1]$, and that $a = 1$. In Fig. 4.5 we show the eigenvalue spectrum for the basic differentiation operator with periodic boundary conditions at $N = 24$ but for different choices of the flux. We note in particular the purely imaginary spectrum for the central flux, consistent with the energy-conserving nature of this discretization. Note also that as soon as $\alpha \neq 1$, a negative real part is associated with some of the eigenvalues, reflecting the dissipative nature of the spatial scheme.

Comparing with the stability regions for the two explicit Runge-Kutta methods in Fig. 4.4, it is clear that we can always choose Δt sufficiently small so that we can ensure $\Delta t \lambda_N$ to be inside the region of stability – a necessary condition for discrete stability. A sufficient condition is a bit more restrictive [326] and the necessary condition turns out to be an excellent guideline.

To turn this into a useful approach, we need to understand how one chooses a stable Δt in a semiautomatic fashion for different values of N. We show in Fig. 4.6 the magnitude of the maximum eigenvalue as a function of N. A couple of things are worth noticing. First, for large values of N, we see that $\max(\lambda_N) \propto N^2$, as expected based on the previous analysis, indicating that we should expect $\Delta t \propto N^{-2}$.

It is instructive to consider another example where homogeneous boundary conditions rather than periodic boundary are enforced on the single domain; that is, it has less relevance for the general multi-element scheme. In Fig. 4.7 we show the corresponding eigenvalue spectrum for the basic differentiation operator with $N = 24$ but for different choices of the flux. We note that for $\alpha > 0$ there is little qualitative difference between these results and those shown in Fig. 4.5. For the pure upwinding case the situation is, however, different. In particular, it appears that for small values of N (e.g., $N < 30$), the growth rate of $\max(\lambda_N)$ is different (e.g., $\max(\lambda_N) \propto N$), but then changes

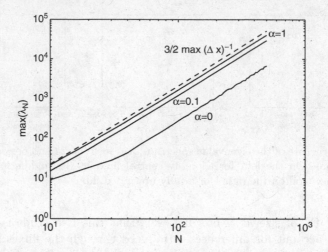

Fig. 4.6. Growth of the maximum eigenvalue of the periodic discrete differentiation operator as a function of N and for different choices of the flux. The dashed line shows the upper bound related to the minimal spacing of the Legendre-Gauss-Lobatto grid. The lower curve, labeled for upwinding ($\alpha = 0$), assumes homogeneous boundary conditions (see text for discussion).

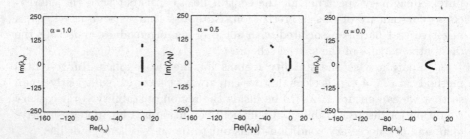

Fig. 4.7. Structure of the eigenvalue spectrum, λ_N, for $N = 24$ and homogeneous boundary conditions. On the left for a purely central flux, in the middle for partially upwinded flux, and on the right for a fully upwinded flux.

for larger values of N for the upwind flux; this is also what is reflected in Fig. 4.6. This is a well-understood phenomenon and is related to a strong non-normality of the discrete differentiation matrix for the pure upwinding case. It is studied in detail in [309], highlighting that the $\mathcal{O}(N)$ growth rate is in fact theoretically correct but finite precision effects makes the eigenvectors highly sensitive, resulting in a practical bound proportional to N^2 rather than N, even for small values of N. It is curiosity rather than something truly useful, as even small changes destroy this effect as illustrated above.

A guideline for stability for the general choice of flux is illustrated in Fig. 4.6 as an upper bound of the type

$$\max(|\lambda_N|) \le \frac{3}{2} \max_i (\Delta_i r)^{-1},$$

where, as previously introduced, $\Delta_i r$ is the grid spacing between the nodes in the standard element, $\mathsf{I} = [-1, 1]$. We can generally expect a timestep limitation like

$$\Delta t \le C(N) \min(\Delta x),$$

where C contains the constant $2/3$ as well as information about the extension of the region of absolute stability (Fig. 4.4).

The extension to more general cases is straightforward. If we consider

$$\frac{\partial u}{\partial t} + a \frac{\partial u}{\partial x} = 0,$$

and a general grid with elements of different lengths, h^k, we have

$$\Delta t \le C \frac{1}{|a|} \min_{k,i} \frac{h^k}{2} (\Delta_i r),$$

as a reasonable bound, also known as the Courant-Friedrichs-Levy (CFL) condition. For the general case of a system

$$\frac{\partial \boldsymbol{u}}{\partial t} + \mathcal{A} \frac{\partial \boldsymbol{u}}{\partial x} = 0,$$

we must ensure that the fastest waves are stable; that is, we obtain a bound like

$$\Delta t \le C \frac{1}{\max(|\lambda(\mathcal{A})|)} \min_{k,i} \frac{h^k}{2} (\Delta_i r),$$

where $\lambda(\mathcal{A})$ represent the real eigenvalues of $\mathcal{A}$, as these are the velocities associated with the characteristic waves of the system.

4.8 Taming the CFL condition

As we have just seen, the timestep associated with a discontinuous Galerkin discretization of a first-order operator has a scaling that asymptotically behaves as $\mathcal{O}\left(h/N^2\right)$ when the order of polynomials, N, is increased or h is decreased. This is primarily because the underlying discretization spaces consists of polynomials defined on each element.

To further highlight the nature of this, we recall the following polynomial inequalities (see e.g., [90, 296] for details and generalizations):

Theorem 4.11. *The inequality*

$$\left| \frac{dq}{dx}(x) \right| \le \frac{N}{\sqrt{1 - x^2}} \|q\|_\infty, \quad -1 < x < 1,$$

holds for every N-th-order polynomial, q.

Theorem 4.12. *The inequality*

$$\left| \frac{dq(x)}{dx} \right| \leq N^2 \, \|q\|_\infty$$

holds for every N-th-order polynomial, q.

These are recognized as Bernstein's and Markov's inequality, respectively. We note that both results are sharp.

Combining these,

$$\left| \frac{dq}{dx}(x) \right| \leq \min \left(N^2, \frac{N}{\sqrt{1 - x^2}} \right) \|q\|_\infty, \quad -1 \leq x \leq 1, \tag{4.6}$$

which highlights that the gradients of the normalized N-th-order polynomials may be up to $O\left(N^2/h\right)$ near the boundary portion of the element and $O\left(N/h\right)$ closer to the center of the element.

These results are suggestive of the main reason why there is a difference between the CFL conditions of a typical finite difference method where a moving interpolation stencil is used, and slope information is generally extracted at element centers, and the DG method where slope information is used throughout the element.

As the small timesteps resulting from the $\mathcal{O}(N^2/h)$ scaling often result in the need to take many timesteps, it is worthwhile discussing whether one can overcome this barrier. The above discussion shows that we must strive to control the high gradients of the polynomials at the boundaries of the interval to have any hope of improving this situation. On the other hand, these polynomials were introduced to guarantee accuracy and robustness, so we must be careful when we attempt to modify them.

In the following, we will discuss three different techniques aimed at addressing the problem with a small timestep introduced by the use of a high-order polynomial basis. As we will see, the ideas are very different and it is likely that the best choice is guided by the specifics of the problems being solved.

4.8.1 Improvements by mapping techniques

Let us first discuss use of a mapping function, $\psi : [-1, 1] \to [-1, 1]$ constructed to control the sharp growth of the polynomials. Hence, if we consider the N-th-order polynomial, q, we have that

$$\frac{dq}{d\xi} = \frac{dr}{d\xi}\frac{dq}{dr}, \quad \frac{d\xi}{dr} = \frac{d\psi}{dr} = \psi'(r),$$

assuming that $\psi : \ r \xrightarrow{} \xi$. Clearly, we would like ψ' to have the property that when r is close to zero, the value is constant, while it should grow like N toward the ends of the domain. Let us consider the following mapping function, known as the Kosloff-Tal-Ezer mapping [206]:

$$\xi = \psi(r, \alpha) = \frac{\arcsin(\alpha r)}{\arcsin \alpha},$$

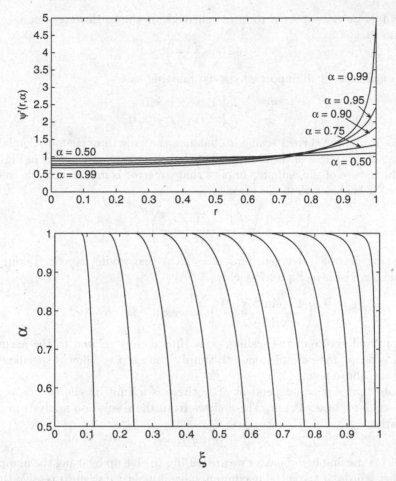

Fig. 4.8. On the top figure, we illustrate the mapping Jacobian associated with the mapping ψ for different values of α. Note that only $[0,1]$ is shown as the function is even around zero. On the bottom figure, we illustrate the effect of the mapping on the grid points at $N = 24$ as a function of α, highlighting that the effect of the mapping is to move the points toward an equidistant grid as α approaches 1.

giving the mapping Jacobian as

$$\frac{dr}{d\xi} = \left(\frac{d\xi}{dr}\right)^{-1} = \frac{1}{\psi'(r,\alpha)}, \quad \frac{d\psi}{dr} = \frac{\alpha}{\arcsin(\alpha)} \frac{1}{\sqrt{1 - (\alpha r)^2}}.$$

In Fig. 4.8 we show the mapping Jacobian associated with ψ for different values of α, emphasizing the growth close to the boundary of the interval. The figure also illustrates the effect of the mapping on the grid points at $N = 24$ as a function of α, highlighting that the effect of the mapping is to move the points toward an equidistant grid as α approaches 1.

So far, we do not seem to have achieved anything. However, let use now assume that

$$\alpha = \cos\left(\frac{\beta}{N}\right), \tag{4.7}$$

and consider the two important special limiting cases

$$\frac{d\psi}{dr} = \frac{2}{\pi}\begin{cases} 1, & r = 0 \\ N/\beta, & r = \pm 1; \end{cases}$$

that is, it has the correct scaling to balance the growth of the polynomials as they reach the edges of the element. This behavior is clearly shown in Fig. 4.8.

The nature of the mapping implies that an error is introduced due to this. In [206] it is shown that this error is

$$\varepsilon = \left(\frac{1 - \sqrt{1 - \alpha^2}}{\alpha}\right)^N,$$

indicating loss of convergence for $\alpha = 1$, as one would expect. Taking the generic form of α in Eq. (4.7) yields

$$\varepsilon = \left(\frac{1 - \sin(\beta N^{-1})}{\cos(\beta N^{-1})}\right)^N \rightarrow \exp(-\beta), \quad N \rightarrow \infty.$$

The optimal choice of the scaling, β, is thus closely related to the accuracy of the scheme; for example, one can simply choose β to allow a specific error using the above result.

Note that if β is independent of N there is a limit on the accuracy; that is, spectral accuracy is lost. This follows from the resolution analysis in [206] indicating

$$\beta = j\pi, \quad j \leq \frac{N}{2},$$

where j is the number of modes we are willing to give up by using the mapping. In other words, if k_m is the maximum wavenumber we wish to resolve then

$$k_m = \frac{N}{2} - j.$$

Another choice of β, advocated in [156], is to choose β such that it balances the basic truncation error for the spatial approximation. For a first order spatial derivative, (i.e., as for a linear wave equation) this leads to the surprising result that

$$\beta = \frac{N}{4},$$

in some contradiction to the arguments presented in the above. It has, nevertheless, been found to be useful for small values of N. In particular, this choice ensures that spectral convergence is maintained.

Another highly successful choice is a conservative choice based on making sure that the mapping error is always dominated by the machine precision; that is,

$$\alpha = \frac{2}{t + t^{-1}}, \quad t = \left(\sqrt[N]{\varepsilon_M}\right)^{-1},$$

where ε_M is the machine zero. This naturally works very well but has the disadvantage of being effective only for N being very high [101].

To illustrate the effect of this mapping on the actual timestep, let us again consider

$$\frac{\partial u}{\partial t} + a\frac{\partial u}{\partial x} = 0 \;\;\Rightarrow\;\; \frac{d}{dt}u_h + \mathcal{L}_h u_h = 0,$$

subject to periodic boundary conditions. We recall that

$$\mathcal{L}_h = \frac{2a}{h}\left(\mathcal{M}^{\psi}\right)^{-1}\left[\mathcal{S} - \frac{2-\gamma}{2}e_N\left(e_N^T - e_0^T\right) + \frac{\gamma}{2}e_0\left(e_0^T - e_N^T\right)\right],$$

where, in a slight abuse of notation, $0 \le \gamma \le 1$ controls the type of the flux. The effect of the mapping enters solely through the mass matrix

$$\mathcal{M}_{ij}^{\psi} = \int_{-1}^{1}\ell_i(r)\ell_j(r)\psi'(r,\alpha)\,dr,$$

which can be computed by quadrature.

In Fig. 4.9 we illustrate how this changes the eigenvalue spectrum, showing that the structure of the spectrum remains essentially the same. Note also, however, that the mapping shrinks the spectral radius of the operator, but the properties are otherwise maintained (e.g., stability and energy conservation for the central flux). In this particular example we have chosen β to allow an asymptotic error of 10^{-4}, yielding $\beta = 9.2$. Decreasing β will yield a higher degree of compression of the spectrum, albeit at the price of an increased minimum error level.

As a final example of the impact of this mapping, we show in Fig. 4.10 the scaling of the maximum eigenvalue with N for different values of β, also providing the connection to the error. This figure should be contrasted with the unmapped results in Fig. 4.6.

As expected, we see in Fig. 4.10 a significant reduction in the maximum eigenvalue, confirming that the use of the local elementwise mapping offers a systematic approach to trade accuracy for speed, through an increased timestep to the extent where the Δt may scale as $\mathcal{O}(N^{-1})$ for large values of N. We can combine this with a timestepping scheme of an order proportional to N; for example, a simple explicit Runge-Kutta scheme such as [185]

$$v = u_h^n,$$
$$v = u_h^n + \frac{1}{k}\Delta t\left(\mathcal{L}_h v\right), \;\; k = s,\ldots,1,$$
$$u_h^{n+1} = v,$$

which, for linear problems, is an s-stage scheme of order s. The stability region of this scheme is illustrated in Fig. 4.11. Furthermore, one can show that the extension of the stability region grows linearly with s [40]. Hence, if $s \propto N$, the timestep is independent of the order the spatial discretization.

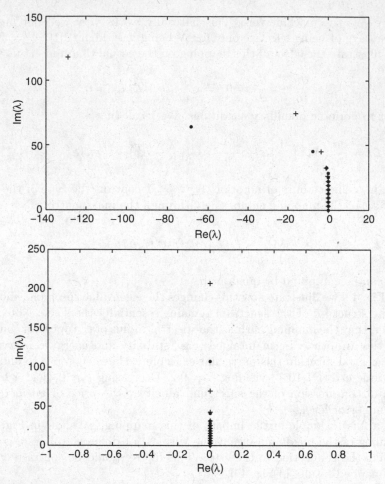

Fig. 4.9. On the top we show the upper half of the eigenvalue spectrum for $\mathcal{L}_h$ for $N = 24$, both for the original operator, (+), and the mapped operator, (·), when using an upwind flux. On the bottom, we show the similar result for the central flux.

4.8.2 Filtering by co-volume grids

Let us consider another way of achieving a similar result (i.e., a timestep independent of the order of approximation, N) by considering the use of a co-volume grid. The motivation for this is illustrated in Fig. 4.12, where we show the upper bound for the polynomials at three biunit one-dimensional elements that each overlap by half their lengths with their neighbors. This upper bound on the scaled gradient of polynomials grows dramatically near each of the element boundaries as expected from the basic estimates discussed previously. However, we have also highlighted the minimum upper bound of all three polynomials in the central element, resulting in an almost constant

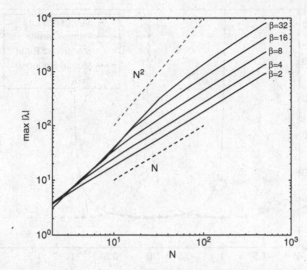

Fig. 4.10. Growth of the maximum eigenvalue of the mapped discrete differentiation operator as a function of N and for different choices of the mapping parameter, β.

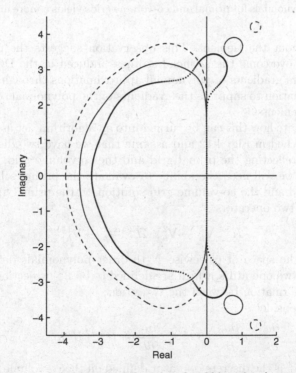

Fig. 4.11. Margins of absolute stability for the fifth-order (points), sixth-order (solid line), and seventh-order (dot-dash line) linear s-th-order RK scheme [185].

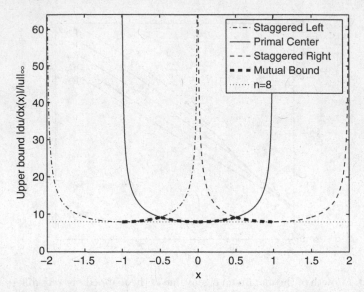

Fig. 4.12. Combining upper bounds on the relative maximal gradient of polynomials eigth-order polynomials for primal and co-volume grids yields a more uniform bound.

curve throughout the element. This observation suggests the possibility of being able to overcome the artificial stiffness induced in the DG scheme by the nonuniform gradients of polynomial approximations through a co-volume grid approximation to suppress the gradients of the polynomial solutions near the ends of elements.

To appreciate how this can be turned into an algorithm, let us consider the situation sketched in Fig. 4.12 and assume that we have two discretizations, Ω_h^1, and Ω_h^2, reflecting the primal grid and the co-volume grid, respectively. Furthermore, we will need the ability to evaluate the primal solution on the co-volume grid and the co-volume grid solution on the primal grid. For this, we define the two operators

$$\Pi^1 : \mathsf{V}_h^1 \to \mathsf{V}_h^2, \ \ \Pi^2 : \mathsf{V}_h^2 \to \mathsf{V}_h^1,$$

where V_h^s is the space of piecewise N-th-order polynomials defined on Ω_h^s; that is, these two operators, which are defined to be L^2-projections, allow one to transfer information between the two grids.

We then consider

$$\frac{\partial u}{\partial t} + a\frac{\partial u}{\partial x} = 0 \ \Rightarrow \ \frac{d\boldsymbol{u}_h}{dt} + \Pi^2 \mathcal{L}_h^2 \Pi^1 \boldsymbol{u}_h = 0.$$

In this case, $\mathcal{L}_h^2$ is the discrete operator defined on the co-volume grid, leading to the following simple algorithm for evaluating the right-hand side:

1. Given a solution, $\boldsymbol{u}_h$ at the primal grid, apply the L^2-projection to recover it on the co-volume grid.

2. Apply the operator at the co-volume grid to recover $d\boldsymbol{u}_h^2/dt$.
3. Project this to the primal grid.

A generalization of this scheme is given as

$$\frac{d\boldsymbol{u}_h}{dt} + \beta \Pi^2 \mathcal{L}_h^2 \Pi^1 \boldsymbol{u}_h + (1-\beta)\mathcal{L}_h^1 \boldsymbol{u}_h = 0, \ \ 0 \le \beta \le 1,$$

effectively combining the original and the co-volume grid scheme. We will discuss the impact of choosing β shortly.

Stability of this new scheme follows directly from the stability of the previous non-co-volume grid scheme and the contractive behavior of the L^2-projection in the following sense [321]:

$$\frac{d}{dt}\|\boldsymbol{u}_h^1\|^2_{\Omega_h^1} \le -\frac{\beta}{2}\sum_{k=1}^{K}(1-\alpha)[\![\Pi^2\boldsymbol{u}_h^2]\!]^2 - \frac{1-\beta}{2}\sum_{k=1}^{K}(1-\alpha)[\![\boldsymbol{u}_h^1]\!]^2;$$

that is, when a dissipative flux is used ($\alpha \ne 1$), there will be decay of energy if there are jumps in the primal solution or jumps in the L^2-projected $\boldsymbol{u}_h^1$ on the co-volume grid.

By requiring that the solution is sufficiently smooth on both discretizations, [321] presents a complete accuracy analysis showing that the scheme is at least $\mathcal{O}(h^N)$, although computational results indicate that this is a conservative estimate.

In the anticipation that the above procedure will allow us to advance at a larger timestep, it is natural to discuss the algorithmic overhead of using the co-volume grid-based filtering. For the generalized advection equation example, the filter requires approximately two grid transfers and two evaluations of the upwind discontinuous Galerkin derivative per stage in the RK-timestepping scheme. Thus, the minimum N required for break even is quite high. We can, however, reduce the computational cost with the following version:

$$\frac{d\boldsymbol{u}_h^1}{dt} = -\mathcal{L}_h^1 \mathcal{F}^\beta \boldsymbol{u}_h^1, \tag{4.8}$$

where $\mathcal{F}^\beta = \beta \Pi^2 \Pi^1 + (1-\beta)$. This formulation only requires one upwind DG derivative evaluation, and the action of the transfers can be consolidated into one operator that uses a three-element stencil to filter the upwind DG gradient in each element. This approach yields a breakeven point for N, which is quite modest. It furthermore emphasized that the co-volume grid approach is a filter – something that we will return to in more detail in Chapter 5.

To fully appreciate the impact that the use of the co-volume filtering can have on the maximum eigenvalue and, thus, the stable timestep, let us consider a well-known problem in more detail. We consider the one-dimensional Maxwell's equations

$$\frac{\partial H}{\partial t} = \frac{\partial E}{\partial x}, \ \frac{\partial E}{\partial t} = \frac{\partial H}{\partial x},$$

Table 4.1. Dependence of the spectral radius of the partially stabilized operator on the β parameter for the $K = 20$, $N = 7$ upwind DG operator on the $[0, 2]$ periodic interval.

β	0	0.2	0.4	0.6	0.8	1.0
Spectral radius	614.8	495.8	376.6	257.3	137.2	129.4
Decrease factor	1.00	1.24	1.63	2.39	4.48	4.75

where H and E represent transverse magnetic and electric fields, respectively. We assume perfectly electrically conducting boundary conditions the ends of the domain; that is,

$$E(x_L, t) = E(x_R, t) = 0.$$

The appearance of the boundary conditions is in contrast to our discussion above where we focused on interior interfaces. The approach above requires one to treat these Dirichlet conditions by extending the solution beyond the domain boundary. In the specific circumstance of Maxwell's equations with homogeneous boundary data, one can use the method of images. For both cases, we extend the primal mesh by two elements, constructed by reflecting the mesh about each end vertex, at each end. The primal solution can be extended by reflection for the zero Neumann condition and by antireflection for the zero Dirichlet condition. The last modification to note is that the DG operator will be applied on the co-volume mesh, and since we have extended the primal grid by two elements at each end, we are able to apply arbitrary boundary condition states for the DG derivative on the co-volume grid without affecting the result after projection back onto the primal mesh. With this approach we can recycle the solvers from the modified DG-FEM approach for advection.

Let us now investigate the order of accuracy and in fact go further and experimentally test the spectral correctness of the discrete modified upwind DG operators. In a first test, we discretize the periodic interval $[0, 2]$ with $K = 20$ elements using seventh-order polynomials in each element. In Fig. 4.13 we show the full spectra of the discretized upwind DG differentiation operator for $\beta = 0, 0.2, 0.4, 0.6, 0.8, 1.0$. It is immediately apparent that the modified operator has a much reduced spectral radius as β increases.

We see that with $\beta = 1$, the normally highly dissipated DG eigenmodes have reduced dissipation and are now in close proximity to the imaginary axis – in particular, near the origin and may be present in time-domain simulations as parasitic solutions. This effect can be mitigated in this specific case by choosing $\beta = 0.8$. In this case, there is a clear separation between the spurious and nonspurious eigenvalues. There is clearly a need for a subtle balance between decreasing the spectral radius of the original DG operator and confusing the spectrum near the imaginary axis.

In Table 4.1 we show the spectral radius for each β value considered. It is clear that this simple modification yields a significant reduction in the spectral

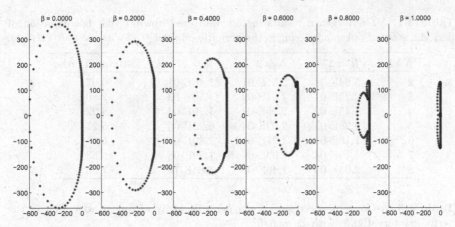

Fig. 4.13. Full spectra for $\beta = 0, 0.2, 0.4, 0.6, 0.8, 1.0$ version of the discrete upwind DG operator on a mesh of $K = 20$ and $N = 7$. Notice how the spectral radius diminishes in the $\beta \to 1$ limit (i.e. the fully co-volume filtered approximation of the derivative).

radius and a corresponding increase in the maximum size of Δt. It seems that using $\beta = 0.8$ is a reasonable compromise between spectral radius deduction and maintaining separation of the nonphysical modes from the imaginary axis.

The decrease in spectral radius is an attractive property of this approach, but it is also necessary to examine the impact of the use of the co-volume grids on the accuracy of the solution. We performed h-convergence tests for $K = 32, 64, 128, 256$, $N = 1, 2, \ldots, 7$ with $\beta = 0$ (i.e., an unfiltered upwind discontinuous Galerkin operator) with $\Delta t = h/2N_p$ and then with $\beta = 1 - 2/N_p$ and $dt = h/4$ using an initial condition

$$E(x, 0) = \exp\left(-40 \sin\left((2\pi x)^2\right)\right), \quad H(x, 0) = 0.$$

The rate of convergence is estimated by assuming the error has form $\epsilon = C_N h^\alpha$.

The results for the unfiltered upwind DG solution of the advection equation are shown in Table 4.2. We clearly see that for $2 \leq N \leq 7$, the order of convergence measured at time $t = 1$ is between N and $N + 1$. It is clear that the $N = 1$ results are preasymptotic for this case. In this case we used a timestep $\Delta t = h/2N_p$, which was experimentally found to be stable for all test cases and close to the maximum Δt.

The corresponding co-volume grid results with $\beta = 1 - 2/N_p$ are shown in Table 4.3. Again, the solution includes relatively narrow exponential pulses, which are not well resolved by the linear elements, even with $K = 512$ elements, and this is reflected in the preasymptotic order of accuracy obtained for the $N = 1$ test. With the exception of the $N = 1$ case, we see that the convergence rates are close to optimal, with rates of nearly $N + 1$. The filtered

Table 4.2. h-Convergence results for an unfiltered upwind DG discretization of periodic array of pulses advecting in the periodic interval $[0, 2]$ with $\Delta t = h/(2(N_p))$.

$N+1$	$h = 2^{-5}$	$h = 2^{-6}$	$h = 2^{-7}$	Estimated order
2	5.64E-01	3.51E-01	1.46E-01	0.97
3	2.23E-01	4.61E-02	4.56E-03	2.80
4	4.45E-02	2.29E-03	1.29E-04	4.22
5	9.28E-03	2.02E-04	6.25E-06	5.27
6	1.10E-03	1.27E-05	2.89E-07	5.95
7	1.10E-04	1.63E-06	1.29E-08	6.53
8	2.03E-05	7.46E-08	4.83E-10	7.68

Table 4.3. h-Convergence results for a periodic array of pulses advecting in the periodic interval $[0, 2]$ with $\Delta t = h/5$.

$N+1$	$h = 2^{-5}$	$h = 2^{-6}$	$h = 2^{-7}$	Estimated order
2	5.53E-01	3.32E-01	1.32E-01	1.03
3	2.27E-01	5.52E-02	6.67E-03	2.54
4	5.99E-02	3.80E-03	2.01E-04	4.11
5	5.08E-03	1.91E-04	7.34E-06	4.72
6	7.34E-04	1.55E-05	4.42E-07	5.35
7	2.42E-04	2.22E-06	1.49E-08	6.70
8	3.33E-05	8.43E-08	4.04E-10	8.16

upwind DG results are comparable with the unfiltered upwind DG results shown in Table 4.2 despite the timesteps being used for the former are much larger than those used for the latter. The method as presented in the above is rather limited in geometric flexibility and in representing solutions of low regularity. However, the example clearly establishes the potential for the method as a way of taming the otherwise very strong condition on Δt for an explicit method.

4.8.3 Local timestepping

Compared to finite elements methods, one of the most appealing features of DG-FEM is the explicit nature of the semidiscrete formulation. Apart for ensuring flexibility in the formulation and simplicity in the implementation, it also opens up for the possibility of doing local timestepping; that is, each element is timestepped with a local timestep, $\Delta t^{(k)}$, chosen to guarantee local stability only. The clear advantage of this is that each element is advanced at the maximum stable timestep rather than being forced to advance at a much smaller timestep due to more stringent constraints elsewhere (e.g., for a highly nonuniform grid or in cases where a different order of approximation is used in different elements).

To appreciate the prospects of such an approach, let us consider a simple example [107]. We assume that the grid consists of elements of two different sizes and/or order of approximation, with two different stable timesteps, $\tau \Delta t^1 = \Delta t^2$, where $\tau \geq 1$, when used with an explicit timestepping method such as RK methods. Let us furthermore assume that a fraction, ν, of the elements can be advanced at Δt^1 whereas the remaining fraction of $(1 - \nu)$ can benefit from the larger timestep, Δt^2.

If we use a globally stable timestep, the total number of timesteps scales as

$$N^G = \frac{1}{\Delta t^1},$$

to advance one timeunit. In contrast to this, if we assume that each element can be advanced at the optimal timestep, we recover

$$N^L = \frac{\nu}{\Delta t^1} + \frac{1 - \nu}{\Delta t^2},$$

resulting in a potential speedup of

$$\sigma = \frac{N^G}{N^L} = \frac{\tau}{1 - \nu + \nu\tau} \geq 1.$$

The most important special case is that of $\tau \gg 1$; that is, there is a dramatic difference between the two stable timesteps. In this case, we recover

$$\lim_{\tau \to \infty} \sigma = \nu^{-1}.$$

Hence, if a small fraction of the elements require a small timestep, the speedup can be very dramatic. It is interesting to observe, however, that in this case, it is the fraction of small elements and not the size of the actual timestep that determines the potential for speedup. The contrast to this is the case of $\nu \ll 1$, where

$$\lim_{\alpha \to 0} \sigma = \tau;$$

that is, the potential speedup is determined solely by the separation in stable timesteps.

However, to realize this potential for speedup, we must exercise some caution. Consider Fig. 4.14 which illustrates a situation where we have three different elements, each of which can be integration in time using a maximum stable timestep, Δt^k.

By considering this situation, one quickly realizes that the challenge lies in finding a way to compute numerical fluxes at the interfaces when they are needed; for example, to advance the left element Δt^1, one would need information from the center element, which may not yet have been advanced. If the emphasis is on steady-state problems, the solution is to simply just take old information from the nearest time slice and use that. This reduces the

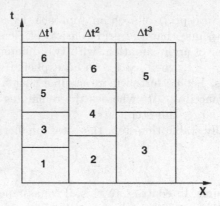

Fig. 4.14. Situation for three elements, each with the potential to be advanced with a local timestep, Δt^k.

formal temporal order of the scheme to first-order, but since we are interested in a steady-state solution, this does not matter.

The challenge lies in finding a way to enable local timestepping in a time-accurate and stable manner. While there are several examples in the literature on ways to address this (e.g., for wave problems in [52] and specialized to Maxwell's equations in [41, 240, 259]), we describe in the following an elegant and entirely general approach, proposed in [107], and named the ADER-DG approach (Arbitrary high-order scheme using derivatives for DG). In its simplest form it can be viewed as a direct extension of the well-known Lax-Wendroff method for conservation laws.

Let us assume that we solve a linear one-dimensional problem of the form

$$\frac{\partial \boldsymbol{u}}{\partial t} = \mathcal{A}\frac{\partial \boldsymbol{u}}{\partial x},$$

subject to appropriate boundary and initial conditions. For simplicity, we assume $\mathcal{A}$ is constant, but this is easily overcome. Returning to the three cell problem in Fig. 4.14, we now express the solution in each cell over the local timestep using a Taylor expansion

$$t \in [t^k, t^k + \Delta t^k] : \boldsymbol{u}^k(x,t) = \sum_{p=0}^{P} \frac{(t-t^k)^p}{p!} \frac{\partial^p \boldsymbol{u}^k(x,t^k)}{\partial t^p}$$

$$= \sum_{p=0}^{P} \frac{(t-t^k)^p}{p!} \mathcal{A}^p \frac{\partial^p \boldsymbol{u}^k(x,t^k)}{\partial x^p},$$

where $P+1$ is the order of the Taylor expansion. Note that we have used the original equation to exchange temporal for spatial derivatives. To match the temporal order of the scheme, P must match the global order of the timestepping scheme used in each element. This approach is the same one pioneered in the classic Lax-Wendroff method, discussed in the context of discontinuous Galerkin methods in [262].

With this expansion of the local solution, we can now express the solution everywhere in time within a timestep as needed to enable local timestepping without losing the temporal accuracy. The approach begins by evaluating an update criterion; that is, an element can be updated if

$$t^k + \Delta t^k \le \min_m (t^m + \Delta t^m),$$

where m runs over all neighboring elements. Referring back to Fig. 4.14, this would indicate that space-time element 1 can be updated first. The values of $u^2(x, t)$ needed to compute the numerical fluxes at the interface between elements 1 and 2 can be obtained using the Taylor expansion in element 2. Completing the timestep in element 1, one can now update space-time element 2 using the same procedure. After that, both elements marked 3 can be updated, as they are not connected and rely only on boundary information and updated information available from element 2. It is easy to see that this procedure can be continued to completion at final time, T.

The clear advantage of this, conceptionally simple, approach is that the temporal order of the scheme relies only on the order of the local time integration scheme (e.g., a RK method used in each cell), and the order of the cellwise Taylor expansion. Stability is ensured by the local stability and no synchronization is needed unless one is interested in global solutions at specific times.

A closely related approach is to entirely avoid the local timestepping method and integrate exactly; that is, consider

$$\frac{\partial \boldsymbol{u}}{\partial t} = \mathcal{A} \frac{\partial \boldsymbol{u}}{\partial x} = \boldsymbol{F}(\boldsymbol{u}),$$

and write

$$t \in [t^k, t^k + \Delta t^k] : \boldsymbol{F}^k(x, t) = \sum_{p=0}^{P} \frac{(t - t^k)^p}{p!} \frac{\partial^p \boldsymbol{F}^k(x, t^k)}{\partial t^p}.$$

We can then integrate exactly in time to recover

$$\boldsymbol{u}^k(x, t^k + \Delta t^k) = \boldsymbol{u}^k(x, t^k) + \sum_{p=0}^{P} \frac{(\Delta t^k)^{p+1}}{(p+1)!} \frac{\partial^p \boldsymbol{F}^k(x, t^k)}{\partial t^p}.$$

For the linear case, we now recover

$$\frac{\partial^p \boldsymbol{F}^k(x, t^k)}{dt^p} = \left(\mathcal{A} \frac{\partial}{\partial x} \right)^{p+1} \boldsymbol{u},$$

which is essentially the same as for the previous approach.

Both the two slightly different formulations above extend to the general nonlinear case through the use of the Cauchy-Kovalevskaya procedure to interchange temporal derivatives for spatial derivatives by the use of the equation and the generalized Leibniz rules for differentiation of products and fractions appearing in the nonlinear flux. Details on this can be found [105, 302] for a general outline and a detailed special example for three-dimensional magnetohydrodynamics in [299].

Another even more dramatic approach to deal with the timestep conditions is to consider a discontinuous Galerkin method with a full space-time basis. This has been considered in detail in [312], where it is shown to perform well. An interesting combination of the ADER approach above with the space-time expansion has been considered in [128, 231], where the Taylor expansion is now applied in both space and time, resulting in a fully explicit space-time DG-FEM with support for local timestepping along the lines discussed above.

4.9 Exercises

1. Consider a function $v(x) \in H^p(\mathsf{I})$. Prove that if that one expresses

$$v_h(r) = \sum_{n=0}^{N} \tilde{v}_n \tilde{P}_n(r), \quad \tilde{v}_n = \int_{-1}^{1} v(r) \tilde{P}_n(r) \, dr.$$

then

$$|\tilde{v}_n| \propto \left(\frac{1}{n(n+1)} \right)^q,$$

and relate q and p.

2. Consider a polynomial, u_h, of order N, defined on $\mathsf{D} = [0, h]$. Express

$$u_h(x) = \sum_{n=0}^{N} \hat{u}_n \tilde{P}_n(r(x)),$$

and prove the inverse trace inequality

$$\|u_h\|_{\partial \mathsf{D}} \leq \frac{N+1}{\sqrt{h}} \|u_h\|_{\mathsf{D}}.$$

3. Implement the mapping technique for increasing the stable timestep and test the accuracy when solving

$$\frac{\partial u}{\partial t} + a \frac{\partial u}{\partial x} = 0,$$

subject to periodic boundary conditions. Consider different choices of β in the mapping parameter and see how it impacts the accuracy of the scheme in terms of $\mathcal{O}(h^\gamma)$ with γ to be determined.

4. The discussion of the mapping technique shows that β should depend on N to maintain spectral convergence. On the other hand, the discussion also shows that to recover an $\mathcal{O}(N)$ scaling of the maximum eigenvalue, β cannot depend on N. A suitable compromise seems to be that $\beta = c\sqrt{N}$.

   With this choice of β, evaluate the performance of the method computationally; for example, how does the maximum eigenvalue scale for β like this and how is the accuracy affected, if at all?

5. Implement the timestepping scheme using co-volume grids for the linear one-dimensional problem

$$\frac{\partial u}{\partial t} + a\frac{\partial u}{\partial x} = 0,$$

   and test stability, accuracy, and efficiency on grids with different element sizes and for different values of β.

6. Implement the local timestepping scheme for the linear problem

$$\frac{\partial u}{\partial t} + a\frac{\partial u}{\partial x} = 0,$$

   and test stability, accuracy, and efficiency on grids with different element sizes.

5

Nonlinear problems

So far, we have focused entirely on linear problems with constant or piecewise constant coefficients. As we have seen, the methods and analysis for these cases is relatively complete.

In this chapter we expand the discussion to include more complex problems – in particular, problems with smoothly varying coefficients and genuinely nonlinear problems. As we will see, this introduces new elements that need attention, and the analysis of the methods for such problems is more complex. In fact, we will often not attempt to give a complete analysis but merely outline the key results. However, the extension to strongly nonlinear problems displays many unique features and the power and robustness of the discontinuous Galerkin methods.

5.1 Conservation laws

Let us recall the basic scalar conservation law

$$\frac{\partial u}{\partial t} + \frac{\partial f(u)}{\partial x} = 0, \quad x \in [L, R] = \Omega, \tag{5.1}$$
$$u(x, 0) = u_0(x),$$

where $f(u)$ is the flux function, assumed to be convex, and we assume that boundary conditions are given at inflow boundaries; that is, where

$$\hat{n} \cdot \frac{\partial f}{\partial u} = \hat{n} \cdot f_u < 0.$$

Here $\hat{n}$ is the outward pointing normal at $\partial\Omega$. A fundamental property of Eq. (5.1), used throughout the sciences as a way of expressing basic physical laws of conservation of mass, momentum, and energy, is that

$$\frac{d}{dt}\int_a^b u(x)dx = f(u(a)) - f(u(b)); \tag{5.2}$$

that is, the change of mass is exactly the difference between what enters and what leaves any volume of interest, $[a, b]$. Furthermore, if the flux vanishes or is periodic at the boundaries, there is conservation of the mass.

The linear constant coefficient case discussed so far is a special case of the above with $f(u) = au$. In that case, the smoothness of the solution, $u(x, t)$, depends on the initial conditions only. However, for general nonlinear cases, this is no longer true as nonsmooth solutions, known as shocks, may develop even if the initial data are smooth.

A well-known example is $f(u) = u^2$ (see e.g., [218, 301] for others and more details). If we define the characteristic path, $X(t)$, as

$$\frac{d}{dt}X(t) = f'(u(X, t)) = 2u(X, t), \quad X(0) = x_0,$$

and take $\psi(t) = u(X, t)$ (i.e., the solution along the path), we have

$$\frac{d\psi}{dt} = \frac{\partial u}{\partial t} + X_t\frac{\partial u}{\partial X} = \frac{\partial u}{\partial t} + \frac{\partial u^2}{\partial X} = 0;$$

that is, $\psi(t)$ is constant along these paths. If these paths do not cross, one can uniquely determine the solution $u(x, t)$ from the initial conditions.

Unfortunately, this assumption is generally not fulfilled. Consider the initial condition of $u(x, 0) = -\sin(\pi x)$ for $x \in [-1, 1]$ and $u(\pm 1, t) = 0$. Then all information to the left of $x = 0$ will move to the right. To the right of $x = 0$, however, it will move to the left. This leaves open the question of what exactly happens where the characteristics meet. Since there is mass conservation, the mass will continue to pile up at $x = 0$ and a shock forms. The solution in this area looses smoothness, and we will therefore introduce the notion of weak solutions to properly define a derivative of a nonsmooth function.

Let us define the C^0 compactly supported test function, $\phi(x, t)$. The solution $u(x, t)$ is a weak solution to the conservation law if it satisfies

$$\int_0^\infty \int_{-\infty}^\infty \left(u(x, t)\frac{\partial \phi}{\partial t} + f(u)\frac{\partial \phi}{\partial x}\right) dx\, dt = 0,$$

for all such smooth test functions and consistent initial conditions

$$\int_{-\infty}^\infty (u(x, 0) - u_0(x))\, \phi(x, 0)dx = 0.$$

Unfortunately, this broader notion of a solution introduces other complications, as is it now easy to find examples of initial conditions that can result in multiple weak solutions; that is, the weak solution is introduced at the expense of uniqueness.

Example 5.1. To illustrate the problem of uniqueness for the weak solutions, consider Burgers' equation

$$\frac{\partial u}{\partial t} + \frac{1}{2}\frac{\partial u^2}{\partial x} = 0, \quad x \in [-\infty, \infty],$$

with the initial conditions

$$u(x,0) = \begin{cases} 0, & x \le 0 \\ 1, & x > 0. \end{cases}$$

A weak solution can clearly be found by using the Rankine-Hugoniot condition to give the shock speed of $\frac{1}{2}$:

$$u_I(x,t) = \begin{cases} 0, & x \le t/2 \\ 1, & x > t/2. \end{cases}$$

However, consider

$$u_{II}(x,t) = \begin{cases} 0, & x \le 0 \\ x/t, & 0 \le x \le t \\ 1, & x \ge t, \end{cases}$$

which is a valid piecewise solution to the equation in strong form, known as a classic solution. Hence, we have two different solutions with the same initial conditions.

This lack of uniqueness leads to the natural question of which solution is the physically relevant one. To answer this, we consider the viscosity solution, $u^\varepsilon(x,t)$, defined as the solution to the equation

$$\frac{\partial u^\varepsilon}{\partial t} + \frac{\partial f(u^\varepsilon)}{\partial x} = \varepsilon \frac{\partial^2 u^\varepsilon}{\partial x^2},$$

with $u^\varepsilon(x,0) = u(x,0)$. The physically relevant solution, $u(x,t)$, is obtained as the limit solution $\lim_{\varepsilon \to 0} u^\varepsilon(x,t) = u(x,t)$ – if it exists. However, proving existence of this limit for general problems remains an open problem.

A necessary condition for the viscosity limit to exist can be recovered through an entropy condition. If we define the convex entropy, $\eta(u)$, such that $\eta''(u) > 0$, and with the entropy flux

$$F(u) = \int_u \eta'(v) f'(v) \, dv,$$

then if $u(x,t)$ satisfies the entropy condition

$$\frac{\partial \eta}{\partial t} + \frac{\partial}{\partial x} F(u) \le 0, \tag{5.3}$$

uniqueness is guaranteed provided $f(u)$ is convex (i.e., $f''(u) > 0$). One easily proves that all viscosity solutions satisfy Eq. (5.3) which becomes an equality for smooth solutions. Furthermore, a weak solution satisfying the entropy condition is also a unique solution [290], provided the flux is convex. This leaves open the question of how to determine whether a solution satisfies an entropy condition.

Perhaps the most celebrated condition, known as the Lax entropy condition, for problems with a convex flux takes the following form [214]:

Theorem 5.2. *Let $u(x,t)$ be a weak solution and S be an (x,t)-curve along which u has a discontinuity. Let $(x_0,t_0) \in S$ and u^- and u^+ be the left and right limits of $u(x,t)$, respectively, at (x_0,t_0) and define*

$$s = \frac{f(u^-) - f(u^+)}{u^- - u^+}.$$

Then $u(x,t)$ satisfies the entropy condition at (x_0,t_0) if and only if

$$f'(u^-) > s > f'(u^+).$$

A discontinuity satisfying this is called a shock and s is the shock velocity.

Note that the shock speed is exactly what is obtained from the Rankine-Hugoniot condition (see Section 2.4).

Hence, we must not only be careful with how to approximate the more complex fluxes and nonlinear terms, but we must also consider that solutions may lose smoothness and, thus, may lose uniqueness, unless some entropy inequality can be established.

For more general fluxes (e.g., nonconvex fluxes), issues like existence and uniqueness of solutions become significantly more complex, even for scalar equations. For systems of equations, for example, the situation is even more complicated, with a largely incomplete theory, and a detailed discussion is well beyond the scope of this text. Good references regarding such issues are [87, 115, 301].

In what remains we will, unless otherwise stated, simply assume that the conservation laws are well-posed; that is, that a unique solution exist and that this solution depend smoothly on the data.

5.2 The basic schemes and their properties

If we consider the scalar conservation law in Eq. (5.1) and follow the basic guidelines from Chapter 2. We assume that

$$x \in \mathsf{D}^k: u_h^k(x,t) = \sum_{i=1}^{N_p} u^k(x_i,t)\ell_i^k(x), \quad f_h^k(u_h(x,t)) = \sum_{i=1}^{N_p} f^k(x_i,t)\ell_i^k(x),$$

are polynomial representations of the local solution and the local flux, respectively, and require the local residual to be orthogonal to all test functions,

$\phi_h \in V_h$. As usual V_h is the space of all piecewise polynomial functions defined on Ω. This results in the local semidiscrete weak formulation

$$\int_{D^k} \left(\frac{\partial u_h^k}{\partial t} \ell_i^k(x) - f_h^k(u_h^k) \frac{d\ell_i^k}{dx} \right) dx = -\int_{\partial D^k} \hat{n} \cdot f^* \ell_i^k(x) \, dx, \qquad (5.4)$$

and the corresponding strong form

$$\int_{D^k} \left(\frac{\partial u_h^k}{\partial t} + \frac{\partial f_h^k(u_h^k)}{\partial x} \right) \ell_i^k(x) \, dx = \int_{\partial D^k} \hat{n} \cdot \left(f_h^k(u_h^k) - f^* \right) \ell_i^k(x) \, dx. \quad (5.5)$$

Furthermore, we have the numerical flux f^*, for which we generally use the monotone Lax-Friedrichs flux

$$f^*(u_h^-, u_h^+) = \{\{f_h(u_h)\}\} + \frac{C}{2} [\![u_h]\!],$$

where $C = \max |f_u|$ is an upper bound on the (local) wave speed.

Using a simpler notation, we recover the semidiscrete schemes

$$\mathcal{M}^k \frac{d}{dt} \boldsymbol{u}_h^k + \mathcal{S} \boldsymbol{f}_h^k = \left[\boldsymbol{\ell}^k(x)(f_h^k - f^*) \right]_{x_l^k}^{x_r^k},$$

for the strong form and

$$\mathcal{M}^k \frac{d}{dt} \boldsymbol{u}_h^k - \mathcal{S}^T \boldsymbol{f}_h^k = -\left[\boldsymbol{\ell}^k(x) f^* \right]_{x_l^k}^{x_r^k},$$

for the weak form. In both cases, we recall the vectors of nodal values

$$\boldsymbol{u}_h^k = [u_h^k(x_1^k), \ldots, u_h^k(x_{N_p}^k)]^T, \quad \boldsymbol{f}_h^k = [f_h^k(x_1^k), \ldots, f_h^k(x_{N_p}^k)]^T.$$

Let us consider the schemes in a little more detail. If we multiply the weak from by a smooth test function, $\phi = 1$, we have

$$\boldsymbol{\phi}_h^T \mathcal{M}^k \frac{d}{dt} \boldsymbol{u}_h^k - \boldsymbol{\phi}_h^T \mathcal{S}^T \boldsymbol{f}_h^k = -\boldsymbol{\phi}_h^T \left[\boldsymbol{\ell}^k(x) f^* \right]_{x_l^k}^{x_r^k} \qquad (5.6)$$

and, thus,

$$\frac{d}{dt} \int_{x_l^k}^{x_r^k} u_h \, dx = f^*(x_l^k) - f^*(x_r^k).$$

This is exactly the discrete equivalent of the fundamental conservation property of the continuous conservation law, Eq. (5.2); that is, the scheme is locally conservative.

If we sum over all the elements, we recover

$$\sum_{k=1}^{K} \frac{d}{dt} \int_{x_l^k}^{x_r^k} u_h \, dx = \sum_{k_e} \hat{\boldsymbol{n}}_e \cdot [\![f^*(x_e^k)]\!],$$

where k_e are the number of interfaces, x_e^k is the interface position, and $\hat{\boldsymbol{n}}_e$ is an outward pointing normal at this edge. If we choose a reasonable flux (e.g., the Lax-Friedrichs flux or some other consistent monotone flux), and assume periodic boundary conditions, we likewise recover global conservation since the flux is unique at an interface by consistency.

Let us now consider the more general local smooth test function

$$x \in \mathsf{D}^k : \quad \phi_h(x, t) = \sum_{i=1}^{N_p} \phi(x_i^k, t) \ell_i^k(x),$$

which we assume vanishes for large values of t. From Eq. (5.6), we recover

$$\left(\phi_h, \frac{\partial}{\partial t} u_h \right)_{\mathsf{D}^k} - \left(\frac{\partial \phi_h}{\partial x}, f_h \right)_{\mathsf{D}^k} = - [\phi_h f^*]_{x_l^k}^{x_r^k}.$$

Integration by parts in time yields

$$\int_0^\infty \left[\left(\frac{\partial}{\partial t} \phi_h, u_h \right)_{\mathsf{D}^k} + \left(\frac{\partial \phi_h}{\partial x}, f_h \right)_{\mathsf{D}^k} - [\phi_h f^*]_{x_l^k}^{x_r^k} \right] dt + (\phi_h(0), u_h(0))_{\mathsf{D}^k} = 0.$$

Summing over all elements yields

$$\int_0^\infty \left[\left(\frac{\partial}{\partial t} \phi_h, u_h \right)_{\Omega, h} + \left(\frac{\partial \phi_h}{\partial x}, f_h \right)_{\Omega, h} \right] dt$$

$$+ (\phi_h(0), u_h(0))_{\Omega, h} = \int_0^\infty \sum_{k_e} \hat{\boldsymbol{n}}_e \cdot [\![\phi_h(x_e^k) f^*(x_e^k)]\!] \, dt.$$

Since ϕ_h is a polynomial representation of a smooth test function, ϕ, it converges as N increases and/or h decreases. Also, the global smoothness of ϕ ensures that the right-hand side vanishes as for the constant test function since the numerical flux is unique. Thus, if $u_h(x, t)$ converges almost everywhere to a function, $u(x, t)$, then this is guaranteed to be a weak solution to the conservation law [45]. This implies, among other things, that shocks will move at the right speed.

Let us again consider the scalar conservation law

$$\frac{\partial u}{\partial t} + \frac{\partial f}{\partial x} = 0,$$

and the local, strong semidiscrete form of the discretization

$$\mathcal{M}^k \frac{d}{dt} \boldsymbol{u}_h^k + \mathcal{S} \boldsymbol{f}_h^k = \left[\boldsymbol{\ell}^k(x)(f_h^k - f^*) \right]_{x_l^k}^{x_r^k}.$$

Now, we multiply with $\boldsymbol{u}^T$ from the left to obtain

$$\frac{1}{2} \frac{d}{dt} \|u_h^k\|_{\mathsf{D}^k}^2 + \int_{\mathsf{D}^k} u_h^k \frac{\partial}{\partial x} f_h^k \, dx = \left[u_h^k(x)(f_h^k - f^*) \right]_{x_l^k}^{x_r^k}.$$

We define the convex entropy function and the associated entropy flux

$$\eta(u) = \frac{u^2}{2}, \quad F'(u) = \eta' f'.$$

First, note that

$$F(u) = \int_u f' u \, du = f(u)u - \int_u f \, du = f(u)u - g(u), \tag{5.7}$$

where we define

$$g(u) = \int_u f(u) \, du.$$

Consider now

$$\int_{\mathsf{D}^k} u_h^k \frac{\partial}{\partial x} f_h^k \, dx = \int_{\mathsf{D}^k} \eta'(u_h^k) f'(u_h^k) \frac{\partial}{\partial x} u_h^k \, dx$$
$$= \int_{\mathsf{D}^k} F'(u_h^k) \frac{\partial}{\partial x} u_h^k \, dx = \int_{\mathsf{D}^k} \frac{\partial}{\partial x} F(u_h^k) \, dx,$$

to recover

$$\frac{1}{2} \frac{d}{dt} \|u_h^k\|_{\mathsf{D}^k}^2 + \left[F(u_h^k) \right]_{x_l^k}^{x_r^k} = \left[u_h^k(x)(f_h^k - f^*) \right]_{x_l^k}^{x_r^k}.$$

At each interface we are left with conditions of the form

$$F(u_h^-) - F(u_h^+) - u_h^-(f_h^- - f^*) + u_h^+(f_h^+ - f^*) \geq 0,$$

to ensure nonlinear stability. Here, u_h^- reflects solutions left of the interface and u_h^+ are the solutions right of the interface. Using Eq. (5.7), we have

$$-g(u_h^-) + g(u_h^+) - f^*(u_h^+ - u_h^-) \geq 0.$$

Now, we use the mean value theorem to obtain

$$g(u_h^+) - g(u_h^-) = g'(\xi)(u_h^+ - u_h^-) = f(\xi)(u_h^+ - u_h^-),$$

for some $\xi \in [u_h^-, u_h^+]$. Combining this, we recover the condition

$$(f(\xi) - f^*)(u_h^+ - u_h^-) \geq 0,$$

which is a standard condition for a monotone flux – known as an E-flux. As already discussed in Chapter 2, the Lax-Friedrichs flux and the other fluxes discussed here satisfy this condition.

Thus, we have established the general result that

$$\frac{1}{2}\frac{d}{dt}\|u_h\|_{\Omega,h} \leq 0. \tag{5.8}$$

This is a very strong result of nonlinear stability for the scheme without any use of stabilization or control of the solutions.

Let us follow [187] and take this argument one step further. We define the consistent entropy flux, $\hat{F}$, as

$$\hat{F}(x) = f^*(x)u(x) - g(x),$$

and consider the local cellwise entropy

$$\frac{1}{2}\frac{d}{dt}\|u_h^k\|_{\mathsf{D}^k}^2 - \int_{\mathsf{D}^k}\frac{\partial u_h^k}{\partial x}f_h^k\,dx + \left[u_h^k f^*\right]_{x_l^k}^{x_r^k}$$

$$= \frac{d}{dt}\int_{\mathsf{D}^k}\eta(u_h^k)\,dx + \hat{F}(x_r^k) - \hat{F}(x_l^k) = 0.$$

The above is recognized as a cell entropy condition, and since the polynomial solution, u_h^k, is smooth, it is a strict equality. Let us also consider an interface

$$\frac{d}{dt}\int_{\mathsf{D}^k}\eta(u_h^k)\,dx + \hat{F}(x_r^k) - \hat{F}(x_l^k)$$

$$= \frac{d}{dt}\int_{\mathsf{D}^k}\eta(u_h^k)\,dx + \hat{F}(x_r^k) - \hat{F}(x_r^{k-1}) + \hat{F}(x_r^{k-1}) - \hat{F}(x_l^k)$$

$$= \frac{d}{dt}\int_{\mathsf{D}^k}\eta(u_h^k)\,dx + \hat{F}(x_r^k) - \hat{F}(x_r^{k-1}) + \Phi^k = 0,$$

where

$$\Phi^k = \hat{F}(x_r^{k-1}) - \hat{F}(x_l^k) = f^*(u_h^- - u_h^+) + g(u_h^+) - g(u_h^-) \geq 0.$$

The latter argument follows directly from the proof of Eq. (5.8) to establish nonlinear stability. Hence, we have

$$\frac{d}{dt}\int_{\mathsf{D}^k}\eta(u_h^k)\,dx + \hat{F}(x_r^k) - \hat{F}(x_r^{k-1}) \leq 0, \tag{5.9}$$

which is a discrete cell entropy condition for the quadratic entropy, η. It can be generalized in a straightforward manner to other convex entropy functions as well as to multidimensional scalar problems. The result is much stronger than nonlinear L^2-stability, as it suffices to guarantee convergence for scalar problems to the unique entropy solution under light additional conditions; for example, the flux is convex and the solution is total variation bounded (see Section 5.6 for more on this).

5.3 Aliasing, instabilities, and filter stabilization

In spite of such strong results, let us consider an example to emphasize that caution is still needed.

Example 5.3. To keep things simple, we consider the problem

$$\frac{\partial u}{\partial t} + \frac{\partial f}{\partial x} = 0, \ x \in [-1, 1],$$

subject to periodic boundary conditions and with the simple initial condition

$$u(x, 0) = \sin(4\pi x).$$

As the flux, we use the variable coefficient function

$$f(u) = a(x)u(x, t), \ a(x) = (1 - x^2)^5 + 1.$$

As usual, we assume that

$$x \in \mathsf{D}^k : \ u_h^k(x, t) = \sum_{i=1}^{N_p} u_h^k(x_i^k, t)\ell_i^k(x).$$

Let us now consider three different ways of implementing this scheme.

The first one follows directly from the general discussion leading to the local scheme

$$\mathcal{M}^k \frac{d}{dt} \boldsymbol{u}_h^k + \mathcal{S}\boldsymbol{f}_h^k = \frac{1}{2} \oint_{x_l^k}^{x_r^k} \hat{\boldsymbol{n}} \cdot [\![f_h^k]\!]\boldsymbol{\ell}^k(x) \, dx,$$

using a central flux. We recall that

$$\boldsymbol{f}_h^k = \left[f_h^k(x_1^k), \ldots, f_h^k(x_{N_p}^k) \right]^T, \ f_h^k(x, t) = \sum_{i=1}^{N_p} f_h^k(x_i^k)\ell_i^k(x),$$

and that $f_h^k(x) = \mathcal{P}_N(a(x)u_h^k(x))$ is the projection of the flux onto the space of piecewise polynomial functions of order N.

In the second approach, we utilize knowledge of the flux and define a new operator, $\mathcal{S}^{k,a}$, as

$$\mathcal{S}_{ij}^{k,a} = \int_{x_l^k}^{x_r^k} \ell_i^k \frac{d}{dx} a(x)\ell_j^k \, dx,$$

leading to the scheme

$$\mathcal{M}^k \frac{d}{dt} \boldsymbol{u}_h^k + \mathcal{S}^{k,a} \boldsymbol{u}_h^k = \frac{1}{2} \oint_{x_l^k}^{x_r^k} \hat{\boldsymbol{n}} \cdot [\![a(x)u_h^k]\!]\boldsymbol{\ell}^k(x) \, dx.$$

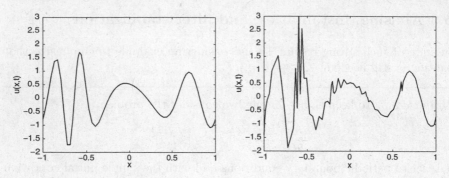

Fig. 5.1. On the left, is shown the results obtained with the first two schemes discussed in the example and the right illustrates the unstable result computed with the last approach. In all cases, we use $K = 5$ elements of 16th order with results shown at $T = 10.5$.

As a third alternative, we consider a slightly modified version of the first scheme and express the flux, f_h^k, as

$$x \in \mathsf{D}^k : f_h^k(x,t) = \sum_{i=1}^{N_p} a(x_i^k) u_h^k(x_i, t) \ell_i^k(x);$$

that is, the flux is obtained by interpolation at the grid points, x_i^k.

In Fig. 5.1 we show the results of the computations using $K = 5$ elements, each of length 0.4 and with a local $N = 16$ order basis. Comparing the results of the three schemes, one realizes that while the two first schemes work well and are stable, the last one fails to maintain stability.

This example highlights that we need to be careful when implementing these methods. Based on this example, one would be inclined to discard the last approach. Indeed, using standard energy methods, one can prove both of the first two methods to be stable and energy conserving, in agreement with previous analysis.

However, the simple computation of the flux in the last scheme as

$$f_h^k(x,t) = \sum_{i=1}^{N_p} a(x_i^k) u_h^k(x_i^k, t) \ell_i^k(x),$$

has significant computational advantages over the two alternatives. For the second formulation, it is a clear disadvantage, both in computational cost and memory, that we need to define a new operator, $\mathcal{S}^a$, for each individual element. The difference between the first and the last scheme may appear more subtle. In the former, we represent the flux as

$$f_h^k(x,t) = \sum_{i=1}^{N_p} f_h^k(x_i^k, t)\ell_i^k(x);$$

that is, we must represent $f(u_h)$ as an N-th-order polynomial. This requires that one projects f on the space of polynomials through the evaluation of the inner product. In this particular case, $a(x)$ itself is a 10th-order polynomial so the product with $u_h(x,t)$ is an $(N+10)$-th-order product, requiring a Gauss-Lobatto quadrature of order $2N+10$ (i.e., with approximately $N+5$ points). For general nonlinear functions $f(u_h)$, the situation can be much worse and the evaluation of this inner product becomes expensive.

In contrast to this approach, in the third approach we express the flux as

$$f_h^k(x,t) = \mathcal{I}_N(a(x)u_h^k(x,t)) = \sum_{i=1}^{N_p} a(x_i^k)u_h^k(x_i^k,t)\ell_i^k(x),$$

where $\mathcal{I}_N v$ refers to an interpolation of order N on element k. In this case, we simply read $a(x)u_h(x,t)$ at the N_p points. This does not suffice to uniquely specify an $(N+10)$-th-order polynomial, and an aliasing error is introduced. To illustrate how this can cause the instability consider

$$\frac{\partial u}{\partial t} + \frac{\partial}{\partial x}(a(x)u) = 0.$$

For simplicity, but without loss of generality, we restrict the attention to one domain with periodic boundary conditions and consider the scheme

$$\frac{\partial u_h}{\partial t} + \frac{\partial}{\partial x}\mathcal{I}_N(au_h) = 0.$$

We rewrite this as

$$\frac{\partial u_h}{\partial t} + \frac{1}{2}\frac{\partial}{\partial x}\mathcal{I}_N(au_h) + \frac{1}{2}\mathcal{I}_N\left(a\frac{\partial u_h}{\partial x}\right)$$

$$+ \frac{1}{2}\mathcal{I}_N\frac{\partial}{\partial x}au_h - \frac{1}{2}\mathcal{I}_N\left(a\frac{\partial u_h}{\partial x}\right)$$

$$+ \frac{1}{2}\frac{\partial}{\partial x}\mathcal{I}_N(au_h) - \frac{1}{2}\mathcal{I}_N\frac{\partial}{\partial x}au_h = 0$$

or simplified as

$$\frac{\partial u_h}{\partial t} + \mathcal{N}_1 + \mathcal{N}_2 + \mathcal{N}_3 = 0. \tag{5.10}$$

The local discontinuous Galerkin scheme becomes

$$\int_{-1}^{1}\left(\frac{\partial u_h}{\partial t} + \mathcal{N}_1 + \mathcal{N}_2 + \mathcal{N}_3\right)\ell_i(x)\,dx = \frac{1}{2}\oint_{-1}^{1}\hat{n}\cdot[\![\mathcal{I}_N(au_h)]\!]\ell_i(x)\,dx \tag{5.11}$$

or

$$\sum_{j=1}^{N_p} \mathcal{M}_{ij} \frac{du_j}{dt} + (\mathcal{N}_1, \ell_i)_\Omega + (\mathcal{N}_2, \ell_i)_\Omega + (\mathcal{N}_3, \ell_i)_\Omega = \frac{1}{2} \oint_{-1}^{1} \hat{n} \cdot [\![\mathcal{I}_N(au_h)]\!] \ell_i(x)\, dx.$$

Multiplying by $u_h(x_i)$ from the left and summing over all points yields

$$\frac{1}{2} \frac{d}{dt} \|u_h\|_\Omega^2 = -(\mathcal{N}_1, u_h)_\Omega - (\mathcal{N}_2, u_h)_\Omega - (\mathcal{N}_3, u_h)_\Omega + \frac{1}{2} \oint_{-1}^{1} \hat{n} \cdot [\![\mathcal{I}_N(au_h)]\!] u_h\, dx.$$

$$(5.12)$$

We consider the three terms individually

$$-2(\mathcal{N}_1, u_h)_\Omega + \oint_{-1}^{1} \hat{n} \cdot [\![\mathcal{I}_N(au_h)]\!] u_h\, dx$$

$$= -\left(\frac{\partial}{\partial x}\mathcal{I}_N(au_h), u_h\right)_\Omega - \left(\mathcal{I}_N\left(a\frac{\partial u_h}{\partial x}\right), u_h\right)_\Omega + \oint_{-1}^{1} \hat{n} \cdot [\![\mathcal{I}_N(au_h)]\!] u_h\, dx$$

$$= \left(\mathcal{I}_N(au_h), \frac{\partial}{\partial x}u_h\right)_\Omega - \left(\mathcal{I}_N\left(a\frac{\partial u_h}{\partial x}\right), u_h\right)_\Omega = 0,$$

where the boundary terms vanish after integration by parts and the volume terms cancel by the property of the interpolation.

For the second term, we have

$$-2(\mathcal{N}_2, u_h)_\Omega = -\left(\mathcal{I}_N\left(\frac{\partial}{\partial x}au_h\right), u_h\right)_\Omega + \left(\mathcal{I}_N\left(a\frac{\partial}{\partial x}u_h\right), u_h\right)_\Omega$$

$$= -\left(\mathcal{I}_N(a\frac{\partial}{\partial x}u_h), u_h\right)_\Omega - (\mathcal{I}_N(a_x u_h), u_h)_\Omega$$

$$+ \left(\mathcal{I}_N\left(a\frac{\partial}{\partial x}u_h\right), u_h\right)_\Omega = -(\mathcal{I}_N(a_x u_h), u_h)_\Omega$$

$$\leq \max_x |a_x| \|u_h\|_\Omega^2.$$

In other words, this term is not the source of the instability, as it is bounded independently of N. Consider the last term of the form

$$-2(\mathcal{N}_3, u_h)_\Omega = -\left(\frac{\partial}{\partial x}\mathcal{I}_N(au_h), u_h\right)_\Omega + \left(\mathcal{I}_N\frac{\partial}{\partial x}au_h, u_h\right)_\Omega$$

$$= \left(\mathcal{I}_N\frac{\partial}{\partial x}au_h - \frac{\partial}{\partial x}\mathcal{I}_N(au_h), u_h\right)_\Omega$$

$$\leq \left\|\mathcal{I}_N\frac{\partial}{\partial x}au_h - \frac{\partial}{\partial x}\mathcal{I}_N(au_h)\right\|_\Omega^2 + \|u_h\|_\Omega^2.$$

The latter term causes no problems. However, the first term is of the form

$$\left\| \mathcal{I}_N \frac{dv}{dx} - \frac{d}{dx}\mathcal{I}_N v \right\|_\Omega^2.$$

To understand this, let us bound it as

$$\left\| \mathcal{I}_N \frac{dv}{dx} - \frac{d}{dx}\mathcal{I}_N v \right\|_\Omega^2 \leq \left\| \frac{dv}{dx} - \mathcal{I}_N \frac{dv}{dx} \right\|_\Omega^2 + \left\| \frac{dv}{dx} - \frac{d}{dx}\mathcal{I}_N v \right\|_\Omega^2.$$

Based on the polynomial approximation results discussed in Section 4.3, both of these terms can be bounded for $v \in H^p(\Omega)$ as

$$\left\| \frac{dv}{dx} - \mathcal{I}_N \frac{dv}{dx} \right\|_\Omega \leq C \frac{h^{\sigma-1}}{N^{p-1}} |v|_{\Omega,p},$$

for $\sigma = \min(N+1, p)$. In a similar way, we have

$$\left\| \frac{dv}{dx} - \frac{d}{dx}\mathcal{I}_N v \right\|_\Omega \leq \|v - \mathcal{I}_N v\|_{\Omega,1} \leq C \frac{h^{\sigma-1}}{N^{p-1}} |v|_{\Omega,p}.$$

Combining it all, we have

$$\frac{1}{2}\frac{d}{dt}\|u_h\|_\Omega \leq C_1 \|u_h\|_\Omega + C_2(h,a)N^{1-p}|u|_{\Omega,p}.$$

This implies that if u is not sufficiently smooth we cannot control the last term and an instability may appear; this is what is manifested in Example 5.3.

The above analysis reveals that the term leading to instability is of the form

$$\left\| \mathcal{I}_N \frac{dv}{dx} - \frac{d}{dx}\mathcal{I}_N v \right\|_\Omega;$$

that is, it reflects that the derivative of an interpolation is not the same as the interpolation of the derivative and this causes the instability. This also means that if u is smooth but underresolved, simply adding resolution may cure the instability, as the commutation error in that case is further reduced. A further discussion of these aspects can be found in [135, 159].

With the added understanding of this aliasing-driven instability, let us now seek a solution that is more practical than to simply add more resolution. A classic technique to deal with relatively weak instabilities is to add some dissipation to the problem. To understand whether this suffices, let us consider the slightly modified problem

$$\frac{\partial u_h}{\partial t} + \frac{\partial}{\partial x}\mathcal{I}_N(au_h) = \varepsilon(-1)^{\tilde{s}+1}\left[\frac{\partial}{\partial x}(1-x^2)\frac{\partial}{\partial x}\right]^{\tilde{s}} u_h. \qquad (5.13)$$

Right now it is, admittedly, not at all clear why the dissipative term takes this particular form or even how to discretize it using a discontinuous Galerkin method. We will return to this issue shortly. However, consider first the right-hand side

$$\varepsilon(-1)^{\tilde{s}+1}\left(u_h, \left[\frac{\partial}{\partial x}(1-x^2)\frac{\partial}{\partial x}\right]^{\tilde{s}} u_h\right)_{\Omega},$$

which is the change introduced into the energy statement after the residual statement, Eq. (5.12). Integration by parts $\tilde{s}$ times yields exactly

$$\varepsilon(-1)^{\tilde{s}+1}\left(u_h, \left[\frac{\partial}{\partial x}(1-x^2)\frac{\partial}{\partial x}\right]^{\tilde{s}} u_h\right)_{\Omega} = -\varepsilon\|u_h^{(\tilde{s})}\|^2_{\Omega} = -\varepsilon|u_h|^2_{\Omega,\tilde{s}}.$$

Including this into the stability analysis, we recover the modified energy statement

$$\frac{1}{2}\frac{d}{dt}\|u_h\|^2_{\Omega} \le C_1\|u_h\|^2_{\Omega} + C_2 N^{2-2p}|u|^2_{\Omega,p} - C_3\varepsilon|u_h|^2_{\Omega,\tilde{s}}.$$

Clearly, choosing $\varepsilon \propto N$, the dissipative term dominates the unstable term and guarantees stability. What remains an open question is how to implement this term in the most efficient way.

To understand this, we first recall that the added term is a trick. It is something we do to stabilize the algorithm and it is not related to the physics of the problem. In other words, we have the freedom to implement this dissipation in any reasonable way.

We therefore include it in a timesplitting fashion; that is, we first advance

$$\frac{\partial u_h}{\partial t} + \frac{\partial}{\partial x}\mathcal{I}_N f(u_h) = 0,$$

one timestep, followed by

$$\frac{\partial u_h}{\partial t} = \varepsilon(-1)^{\tilde{s}+1}\left[\frac{\partial}{\partial x}(1-x^2)\frac{\partial}{\partial x}\right]^{\tilde{s}} u_h. \tag{5.14}$$

This is only an order $\mathcal{O}(\Delta t)$ accurate approximation to the directly stabilized scheme, Eq. (5.13), but that does not really matter. If we, furthermore, restrict the attention to this second problem and advance it in time using a forward Euler method, we recover

$$u_h^* = u_h(t+\Delta t) = u_h(t) + \varepsilon\Delta t(-1)^{\tilde{s}+1}\left[\frac{\partial}{\partial x}(1-x^2)\frac{\partial}{\partial x}\right]^{\tilde{s}} u_h(t). \tag{5.15}$$

So far, not much seems to have been achieved. Recall now, however, that we can express

$$u_h(x,t) = \sum_{n=1}^{N_p} \hat{u}_n(t) \tilde{P}_{n-1}(x).$$

Since $\tilde{P}_n(x)$ satisfies the Sturm-Liouville equation

$$\frac{d}{dx}(1-x^2)\frac{d}{dx}\tilde{P}_n + n(n+1)\tilde{P}_n = 0,$$

we obtain

$$u_h^*(x,t) \simeq u_h(x,t) + \varepsilon \Delta t (-1)^{\tilde{s}+1} \sum_{n=1}^{N_p} \hat{u}_n(t)(n(n-1))^{\tilde{s}} \tilde{P}_{n-1}(x)$$

$$\simeq \sum_{n=1}^{N_p} \sigma\left(\frac{n-1}{N}\right) \hat{u}_n(t) \tilde{P}_{n-1}(x), \quad \varepsilon \propto \frac{1}{\Delta t N^{2\tilde{s}}}.$$

We have introduced the filter function, $\sigma(\eta)$, which must have a few obvious properties; that is,

$$\sigma(\eta) \begin{cases} = 1, & \eta = 0 \\ \leq 1, & 0 \leq \eta \leq 1 \\ = 0, & \eta > 1, \end{cases} \quad \eta = \frac{n-1}{N}.$$

The first condition ensures that mean values are maintained and the second condition expresses the dissipation of the high modes.

The above derivation indicates that

$$\sigma(\eta) = 1 - \alpha\eta^{2\tilde{s}},$$

suffices for stabilization. A popular alternative to this is

$$\sigma(\eta) = \exp(-\alpha\eta^{2\tilde{s}}),$$

known as an exponential filter. Both filters are filters of order $2\tilde{s}$. In fact, the first is simply a leading order approximation to the second one.

The actual choice of α is somewhat arbitrary; that is, choosing $\alpha = 0$ means no dissipation whereas taking α to be large results in increasing dissipation. A choice often used is $\alpha \simeq -\log(\varepsilon_M) \simeq 36$, where ε_M is the machine precision in double precision. This choice means that $\sigma(1) \simeq \varepsilon_M$ in the exponential filter.

There is one very important difference between solving Eq. (5.14) and applying the filter, σ, to the expansion coefficients. In the latter case, no timestepping is required and, thus, there is no stability constraint as would be the case when solving Eq. (5.14). This makes the use of the filter a highly advantageous way of introducing the dissipation needed for stabilization.

In the formulation developed in Chapter 3, the process of filtering reduces to a matrix multiplication with a filter matrix, $\mathcal{F}$, defined as

$$\mathcal{F} = \mathcal{V}\Lambda\mathcal{V}^{-1},$$

where the diagonal matrix, Λ, has the entries

$$\Lambda_{ii} = \sigma\left(\frac{i-1}{N}\right), \quad i = 1, \ldots, N_p.$$

A script for defining this is shown in Filter1D.m, assuming that we use a filter of the form

$$\sigma(\eta) = \begin{cases} 1, & 0 \le \eta \le \eta_c = \frac{N_c}{N} \\ \exp(-\alpha((\eta - \eta_c)/(1 - \eta_c))^s), & \eta_c \le \eta \le 1, \end{cases} \tag{5.16}$$

To simplify matters, we define $s = 2\tilde{s}$; that is, it must be even. As discussed above we use $\alpha = -\log(\varepsilon_M)$ and N_c represents a cutoff below which the low modes are left untouched.

| Filter1D.m |

```
function [F] = Filter1D(N,Nc,s)

% function [F] = Filter1D(N,Nc,s)
% Purpose : Initialize 1D filter matrix of size N.
%           Order of exponential filter is (even) s with cutoff at Nc;

Globals1D;
filterdiag = ones(N+1,1);
alpha = -log(eps);

% Initialize filter function
for i=Nc:N
    filterdiag(i+1) = exp(-alpha*((i-Nc)/(N-Nc))^s);
end;

F = V*diag(filterdiag)*invV;
return;
```

To illustrate the dependence of the filter function on the choice of the parameters α, N_c, and s, we show it in Fig. 5.2 for a number of variations. The general trend is that increasing s and/or N_c decreases the dissipation while increasing α has the opposite effect.

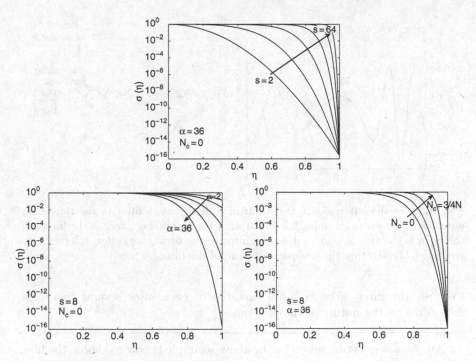

Fig. 5.2. Examples of how the filter function, $\sigma(\eta)$, varies as a function of the three parameters: the order s, the cutoff, $N_c = N\eta_c$, and the maximum damping parameter, α.

Example 5.4. To further understand the filter, let us return to Example 5.3 and repeat the unstable computation with a weak filter applied – in this case, we use $\eta_c = 0.5$, $s = 32$, and $\alpha = 36$.

The result is shown in Fig. 5.3. Comparing with the results in Fig. 5.1, the solutions are indistinguishable, although the filtered solution is obtained at a significantly reduced cost.

A tale of caution is also illustrated in Fig. 5.3. We show $\|u\|_{\Omega,h}^2$ as a function of time for different filters. This highlights that one can recover a very good solution over a long time. However, one must be careful when choosing the filter, as it adds dissipation. Too much filtering (e.g., too low a value of s), may destroy the solution.

In this example, the filter matrix is applied to the solution vector at every stage of the RK scheme. However, this is not required and it often suffices

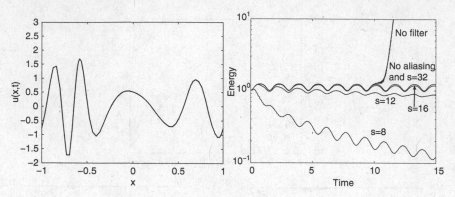

Fig. 5.3. On the left is shown the result of adding a weak filter to the third computation discussed in Example 5.3, illustrating the stabilizing effect of the filter. On the right is shown the temporal development of the total energy for different filter strengths, highlighting the dissipative nature of the filter.

to apply the filter after each full timestep or even after several timesteps, depending on the nature of the problem.

An obvious question raised by the above example is how to choose the filter variables, s, N_c, and α for a particular problem. The best answer to this is

Filter as little as possible
.. but as much as is needed.

This is not quantitative but is perhaps the best guideline there is. Clearly, the amount of filtering needed is highly problem dependent; for example, problems with underresolved features/shocks and/or highly nonlinear terms are likely to suffer more from instabilities than simpler types of problem. For specific problems, it is, nevertheless, often an easy matter to choose suitable parameters (e.g., start by filtering a small amount and then increase the amount of filtering over a few runs until the scheme is stabilized). A good set of starting parameters is $\alpha = 36$, $N_c = 0$, and $s = 16$ for a reasonably resolved computation.

While it is now clear that the simplest and fastest way of computing the fluxes may require one to add some dissipation to the scheme, it is natural to question whether there are other ways than through filtering. Indeed, a second look at the operator associated with filtering,

$$\left[\frac{d}{dx}(1 - x^2)\frac{d}{dx} \right]^{\tilde{s}} = \left[(1 - x^2)\frac{d^2}{dx^2} - 2x\frac{d}{dx} \right]^{\tilde{s}},$$

reveals that the dissipation is added in a nonuniform way; for example, it is strongest around $x = 0$ and weaker around $x = \pm 1$, leading to the possibility of over-dissipating in the center to ensure stability throughout the element.

This could suggest that we instead use a term like

$$(-1)^{\tilde{s}+1}\left[\frac{d^2}{dx^2}\right]^{\tilde{s}},$$

as a dissipative operator. This term cannot be implemented directly through the filter so we need to consider the modified equation

$$\frac{\partial u}{\partial t}+\frac{\partial f(u)}{\partial x}=\varepsilon(-1)^{\tilde{s}+1}\left[\frac{d^2}{dx^2}\right]^{\tilde{s}}u.$$

While this is dissipative and sufficiently strong to stabilize the scheme, a central problem is that the order of the equation is changed. This impacts the number of boundary conditions needed to ensure well-posedness. As an example, assume $f(u) = au$. The unaltered equation will need only one boundary condition while the stabilized version requires $2\tilde{s}$ boundary conditions. As this results from our choice of numerical scheme, it is not easy to see how to choose these conditions without impacting the solution.

For the more complicated dissipative term, we have

$$\left(u,(-1)^{\tilde{s}+1}\left[\frac{d}{dx}(1-x^2)\frac{d}{dx}\right]^{\tilde{s}}u\right)_{\Omega}=-\left\|\left[\frac{d}{dx}(1-x^2)\frac{d}{dx}\right]^{\tilde{s}/2}u\right\|_{\Omega}^2,$$

which implies that no new boundary conditions are needed due to the singular operator.

Returning to the question of whether $u_h(x,t)$ converges to a weak solution of the conservation law, we consider a term like

$$\left(\phi_h,(-1)^{\tilde{s}+1}\frac{1}{N^{2\tilde{s}-1}}\left[\frac{d}{dx}(1-x^2)\frac{d}{dx}\right]^{\tilde{s}}u_h\right)_{\Omega}.$$

To maintain convergence to the weak solution, we must ensure that this term vanishes as $N \to \infty$.

Recall first that

$$\phi_h(x,t)=\sum_{n=1}^{N_p}\hat{\phi}_n(t)\tilde{P}_{n-1}(x),\quad u_h(x,t)=\sum_{n=1}^{N_p}\hat{u}_n(t)\tilde{P}_{n-1}(x).$$

Inserting this yields

$$-\frac{1}{N^{2\tilde{s}-1}}\sum_{n=1}^{N_p}\hat{\phi}_n(t)\hat{u}_n(t)n^{\tilde{s}}(n-1)^{\tilde{s}},$$

since $\tilde{P}_n(x)$ satisfies the Sturm-Liouville problem.

Furthermore, if $u_h(x,t)$ is bounded; that is, the scheme is assumed to be stable, we know from Theorem 4.1 that

$$|\hat{u}_n\hat{\phi}_n|\le Cn^{-3},$$

provided ϕ_h is at least continuous. Then

$$\frac{1}{N^{2\tilde{s}-1}} \sum_{n=1}^{N_p} \frac{n^{\tilde{s}}(n-1)^{\tilde{s}}}{n^3} \leq \frac{C}{N}.$$

Thus, the stabilizing term vanishes and guarantees that if $u_h(x,t)$ converges to $u(x,t)$ for increasing N and/or decreasing h, then the solution is a weak solution to the original conservation law (see [45] for more details on this).

It is worth emphasizing that other kinds of dissipative terms have the same characteristics [235, 236] although these terms generally need to be added directly to the equation rather than through the use of the filter matrix.

5.4 Problems on nonconservative form

In most of the discussions so far we have considered problems in conservation form; that is

$$\frac{\partial u}{\partial t} + \frac{\partial f}{\partial x} = 0.$$

However, one often encounters closely related problems in nonconservative form as

$$\frac{\partial u}{\partial t} + a(x,t)\frac{\partial u}{\partial x} = 0, \qquad (5.17)$$

with a prominent example being the Hamilton-Jacobi equation encountered in control theory, level set methods, and many other types of application.

It is worthwhile briefly to discuss how these kinds of problem fit into the discontinuous Galerkin formulation. First, we can rewrite Eq. (5.17) in inhomogeneous conservation form as

$$\frac{\partial u}{\partial t} + \frac{\partial au}{\partial x} - \frac{\partial a}{\partial x}u = 0,$$

highlighting that $|a_x| \leq C$ suffices to ensure well-posedness of the problem. If a is sufficiently smooth, this is a reasonable approach. However, care must be exercised in case steady-state solutions are required, as these must now satisfy a balance equation and the discrete scheme must support such solutions. This particular complication is discussed in detail in [327, 328] and references therein.

A closely related approach is to borrow the results from Example 2.5 and consider the direct discretization of Eq. (5.17) but with a numerical flux given as

$$\hat{n} \cdot (au)^* = \frac{2a^+a^-}{a^+ + a^-}u^-,$$

which, in the case where a is smooth, reduces to a flux based on $f = au$ $(a > 0)$.

A third option, often used in the context of Hamilton-Jacobi equations, is to introduce the new variable $v = \frac{\partial u}{\partial x}$ and differentiate Eq. (5.17) to obtain

$$\frac{\partial v}{\partial t} + \frac{\partial av}{\partial x} = 0,$$

which can now be solved using a standard discretization. To recover a boundary condition on v, we use that if $u(x,t) = g(t)$ at an inflow boundary condition, Eq. (5.17) implies

$$v(x,t) = \frac{\partial u}{\partial x} = -\frac{1}{a}\frac{dg(t)}{dt}$$

for $v(x,t)$ at the boundary. Once $v(x,t)$ is computed, $u(x,t)$ can be recovered directly by integration. In the context of Hamilton-Jacobi equations, this approach is discussed further in [178, 216]. A different approach for Hamilton-Jacobi equations, focusing on solving these directly rather than by reduction to conservation form, is introduced in [56]. This results in methods similar to the one discussed above for Eq. (5.17) but with a more complex dissipative numerical flux.

5.5 Error estimates for nonlinear problems with smooth solutions

Provided the solutions u and fluxes f are sufficiently smooth and that monotone fluxes are used, there are no essential differences in the results between the error analysis for the linear and nonlinear problems. We have already established nonlinear stability in Section 5.2 and the error analysis, following the approach in Section 4.5, yields the expected result [337].

Theorem 5.5. *Assume that the flux $f \in C^3$ and the exact solution u is sufficiently smooth with bounded derivatives. Let u_h be a piecewise polynomial semidiscrete solution of the discontinuous Galerkin approximation to the one-dimensional scalar conservation law; then*

$$\|u(t) - u_h(t)\|_{\Omega,h} \le C(t)h^{N+\nu},$$

provided a regular grid of $h = \max h^k$ is used. The constant C depends on u, N, and time t, but not on h. If a general monotone flux is used, $\nu = \frac{1}{2}$, resulting in suboptimal order, while $\nu = 1$ in the case an upwind flux is used.

It is worth noting the suboptimal convergence for a general flux, i.e., a global Lax-Friedrichs flux. The reason for this, as discussed in Section 4.5, is that to recover the optimal rate, one relies on an upwinded special projection that allows the superconvergence result. Hence, to maintain this, one must use upwind/streamline numerical fluxes such as Godonov, Engquist-Osher, Roe fluxes, etc. [79, 218].

The extension of the above result to the system case introduces an additional complication since the upwind direction is no longer immediately identifiable. Hence, the general result for smooth solutions to one-dimensional symmetrizable systems of conservation laws follows the result in Theorem 5.5 with $\nu = \frac{1}{2}$ [338].

Exceptions to this are cases where one can apply flux splitting to derive the exact upwind fluxes, similar to the linear systems discussed in detail in Section 2.4. The equivalent nonlinear condition is that the flux is homogeneous of degree 1 (i.e., $\boldsymbol{f}(\boldsymbol{u}) = \boldsymbol{f_u}(\boldsymbol{u})\boldsymbol{u}$ and the flux Jacobian, $\boldsymbol{f_u}(\boldsymbol{u})$, is symmetrizable). In such cases, which include many important problems such as the Euler equations of gas dynamics and the equations of magnetohydrodynamics, optimal convergence rates can be recovered by doing strict upwinding. The main disadvantage of doing this is the associated cost for nonlinear problems.

5.6 Problems with discontinuous solutions

While we have discussed various aspects of nonlinear problems, we have avoided one critical issue: what happens when discontinuities and shocks develop in the solution? In particular, we need to understand what we can expect regarding accuracy and stability in such situations.

Before considering the full problem, let us illustrate a fundamental problem associated with the appearance of a shock. In Fig. 5.4 we show the polynomial representation of the simple discontinuity

$$u(x) = -\text{sign}(x), \quad x \in [-1, 1],$$

and plots of the pointwise error for increasing number of terms, N, in the expansion. The results in Fig. 5.4 illustrate three unfortunate effects of the shock:

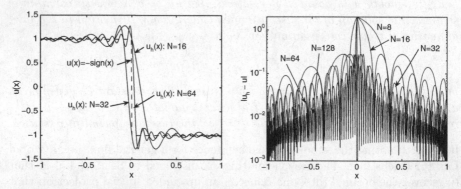

Fig. 5.4. On the left is shown the approximation of a sign function using a Legendre expansion with increasing number of terms, illustrating the Gibbs phenomenon. On the right is shown the pointwise error of the expansion.

- The reduction to first order pointwise accuracy away from the point of discontinuity.
- The loss of pointwise convergence at the point of discontinuity.
- The introduction of artificial and persistent oscillations around the point of discontinuity.

This phenomenon, classic by now, is known as the Gibbs phenomenon and its behavior is well understood (see [135, 137]). The issue at hand is how to deal with this. After all, it is the use of the high-order basis on the elements that gives the high-order accuracy of the scheme for smooth solutions, as discussed in Chapter 4.

Let us first return to the simpler case of

$$\frac{\partial u}{\partial t} + a(x)\frac{\partial u}{\partial x} = 0,$$

discussed in some detail previously. For this case we have established stability, possibly with the use of a filter, depending on the details of the flux computation.

While we return to the impact of the filter shortly, let us attempt to understand whether the combination of the Gibbs oscillation with propagation destroys the solution globally – or rather reduces the accuracy to first order with no hope of recovering a high-order accurate solution.

We write the problem on the skew-symmetric form (term $\mathcal{N}_1$ in Eq. (5.10))

$$\frac{\partial u}{\partial t} + \frac{1}{2}a\frac{\partial u}{\partial x} + \frac{1}{2}\frac{\partial au}{\partial x} - \frac{1}{2}a_x u = \frac{\partial u}{\partial t} + \mathcal{L}u = 0,$$

and assume that $a(x)$ is smooth. Both $a(x)$ and $u(x,t)$ are considered periodic for simplicity. However, we assume that the initial condition, $u(x,0)$, is nonsmooth to introduce the Gibbs phenomenon.

Introduce the adjoint problem

$$\frac{\partial v}{\partial t} - \mathcal{L}^* v = 0,$$

where $(\mathcal{L}u, v)_\Omega = (u, \mathcal{L}^* v)_\Omega$. For solving the adjoint problem we assume smooth initial conditions.

We immediately have that

$$\frac{d}{dt}(u,v)_\Omega = 0 \quad \Rightarrow \quad (u(t), v(t))_\Omega = (u(0), v(0))_\Omega.$$

Assuming exact integration of all terms and central fluxes (i.e., no aliasing or dissipation), we obtain a similar statement for the semidiscrete scheme

$$(u_h(t), v_h(t))_{\Omega,h} = (u_h(0), v_h(0))_{\Omega,h}.$$

Before proceeding, we need to be a bit careful about exactly what $u_h(0)$ means, given that $u(0)$ is assumed to be discontinuous. If we simply read this function at the grid points, we have forever lost the information of the location of the shock and the solution cannot be better than first order. We therefore assume that $u_h(0)$ is understood to be the polynomial interpolation of the projection of $u(0)$. In that case, there is no quantization error caused by the grid.

Now, consider

$$(u_h(0), v_h(0))_{\Omega,h} = (u(0), v(0))_\Omega + (u_h(0) - u(0), v_h(0))_{\Omega,h}$$
$$+ (u(0), v_h(0) - v(0))_{\Omega,h}.$$

First, one realizes that the second term on the right-hand side vanishes due to Galerkin orthogonality. The last term can be controlled by the Cauchy-Schwarz inequality and the smoothness of $v(0)$ to obtain

$$(u_h(0), v_h(0))_{\Omega,h} \leq (u(0), v(0))_\Omega + C(u)h^{N+1}N^{-q}|v(0)|_{\Omega,q}.$$

Finally, since the approximation of the dual problem is stable and $v(0)$ smooth, we have

$$\|v(t) - v_h(t)\|_{\Omega,h} \leq C(t)\frac{h^{N+1}}{N^q}|v(t)|_{\Omega,q};$$

that is, we can safely exchange v for v_h. This results in

$$(u_h(t), v(t))_{\Omega,h} = (u(t), v(t))_\Omega + \varepsilon, \tag{5.18}$$

where ε is very small and depends only on the smoothness of $v(x,t)$.

Equation (5.18) is interesting, as it highlights, at least for the case of a variable coefficient problem with nonsmooth initial conditions, the possibility of recovering a high-order accurate solution, $u_h(x,t)$. The catch is that the accuracy is not seen directly in the solution $u_h(x,t)$, but, rather, there exists some smooth function, $v(x,t)$, that allows one to extract a highly accurate solution even after propagation in time.

While it is a surprising result, it is also an encouraging result. It clarifies that the oscillations may look bad, but they do not destroy the attractive basic properties of the schemes – in particular, the properties related to propagation. In fact, the highly oscillatory result contains the information needed to reconstruct a spectrally accurate solution.

Similar results to the one above are not known with rigor for general nonlinear problems although some experiments show similar properties for certain problems (e.g., Burgers' equation [287]). The retainment of the high-order information was also argued much earlier in [239]. It is worth noticing that there are currently no general results showing that the higher-order information is destroyed by the interaction of oscillations and the propagation (e.g., through interacting shocks). It is, however, likely that the solutions with improved accuracy has to be recovered through some process, as the direct solution is first order accurate in a pointwise sense, in agreement with the simpler analysis of the linear problem. This has been demonstrated also for much more complex problems [98, 99, 100].

5.6.1 Filtering

In light of the above, it is reasonable to consider ways to recover some of the accuracy hidden in the oscillatory solutions. As it turns out, we already have such a tool at our disposal.

Consider

$$u_h(x) = \sum_{n=1}^{N_p} \hat{u}_n \tilde{P}_{n-1}(x), \quad \hat{u}_n = \int_{-1}^{1} u(x) \tilde{P}_{n-1}(x) \, dx.$$

If $u(x)$ has a discontinuity in $[-1, 1]$, the analysis in Section 4.5 shows that

$$\hat{u}_n \simeq \frac{1}{n}.$$

Thus, a manifestation of the Gibbs phenomenon, or rather lack of regularity, is a slow decay of the expansion coefficients. This suggests that we could attempt to modify the expansion coefficients to decay faster in the hope of recovering a more rapidly convergent expansion and, thus, a more accurate approximation.

In Fig. 5.5 we illustrate the impact of using the exponential filter

$$\sigma(\eta) = \exp(-\alpha \eta^s)$$

to obtain the filtered expansion

$$u_h^F(x) = \sum_{n=1}^{N_p} \sigma\left(\frac{n-1}{N}\right) \hat{u}_n \tilde{P}_{n-1}(x).$$

In Section 5.3 this was used to recover stability by controlling the impact of the aliasing errors.

We consider the sequence of functions

$$u^{(0)} = \begin{cases} -\cos(\pi x), & -1 \leq x \leq 0 \\ \cos(\pi x), & 0 < x \leq 1, \end{cases} \quad u^{(i)} = \int_{-1}^{x} u^{(i-1)}(s) \, ds,$$

which is constructed such that $u^{(p)} \in H^p[-1, 1]$.

In Fig. 5.5 we show the pointwise error associated with three different filter orders for four test functions (i.e., $u^{(0)} - u^{(3)}$). A number of observations are worth making. For a function of fixed regularity (rows in Fig. 5.5), filtering can dramatically improve the accuracy of the expansion away from the point of discontinuity. Also, increasing N and s decreases the size of the region around the nonsmooth point, where the impact of filtering is less effective. On the other hand, the order of the filter impacts the accuracy unfavorably, as illustrated

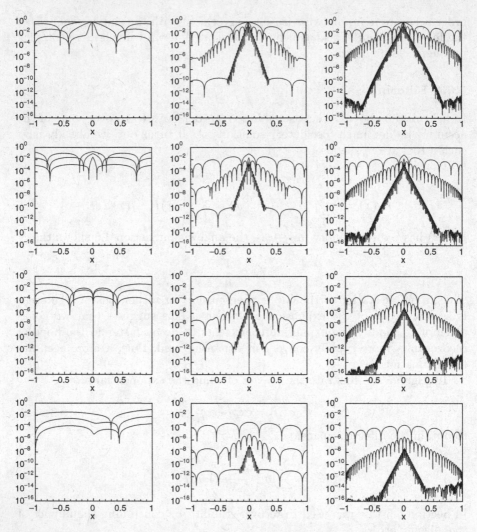

Fig. 5.5. In the left column is shown the pointwise error after the exponential filter with $s = 2$ has been applied to the Legendre expansions of the four test functions. We use $N = 16$, $N = 64$, and $N = 256$ for each function. The middle column shows similar results for $s = 6$ and the right column displays the results for $s = 10$.

for the $s = 2$ filter (first column in Fig. 5.5), which limits the convergence rate regardless of the regularity of the function being approximated. However, if s is sufficiently large, this does not adversely impact the pointwise error when filtering smooth functions. The results in Fig. 5.5 highlight that the filter does not only help to stabilize the computation but also recovers a much higher convergence rate away from the point of discontinuity.

Hence, a possible model for computing with discontinuous solutions is to simply stabilize the computation with the filter as much (or rather as little) as is needed; that is, compute with the oscillations that clearly are not noise, and then postprocess the solution when needed. To illustrate the prospects of this approach, let us consider an example.

Example 5.6. We solve Burgers' equation

$$\frac{\partial u}{\partial t} + \frac{\partial u^2}{\partial x} = 0, \ x \in [-1,1],$$

with the discontinuous initial condition

$$u_0(x) = u(x,0) = \begin{cases} 2, & x \leq -0.5 \\ 1, & x > -0.5. \end{cases}$$

Using the Rankine-Hugoniot conditions, we easily find that the shock propagates with the constant speed of 3; that is, the exact solution is

$$u(x,t) = u_0(x - 3t),$$

which we also use to define the appropriate boundary conditions.

The Burgers' equation is solved using a discontinuous Galerkin method on strong form with aliasing in the computation of the flux and we add a filter to ensure stability.

We use $K = 20$ equidistant elements, each with an $N = 8$ order polynomial basis and a Lax-Friedrichs flux to connect the elements. To stabilize the computation, we use an exponential filter with $\alpha = 36$ and $N_c = 0$ (see Eq. (5.16)), applied at every stage of the RK method.

In Fig. 5.6 we show the results obtained using three different filters of decreasing strength. As we have already seen, using too low a filter order results in an overly dissipated solution, similar to what we observe with a second order filter. While the center of the shock is in the right place, the shock is severely smeared out, although the solution is monotone. We also observe a faceting of the solution, which is a characteristic of too strong filtering (see [192] for a discussion of this). Nevertheless, away from the location of the shock, we see better accuracy, improving with the distance to the shock.

Decreasing the strength of the filter results in a profound improvement of the quality of the solution. Both with the sixth-order filter and the tenth-order filter we see an excellent resolution of the shock as well as very high accuracy away from the shock location. It is noteworthy that we find such high accuracy even in regions where the shock has passed through. This confirms that the

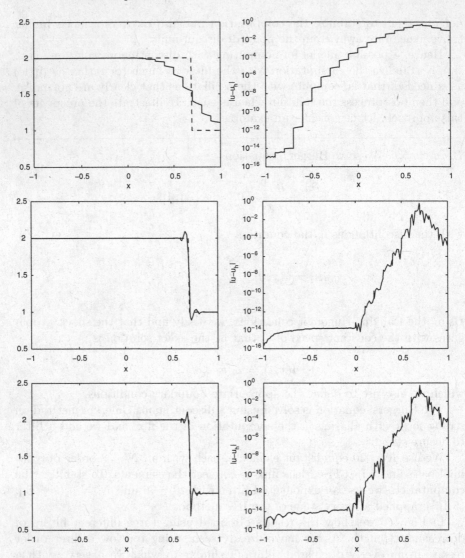

Fig. 5.6. Solution of Burgers' equation with a shock using a filtering approach. In all examples, we use $K = 20$ and $N = 8$ while varying the strength of the filter. The left column shows the computed results and the right column displays the pointwise error. Results are shown for $T = 0.4$. In the top row, we use a second order exponential filter; in the middle row a sixth order filter; and in the bottom row, a tenth-order filter.

effect of the oscillations do not destroy the accuracy in the smooth regions of the solution.

Since the analysis in the above suggests that the oscillatory solutions contain the information needed to reconstruct a spectrally accurate solution, it is natural to attempt to develop such reconstruction techniques. During the last decade, a number of techniques for recovering a solution with a higher pointwise accuracy have been developed.

One can simply filter the computed solution in the hope of recovering improved accuracy. The analysis of this approach is not complete for a polynomial basis [162], but the analysis [313] for a simpler periodic case confirms that spectral accuracy can be recovered everywhere except at the point(s) of discontinuity. In a similar spirit, one can filter in physical space, often known as mollification, and achieve comparable results [138, 298].

Similar results can be obtained (i.e., exponential convergence away from the points of discontinuity), by reexpanding the computed solutions using rational functions (e.g., Padé forms). In this case, one seeks two local polynomials, R_M and Q_L of order M and L, respectively, defined on D^k, such that

$$x \in \mathsf{D}^k: \quad u_h^k(x) = \frac{R_M(x)}{Q_L(x)},$$

where $N \geq M + L + 1$. There are several ways of defining in which sense the two expressions are the same, with the most obvious one being

$$\forall m \in [0, M + L]: \quad \int_{\mathsf{D}^k} \left(u_h^k Q_L - R_M \right) \tilde{P}_m \, dx = 0.$$

For general functions, one can use a quadrature to evaluate the inner product.

The advantage of the Padé reconstruction approach over straightforward filtering is a dramatic reduction of the Gibbs oscillations, a reduced smearing of the shock, and a much more rapid convergence even close to the point of discontinuity.

Example 5.7. We solve Burgers' equation

$$\frac{\partial u}{\partial t} + \frac{\partial u^2}{\partial x} = 0, \quad x \in [-1, 1],$$

with the smooth initial condition

$$u_0(x) = u(x, 0) = 0.5 + \sin(\pi x).$$

Burgers' equation is solved using only one domain with $N = 256$, although this is not important in this context. The equation is solved on strong form and filtering is used to ensure stability.

The direct computational result is shown in Fig. 5.7, where we recognize the Gibbs oscillations around the shock. We also show in Fig. 5.7 an example of a good reconstruction obtained by using the Padé reconstruction technique.

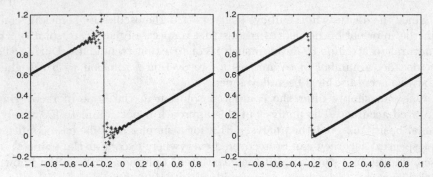

Fig. 5.7. On the left is shown the purely polynomial solution of Burgers' equation with $N = 256$ and the right shows the Padé-Legendre reconstructed solution with $M = 20$ and $L = 8$.

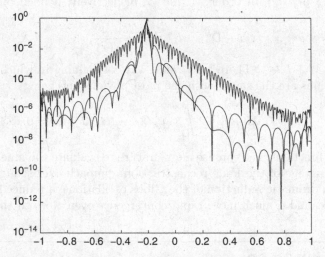

Fig. 5.8. Pointwise error for the reconstructed solution to Burgers' equation. We use $N = 256$ and $M = 20$ for the numerator and the three curves represent, from top to bottom, $L = 0$, $L = 4$, and $L = 8$, respectively.

As evidence of the possibility of dramatically increasing the pointwise accuracy away from the shock, we show in Fig. 5.8 the pointwise error for $M = 20$ and for increasing values of L (i.e., by increasing the order of denominator in the rational approximation). This clearly confirms the enhanced convergence in the piecewise smooth regions away from the point of discontinuity.

These Padé-based expansion techniques are discussed at length in [108, 160, 161] for Legendre-Padé-reconstruction, relevant to the present discussion.

Whereas the above techniques offer significant improvements, they are unable to recover spectral accuracy all the way up to and including the point of discontinuity. To achieve this, one needs to also introduce knowledge of the location of the discontinuity. If, however, this is available or computable with sufficient accuracy, reprojection techniques, also known as Gibbs reconstruction methods, enable the recovery of the full accuracy everywhere. The basic idea of these techniques is to isolate smooth regions, $x \in [a, b]$, between the discontinuities and reexpand these as

$$x \in [a, b]: \quad u_h(x) = \sum_{n=1}^{N_p} \hat{u}_n \tilde{P}_{n-1}(r(x)) = \sum_{m=0}^{M} \hat{v}_m \tilde{P}_m^{(\alpha,\alpha)}(r(x)),$$

where $\tilde{P}_m^{(\alpha,\alpha)}(r)$ are the normalized symmetric Jacobi polynomials, also known as Gegenbauer polynomials. The new expansion coefficients, $\hat{v}_m$, are found through projection, using the orthonormality of the Gegenbauer basis in a weighted norm. Under certain conditions on M and α, both of which are functions of N, one can show spectral decay of $\hat{v}_m$ and, thus, full recovery of pointwise spectral accuracy for $x \in [a, b]$. For the details, we refer to the review paper [137] and references therein. Large-scale applications of this technique are considered in [98, 99, 100].

5.6.2 Limiting

While the use of a filter suffices to stabilize the computations and the postprocessing enables one to obtain accurate solutions, even for problems with shocks, this approach falls short in one particular area: It does not eliminate the artificial oscillations during the computations unless a second order filter is used. However, as we have seen in the simple tests above, doing this leads to a severe smearing of the shocks.

For certain applications, this may not pose a problem, but unfortunately, it does present problems for many others; for example, if the solution is a temperature or a density, the oscillations may cause these fields to take unphysical values. Furthermore, there could be examples where a slightly higher value of a field variable would change the solution dramatically (e.g., in a combustion process where things could ignite artificially).

The ultimate solution would naturally be if one could develop a process that would eliminate the oscillations entirely, without adversely impacting the accuracy of the computation. To understand what is required to achieve this, we follow the discussion in [60] loosely and return to the continuous viscosity equation

$$\frac{\partial}{\partial t} u^\varepsilon + \frac{\partial}{\partial x} f(u^\varepsilon) = \varepsilon \frac{\partial^2}{\partial x^2} u^\varepsilon. \tag{5.19}$$

Apart from solution uniqueness, we must also ensure uniform boundedness to guarantee existence of the solution; that is,

$$\|u\|_{L^1} \leq C, \quad \|u\|_{L^1} = \int_\Omega |u| \, dx.$$

To see that this holds for solutions to Eq. (5.19), we define $\eta(u) = |u|$. Assuming simple periodic boundaries, we note that

$$-\int_\Omega (\eta'(u_x))_x u_t \, dx = \int_\Omega \frac{u_x}{|u_x|} u_{xt} \, dx = \frac{d}{dt} \int_\Omega |u_x| dx = \frac{d}{dt} \|u_x\|_{L^1}.$$

Now, multiply Eq. (5.19) by $-(\eta'(u_x^\varepsilon))_x$ and integrate over the domain to obtain

$$\frac{d}{dt} \|u_x^\varepsilon\|_{L^1} + \int_\Omega -(\eta'(u_x^\varepsilon))_x f_x(u^\varepsilon) \, dx = \int_\Omega -(\eta'(u_x^\varepsilon))_x \varepsilon u_{xx}^\varepsilon \, dx.$$

We realize that

$$-\int_\Omega (\eta'(u_x^\varepsilon))_x f_x(u^\varepsilon) \, dx = -\int_\Omega \eta''(u_x^\varepsilon) u_{xx}^\varepsilon f'(u^\varepsilon) u_x^\varepsilon \, dx = 0,$$

since $\eta''(u)u \equiv 0$. Furthermore, we have

$$-\int_\Omega (\eta'(u_x^\varepsilon))_x \varepsilon u_{xx}^\varepsilon \, dx = -\int_\Omega \varepsilon \eta''(u_x^\varepsilon)(u_{xx}^\varepsilon)^2 \, dx \le 0,$$

establishing uniform boundedness

$$\frac{d}{dt} \|u_x^\varepsilon\|_{L^1} \le 0.$$

To translate this into the discrete case, let us first consider the semidiscrete case with a constant basis (i.e., a first-order finite volume scheme). In this case, we have the one-dimensional method

$$h \frac{du_h^k}{dt} + f^*(u_h^k, u_h^{k+1}) - f^*(u_h^k, u_h^{k-1}) = 0, \qquad (5.20)$$

where we assume that the K cells are located equidistantly with a spacing of h and $f^*(a,b)$ is a monotone flux.

To mimic the continuous case, we define

$$v_h^k = -\frac{1}{h} \left[\eta' \left(\frac{u_h^{k+1} - u_h^k}{h} \right) - \eta' \left(\frac{u_h^k - u_h^{k-1}}{h} \right) \right]$$

as a first-order approximation to $-(\eta'(u_x))_x$. Multiplying Eq. (5.20) by this and summing all terms yields

$$\frac{d}{dt} |u_h|_{TV} + \sum_{k=1}^K v_h^k \left(f^*(u_h^k, u_h^{k+1}) - f^*(u_h^k, u_h^{k-1}) \right) = 0,$$

where the discrete total variation norm is defined as

$$|u_h|_{TV} = \sum_{k=1}^{K} |u_h^{k+1} - u_h^k|.$$

Consider the term

$$v_h^k \left(f^*(u_h^k, u_h^{k+1}) - f^*(u_h^k, u_h^{k-1}) \right)$$
$$= v_h^k (f^+(u_h^k) - f^+(u_h^{k-1}) + f^-(u_h^{k+1}) - f^-(u_h^k)),$$

where we have split the flux into increasing and decreasing components; that is, for the Lax-Friedrichs flux,

$$f^*(a, b) = \frac{f(a) + f(b)}{2} + \hat{\boldsymbol{n}}\frac{C}{2}(a - b),$$

then

$$f^*(a, b) = \begin{cases} f^+(a) + f^-(b), & \hat{\boldsymbol{n}} = 1 \\ f^-(a) + f^+(b), & \hat{\boldsymbol{n}} = -1, \end{cases}$$

where

$$f^+(a) = \frac{1}{2}(f(a) + Ca), \quad f^-(b) = \frac{1}{2}(f(b) - Cb).$$

Since f^+ and $-f^-$ are nondecreasing due to monotonicity, one easily shows that

$$v_h^k(f^+(u_h^k) - f^+(u_h^{k-1}) + f^-(u_h^{k+1}) - f^-(u_h^k)) \geq 0,$$

by considering the few possible combinations. Hence, we obtain

$$\frac{d}{dt}|u_h|_{TV} \leq 0,$$

and, thus, uniform boundedness. Similar results can be obtained with other monotone fluxes.

To go beyond this simple case, we must consider the two key questions of what happens when a higher-order basis is used and how does the time integration come into play.

Postponing the latter question for a while, we first recall that a higher-order basis may introduce oscillations and, thus, violate the bound on the total variation. This severely complicates the analysis and a complete analysis quickly becomes very technical.

To appreciate the challenges in this, let us consider a high-order scheme but require that the local means, or cell averages, $\bar{u}_h^k$ are uniformly bounded. In this case, we consider

$$h\frac{d\bar{u}_h^k}{dt} + f^*(u_r^k, u_l^{k+1}) - f^*(u_l^k, u_r^{k-1}) = 0,$$

where we have introduced the notation of u_l^k and u_r^k as the left and right limit value of u_h^k, respectively. If we use a first-order forward Euler method to integrate in time, we have

$$\frac{h}{\Delta t}\left(\bar{u}^{k,n+1} - \bar{u}^{k,n}\right) + f^*(u_r^{k,n}, u_l^{k+1,n}) - f^*(u_l^{k,n}, u_r^{k-1,n}) = 0,$$

and with a monotone flux, one can show that [60]

$$|\bar{u}^{n+1}|_{TV} - |\bar{u}^n|_{TV} + \Phi = 0,$$

where

$$\Phi = \sum_{k=1}^{K}\left(\eta'(\bar{u}^{k+1/2,n}) - \eta'(\bar{u}^{k+1/2,n})\right)\left(p(u^{k+1,n}) - p(u^{k,n})\right)$$

$$+\frac{\Delta t}{h}\sum_{k=1}^{K}\left(\eta'(\bar{u}^{k-1/2,n}) - \eta'(\bar{u}^{k+1/2,n})\right)\left(f^+(u_r^{k,n}) - f^+(u_r^{k-1,n})\right)$$

$$-\frac{\Delta t}{h}\sum_{k=1}^{K}\left(\eta'(\bar{u}^{k+1/2,n}) - \eta'(\bar{u}^{k-1/2,n})\right)\left(f^-(u_l^{k+1,n}) - f^-(u_l^{k,n})\right).$$

Here $\eta(u) = |u|$,

$$\eta'(\bar{u}^{k+1/2,n}) = \eta'\left(\frac{\bar{u}^{k+1,n} - \bar{u}^{k,n}}{h}\right),$$

and

$$p(u^{k,n}) = \bar{u}^k - \frac{\Delta t}{h}f^+(u_r^{k,n}) + \frac{\Delta t}{h}f^-(u_l^{k,n}).$$

By the properties of $\eta'(u) = \text{sign}(u)$, we see that the solution is total variation diminishing in the mean (TVDM) (i.e., $\Phi \geq 0$), if

$$\text{sign}(\bar{u}^{k+1,n} - \bar{u}^{k,n}) = \text{sign}(p(u^{k+1,n}) - p(u^{k,n})), \qquad (5.21)$$

$$\text{sign}(\bar{u}^{k,n} - \bar{u}^{k-1,n}) = \text{sign}(u_r^{k,n} - u_r^{k-1,n}), \qquad (5.22)$$

$$\text{sign}(\bar{u}^{k+1,n} - \bar{u}^{k,n}) = \text{sign}(u_l^{k+1,n} - u_l^{k,n}). \qquad (5.23)$$

Unfortunately, there is no guarantee that the numerical solution is restricted to behave like this, so to ensure total variation stability it must be enforced directly. This is the role of the limiter – also known as a slope limiter, Π. When designing a slope limiter, it must have the following properties:

- It does not violate conservation.
- It ensures that the three constraints are satisfied.
- It does not change the formal accuracy of the method.

While the first two are easy to satisfy, the third property causes problems. To see this, let us consider a simple example.

Example 5.8. We solve the simple problem

$$\frac{\partial u}{\partial t} + \frac{\partial u}{\partial x} = 0, \ x \in [-1, 1],$$

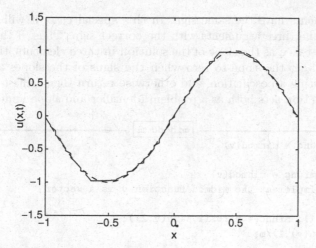

Fig. 5.9. Solution to linear wave equation using $N = 1$ and $K = 50$ and a simple slope limiter. The dashed line is the exact solution at $T = 10$ and the solid line is the computed solution.

with periodic boundary conditions and initial conditions, $u(x, 0) = \sin(\pi x)$.

Figure 5.9 shows the solution computed for this problem using 50 elements, each with an $N = 1$ basis and a basic slope limiter. The loss of accuracy around the smooth extrema is noticeable and is caused by the slope limiter.

This example highlights a significant problem with slope limiters, as they reduce the accuracy in smooth regions of the solution. Unfortunately, this cannot be avoided. Consider an approximation, u_h, to a smooth solution, $u(x)$. If $u(x)$ locally is monotone, then u_h will behave locally like a straight line and it is easy to see that Eqs. (5.21)-(5.23) are trivially obeyed. However, if $u(x)$ has a local smooth extrema, then the conditions are not necessarily fulfilled. In this case, the solution is wrongfully identified as having an oscillation and the limiter will alter the local solution and reduce the formal accuracy to first order. This is a general property of slope limiters if the solution is required to be TVDM. Naturally, for the simple example above, there is no need to use a slope limiter. However, a similar effect will occur in any computation of problems with shocks and regions of smooth behavior.

To find a practical way of modifying the local mean of the solution to guarantee the TVDM property, we define the minmod function

$$m(a_1, \ldots, a_m) = \begin{cases} s \min_{1 \le i \le m} |a_i|, & |s| = 1 \\ 0, & \text{otherwise,} \end{cases} \qquad s = \frac{1}{m} \sum_{i=1}^{m} \text{sign}(a_i).$$

$$(5.24)$$

To appreciate what this function does, consider a situation where we have three arguments, (a_1, a_2, a_3). Then $m(a_1, a_2, a_3)$ will return a zero unless the three arguments have the same sign. In this special case, it will return the smallest of the three arguments with the correct sign. Thus, if the three arguments are taken as the slope of the solution in three elements, the minmod function will set the slope to zero when the signs of the slopes are not the same, indicating an oscillation, and otherwise return the smallest slope. The minmod function acts both as a problem identifier and slope limiter.

$\boxed{\texttt{minmod.m}}$

```
function mfunc = minmod(v)

% function mfunc = minmod(v)
% Purpose: Implement the midmod function v is a vector

m = size(v,1); mfunc = zeros(1,size(v,2));
s = sum(sign(v),1)/m;

ids = find(abs(s)==1);
if(~isempty(ids))
  mfunc(ids) = s(ids).*min(abs(v(:,ids)),[],1);
end
return;
```

We can now define the interface fluxes as

$$v_l^k = \bar{u}_h^k - m(\bar{u}_h^k - u_l^k, \bar{u}_h^k - \bar{u}_h^{k-1}, \bar{u}_h^{k+1} - \bar{u}_h^k), \tag{5.25}$$
$$v_r^k = \bar{u}_h^k + m(u_r^k - \bar{u}_h^k, \bar{u}_h^k - \bar{u}_h^{k-1}, \bar{u}_h^{k+1} - \bar{u}_h^k). \tag{5.26}$$

Inserting these into Eqs. (5.21)-(5.23) reveal that they suffice to guarantee the TVDM property of the solution for sufficiently small timestep, Δt.

This can be explored to define families of slope limiters with slightly different properties. As a first step, assume that the solution is represented by a piecewise linear solution; that is,

$$u_h^k(x) = \bar{u}_h^k + (x - x_0^k)(u_h^k)_x,$$

where x_0^k represents the center coordinate of D^k. We define the slope limited solution

$$\Pi^1 u_h^k(x) = \bar{u}_h^k + (x - x_0^k) m \left((u_h^k)_x, \frac{\bar{u}_h^{k+1} - \bar{u}_h^k}{h/2}, \frac{\bar{u}_h^k - \bar{u}_h^{k-1}}{h/2} \right), \tag{5.27}$$

which can be shown to satisfy Eqs. (5.21)-(5.23). A slightly more dissipative limiter is the classic MUSCL (Monotone Upstream-centered Scheme for Conservation Laws) limiter [218, 301].

$$\Pi^1 u_h^k(x) = \bar{u}_h^k + (x - x_0^k) m \left((u_h^k)_x, \frac{\bar{u}_h^{k+1} - \bar{u}_h^k}{h}, \frac{\bar{u}_h^k - \bar{u}_h^{k-1}}{h} \right), \tag{5.28}$$

```
                       ┌─────────────────┐
─────────────────────── │ SlopeLimitLin.m │ ───────────────────────
                       └─────────────────┘
function ulimit = SlopeLimitLin(ul,xl,vm1,v0,vp1);

% function ulimit = SlopeLimitLin(ul,xl,vm1,v0,vp1);
% Purpose: Apply slopelimited on linear function ul(Np,1) on x(Np,1)
%          (vm1,v0,vp1) are cell averages left, center, and right

Globals1D;

% Compute various geometric measures
ulimit = ul; h = xl(Np,:)-xl(1,:);
x0 = ones(Np,1)*(xl(1,:) + h/2);

hN = ones(Np,1)*h;

% Limit function
ux = (2./hN).*(Dr*ul);

ulimit = ones(Np,1)*v0+(xl-x0).*(ones(Np,1)*...
               minmod([ux(1,:); (vp1-v0)./h; (v0-vm1)./h]));
return
```

An implementation of the slope limiter on one element is shown in Slope-
LimitLin.m and the script for slope limiting of the whole solution is shown in
SlopeLimit1.m.

```
                       ┌──────────────┐
─────────────────────── │ SlopeLimit1.m │ ───────────────────────
                       └──────────────┘
function ulimit = SlopeLimit1(u);

% function ulimit = SlopeLimit1(u);
% Purpose: Apply slopelimiter (Pi^1) to u

Globals1D;
ulimit = zeros(Np,K);

% Compute modal coefficients
uh = invV*u;

% Extract linear polynomial
ul = uh; ul(3:Np) = 0;ul = V*ul;

% Extract cell averages
uh(2:Np,:)=0; uavg = V*uh; v = uavg(1,:);

% Find cell averages in neighborhood of each element
vk = v; vkm1 = [v(1),v(1:K-1)]; vkp1 = [v(2:K),v(K)];

% Limit function in all cells
```

```
ulimit = SlopeLimitLin(ul,x,vkm1,vk,vkp1);
return
```

As discussed above, the use of a simple slope limiter has the disadvantage that it destroys the high-order accuracy in smooth regions. There are, however, a couple of things one can do to avoid this. If we assume that the computed solution, u_h, is a piecewise N-th-order polynomial, it is natural to only limit in those elements where oscillations are detected. A procedure [60] which appears to work well for this is as follows:

- Compute the limited edge values, v_l^k and v_r^k using Eqs. (5.25)-(5.26).
- If $v_l^k = u_h^k(x_l^k)$ and $v_r^k = u_h^k(x_r^k)$, there is no need for limiting and the local solution is not altered.
- If limiting is needed, compute the limited version of u_h^k as $\Pi^1 \tilde{u}_h^k$, where $\tilde{u}_h^k$ is the linear approximation to u_h^k; that is,

$$\tilde{u}_h^k = \bar{u}_h^k + (x - x_0^k)(u_h^k)_x.$$

This type of limiter, referred to as a generalized slope limiter, Π^N, is implemented in SlopeLimitN.m and generally yields improved results in smooth regions of the solution.

───────────────────────── SlopeLimitN.m ─────────────────────────

```
function ulimit = SlopeLimitN(u);

% function ulimit = SlopeLimitN(u);
% Purpose: Apply slopelimiter
%          (Pi^N) to u assuming u an N'th order polynomial

Globals1D;

% Compute cell averages
uh = invV*u; uh(2:Np,:)=0; uavg = V*uh; v = uavg(1,:);

% Apply slope limiter as needed.
ulimit = u; eps0=1.0e-8;

% find end values of each element
ue1 = u(1,:); ue2 = u(end,:);

% find cell averages
vk = v; vkm1 = [v(1),v(1:K-1)]; vkp1 = [v(2:K),v(K)];

% Apply reconstruction to find elements in need of limiting
ve1 = vk - minmod([(vk-ue1);vk-vkm1;vkp1-vk]);
ve2 = vk + minmod([(ue2-vk);vk-vkm1;vkp1-vk]);
ids = find(abs(ve1-ue1)>eps0 | abs(ve2-ue2)>eps0);
```

```
% Check to see if any elements require limiting
if(~isempty(ids))
  % create piecewise linear solution for limiting on specified elements
  uhl = invV*u(:,ids); uhl(3:Np,:)=0; ul = V*uhl;

  % apply slope limiter to selected elements
  ulimit(:,ids) = SlopeLimitLin(ul,x(:,ids),vkm1(ids),vk(ids),
  vkp1(ids));
end
return;
```

This may improve the accuracy in smooth regions of the solution, but it does not overcome the loss of accuracy around local extrema. One way to address this is to relax the condition on the decay of the total variation and require that the total variation of the mean is just bounded, called the TVBM condition. Following [285], this can be achieved by slightly modifying the definition of the minmod function, $m(\cdot)$, as

$$\bar{m}(a_1,\ldots,a_m) = m\left(a_1, a_2 + Mh^2\text{sign}(a_2),\ldots,a_m + Mh^2\text{sign}(a_m)\right), \quad (5.29)$$

where M is a constant that should be an upper bound on the second derivative at the local extrema. This is naturally not easy to estimate a priori. Too small a value of M implies higher local dissipation and order reduction, whereas too high a value of M reintroduces the oscillations.

—————————————————————— minmodB.m ——————————————

```
function mfunc = minmodB(v,M,h)

% function mfunc = minmodB(v,M,h)
% Purpose: Implement the TVB modified midmod function. v is a vector

mfunc = v(1,:);
ids = find(abs(mfunc) > M*h.^2);

if(size(ids,2)>0)
  mfunc(ids) = minmod(v(:,ids));
end
return
```

To implement this, one simply exchanges the use of the minmod function in SlopeLimitLin.m with the above approach, implemented in minmodB.m.

Example 5.9. We repeat the solution of the linear problem

$$\frac{\partial u}{\partial t} + \frac{\partial u}{\partial x} = 0, \quad x \in [-1,1],$$

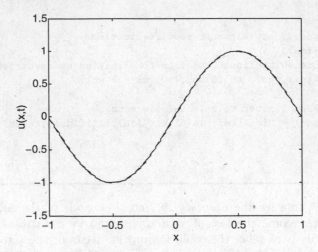

Fig. 5.10. Solution to the linear wave equation using $N = 1$ and $K = 50$ and a TVBM slope limiter with $M = 20$. The dashed line is the exact solution at $t = 10$ and the solid line the computed solution. To appreciate the effect of the TVBM limiter, compare the results in Fig. 5.9.

with periodic boundary conditions and initial conditions, $u(x,0) = \sin(\pi x)$, using the modified slope limiter based on $\bar{m}$. In Fig. 5.10 we show the solution computed for this problem using 50 elements, each with an $N = 1$ basis. The modified minmod function is used with $M = 20$ to restore high-order accuracy around the smooth extrema.

To illustrate the performance of the different limiters and compare the results with those obtained with the use of filtering, let us consider another example.

Example 5.10. We again solve Burgers' equation

$$\frac{\partial u}{\partial t} + \frac{\partial u^2}{\partial x} = 0, \ x \in [-1, 1],$$

with the discontinuous initial condition

$$u_0(x) = u(x,0) = \begin{cases} 2, & x \le -0.5 \\ 1, & x > -0.5. \end{cases}$$

Recall from Example 5.6 that the exact solution is given as

$$u(x,t) = u_0(x - 3t).$$

The Burgers' equation is solved using a DG method on strong form with aliasing in the computation of the flux. No filtering is applied, but the solution is limited using different types of limiters.

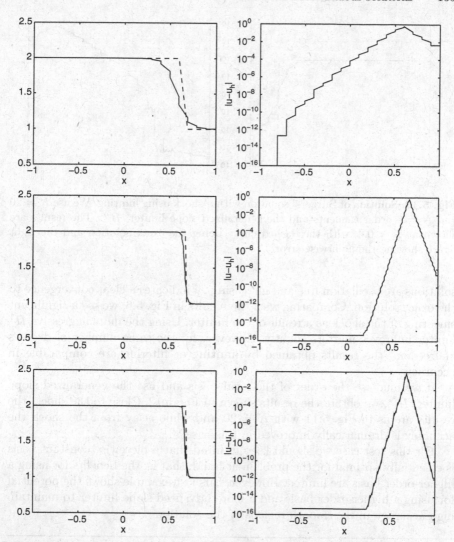

Fig. 5.11. Solution of Burgers' equation with a shock using limiting. In all examples, we use linear elements and propagate the solution until $T = 0.4$. The dashed curves are the exact solution. The left column shows the computed results and the right column displays the pointwise error. In the first row, we show the results with $K = 20$ and a MUSCL limiter, Eq. (5.28). The second row shows the results obtained with $K = 100$, while in the third row we show the results recovered with $K = 100$ but the more aggressive Π^1 limiter, Eq. (5.27).

We first consider the case where the local basis is piecewise linear (i.e., $N = 1$), and we use a Lax-Friedrichs flux to connect the elements.

In Fig. 5.11 we show the results of using different number of elements and limiter types, all computed with a local first-order basis. As expected, the

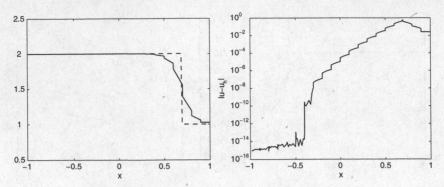

Fig. 5.12. Solution of Burgers' equation with a shock using limiting. We use $K = 20$ and $N = 8$ order elements and the generalized slope limiter, Π^N. The results are shown at $T = 0.4$, with the dashed curve being the exact solution and the right figure showing the pointwise error.

solutions are oscillation free and increasing K indicates a clear convergence to the exact solution. Comparing with the results in Fig. 5.6, we see a significant smearing of the shock as a result of the limiter. Using the more aggressive Π^1 limiter improves matters slightly. However, for the same number of degrees of freedom, the results obtained by limiting or filtering are comparable in accuracy.

If we increase the order of the local basis and use the generalized slope limiter, Π^N, we obtain the results shown in Fig. 5.12. Close to the shock, the results are as in Fig. 5.11 with $K = 20$ and while away from the shock the accuracy is dramatically improved in this case.

For this test case we should keep in mind that a piecewise constant basis is essentially optimal for the problem at hand; that is, the benefits for using a higher-order basis are limited. However, this last example shows the potential for using a higher-order basis and a more advanced slope limiter to maintain high accuracy in the smooth regions of the solution.

Another attempt to derive a slope limiter, which does not impact the high-order accuracy around extrema, has been made in [28]. This approach utilizes a hierarchical slope limiting of the modal coefficients in the following way:

- Compute $\hat{\boldsymbol{u}}^k = \mathcal{V}^{-1}\boldsymbol{u}^k$.
- For $n = N, \dots, 1$, compute

$$\sqrt{(2n+1)(2n+3)}\hat{v}_n^k = m\left(\sqrt{(2n+1)(2n+3)}\hat{u}_n^k, \hat{u}_{n-1}^{k+1} - \hat{u}_{n-1}^k, \hat{u}_{n-1}^k - \hat{u}_{n-1}^{k-1}\right).$$

- Repeat until $\hat{v}_n^k = \hat{u}_n^k$ (i.e., no limiting on that coefficient) or $n = 1$, in which case the solution is reduced to the cell-averaged solution.

The constant in the limiter comes from the identity

$$2\tilde{P}_n = \sqrt{\gamma_n \gamma_{n+1}} \tilde{P}'_{n+1} - \sqrt{\gamma_n \gamma_{n-1}} \tilde{P}'_{n-1},$$

i.e., it is a leading term approximation of the derivative. Generally, this limiter works well although there is no proof of TVDM/TVBM properties and there are cases where it destroys the formal accuracy of the scheme [120, 212]. A recent extension is discussed in [209].

5.7 Strong stability-preserving Runge-Kutta methods

A subtlety that may have escaped even the careful reader is the fact that all analysis and results on limiting were obtained under the assumption that a first-order forward Euler method was used to integrate in time. As one of the strengths of the DG formulation is the ability to compute with high-order accuracy, a requirement of first order in time is clearly not satisfactory. The question becomes whether one can design high-order temporal integration schemes that maintain the TVDM or TVBM property, provided this can be shown for the first-order forward Euler method.

Consider the semidiscrete scheme

$$\frac{d}{dt} u_h = \mathcal{L}_h(u_h, t),$$

and assume that we can establish the TV property using a forward Euler method; that is,

$$u_h^{n+1} = u_h^n + \Delta t \mathcal{L}_h(u_h^n, t^n), \quad |u_h^{n+1}|_{TV} \le |u_h^n|_{TV}.$$

Let us consider an explicit RK method with s stages of the form

$$\begin{cases} v^{(0)} = u_h^n \\ i = 1, \dots, s: \ v^{(i)} = \sum_{j=0}^{i-1} \alpha_{ij} v^{(j)} + \beta_{ij} \Delta t \mathcal{L}_h(v^{(j)}, t^n + \gamma_j \Delta t) \ . \\ u_h^{n+1} = v^{(s)} \end{cases} \quad (5.30)$$

Clearly, we must find $(\alpha_{ij}, \beta_{ij}, \gamma_j)$ such that the order conditions (see, e.g., [40, 143, 144]) are satisfied, and if additional degrees of freedom are available, we can attempt to optimize the scheme in some way. For consistency, we must have

$$\sum_{l=0}^{i-1} \alpha_{il} = 1.$$

A closer look at the form of the RK method in Eq. (5.30) reveals that if $(\alpha_{ij}, \beta_{ij})$ are all positive, the RK method is simply a convex combination of forward Euler steps since we can write the stages as

$$v^{(i)} = \sum_{j=0}^{i-1} \alpha_{ij} \left(v^{(j)} + \frac{\beta_{ij}}{\alpha_{ij}} \Delta t \mathcal{L}_h(v^{(j)}, t^n + \gamma_j \Delta t) \right).$$

If the scheme is TVD/TVB for Δt_E using an Euler method, we can directly rely on that result at high order also, provided we use a maximum timestep

$$\Delta t_{RK} \leq \min_{ij} \frac{\alpha_{ij}}{\beta_{ij}} \Delta t_E.$$

When optimizing the scheme, the objective should be to maximize the fraction in front of Δt_E to minimize the cost of the time integration.

Methods of this kind are known as strong stability-preserving Runge-Kutta (SSP-RK) or TVD-RK methods and can be used with advantage for problems with strong shocks and discontinuities, as they guarantee that no additional oscillations are introduced as part of the time-integration process.

For a second-order two-stage SSP-RK scheme, the optimal scheme is [139, 140]

$$v^{(1)} = u_h^n + \Delta t \mathcal{L}_h(u_h^n, t^n), \tag{5.31}$$
$$u_h^{n+1} = v^{(2)} = \frac{1}{2} \left(u_h^n + v^{(1)} + \Delta t \mathcal{L}_h(v^{(1)}, t^n + \Delta t) \right),$$

and the optimal third-order three-stage SSP-RK scheme is given as

$$v^{(1)} = u_h^n + \Delta t \mathcal{L}_h(u_h^n, t^n),$$
$$v^{(2)} = \frac{1}{4} \left(3u_h^n + v^{(1)} + \Delta t \mathcal{L}_h(v^{(1)}, t^n + \Delta t) \right), \tag{5.32}$$
$$u_h^{n+1} = v^{(3)} = \frac{1}{3} \left(u_h^n + 2v^{(2)} + 2\Delta t \mathcal{L}_h \left(v^{(2)}, t^n + \frac{1}{2} \Delta t \right) \right).$$

Both schemes are optimal in the sense that the maximum timestep is the same as that of the forward Euler method.

Unfortunately, one can show [139] that it is not possible to construct fourth-order four-stage SSP-RK schemes where all coefficients are positive. However, one can derive a fourth-order scheme by allowing a fifth stage [292]. The optimal scheme is given as

$$\begin{aligned} v^{(1)} =\ & u_h^n + 0.39175222700392 \Delta t \mathcal{L}_h(u_h^n, t^n), \\ v^{(2)} =\ & 0.44437049406734 u_h^n + 0.55562950593266 v^{(1)} \\ & + 0.36841059262959 \Delta t \mathcal{L}_h(v^{(1)}, t^n + 0.39175222700392 \Delta t), \\ v^{(3)} =\ & 0.62010185138540 u_h^n + 0.37989814861460 v^{(2)} \\ & + 0.25189177424738 \Delta t \mathcal{L}_h(v^{(2)}, t^n + 0.58607968896780 \Delta t), \\ v^{(4)} =\ & 0.17807995410773 u_h^n + 0.82192004589227 v^{(3)} \\ & + 0.54497475021237 \Delta t \mathcal{L}_h(v^{(3)}, t^n + 0.47454236302687 \Delta t), \end{aligned} \tag{5.33}$$

$$u_h^{n+1} = v^{(5)} = 0.00683325884039u_h^n + 0.51723167208978v^{(2)}$$
$$+ 0.12759831133288v^{(3)} + 0.34833675773694v^{(4)}$$
$$+ 0.08460416338212\Delta t \mathcal{L}_h(v^{(3)}, t^n + 0.47454236302687\Delta t)$$
$$+ 0.22600748319395\Delta t \mathcal{L}_h(v^{(4)}, t^n + 0.93501063100924\Delta t).$$

The additional work is partially offset by the maximum timestep being approximately 50% larger than for the forward Euler method. Other SSP-RK methods are known (e.g., low-storage forms and SSP multistep schemes). We refer to [140] for an overview of these methods and more references.

If a filter or limiter is used, it must be applied at each stage of the SSP-RK scheme; for example,

$$v^{(i)} = \Pi^p \left(\sum_{l=0}^{i-1} \alpha_{il} v^{(l)} + \beta_{il} \Delta t \mathcal{L}_h(v^{(l)}, t^n + \gamma_l \Delta t) \right).$$

To illustrate the problems that may arise by using a poorly chosen time-integration method, let us consider a well-known example, taken from [139].

Example 5.11. We solve Burgers' equation

$$\frac{\partial u}{\partial t} + \frac{\partial u^2}{\partial x} = \frac{\partial u}{\partial t} + \mathcal{L}(u) = 0, \quad x \in [-1, 1],$$

with the discontinuous initial condition

$$u_0(x) = u(x, 0) = \begin{cases} 2, & x \le -0.5 \\ 1, & x > -0.5 \end{cases}$$

and the exact solution

$$u(x, t) = u_0(x - 3t).$$

Burgers' equation is solved using a DG method on strong form with a MUSCL limiter and Lax-Friedrichs fluxes. We use a linear local basis and $K = 100$ elements.

As the first time-integration method, we consider the second-order, two-stage RK method of the form

$$v^{(1)} = u_h^n - 20\Delta \mathcal{L}_h(u_h^n),$$
$$u_h^{n+1} = u_h^n + \frac{\Delta t}{40} \left(41\mathcal{L}_h(u_h^n) - \mathcal{L}_h(v^{(1)}) \right).$$

This is perhaps not a standard second-order RK method, but it is certainly a valid one. As an alternative, we use the second-order SSP-RK method given in Eq. (5.31). Both methods are run at the same timestep although the latter is stable at a larger timestep.

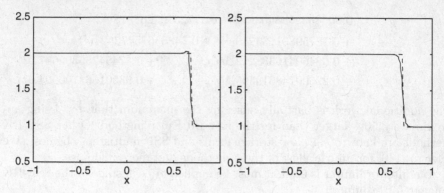

Fig. 5.13. Solution of Burgers' equation with a shock using limiting. We use $K = 100$ and $N = 1$ order elements and the MUSCL slope limiter. The results are shown at $T = 0.4$. On the left is shown the result computed using a generic second-order RK method and the figure on the right shows the nonoscillatory results obtained by the use of the SSP-RK2 method.

The results in Fig. 5.13 highlight the possibility of generating spurious oscillations in the solution solely by failing to use a suitable timestepping method.

5.8 A few general results

With all the pieces in place, one can establish a number of more general results regarding convergence for nonlinear scalar conservation laws with convex fluxes. For completeness, we will summarize a few of them here without proofs.

Theorem 5.12. *Assume that the limiter,* Π, *ensures the TVDM property; that is,*

$$v_h = \Pi(u_h) \quad \Rightarrow \quad |v_h|_{TV} \leq |u_h|_{TV},$$

and that the SSP-RK method is consistent.
    Then the DG-FEM with the SSP-RK solution is TVDM as

$$\forall n : \ |u_h^n|_{TV} \leq |u_h^0|_{TV}.$$

This is an almost immediate consequence of the results we have discussed above.
    Furthermore, we have similar results as follows:

Theorem 5.13. *Assume that the limiter,* Π, *ensures the TVBM property and that the SSP-RK method is consistent.*

Then the DG-FEM with the SSP-RK solution is TVBM as

$$\forall n : \quad |u_h^n|_{TV} \le |u_h^0|_{TV} + CM,$$

where the constant, C, depends only on the order of approximation, N, and M is the constant in the TVBM limiter, Eq. (5.29).

These results allow one to prove the fundamental convergence theorem [60]

Theorem 5.14. *Assume that the slope limiter, Π, ensures that u_h is TVDM or TVBM and that the SSP-RK method is consistent.*

Then there is a subsequence, $\{\bar{u}_h'\}$, of the sequence $\{\bar{u}_h\}$ generated by the scheme that converges in $L^\infty(0, T; L^1)$ to a weak solution of the scalar conservation law.

Moreover, if a TVBM limiter is used, the weak solution is the entropy solution and the whole sequence converges.

Finally, if the generalized slope limiter guarantees that

$$\|\bar{u}_h - \Pi\bar{u}_h\|_{L^1} \le Ch|\bar{u}_h|_{TV},$$

then the above results hold not only for the sequence of cell averages, $\{\bar{u}_h\}$, but also for the sequence of functions, $\{u_h\}$.

If one further makes assumptions of smoothness, the semidiscrete results discussed in Section 5.5 are extended to the fully discrete case in [337, 338], based on a second-order SSP-RK method, resulting in the expected additional error term of $\mathcal{O}(\Delta t^2)$.

5.9 The Euler equations of compressible gas dynamics

To conclude this chapter, let us consider a more elaborate example to see how all the pieces come together. We consider the one-dimensional equations of gas dynamics, known as the Euler equations. These are a set of three coupled nonlinear conservation laws given as

$$\frac{\partial \rho}{\partial t} + \frac{\partial \rho u}{\partial x} = 0,$$
$$\frac{\partial \rho u}{\partial t} + \frac{\partial (\rho u^2 + p)}{\partial x} = 0,$$
$$\frac{\partial E}{\partial t} + \frac{\partial (E + p)u}{\partial x} = 0,$$

where we have the conserved variables of density, ρ, momentum, ρu, and energy, E. The energy and the pressure are related through the ideal gas law as

$$p = (\gamma - 1)\left(E - \frac{1}{2}\rho u^2\right), \quad c = \sqrt{\frac{\gamma p}{\rho}},$$

and the local speed of sound c. In both cases, γ is a constant related to the type of fluid. We take it to be $\gamma = 7/5$, which is typical for atmospheric gases.

In EulerRHS1D.m, we show the implementation needed to evaluate the right-hand side of the Euler equations, using a Lax-Friedrichs flux.

──────────────── EulerRHS1D.m ────────────────

```
function [rhsrho, rhsrhou, rhsEner] = EulerRHS1D(rho, rhou ,Ener)

% function [rhsrho, rhsrhou, rhsEner] = EulerRHS1D(rho, rhou ,Ener)
% Purpose  : Evaluate RHS flux in 1D Euler

Globals1D;

% compute maximum velocity for LF flux
gamma = 1.4;
pres = (gamma-1.0)*(Ener - 0.5*(rhou).^2./rho);
cvel = sqrt(gamma*pres./rho); lm = abs(rhou./rho)+cvel;

% Compute fluxes
rhof = rhou; rhouf=rhou.^2./rho+pres; Enerf=(Ener+pres).*rhou./rho;

% Compute jumps at internal faces
drho  =zeros(Nfp*Nfaces,K);  drho(:)  =  rho(vmapM)- rho(vmapP);
drhou =zeros(Nfp*Nfaces,K);  drhou(:)  =  rhou(vmapM)- rhou(vmapP);
dEner =zeros(Nfp*Nfaces,K);  dEner(:)  = Ener(vmapM)- Ener(vmapP);
drhof =zeros(Nfp*Nfaces,K);  drhof(:)  =  rhof(vmapM)- rhof(vmapP);
drhouf=zeros(Nfp*Nfaces,K);  drhouf(:) =rhouf(vmapM)-rhouf(vmapP);
dEnerf=zeros(Nfp*Nfaces,K);  dEnerf(:) =Enerf(vmapM)-Enerf(vmapP);
LFc   =zeros(Nfp*Nfaces,K);  LFc(:)    =max(lm(vmapP),lm(vmapM));

% Compute fluxes at interfaces
drhof(:) = nx(:).*drhof(:)/2.0-LFc(:)/2.0.*drho(:);
drhouf(:)=nx(:).*drhouf(:)/2.0-LFc(:)/2.0.*drhou(:);
dEnerf(:)=nx(:).*dEnerf(:)/2.0-LFc(:)/2.0.*dEner(:);

% Boundary conditions for Sod's problem
rhoin   = 1.000;   rhouin  = 0.0;
pin     = 1.000;   Enerin  = pin/(gamma-1.0);
rhoout  = 0.125;   rhouout = 0.0;
pout    = 0.100;   Enerout = pout/(gamma-1.0);

% Set fluxes at inflow/outflow
rhofin =rhouin; rhoufin=rhouin.^2./rhoin+pin;
Enerfin=(pin/(gamma-1.0)+0.5*rhouin^2/rhoin+pin).*rhouin./rhoin;
lmI=lm(vmapI)/2; nxI=nx(mapI);
drho (mapI)=nxI*(rhof (vmapI)-rhofin )/2.0-lmI*(rho(vmapI) -rhoin);
```

```
drhou(mapI)=nxI*(rhouf(vmapI)-rhoufin)/2.0-lmI*(rhou(vmapI)-rhouin);
dEner(mapI)=nxI*(Enerf(vmapI)-Enerfin)/2.0-lmI*(Ener(vmapI)-Enerin);

rhofout=rhouout; rhoufout=rhouout.^2./rhoout+pout;
Enerfout=(pout/(gamma-1.0)+0.5*rhouout^2/rhoout+pout).*rhouout./rhoout;
lm0=lm(vmap0)/2; nx0=nx(map0);
drho (map0)=nx0*(rhof(vmap0) - rhofout)/2.0-lm0*(rho (vmap0)- rhoout);
drhou(map0)=nx0*(rhouf(vmap0)-rhoufout)/2.0-lm0*(rhou(vmap0)-rhouout);
dEner(map0)=nx0*(Enerf(vmap0)-Enerfout)/2.0-lm0*(Ener(vmap0)-Enerout);

% compute right hand sides of the PDE's
rhsrho  = -rx.*(Dr*rhof)  + LIFT*(Fscale.*drhof);
rhsrhou = -rx.*(Dr*rhouf) + LIFT*(Fscale.*drhouf);
rhsEner = -rx.*(Dr*Enerf) + LIFT*(Fscale.*dEnerf);
return
```

The temporal integration is done using a third order SSP-RK method. We use the slope limiter, Π^1, at each stage. This is implemented in Euler1D.m. The final driver routine, setting initial conditions and the order of the approximation, is given in EulerDriver1D.m.

──────────────────────── Euler1D.m ────────────────────────
```
function [rho,rhou,Ener] = Euler1D(rho, rhou, Ener, FinalTime)

% function [rho, rhou, Ener] = Euler1D(rho, rhou, Ener, FinalTime)
% Purpose  : Integrate 1D Euler equations until FinalTime starting with
%            initial conditions [rho, rhou, Ener]

Globals1D;

% Parameters
gamma = 1.4; CFL = 1.0; time = 0;

% Prepare for adaptive time stepping
mindx = min(x(2,:)-x(1,:));

% Limit initial solution
rho =SlopeLimitN(rho); rhou=SlopeLimitN(rhou); Ener=SlopeLimitN(Ener);

% outer time step loop
while(time<FinalTime)

  Temp = (Ener - 0.5*(rhou).^2./rho)./rho;
  cvel = sqrt(gamma*(gamma-1)*Temp);
  dt = CFL*min(min(mindx./(abs(rhou./rho)+cvel)));

  if(time+dt>FinalTime)
```

```
   dt = FinalTime-time;
 end

% 3rd order SSP Runge-Kutta

% SSP RK Stage 1.
[rhsrho,rhsrhou,rhsEner]  = EulerRHS1D(rho, rhou, Ener);
rho1  = rho  + dt*rhsrho;
rhou1 = rhou + dt*rhsrhou;
Ener1 = Ener + dt*rhsEner;

% Limit fields
rho1  = SlopeLimitN(rho1); rhou1 = SlopeLimitN(rhou1);
Ener1 = SlopeLimitN(Ener1);

% SSP RK Stage 2.
[rhsrho,rhsrhou,rhsEner]  = EulerRHS1D(rho1, rhou1, Ener1);
rho2  = (3*rho  + rho1  + dt*rhsrho )/4;
rhou2 = (3*rhou + rhou1 + dt*rhsrhou)/4;
Ener2 = (3*Ener + Ener1 + dt*rhsEner)/4;

% Limit fields
rho2  = SlopeLimitN(rho2); rhou2 = SlopeLimitN(rhou2);
Ener2 = SlopeLimitN(Ener2);

% SSP RK Stage 3.
[rhsrho,rhsrhou,rhsEner]  = EulerRHS1D(rho2, rhou2, Ener2);
rho  = (rho  + 2*rho2  + 2*dt*rhsrho )/3;
rhou = (rhou + 2*rhou2 + 2*dt*rhsrhou)/3;
Ener = (Ener + 2*Ener2 + 2*dt*rhsEner)/3;

% Limit solution
rho =SlopeLimitN(rho); rhou=SlopeLimitN(rhou);
Ener=SlopeLimitN(Ener);

% Increment time and adapt timestep
time = time+dt;
end
return
```

```
                          ──────┤ EulerDriver1D.m ├──────
% Driver script for solving the 1D Euler equations
Globals1D;

% Polynomial order used for approximation
N = 6;
```

```
% Generate simple mesh
[Nv, VX, K, EToV] = MeshGen1D(0.0, 1.0, 250);

% Initialize solver and construct grid and metric
StartUp1D;
gamma = 1.4;

% Set up initial conditions -- Sod's problem
MassMatrix = inv(V')/V;
cx = ones(Np,1)*sum(MassMatrix*x,1);

rho = ones(Np,K).*( (cx<0.5) + 0.125*(cx>=0.5));
rhou = zeros(Np,K);
Ener = ones(Np,K).*((cx<0.5) + 0.1*(cx>=0.5))/(gamma-1.0);
FinalTime = 0.2;

% Solve Problem
[rho,rhou,Ener] = Euler1D(rho,rhou,Ener,FinalTime);
```

We use this to solve a classic test problem, known as Sod's problem. The problem is set in $x \in [0, 1]$ with the initial conditions

$$
\rho(x,0) = \begin{cases} 1.0, & x < 0.5 \\ 0.125, & x \geq 0.5, \end{cases} \quad \rho u(x,0) = 0 \quad E(x,0) = \frac{1}{\gamma - 1} \begin{cases} 1, & x < 0.5 \\ 0.1, & x \geq 0.5. \end{cases}
$$

The problem has an exact solution by solving Riemann problems.

In Fig. 5.14 we show the computed results with $N = 1$ and $K = 250$ elements, compared with the exact solution at $T = 0.2$. We observe an excellent resolution of the shocks but also some smearing of the contact discontinuities. In Fig. 5.15 we illustrate the general convergence by using $K = 500$ elements, showing decreased smearing of contacts.

5.10 Exercises

1. Consider a smooth nonlinear problem, known as the shallow water system, as

$$
\frac{\partial h}{\partial t} + \frac{\partial hu}{\partial x} = 0,
$$
$$
\frac{\partial u}{\partial t} + G\frac{\partial h}{\partial x} + u\frac{\partial u}{\partial x} = 0,
$$

where $h(x,t)$ is the water height and $u(x,t)$ is the velocity. This model can be used to model water waves on shallow water.

It has an exact solution of the form

$$
h(x,t) = \xi^2, \quad u(x,t) = 2\sqrt{G}\xi - 2\sqrt{GH},
$$

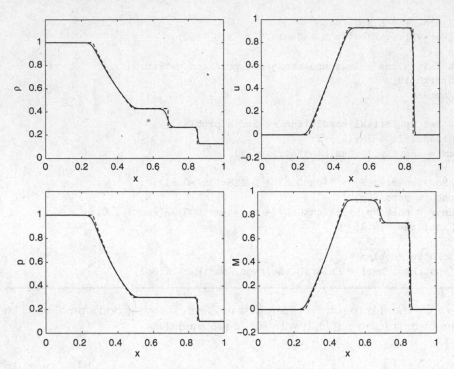

Fig. 5.14. Solution of Sod's shock tube problem at $T = 0.2$ with $K = 250$ linear elements and the MUSCL TVBM limiter. Shown is the computed density (ρ), the velocity, (u), the pressure, (p), and the local Mach number (M). The dashed lines represent the exact solution.

where

$$\xi(x,t) = \frac{x + 2\sqrt{GH}t}{1 + 3\sqrt{G}t}.$$

H is the steady-state water height and G is the constant of gravity.

a) Design and implement a DG-FEM method for this problem. Motivate the choice of numerical flux and general formulation of the scheme.

b) Determine the fastest wave velocity of the system. What role does that play in determining the stable timestep for an RK method?

c) Validate the accuracy of the code; that is, show that $\|\varepsilon\|_{\Omega,h} \leq Ch^s$. What is s and can you determine how C depends on time and N?

2. Consider the linear wave problem

$$\frac{\partial u}{\partial t} + a(x)\frac{\partial u}{\partial x} = 0, \quad x \in [-1, 1],$$

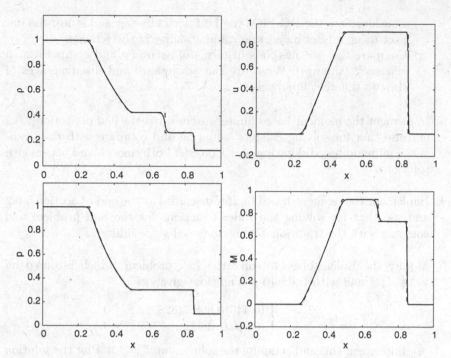

Fig. 5.15. Solution of Sod's shock tube problem at $T = 0.2$ with $K = 500$ linear elements and the MUSCL TVBM limiter. Shown is the computed density (ρ), the velocity, (u), the pressure, (p), and the local Mach number (M). The dashed lines represent the exact solution.

where

$$a(x) = \frac{1}{\pi}\sin(\pi x - 1).$$

If the initial condition is

$$u(x,0) = f(x) = \sin(x),$$

the exact solution to the linear problem is

$$u(x,t) = f\left(2\tan^{-1}\left[e^{-t}\tan\left(\frac{\pi x - 1}{2}\right)\right] + 1\right).$$

a) Confirm that $u(x,t)$ is an exact solution and that the solution asymptotes to a constant. Find the constant.

b) Formulate a DG-FEM method with exact integration for solving the problem and implement the scheme. Use the exact solution as the boundary condition at $x = -1$ and confirm the accuracy of the scheme. Is it as expected?

c) Consider now the DG-FEM method with aliasing and study the impact using a filter on accuracy and stability of the scheme.

d) Compare the two methods in terms of accuracy and computational efficiency (timings). What are the advantages and disadvantages of the two different approaches.

3. Implement the moment-based limiter discussed at the end of Section 5.6.2 and use that for solving Burgers' equation and compare with the traditional minmod based slope limiters. Consider both smooth and nonsmooth solutions.

4. Implement the moment-based limiter discussed at the end of Section 5.6.2 and use that for solving the Euler equations for the Sod problem and compare with the traditional minmod-based slope limiters.

5. Modify the Euler solver to solve the Lax problem, which is posed on $x \in [-5, 5]$ and with the initial conditions given as

$$(\rho, u, p) = \begin{cases} (0.445, 0.698, 3.528), & x \leq 0 \\ (0.5, 0.0, 0.571), & x > 0. \end{cases}$$

a) Implement this and compute the solution at $T = 1.3$. Plot the solution for different resolutions.

b) Compare the performance of the different limiters for this problem.

6. The Sod problem is not very well suited to display the advantages of high-order methods, as the solutions are essentially piecewise linear. Where high-order methods have a much better potential to shine is for solving problems combining shocks with waves. An example of this is the shock-density wave problem given below.

As is clear from the above, limiting is at the heart of the success of these methods, in particular for problems combining smooth regions with shocks.

a) Consider the shock-density wave problem posed on $x \in [-5, 5]$. The initial conditions are

$$(\rho, u, p) = \begin{cases} (3.857143, 2.629369, 10.333333), & x < -4 \\ (1 + 0.2\sin(5x), 0, 1), & x \geq 4. \end{cases}$$

Implement this and compute the solution at $T = 1.8$.

b) Compare the solutions obtained with Π^1, Π^N, and the TVB limiter for different values of M. A good starting point is $K = 300$ elements.

6

Beyond one dimension

We have so far focused almost entirely on the formulation, analysis, and implementation of the discontinuous Galerkin schemes (DG-FEM) for one-dimensional problems. In this chapter we consider in more detail the extension to multidimensional problems and the challenges introduced by this. We will quickly realize that what may have seemed unnecessarily complicated in the one-dimensional case now enables us to expand the formulation to multiple dimensions with only minor changes.

Although we focus our treatment on methods for solving problems in two dimensions with spatial discretizations based on triangles, it is worth emphasizing that any reasonable element type could be used (e.g., quadrilaterals, hybrid triangle-quadrilateral grids, or even overlapping grids).

As in previous chapters, we focus on solving conservation laws

$$
\begin{aligned}
&\frac{\partial u(\boldsymbol{x}, t)}{\partial t} + \nabla \cdot \boldsymbol{f}(u(\boldsymbol{x}, t), \boldsymbol{x}, t) = 0, \ \ \boldsymbol{x} \in \Omega \in \mathsf{R}^2, \\
&u(\boldsymbol{x}, t) = g(\boldsymbol{x}, t), \ \ \boldsymbol{x} \in \partial \Omega_i, \\
&u(\boldsymbol{x}, 0) = f(\boldsymbol{x}).
\end{aligned}
\tag{6.1}
$$

For simplicity and with minimal loss of generality, we restrict most of the discussion to the scalar case. We will later point out the minimal generalizations needed to solve the general system case. We assume that the correct number of boundary conditions are available at all inflow points of the boundary, $\partial \Omega_i$ (i.e., where the eigenvalues of the flux-Jacobian, $\hat{\boldsymbol{n}} \cdot \boldsymbol{f}_u$, are negative along the outward-pointing normal, $\hat{\boldsymbol{n}}$).

To secure geometric flexibility and pave the way for the DG-FEM formulation, we assume that Ω can be triangulated using K elements,

$$
\Omega \simeq \Omega_h = \bigcup_{k=1}^{K} \mathsf{D}^k,
$$

where D^k is a straight-sided triangle and the triangulation is assumed to be geometrically conforming; for example, $\partial \Omega$ is approximated by a piecewise linear polygon with each line segment being a face of a triangle. In Chapter 9 we will extend this further to also include curvilinear representations of surfaces.

Let us follow the one-dimensional approach and assume that we can approximate $u(\boldsymbol{x}, t)$ using

$$u(\boldsymbol{x}, t) \simeq u_h(\boldsymbol{x}, t) = \bigoplus_{k=1}^{K} u_h^k(\boldsymbol{x}, t) \in \mathsf{V}_h = \bigoplus_{k=1}^{K} \left\{ \psi_n(\mathsf{D}^k) \right\}_{n=1}^{N_p}.$$

Here $\psi_n(\mathsf{D}^k)$ is a two-dimensional polynomial basis defined on element D^k. The local function, $u_h^k(\boldsymbol{x}, t)$, can be expressed by

$$\boldsymbol{x} \in \mathsf{D}^k: \quad u_h^k(\boldsymbol{x}, t) = \sum_{i=1}^{N_p} u_h^k(\boldsymbol{x}_i^k, t) \ell_i^k(\boldsymbol{x}),$$

where $\ell_i(\boldsymbol{x})$ is the multidimensional Lagrange polynomial defined by some grid points, $\boldsymbol{x}_i$, on the element D^k.

Recalling the general discussion in Chapter 2 we require the residual to be orthogonal to all test functions, $\phi_h \in \mathsf{V}_h$, resulting in the local statements

$$\int_{\mathsf{D}^k} \left[\frac{\partial u_h^k}{\partial t} \ell_i^k(\boldsymbol{x}) - \boldsymbol{f}_h^k \cdot \nabla \ell_i^k(\boldsymbol{x}) \right] d\boldsymbol{x} = - \int_{\partial \mathsf{D}^k} \hat{\boldsymbol{n}} \cdot \boldsymbol{f}^* \ell_i^k(\boldsymbol{x}) \, d\boldsymbol{x},$$

and

$$\int_{\mathsf{D}^k} \left[\frac{\partial u_h^k}{\partial t} + \nabla \cdot \boldsymbol{f}_h^k \right] \ell_i^k(\boldsymbol{x}) \, d\boldsymbol{x} = \int_{\partial \mathsf{D}^k} \hat{\boldsymbol{n}} \cdot \left[\boldsymbol{f}_h^k - \boldsymbol{f}^* \right] \ell_i^k(\boldsymbol{x}) \, d\boldsymbol{x},$$

as the weak and strong form, respectively, of the nodal discontinuous Galerkin method in two spatial dimensions.

As we have already discussed, the missing piece is the specification of the numerical flux, $\boldsymbol{f}^*$. In most of the subsequent discussion, we primarily consider the local Lax-Friedrichs flux

$$\boldsymbol{f}^*(a, b) = \frac{\boldsymbol{f}(a) + \boldsymbol{f}(b)}{2} + \frac{C}{2} \hat{\boldsymbol{n}}(a - b),$$

where (a, b) are the interior and exterior solution value, respectively, C is the local maximum of the directional flux Jacobian; that is,

$$C = \max_{u \in [a, b]} \left| \hat{n}_x \frac{\partial f_1}{\partial u} + \hat{n}_y \frac{\partial f_2}{\partial u} \right|,$$

where $\boldsymbol{f} = (f_1, f_2)$. As for the one-dimensional case, many details remain to be discussed for special problems and we will return to these later. However, at this point, the generic framework is settled and we can begin to consider how to actually transform this into an efficient computational method.

6.1 Modes and nodes in two dimensions

By leaning on the experience gained for the one-dimensional approximation in Section 3.1, we continue to develop the tools needed for the polynomial interpolation on triangles.

We assume that the local solution is expressed as

$$x \in \mathsf{D}^k : \quad u_h^k(x,t) = \sum_{i=1}^{N_p} u_h^k(x_i,t)\ell_i^k(x) = \sum_{n=1}^{N_p} \hat{u}_n^k(t)\psi_n(x).$$

Here, $\ell_i^k(x)$ is the multidimensional Lagrange polynomial based on the grid points, x_i, and $\{\psi_n(x)\}_{n=1}^{N_p}$ is a genuine two-dimensional polynomial basis of order N.

In contrast to the one-dimensional case, we first realize that N_p does not represent the order, N, of the polynomial, u_h^k, but, rather, the number of terms in the local expansion. These two are related by

$$N_p = \frac{(N+1)(N+2)}{2},$$

for a polynomial of order N in two variables.

As sketched in Fig. 6.1, we introduce a mapping, Ψ, connecting the general straight-sided triangle, $x \in \mathsf{D}^k$, with the standard triangle, defined as

$$\mathsf{I} = \{r = (r,s)|(r,s) \geq -1; r+s \leq 0\}.$$

To connect the two triangles, assume that D^k is spanned by the three vertices, (v^1, v^2, v^3), counted counterclockwise. We define the barycentric coordinates, $(\lambda^1, \lambda^2, \lambda^3)$, with the properties that

$$0 \leq \lambda^i \leq 1, \quad \lambda^1 + \lambda^2 + \lambda^3 = 1. \tag{6.2}$$

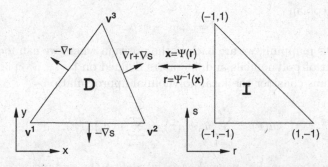

Fig. 6.1. Notation for the mapping between two simplices.

Any point in the triangle, spanned by the three vertices, can now be expressed as

$$\boldsymbol{x} = \lambda^2 \boldsymbol{v}^1 + \lambda^3 \boldsymbol{v}^2 + \lambda^1 \boldsymbol{v}^3.$$

In a similar fashion, we can express points in I as

$$\begin{pmatrix} r \\ s \end{pmatrix} = \lambda^2 \begin{pmatrix} -1 \\ -1 \end{pmatrix} + \lambda^3 \begin{pmatrix} 1 \\ -1 \end{pmatrix} + \lambda^1 \begin{pmatrix} -1 \\ 1 \end{pmatrix}.$$

Combining this with Eq. (6.2), we recover

$$\lambda^1 = \frac{s+1}{2}, \quad \lambda^2 = -\frac{r+s}{2}, \quad \lambda^3 = \frac{r+1}{2},$$

and, thus, the direct mapping

$$\boldsymbol{x} = -\frac{r+s}{2} \boldsymbol{v}^1 + \frac{r+1}{2} \boldsymbol{v}^2 + \frac{s+1}{2} \boldsymbol{v}^3 = \Psi(\boldsymbol{r}). \tag{6.3}$$

It is important to observe that the mapping is linear in $\boldsymbol{r}$. This has the consequence that any two straight-sided triangles are connected through an affine mapping; that is, it has a constant transformation Jacobian. The metric for the mapping can be found directly since

$$\frac{\partial \boldsymbol{x}}{\partial \boldsymbol{r}} \frac{\partial \boldsymbol{r}}{\partial \boldsymbol{x}} = \begin{bmatrix} x_r & x_s \\ y_r & y_s \end{bmatrix} \begin{bmatrix} r_x & r_y \\ s_x & s_y \end{bmatrix} = \begin{bmatrix} 1 & 0 \\ 0 & 1 \end{bmatrix}.$$

Here, we have used the standard notation of a_b to mean a differentiated with respect to b. From Eq. (6.3) we have

$$(x_r, y_r) = \boldsymbol{x}_r = \frac{\boldsymbol{v}^2 - \boldsymbol{v}^1}{2}, \quad (x_s, y_s) = \boldsymbol{x}_s = \frac{\boldsymbol{v}^3 - \boldsymbol{v}^1}{2},$$

yielding

$$r_x = \frac{y_s}{J}, \quad r_y = -\frac{x_s}{J}, s_x = -\frac{y_r}{J}, \quad s_y = \frac{x_r}{J}, \tag{6.4}$$

with the Jacobian

$$J = x_r y_s - x_s y_r. \tag{6.5}$$

Through this mapping, we are back in the position where we can focus on the development of polynomials and operators defined on I.

Let us thus consider the local polynomial approximation

$$u(\boldsymbol{r}) = \sum_{n=1}^{N_p} \hat{u}_n \psi_n(\boldsymbol{r}) = \sum_{i=1}^{N_p} u(\boldsymbol{r}_i)\ell_i(\boldsymbol{r}),$$

where, as in the one-dimensional case, we define the expansion coefficients, $\hat{u}_n$, through an interpolation to avoid problems with having to evaluate multidimensional integrals.

This yields the expression

$$\mathcal{V}\hat{\boldsymbol{u}} = \boldsymbol{u},$$

where, as in the one-dimensional case, $\hat{\boldsymbol{u}} = [\hat{u}_1, \ldots, \hat{u}_{N_p}]^T$ are the N_p expansion coefficients and $\boldsymbol{u} = [u(\boldsymbol{r}_1), \ldots, u(\boldsymbol{r}_{N_p})]^T$ represents the N_p grid point values.

To ensure stable numerical behavior of the generalized Vandermonde matrix $\mathcal{V}$ with the entries

$$\mathcal{V}_{ij} = \psi_j(\boldsymbol{r}_i),$$

we need to address two issues:

- Identify an orthonormal polynomial basis, $\psi_j(\boldsymbol{r})$, defined on the triangle I.
- Identify families of points that lead to good behavior of the interpolating polynomial defined on I. Such points can be viewed as a multidimensional generalization of the Legendre-Gauss-Lobatto points although they do not need to have a quadrature formula associated with them.

The first issue is, at least in principle, easy to resolve. Consider the canonical basis

$$\psi_m(\boldsymbol{r}) = r^i s^j, \quad (i, j) \geq 0; \quad i + j \leq N,$$

$$m = j + (N+1)i + 1 - \frac{i}{2}(i-1), \quad (i, j) \geq 0; \quad i + j \leq N,$$

which spans the space of N-dimensional polynomials in two variables, (r, s). Based on the one-dimensional discussion in Section 3.1 we do not expect this to be a good choice. However, it is a complete polynomial basis and it can be orthonormalized through a Gram-Schmidt process. The resulting basis is

$$\psi_m(\boldsymbol{r}) = \sqrt{2} P_i(a) P_j^{(2i+1,0)}(b)(1-b)^i, \tag{6.6}$$

where

$$a = 2\frac{1+r}{1-s} - 1, \quad b = s,$$

and $P_n^{(\alpha,\beta)}(x)$ is the n-th order Jacobi polynomial. Recall that $\alpha = \beta = 0$ is the Legendre polynomial, and the more general polynomials are discussed in Appendix A.

A library routine, Simplex2DP.m, to evaluate this more complicated basis is likewise introduced in Appendix A. Evaluating the basis is achieved as

>> P = Simplex2DP(a,b,i,j)

with the meaning of (a, b) and (i, j) as above. The mapping of (a, b) from (r, s) is done using rstoab.m.

To get an idea of how this basis behaves, we show in Fig. 6.2 the behavior of the basis for a few choices of (i, j). As expected, the basis resembles Legendre polynomials along the edges and is also, naturally, oscillatory inside the triangle.

Fig. 6.2. Behavior of the first few orthonormal basis functions defined on I. Each row corresponds to all basis functions of the same order.

```
┌─────────┐
│ rstoab.m │
└─────────┘
function [a,b] = rstoab(r,s)

% function [a,b] = rstoab(r,s)
% Purpose : Transfer from (r,s) -> (a,b) coordinates in triangle

Np = length(r); a = zeros(Np,1);
for n=1:Np
  if(s(n) ~= 1)
    a(n) = 2*(1+r(n))/(1-s(n))-1;
  else a(n) = -1; end
end
b = s;
return;
```

With the identification of an orthonormal polynomial basis, we are left with the task of having to identify N_p points on I, leading to well-behaved interpolations. As we saw for the one-dimensional case, this choice is important since a poorly chosen set can result in computational problems due to ill-conditioning. In Section 3.1 this led us to the identification of the Legendre-Gauss-Lobatto quadrature points as a suitable choice.

However, we cannot directly use these nodes on the triangle, as a tensor product in (a, b) would lead to $(N + 1)^2$ points, asymmetrically distributed with a clustering at one vertex, resulting in severely ill-conditioned operators. In other words, we have to be more sophisticated.

One can approach this challenge of finding exactly N_p points for interpolation on the triangle in several different ways [55, 151, 300], leading to nodal distributions with very similar behavior.

A slight disadvantage of most known nodal sets is that their computation requires substantial initial effort, after which they are tabulated. In the following, we develop a simple and constructive approach for the computation of a well-behaved family of nodal points of any order.

We begin by seeking inspiration from the one dimensional case. Recall both the equidistant grid

$$r_i^e = -1 + \frac{2i}{N}, \quad i \in [0, \ldots, N],$$

and the Legendre-Gauss-Lobatto grid, termed r_i^{LGL}, from Section 3.1. These two nodal sets can formally be connected as

$$w(r) = \sum_{i=1}^{N_p} (r_i^{LGL} - r_i^e) \ell_i^e(r), \tag{6.7}$$

where $\ell_i^e(r)$ are the Lagrange polynomials based on r_i^e. Thus, $w(r)$ is an N-th-order polynomial approximation to a function that maps the equidistant points to the Legendre-Gauss-Lobatto points; that is, it serves the purpose of mapping "bad points" to "good points". We show this function in Fig. 6.3 and observe in particular that there is limited variation of this function with N.

To utilize this, consider the symmetric equilateral triangle shown in Fig. 6.4. As in the one-dimensional case, we define an equidistant grid on this triangle using the barycentric coordinates

$$(i, j) \geq 0, \, i + j \leq N : \quad (\lambda^1, \lambda^3) = \left(\frac{i}{N}, \frac{j}{N} \right), \quad \lambda^2 = 1 - \lambda^1 - \lambda^3.$$

Using these nodes will lead to ill-conditioned operators at high order as we have already discussed. However, we can now utilize the one-dimensional mapping function $w(r)$ to blend [134] the edge mapping into the triangle, thus mapping the two-dimensional grid in a fashion similar to the successful one-dimensional mapping, Eq. (6.7). To keep things simple, we express everything in the barycentric coordinates and blend along the normal to the edge. Thus, for the first edge (connecting vertices v^1 and v^2), we define the normal warping function:

$$\boldsymbol{w}^1(\lambda^1, \lambda^2, \lambda^3) = w(\lambda^3 - \lambda^2) \begin{pmatrix} 1 \\ 0 \end{pmatrix}$$

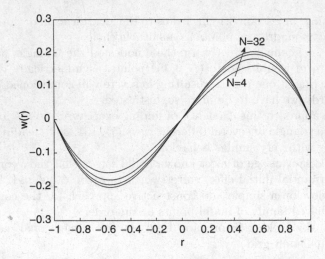

Fig. 6.3. The mapping function, $w(r)$, for different orders of approximation, $N = 4, 8, 16, 32$.

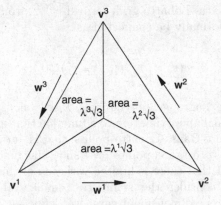

Fig. 6.4. Equilateral triangle with vertex coordinates, $(\boldsymbol{v}^1, \boldsymbol{v}^2, \boldsymbol{v}^3)$, and barycentric coordinates, $(\lambda^1, \lambda^2, \lambda^3)$.

and

$$b^1(\lambda^1, \lambda^2, \lambda^3) = \left(\frac{2\lambda^3}{2\lambda^3 + \lambda^1} \right) \left(\frac{2\lambda^2}{2\lambda^2 + \lambda^1} \right).$$

Note that the apparent singularity for $\lambda^3 = \lambda^1 = 0$ ($\boldsymbol{x} = \boldsymbol{v}^1$) and $\lambda^2 = \lambda^1 = 0$ ($\boldsymbol{x} = \boldsymbol{v}^2$) coincide exactly with points where $w(r) = 0$. Redefining $w(r)$ to account for this (see Warpfactor.m) yields

$$\tilde{w}(r) = \frac{w(r)}{1 - r^2},$$

to recover

$$\boldsymbol{w}^1(\lambda^1, \lambda^2, \lambda^3) = \tilde{w}(\lambda^3 - \lambda^2) \begin{pmatrix} 1 \\ 0 \end{pmatrix},$$

and

$$b^1(\lambda^1, \lambda^2, \lambda^3) = 4\lambda^3\lambda^2.$$

For the two other edges we likewise define the normal mapping functions (see Fig. 6.4)

$$\boldsymbol{w}^2(\lambda^1, \lambda^2, \lambda^3) = \tilde{w}(\lambda^1 - \lambda^3)\frac{1}{2}\begin{pmatrix} -1 \\ \sqrt{3} \end{pmatrix},$$

$$\boldsymbol{w}^3(\lambda^1, \lambda^2, \lambda^3) = \tilde{w}(\lambda^2 - \lambda^1)\frac{1}{2}\begin{pmatrix} -1 \\ -\sqrt{3} \end{pmatrix},$$

and the blending functions

$$b^2(\lambda^1, \lambda^2, \lambda^3) = 4\lambda^3\lambda^1, \quad b^3(\lambda^1, \lambda^2, \lambda^3) = 4\lambda^2\lambda^1.$$

Warpfactor.m

```
function warp = Warpfactor(N, rout)

% function warp = Warpfactor(N, rout)
% Purpose  : Compute scaled warp function at order N based on
%            rout interpolation nodes

% Compute LGL and equidistant node distribution
LGLr = JacobiGL(0,0,N); req  = linspace(-1,1,N+1)';

% Compute V based on req
Veq = Vandermonde1D(N,req);

% Evaluate Lagrange polynomial at rout
Nr = length(rout); Pmat = zeros(N+1,Nr);
for i=1:N+1
  Pmat(i,:) = JacobiP(rout, 0, 0, i-1)';
end;
Lmat = Veq'\Pmat;

% Compute warp factor
warp = Lmat'*(LGLr - req);

% Scale factor
zerof = (abs(rout)<1.0-1.0e-10); sf = 1.0 - (zerof.*rout).^2;
warp = warp./sf + warp.*(zerof-1);
return;
```

We can form the two-dimensional version of $w(x)$,

$$w(\lambda^1, \lambda^2, \lambda^3) = b^1 \boldsymbol{w}^1 + b^2 \boldsymbol{w}^2 + b^3 \boldsymbol{w}^3,$$

as the transformation of the equidistant grid into a grid that is better suited for interpolation. The process is illustrated in Fig. 6.5, where we show the three individual mappings, $b^i \boldsymbol{w}^i$, as well as the action of the full mapping. This highlights how the equidistant nodes are pushed toward the vertices to improve on the interpolation properties, in a way mimicking the successful one-dimensional case.

Although one can use this construction directly, there are some benefits in considering the slightly generalized warping function

$$\boldsymbol{w}(\lambda^1, \lambda^2, \lambda^3) = \left(1 + \left(\alpha\lambda^1\right)^2\right)b^1\boldsymbol{w}^1 + \left(1 + \left(\alpha\lambda^2\right)^2\right)b^2\boldsymbol{w}^2 + \left(1 + \left(\alpha\lambda^3\right)^2\right)b^3\boldsymbol{w}^3.$$

In this case, α can be used to optimize the distribution of the nodes. As discussed in Section 3.1 a suitable measure of the quality of the interpolant is the Lebesque constant defined as

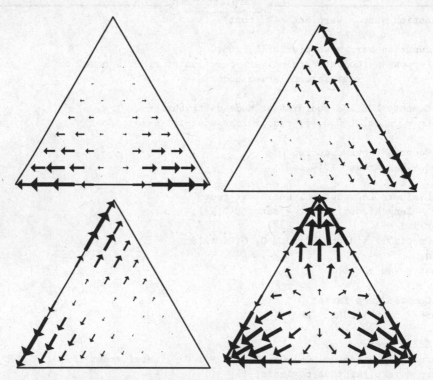

Fig. 6.5. Mapping and blending for the three edges as well as as for the combined equilateral triangle. The arrows represent the amount by which the nodes in an eighth-order triangle is moved from the equidistant nodal set.

Table 6.1. Lebesque constants for the α-optimized grids, for the $\alpha = 0$ grid, and for the equidistant grid defined on the equilateral triangle.

N	α	Λ Optimized	Λ $\alpha = 0$	Λ Equidistant
1	0.0000	1.00	1.00	1.00
2	0.0000	1.67	1.67	1.67
3	1.4152	2.11	2.11	2.27
4	0.1001	2.66	2.66	3.47
5	0.2751	3.12	3.14	5.45
6	0.9808	3.70	3.82	8.75
7	1.0999	4.27	4.55	14.35
8	1.2832	4.96	5.69	24.01
9	1.3648	5.74	7.02	40.92
10	1.4773	6.67	9.16	70.89
11	1.4959	7.90	11.83	124.53
12	1.5743	9.36	16.06	221.41
13	1.5770	11.47	21.17	397.70
14	1.6223	13.97	30.33	720.70
15	1.6258	17.65	42.48	1315.9

$$\Lambda = \max_{\boldsymbol{x}} \sum_{i=1}^{N_p} |\ell_i(\boldsymbol{x})|,$$

and it is natural to choose α to minimize this. The results of this is shown in Table 6.1, comparing also with the Lebesque constants for the equidistant grid. We see that the Lebesque constant is indeed the smallest for the optimized grid. Even the results for the grid with $\alpha = 0$ are likewise dramatically better than the equidistant grid, in particular at high order. In Table 6.1 is also given the optimal values of α up to a 15th order basis, which has a total of 136 nodes on a single triangle.

In Nodes2D.m, the α-optimized nodes are computed following the approach discussed above. If one requires elements of order higher than 15, one can either guess a value or take $\alpha = 0$, which, as illustrated in Table 6.1, still leads to nodal sets with a reasonable Lagrange interpolant. A few examples of the resulting nodal sets are shown in Fig. 6.6, highlighting the symmetric nature of the nodal sets.

—————————————— Nodes2D.m ——————————————

```
function [x,y] = Nodes2D(N);

% function [x,y] = Nodes2D(N);
% Purpose  : Compute (x,y) nodes in equilateral triangle for
%            polynomial of order N

alpopt = [0.0000 0.0000 1.4152 0.1001 0.2751 0.9800 1.0999 ...
          1.2832 1.3648 1.4773 1.4959 1.5743 1.5770 1.6223 1.6258];
```

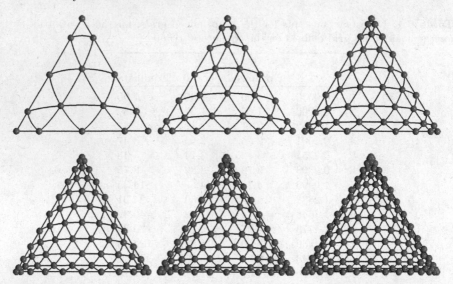

Fig. 6.6. Examples of α-optimized nodal sets on the equilateral triangle. Top row is for orders 4, 6, and 8 and the bottom row shows orders 10, 12, and 14.

```
% Set optimized parameter, alpha, depending on order N
if (N<16)
   alpha = alpopt(N);
else
   alpha = 5/3;
end;

% total number of nodes
Np = (N+1)*(N+2)/2;

% Create equidistributed nodes on equilateral triangle
L1 = zeros(Np,1); L2 = zeros(Np,1); L3 = zeros(Np,1);
sk = 1;
for n=1:N+1
  for m=1:N+2-n
    L1(sk) = (n-1)/N; L3(sk) = (m-1)/N;
    sk = sk+1;
  end
end
L2 = 1.0-L1-L3;
x = -L2+L3; y = (-L2-L3+2*L1)/sqrt(3.0);

% Compute blending function at each node for each edge
blend1 = 4*L2.*L3; blend2 = 4*L1.*L3; blend3 = 4*L1.*L2;
```

```
% Amount of warp for each node, for each edge
warpf1 = Warpfactor(N,L3-L2); warpf2 = Warpfactor(N,L1-L3);
               warpf3 = Warpfactor(N,L2-L1);

% Combine blend & warp
warp1 = blend1.*warpf1.*(1 + (alpha*L1).^2);
warp2 = blend2.*warpf2.*(1 + (alpha*L2).^2);
warp3 = blend3.*warpf3.*(1 + (alpha*L3).^2);

% Accumulate deformations associated with each edge
x = x + 1*warp1 + cos(2*pi/3)*warp2 + cos(4*pi/3)*warp3;
y = y + 0*warp1 + sin(2*pi/3)*warp2 + sin(4*pi/3)*warp3;
return;
```

One last issue to attend to is the fact that the nodes are computed in the equilateral triangle, Fig. 6.4, while the orthonormal basis lives in I. Thus, we need to map the computed nodes into I using an affine mapping as discussed previously. This is done by xytors.m.

───────────────────────── | xytors.m | ─────────────────────────

```
function [r,s] = xytors(x,y)

% function [r,s] = xytors(x, y)
% Purpose : From (x,y) in equilateral triangle to (r,s) coordinates
%           in standard triangle

L1 = (sqrt(3.0)*y+1.0)/3.0;
L2 = (-3.0*x - sqrt(3.0)*y + 2.0)/6.0;
L3 = ( 3.0*x - sqrt(3.0)*y + 2.0)/6.0;

r = -L2 + L3 - L1; s = -L2 - L3 + L1;
return;
```

This now completes the required developments; that is, we have identified an orthonormal basis and a way to construct nodal sets, which are good for interpolation on the triangle. This allows for the construction of a well-behaved Vandermonde matrix for operations on the triangle and extends to general triangles by an affine mapping. In a way very similar to the one-dimensional case, we can construct local approximations as

$$u(\boldsymbol{r}) \simeq u_h(\boldsymbol{r}) = \sum_{n=1}^{N_p} \hat{u}_n \psi_n(\boldsymbol{r}) = \sum_{i=1}^{N_p} u(\boldsymbol{r}_i)\ell_i(\boldsymbol{r}),$$

where $\boldsymbol{r}_i$ is a two-dimensional nodal set and $\psi_n(\boldsymbol{r})$ is the orthonormal two-dimensional polynomial basis. We recall that the number of terms in the expansions is

$$N_p = \frac{(N+1)(N+2)}{2},$$

for an N-th-order polynomial in two variables.

A central result of all this work is the stable construction of the Vandermonde matrix, $\mathcal{V}$, which establishes the connections

$$\boldsymbol{u} = \mathcal{V}\hat{\boldsymbol{u}}, \quad \mathcal{V}^T \boldsymbol{l}(\boldsymbol{r}) = \tilde{\boldsymbol{\psi}}(\boldsymbol{r}), \quad \mathcal{V}_{ij} = \psi_j(\boldsymbol{r}_i).$$

A script for initializing $\mathcal{V}$ is given in Vandermonde2D.m

With $\mathcal{V}$, we can transform directly between the modal representation, using $\hat{u}_n$ as the unknowns, and the nodal form, using $u(\boldsymbol{r}_i)$. Furthermore, this offers a way of evaluating the genuinely two-dimensional Lagrange polynomials, $\ell_i(\boldsymbol{x})$, for which no explicit expression is known. An example of how these may look is shown in Fig. 6.7 for the order $N = 4$ with a total of 15 basis functions.

Fig. 6.7. Example of multidimensional Lagrange polynomials of fourth-order based on the nodes in Fig. 6.6.

—————————————————— | Vandermonde2D.m | ——————————————————

```
function [V2D] = Vandermonde2D(N, r, s);

% function [V2D] = Vandermonde2D(N, r, s);
% Purpose : Initialize the 2D Vandermonde Matrix,
%              V_{ij} = phi_j(r_i, s_i);

V2D = zeros(length(r),(N+1)*(N+2)/2);

% Transfer to (a,b) coordinates
[a, b] = rstoab(r, s);

% build the Vandermonde matrix
sk = 1;
for i=0:N
  for j=0:N - i
    V2D(:,sk) = Simplex2DP(a,b,i,j);
    sk = sk+1;
  end
end
return;
```

6.2 Elementwise operations

With the local approximation in place, we continue the development of the computational components. Following the one-dimensional approach, we need the mass matrix

$$\mathcal{M}_{ij}^k = \int_{\mathsf{D}^k} \ell_i^k(\boldsymbol{x})\ell_j^k(\boldsymbol{x}) \, d\boldsymbol{x} = J^k \int_{\mathsf{I}} \ell_i(\boldsymbol{r})\ell_j(\boldsymbol{r}) \, d\boldsymbol{r},$$

where we have utilized that the transformation Jacobian, J^k, is a positive constant, provided D^k is a straightsided triangle (cf Eq. (6.5)).

Since $\mathcal{V}$ is constructed using an orthonormal basis, we recover, exactly as in the one-dimensional case, that

$$\mathcal{M}^k = J^k (\mathcal{V}\mathcal{V}^T)^{-1}.$$

Let us now consider the evaluation of the stiffness matrices or, to follow the approach in Section 3.2, the differentiation matrices. Using the chain rule, we have

$$\frac{\partial}{\partial x} = \frac{\partial r}{\partial x}\mathcal{D}_r + \frac{\partial s}{\partial x}\mathcal{D}_s, \quad \frac{\partial}{\partial y} = \frac{\partial r}{\partial y}\mathcal{D}_r + \frac{\partial s}{\partial y}\mathcal{D}_s,$$

and we obtain the metric constants from Eq. (6.4).

To define the differentiation matrices, $\mathcal{D}_r$ and $\mathcal{D}_s$, we need

$$\mathcal{V}_{r,(i,j)} = \left.\frac{\partial\psi_j}{\partial r}\right|_{r_i} , \quad \mathcal{V}_{s,(i,j)} = \left.\frac{\partial\psi_j}{\partial s}\right|_{r_i} .$$

If we recall the orthonormal basis

$$m \in [1,\dots,N_p]: \quad \psi_m(\boldsymbol{r}) = \sqrt{2}P_i(a)P_j^{(2i+1,0)}(b)(1-b)^i,$$

where

$$a = 2\frac{1+r}{1-s} - 1, \quad b = s,$$

we directly recover

$$\frac{\partial\psi_j}{\partial r} = \frac{\partial a}{\partial r}\frac{\partial\psi_j}{\partial a}, \quad \frac{\partial\psi_j}{\partial s} = \frac{\partial a}{\partial s}\frac{\partial\psi_j}{\partial a} + \frac{\partial\psi_j}{\partial b}.$$

This is implemented in GradSimplex2DP.m to enable the initialization of $\mathcal{V}_r$ and $\mathcal{V}_s$, as shown in GradVandermonde2D.m.

GradSimplex2DP.m

```
function [dmodedr, dmodeds] = GradSimplex2DP(a,b,id,jd)

% function [dmodedr, dmodeds] = GradSimplex2DP(a,b,id,jd)
% Purpose: Return the derivatives of the modal basis (id,jd)
%          on the 2D simplex at (a,b).

fa = JacobiP(a, 0, 0, id);      dfa = GradJacobiP(a, 0, 0, id);
gb = JacobiP(b, 2*id+1,0, jd); dgb = GradJacobiP(b, 2*id+1,0, jd);

% r-derivative
% d/dr = da/dr d/da + db/dr d/db = (2/(1-s)) d/da = (2/(1-b)) d/da
dmodedr = dfa.*gb;
if(id>0)
  dmodedr = dmodedr.*((0.5*(1-b)).^(id-1));
end

% s-derivative
% d/ds = ((1+a)/2)/((1-b)/2) d/da + d/db
dmodeds = dfa.*(gb.*(0.5*(1+a)));
if(id>0)
  dmodeds = dmodeds.*((0.5*(1-b)).^(id-1));
end
```

```
tmp = dgb.*((0.5*(1-b)).^id);
if(id>0)
  tmp = tmp-0.5*id*gb.*((0.5*(1-b)).^(id-1));
end
dmodeds = dmodeds+fa.*tmp;

% Normalize
dmodedr = 2^(id+0.5)*dmodedr; dmodeds = 2^(id+0.5)*dmodeds;
return;
```

| GradVandermonde2D.m |

```
function [V2Dr,V2Ds] = GradVandermonde2D(N,r,s)

% function [V2Dr,V2Ds] = GradVandermonde2D(N,r,s)
% Purpose : Initialize the gradient of the modal basis (i,j)
%           at (r,s) at order N

V2Dr = zeros(length(r),(N+1)*(N+2)/2); V2Ds = zeros(length(r),
(N+1)*(N+2)/2);

% find tensor-product coordinates
[a,b] = rstoab(r,s);

% Initialize matrices
sk = 1;
for i=0:N
  for j=0:N-i
    [V2Dr(:,sk),V2Ds(:,sk)] = GradSimplex2DP(a,b,i,j);
    sk = sk+1;
  end
end
return;
```

The differentiation matrices follow directly from this,

$$\mathcal{D}_r \mathcal{V} = \mathcal{V}_r, \quad \mathcal{D}_s \mathcal{V} = \mathcal{V}_s,$$

and the corresponding stiffness matrices are

$$\mathcal{S}_r = \mathcal{M}^{-1}\mathcal{D}_r, \quad \mathcal{S}_s = \mathcal{M}^{-1}\mathcal{D}_s.$$

The differentiation matrices are initialized in Dmatrices2D.m.

—————————————————— | Dmatrices2D.m | ——————————————————

```
function [Dr,Ds] = Dmatrices2D(N,r,s,V)

% function [Dr,Ds] = Dmatrices2D(N,r,s,V)
% Purpose : Initialize the (r,s) differentiation matrices
%           on the simplex, evaluated at (r,s) at order N

[Vr, Vs] = GradVandermonde2D(N, r, s);
Dr = Vr/V; Ds = Vs/V;
return;
```

To mimic the one-dimensional formulation of the local elementwise operators, we also define the two-dimensional filter operator. Recall the one-dimensional case discussed in detail in Section 5.3. We extend this directly to the two-dimensional case through

$$\mathcal{F} = \mathcal{V}\Lambda\mathcal{V}^{-1},$$

where the diagonal spectral filter matrix is

$$\Lambda_{mm} = \sigma\left(\frac{i+j}{N}\right), \quad m = j + (N+1)i + 1 - \frac{i}{2}(i-1), \ (i,j) \geq 0; \ i+j \leq N,$$

where $(i,j) \geq 0, i+j \leq N$ represent the order of the basis. As in the one-dimensional case, it is reasonable to use an exponential filter although others are possible [42, 159, 162]. The setup of the filter matrix is illustrated in Filter2D.m.

—————————————————— | Filter2D.m | ——————————————————

```
function [F] = Filter2D(Norder,Nc,sp)

% function [F] = Filter2D(Norder,sp)
% Purpose : Initialize 2D filter matrix of order sp and cutoff Nc

Globals2D;

filterdiag = ones((Norder+1)*(Norder+2)/2,1);
alpha = -log(eps);

% build exponential filter
sk = 1;
for i=0:Norder
  for j=0:Norder-i
    if (i+j>=Nc)
      filterdiag(sk) = exp(-alpha*((i+j - Nc)/(Norder-Nc))^sp);
    end
    sk = sk+1;
  end
end
```

```
end

F = V*diag(filterdiag)*invV;
return;
```

Until now, the development of the local operators has followed the one-dimensional approach in Section 3.2 closely. There is, however, one particular place in the formulation where the multidimensional case becomes slightly more complex.

The evaluation of the right-hand side of the DG-FEM requires one to compute a surface integral like

$$\int_{\partial D^k} \hat{\boldsymbol{n}} \cdot \boldsymbol{g}_h \ell_i^k(\boldsymbol{x}) \, d\boldsymbol{x},$$

where $\boldsymbol{g}_h$ is a polynomial trace function, composed of the numerical flux or the jump in flux, depending on whether we use the weak or the strong form.

In the one-dimensional case, this operation is particularly straightforward, as it only involves the end points of the element. To extend this to the triangle, let us first split the integral into the three individual edge components, each of the type

$$\int_{\text{edge}} \hat{\boldsymbol{n}} \cdot \boldsymbol{g}_h \ell_i^k(\boldsymbol{x}) \, d\boldsymbol{x} = \sum_{j=1}^{N+1} \hat{\boldsymbol{n}} \cdot \boldsymbol{g}_j \int_{\text{edge}} \ell_j^k(\boldsymbol{x}) \ell_i^k(\boldsymbol{x}) \, d\boldsymbol{x}.$$

Here, $\boldsymbol{x}$ is assumed to be the trace along the edge where there are exactly $N+1$ nodal points. Recall that as all triangles are assumed to be straight-sided, the outward pointing normal, $\hat{\boldsymbol{n}}$, is a constant along a face.

Thus, we need to compute edge-mass matrices of the form

$$\mathcal{M}_{ij}^{k,e} = \int_{\text{edge}} \ell_j^k(\boldsymbol{x}) \ell_i^k(\boldsymbol{x}) \, d\boldsymbol{x}.$$

At first, this appears to be a full matrix of size $N_p \times (N+1)$. However, recall now that $\ell_i(\boldsymbol{x})$ is a polynomial of order N, including along the edge. Thus, if $\boldsymbol{x}_i$ (the particular grid point where $\ell_i(\boldsymbol{x}_i) = 1$) does not reside on the edge, $\ell_i(\boldsymbol{x})$ is an N-th-order polynomial, taking the value of zero at $N + 1$ grid points; that is, it is exactly zero along the edge. The mass matrices therefore only have nonzero entries in those rows, i, where $\boldsymbol{x}_i$ resides on the edge. If we define the Vandermonde matrix, $\mathcal{V}^{1D}$, corresponding to the one-dimensional interpolation along the edge, we have

$$\mathcal{M}^1 = J^1 (\mathcal{V}^{1D}(\mathcal{V}^{1D})^T)^{-1},$$

where J^1 is the transformation Jacobian along the face – the ratio between the length of the face in D^k and in I, respectively.

This special property of the nodal basis, allowing one to only use information along the edges to compute the surface integrals, is a clear advantage of the nodal basis over a modal representation where all information is needed to evaluate the solution pointwise.

To specify a simple way to implement this surface integral, we define a matrix, Fmask, of size $Nf_p \times 3$, which, in each column, contains the local numbering for the Nf_p nodes along each of the three faces on the triangle. We can then use Fmask to extract the edge coordinate and, subsequently, form the local Vandermonde matrix and the mass matrix corresponding to the edge.

As an example, along edge 1, this could be done as

```
>> faceR = r(Fmask(:,1));
>> vdmEdge = Vandermonde1D(N, faceR);
>> massEdge1 = inv(vdmEdge*vdmEdge');
```

Define the matrix, $\mathcal{E}$, of size $N_p \times 3Nf_p$. The purpose of this is to compute the surface integral on l; that is,

$$(\mathcal{E}[\boldsymbol{g}^1, \boldsymbol{g}^2, \boldsymbol{g}^3]^T) = \int_{\partial l} \hat{\boldsymbol{n}} \cdot \boldsymbol{g}_h \ell_i(\boldsymbol{r}) \, d\boldsymbol{r}.$$

We can form $\mathcal{E}$ directly by simply inserting the elements of three face-mass matrices in the positions corresponding to the face nodes in the element.

This is all implemented in Lift2D.m, which returns the matrix LIFT of size $N_p \times 3Nf_p$, as

$$\text{LIFT} = \mathcal{M}^{-1}\mathcal{E}.$$

Lift2D.m

```
function [LIFT] = Lift2D()

% function [LIFT] = Lift2D()
% Purpose  : Compute surface to volume lift term for DG formulation

Globals2D;
Emat = zeros(Np, Nfaces*Nfp);

% face 1
faceR = r(Fmask(:,1));
V1D = Vandermonde1D(N, faceR);
massEdge1 = inv(V1D*V1D');
Emat(Fmask(:,1),1:Nfp) = massEdge1;

% face 2
faceR = r(Fmask(:,2));
V1D = Vandermonde1D(N, faceR);
massEdge2 = inv(V1D*V1D');
```

```
Emat(Fmask(:,2),Nfp+1:2*Nfp) = massEdge2;

% face 3
faceS = s(Fmask(:,3));
V1D = Vandermonde1D(N, faceS);
massEdge3 = inv(V1D*V1D');
Emat(Fmask(:,3),2*Nfp+1:3*Nfp) = massEdge3;

% inv(mass matrix)*\I_n (L_i,L_j)_{edge_n}
LIFT = V*(V'*Emat);
return
```

To simplify the notation and discretization of general partial differential equations, we define local operations based on vector algebra. Thus, in Grad2D.m, we define the two-dimensional gradient operator as

$$\boldsymbol{v} = (v_x, v_y) = \nabla u.$$

—————————————————— Grad2D.m ——————————————————
```
function [ux,uy] = Grad2D(u);

% function [ux,uy] = Grad2D(u);
% Purpose: Compute 2D gradient field of scalar u

Globals2D;

ur = Dr*u; us = Ds*u;
ux = rx.*ur + sx.*us; uy = ry.*ur + sy.*us;
return
```

In Div2D.m, we have the local divergence of a two-dimensional vector field as

$$v = \nabla \cdot \boldsymbol{u}.$$

—————————————————— Div2D.m ——————————————————
```
function [divu] = Div2D(u,v);

% function [divu] = Div2D(u,v);
% Purpose: Compute the 2D divergence of the vectorfield (u,v)

Globals2D;

ur = Dr*u; us = Ds*u; vr = Dr*v; vs = Ds*v;
divu = rx.*ur + sx.*us + ry.*vr + sy.*vs;
return;
```

Finally, we define the two-dimensional version of the curl-operator in Curl2D.m as

$$v = (v_x, v_y, v_z) = \nabla \times u = \left(\frac{\partial u_z}{\partial y}, -\frac{\partial u_z}{\partial x}, \frac{\partial u_y}{\partial x} - \frac{\partial u_x}{\partial y} \right).$$

$\boxed{\text{Curl2D.m}}$

```
function [vx,vy,vz] = Curl2D(ux,uy,uz);

% function [vx,vy,vz] = Curl2D(ux,uy,uz);
% Purpose: Compute 2D curl-operator in (x,y) plane

Globals2D;

uxr = Dr*ux; uxs = Ds*ux; uyr = Dr*uy; uys = Ds*uy;
vz =   rx.*uyr + sx.*uys - ry.*uxr - sy.*uxs;
vx=[]; vy=[];

if (~isempty(uz));
  uzr = Dr*uz; uzs = Ds*uz;
  vx =   ry.*uzr + sy.*uzs; vy = -rx.*uzr - sx.*uzs;
end;
return
```

6.3 Assembling the grid

With all local operations in place, we now consider the actual computation of the metric coefficients and the assembly of the global grid structure.

Similar to the one-dimensional case, we assume that we have the following information:

- Two row vectors $(\mathbf{VX}, \mathbf{VY})$ with the coordinates of the N_v vertices. These coordinate sets are numbered from 1 to N_v.
- An integer matrix, EToV, of size $K \times 3$ with the three vertex numbers in each row forming one element of a total of K. It is always assumed that the three vertices are ordered counterclockwise such that the transformation Jacobians, Eq. (6.5), are all positive.

Any grid generator will deliver this information to ensure a meaningful triangulation of the computational domain.

With the commands

```
>> va = EToV(:,1)'; vb = EToV(:,2)'; vc = EToV(:,3)';
>> x = 0.5*(-(r+s)*VX(va)+(1+r)*VX(vb)+(1+s)*VX(vc));
>> y = 0.5*(-(r+s)*VY(va)+(1+r)*VY(vb)+(1+s)*VY(vc));
```

we combine the mapping in Eq. (6.3) with the local (r, s) coordinates of the nodes in an element of order N and the vertex coordinates to compute the two arrays (x, y), each of size $N_p \times K$ with the spatial coordinates of the computational grid. We can now compute (x_r, x_s, y_r, y_s) by using $\mathcal{D}_r$ and $\mathcal{D}_s$ and, thus, the full metric for the mappings of the individual elements. This is all implemented in GeometricFactors2D.m.

| GeometricFactors2D.m |

```
function [rx,sx,ry,sy,J] = GeometricFactors2D(x,y,Dr,Ds)

% function [rx,sx,ry,sy,J] = GeometricFactors2D(x,y,Dr,Ds)
% Purpose  : Compute the metric elements for the local mappings
%              of the elements

% Calculate geometric factors
xr = Dr*x; xs = Ds*x; yr = Dr*y; ys = Ds*y; J = -xs.*yr + xr.*ys;
rx = ys./J; sx =-yr./J; ry =-xs./J; sy = xr./J;
return;
```

An $(N + 1) \times 3$ array with the local node numbering at the three edges is found as:

```
>> fmask1 = find( abs(s+1) < NODETOL)';
>> fmask2 = find( abs(r+s) < NODETOL)';
>> fmask3 = find( abs(r+1) < NODETOL)';
>> Fmask = [fmask1;fmask2;fmask3]';
>> Fx = x(Fmask(:), :); Fy = y(Fmask(:), :);
```

where the latter extracts the (x, y)-coordinates of those nodes at the boundary of the elements; that is, they are arrays of size $3Nf_p \times K$.

Computing the outward pointing vectors at the surfaces of the elements follows directly from the properties of the mapping (see Fig. 6.1); that is,

$$\hat{n}^1 = -\frac{\nabla s}{\|\nabla s\|}, \quad \hat{n}^2 = \frac{\nabla r + \nabla s}{\|\nabla r + \nabla s\|}, \quad \hat{n}^3 = -\frac{\nabla r}{\|\nabla r\|}.$$

Here, $\| \cdot \|$ is the Euclidian length of the vector. These can be computed directly using Eq. (6.4) from the physical coordinates, (x, y). The Jacobian for the mapping along the edge is computed as

$$\|\hat{n}^1\| = 1 = \frac{1}{J^1} \sqrt{y_r^2 + x_r^2},$$

and likewise along the two other edges. Computation of the normals is collected in Normals2D.m.

```
┌─────────────┐
│ Normals2D.m │
└─────────────┘
function [nx, ny, sJ] = Normals2D()

% function [nx, ny, sJ] = Normals2D()
% Purpose : Compute outward pointing normals at elements faces and
%           surface Jacobians

Globals2D;
xr = Dr*x; yr = Dr*y; xs = Ds*x; ys = Ds*y; J = xr.*ys-xs.*yr;

% interpolate geometric factors to face nodes
fxr = xr(Fmask, :); fxs = xs(Fmask, :); fyr = yr(Fmask, :);
fys = ys(Fmask, :);

% build normals
nx = zeros(3*Nfp, K); ny = zeros(3*Nfp, K);
fid1 = (1:Nfp)'; fid2 = (Nfp+1:2*Nfp)'; fid3 = (2*Nfp+1:3*Nfp)';

% face 1
nx(fid1, :) =  fyr(fid1, :); ny(fid1, :) = -fxr(fid1, :);

% face 2
nx(fid2, :) =  fys(fid2, :)-fyr(fid2, :);
ny(fid2, :) = -fxs(fid2, :)+fxr(fid2, :);

% face 3
nx(fid3, :) = -fys(fid3, :); ny(fid3, :) =  fxs(fid3, :);

% normalise
sJ = sqrt(nx.*nx+ny.*ny); nx = nx./sJ; ny = ny./sJ;
return;
```

We are now left to combine the K elements into a continuous region of elements by computing the connectivity of the elements. As in the one-dimensional case, we form a sparse integer matrix, FToV, of size $3K \times N_v$ corresponding to the total number of edges and vertices, respectively. The value 1 is inserted in the two columns corresponding to the global vertex numbers forming the face.

The matrix FToF with information about face connectivities is constructed as

$$\text{FToF} = (\text{FToV})(\text{FToV})^T - 2I,$$

where the self-references have been removed.

As in one dimension, each row now has only zeros, or, in most cases, a single instance of one. This latter case then directly tells which element (column number mod 3) and face number a particular row corresponds to, thus allowing one to directly establish a face-to-face connectivity matrix. In case the whole row is zero, there is no connectivity to this particular face, which is then assigned to be a boundary face.

This is implemented in Connect2D.m, which gives as output two arrays of size $K \times 3$ with the information in terms of element-to-element (EToE) and element-to-face (EToF) connectivity. Each of these two arrays holds, in position (i, j), the global element number (EToE) and the local face number (EToF) to which face j on element i connects. Note that if $EToE(i, j) = i$, then face j of element i self-connects and is therefore taken to be a boundary face.

---------------------------------| Connect2D.m |--------------------------------

```
function [EToE, EToF] = Connect2D(EToV)

% function [EToE, EToF] = Connect2D(EToV)
% Purpose  : Build global connectivity arrays for grid based on
%            standard EToV input array from grid generator

Nfaces = 3;

% Find number of elements and vertices
K = size(EToV,1); Nv = max(max(EToV));

% Create face to node connectivity matrix
TotalFaces = Nfaces*K;

% List of local face to local vertex connections
vn = [[1,2];[2,3];[1,3]];

% Build global face to node sparse array
SpFToV = spalloc(TotalFaces, Nv, 2*TotalFaces);
sk = 1;
for k=1:K
  for face=1:Nfaces
    SpFToV( sk, EToV(k, vn(face,:))) = 1;
    sk = sk+1;
  end
end

% Build global face to global face sparse array
SpFToF = SpFToV*SpFToV' - 2*speye(TotalFaces);

% Find complete face to face connections
[faces1, faces2] = find(SpFToF==2);

% Convert face global number to element and face numbers
element1 = floor( (faces1-1)/Nfaces ) + 1;
face1    =   mod( (faces1-1), Nfaces ) + 1;
element2 = floor( (faces2-1)/Nfaces ) + 1;
face2    =   mod( (faces2-1), Nfaces ) + 1;
```

```
% Rearrange into Nelements x Nfaces sized arrays
ind = sub2ind([K, Nfaces], element1, face1);

EToE = (1:K)'*ones(1,Nfaces); EToF = ones(K,1)*(1:Nfaces);
EToE(ind) = element2; EToF(ind) = face2;
return;
```

We finally construct two vectors, $vmapM$ and $vmapP$, each of length $3KNf_p$ (i.e., the total number of individual nodes along the faces of all triangles). Assuming a global numbering of these nodes, the connectivity tables are used to connect the individual nodes along the edges directly, with $vmapM$ corresponding to the interior node, u^-, and $vmapP$ corresponding to the exterior node, u^+. Vertices which are not connected to anything will be assembled in the vector $vmapB$, reflecting all boundary nodes. This is completed in BuildMaps2D.m.

―――――――――――――――――――――― BuildMaps2D.m ――――――――――――――

```
function [mapM, mapP, vmapM, vmapP, vmapB, mapB] = BuildMaps2D()

% function [mapM, mapP, vmapM, vmapP, vmapB, mapB] = BuildMaps2D
% Purpose: Connectivity and boundary tables in the K #
%          of Np elements

Globals2D;

% number volume nodes consecutively
nodeids = reshape(1:K*Np, Np, K);
vmapM   = zeros(Nfp, Nfaces, K); vmapP = zeros(Nfp, Nfaces, K);
mapM    = (1:K*Nfp*Nfaces)';     mapP  = reshape(mapM, Nfp, Nfaces, K);

% find index of face nodes with respect to volume node ordering
for k1=1:K
  for f1=1:Nfaces
    vmapM(:,f1,k1) = nodeids(Fmask(:,f1), k1);
  end
end

one = ones(1, Nfp);
for k1=1:K
  for f1=1:Nfaces
    % find neighbor
    k2 = EToE(k1,f1); f2 = EToF(k1,f1);

    % reference length of edge
    v1 = EToV(k1,f1); v2 = EToV(k1, 1+mod(f1,Nfaces));
    refd = sqrt( (VX(v1)-VX(v2))^2 + (VY(v1)-VY(v2))^2 );
```

```
    % find find volume node numbers of left and right nodes
    vidM = vmapM(:,f1,k1); vidP = vmapM(:,f2,k2);
    x1 = x(vidM); y1 = y(vidM); x2 = x(vidP); y2 = y(vidP);
    x1 = x1*one; y1 = y1*one;  x2 = x2*one;  y2 = y2*one;

    % Compute distance matrix
    D = (x1 -x2').^2 + (y1-y2').^2;
    [idM, idP] = find(sqrt(abs(D))<NODETOL*refd);
    vmapP(idM,f1,k1) = vidP(idP);
    mapP(idM,f1,k1) = idP + (f2-1)*Nfp+(k2-1)*Nfaces*Nfp;
  end
end

% reshape vmapM and vmapP to be vectors and create boundary
% node list
vmapP = vmapP(:); vmapM = vmapM(:); mapP = mapP(:);
mapB = find(vmapP==vmapM); vmapB = vmapM(mapB);
return
```

Everything discussed in the above must be done for any problem being solved with a two-dimensional DG approach. We have assembled these components in the startup scripts StartUp2D.m, which also contains the definition of a few constants that we will use later. Finally, as in the one-dimensional case, we collect a number of constants and global variables in Globals2D.m.

───────────────────────────── StartUp2D.m ─────────────────────────────

```
% Purpose : Setup script, building operators, grid, metric, and
%                connectivity tables.
% Definition of constants
Nfp = N+1; Np = (N+1)*(N+2)/2; Nfaces=3; NODETOL = 1e-12;

% Compute nodal set
[x,y] = Nodes2D(N); [r,s] = xytors(x,y);

% Build reference element matrices
V = Vandermonde2D(N,r,s); invV = inv(V);
MassMatrix = invV'*invV;
[Dr,Ds] = Dmatrices2D(N, r, s, V);

% build coordinates of all the nodes
va = EToV(:,1)'; vb = EToV(:,2)'; vc = EToV(:,3)';
x = 0.5*(-(r+s)*VX(va)+(1+r)*VX(vb)+(1+s)*VX(vc));
y = 0.5*(-(r+s)*VY(va)+(1+r)*VY(vb)+(1+s)*VY(vc));

% find all the nodes that lie on each edge
fmask1   = find( abs(s+1) < NODETOL)';
```

```
fmask2   = find( abs(r+s) < NODETOL)';
fmask3   = find( abs(r+1) < NODETOL)';
Fmask   = [fmask1;fmask2;fmask3]';
Fx = x(Fmask(:), :); Fy = y(Fmask(:), :);

% Create surface integral terms
LIFT = Lift2D();

% calculate geometric factors
[rx,sx,ry,sy,J] = GeometricFactors2D(x,y,Dr,Ds);

% calculate geometric factors
[nx, ny, sJ] = Normals2D();
Fscale = sJ./(J(Fmask,:));

% Build connectivity matrix
[EToE, EToF] = tiConnect2D(EToV);

% Build connectivity maps
BuildMaps2D;

% Compute weak operators (could be done in preprocessing to
% save time)
[Vr, Vs] = GradVandermonde2D(N, r, s);
Drw = (V*Vr')/(V*V'); Dsw = (V*Vs')/(V*V');
```

Globals2D.m

```
% Purpose: declare global variables

global Np Nfp N K
global r s
global Dr Ds LIFT Drw Dsw MassMatrix
global Fx Fy nx ny jac Fscale J
global vmapM vmapP vmapB mapB Fmask
global BCType mapI mapO mapW mapF mapC mapS mapM mapP mapD mapN
global vmapI vmapO vmapW vmapO vmapC vmapS vmapD vmapN
global rx ry sx sy J sJ
global rk4a rk4b rk4c
global Nfaces EToE EToF EToV
global V invV
global x y NODETOL VX VY

% Some curved mesh and cubature quadrature specific data
global cub gauss straight curved

In = 1; Out = 2; Wall = 3; Far = 4; Cyl = 5; Dirichlet = 6;
Neuman = 7; Slip = 8;
```

```
% Low storage Runge-Kutta coefficients
rk4a = [                 0.0 ...
           -567301805773.0/1357537059087.0 ...
          -2404267990393.0/2016746695238.0 ...
          -3550918686646.0/2091501179385.0  ...
          -1275806237668.0/842570457699.0];
rk4b = [ 1432997174477.0/9575080441755.0 ...
           5161836677717.0/13612068292357.0 ...
           1720146321549.0/2090206949498.0  ...
           3134564353537.0/4481467310338.0  ...
           2277821191437.0/14882151754819.0];
rk4c = [                 0.0  ...
           1432997174477.0/9575080441755.0 ...
           2526269341429.0/6820363962896.0 ...
           2006345519317.0/3224310063776.0 ...
           2802321613138.0/2924317926251.0];
```

6.4 Timestepping and boundary conditions

To complete the discussion of the two-dimensional schemes, let us finally consider two minor issues, mainly of a practical character. In Section 4.7 we discussed at length how to choose the timestep to ensure a discretely stable computation. Let us return to this for the multidimensional schemes.

To understand the scaling, similar to what we did in Section 4.7, consider the problem

$$\frac{\partial u}{\partial t} + \boldsymbol{v} \cdot \nabla u = 0,$$

where the computational domain is the reference triangle, I, and $\boldsymbol{v} = (\cos\theta, \sin\theta)$; that is, it is the advection velocity with θ reflecting the angular dependence.

When we discretize this equation, assuming homogeneous boundary conditions and a central flux, we recover

$$\frac{d\boldsymbol{u}_h}{dt} = -\left(v_x \mathcal{D}_x \boldsymbol{u}_h + v_y \mathcal{D}_y \boldsymbol{u}_h - \frac{1}{2} \oint_{\partial \mathsf{I}} (\hat{\boldsymbol{n}} \cdot \boldsymbol{v})(\hat{\boldsymbol{n}} \cdot [\![\boldsymbol{u}_h]\!]) \ell(\boldsymbol{r}) \, d\boldsymbol{r} \right) = \mathcal{A}\boldsymbol{u}_h,$$

where the eigenvalues of $\mathcal{A}$ need to be understood.

In Fig. 6.8 we show the maximum of the eigenvalues as a function of both N and θ. We note that the maximum eigenvalue appears around $\theta = 45°$, which is natural, as that reflects the shortest length in the triangle, I (see Fig. 6.1).

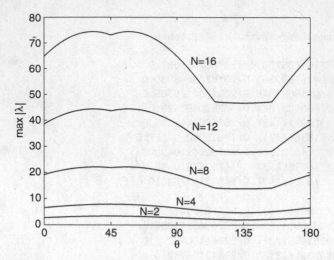

Fig. 6.8. Behavior of maximum eigenvalue of two-dimensional advection operator with advection velocity, $\boldsymbol{v} = (\cos\theta, \sin\theta)$, as a function of the angle θ and the order of approximation N.

As for the one-dimensional case, the eigenvalues in the standard element scales as

$$\max|\lambda| \simeq CN^2.$$

The value of the constant depends on several things such as the type of numerical flux, as discussed in Section 4.7. Based on numerical experiments, we found an upper bound for the one-dimensional case as

$$\max|\lambda|_{1D} \le \frac{3}{2}\max(\Delta r_i)^{-1},$$

where Δr_i reflects the spacing between grid points in the one-dimensional standard interval, $[-1, 1]$. In a similar fashion, we find a scaling on I of the form

$$\max|\lambda|_{2D} \le 2\max|\lambda|_{1D}.$$

We use the same scaling philosophy as in the one-dimensional case, although there remains one issue to consider for the general triangle. For the one-dimensional case, the geometric scaling of the maximum eigenvalue is straightforward, as it scales directly as $2/h$, with h being the length of the smallest element. The extension of this to a triangle requires a few more considerations; that is, we cannot simply take the shortest face length because that would not identify a bad triangle with two small angles and three long edges.

For the characteristic length of the triangle, we use the radius of the inscribed circle, found as

$$r_D(\boldsymbol{x}) = \frac{A}{s},$$

where s is half the triangle perimeter and A is the area of the triangle. With this, we get the scaling for an advection problem like

$$\mathcal{L}u = \boldsymbol{v} \cdot \nabla u,$$

as

$$\Delta t \leq C \left(\frac{2}{3} \min \Delta r_i \right) \min_\Omega \left(\frac{r_D}{|\boldsymbol{v}|} \right),$$

where C is of order 1. A routine to compute r_D is given in dtscale2D.m.

---
dtscale2D.m

```
function dtscale = dtscale2D;

% function dtscale = dtscale2D;
% Purpose : Compute inscribed circle diameter as characteristic
%           for grid to choose timestep

Globals2D;

% Find vertex nodes
vmask1   = find( abs(s+r+2) < NODETOL)';
vmask2   = find( abs(r-1)   < NODETOL)';
vmask3   = find( abs(s-1)   < NODETOL)';
vmask  = [vmask1;vmask2;vmask3]';
vx = x(vmask(:), :); vy = y(vmask(:), :);

% Compute semi-perimeter and area
len1 = sqrt((vx(1,:)-vx(2,:)).^2+(vy(1,:)-vy(2,:)).^2);
len2 = sqrt((vx(2,:)-vx(3,:)).^2+(vy(2,:)-vy(3,:)).^2);
len3 = sqrt((vx(3,:)-vx(1,:)).^2+(vy(3,:)-vy(1,:)).^2);
sper = (len1 + len2 + len3)/2.0;
Area = sqrt(sper.*(sper-len1).*(sper-len2).*(sper-len3));

% Compute scale using radius of inscribed circle
dtscale = Area./sper;
return;
```

Another multidimensional issue that requires some special attention is the handling of different type of boundary conditions and how one imposes them.

In BuildMaps2D.m, we create the vector of nodes, $\boldsymbol{vmapB}$, which are not connected to anything, designating these to be boundary nodes. If all these are to be treated in a similar fashion, this is all that is needed. If, however, some parts of the boundary are to be treated differently than others, we will need to be able to build vectors of these nodes.

We assume that we have a matrix, BCType, of size $K \times 3$, corresponding to one entry for each face in the grid. Most of the entries in this matrix will be zero, with only those located at the boundaries having different values – this information is obtained from the grid generator.

Let us, for simplicity, consider two different types of boundary, "In" and "Out", and associate the value of 1 with "In" and the value of 2 with "Out". Also assume that these values are in the correct places in BCType. The task is then to use this information to build the two maps, ***vmapI*** and ***vmapO***, as the vectors of globally numbered nodes residing on the "In" and "Out" boundaries, respectively.

A small section of code, used to build the required vectors of nodes, is shown in BuildBCMaps2D.m, which can be executed after BuildMaps2D.m in StartUp2D.m. The computed mapping vectors are stored in Globals2D.m and can now be used as needed. Detailed examples of their use is discussed in Section 6.6.

┌──────────────────┐
│ BuildBCMaps2D.m │
└──────────────────┘

```
function BuildBCMaps2D()

% function BuildMaps2DBC
% Purpose: Build specialized nodal maps for various types of
%          boundary conditions, specified in BCType.

Globals2D;

% create label of face nodes with boundary types from BCType
bct    = BCType';
bnodes = ones(Nfp, 1)*bct(:)';
bnodes = bnodes(:);

% find location of boundary nodes in face and volume node lists
mapI = find(bnodes==In);        vmapI = vmapM(mapI);
mapO = find(bnodes==Out);       vmapO = vmapM(mapO);
mapW = find(bnodes==Wall);      vmapW = vmapM(mapW);
mapF = find(bnodes==Far);       vmapF = vmapM(mapF);
mapC = find(bnodes==Cyl);       vmapC = vmapM(mapC);
mapD = find(bnodes==Dirichlet); vmapD = vmapM(mapD);
mapN = find(bnodes==Neuman);    vmapN = vmapM(mapN);
mapS = find(bnodes==Slip);      vmapS = vmapM(mapS);
return;
```

6.5 Maxwell's equations

As a first example, let us consider the solution of the two-dimensional vacuum Maxwell's equations in what is known as transverse magnetic form (TM). These are given as

$$\mu \frac{\partial \tilde{H}^x}{\partial \tilde{t}} = -\frac{\partial \tilde{E}^z}{\partial \tilde{y}},$$

$$\mu \frac{\partial \tilde{H}^y}{\partial \tilde{t}} = \frac{\partial \tilde{E}^z}{\partial \tilde{x}}, \qquad (6.8)$$

$$\varepsilon \frac{\partial \tilde{E}^z}{\partial \tilde{t}} = \frac{\partial \tilde{H}^y}{\partial \tilde{x}} - \frac{\partial \tilde{H}^x}{\partial \tilde{y}}.$$

Here, we have the two magnetic fields, $(\tilde{H}^x, \tilde{H}^y)$, and the electric field, $\tilde{E}^z$, all functions of $(\tilde{x}, \tilde{y}, \tilde{t})$. All fields and units are dimensional. Furthermore, we have the magnetic permeability, $\mu(\tilde{\boldsymbol{x}})$, and the electric permittivity, $\varepsilon(\tilde{\boldsymbol{x}})$, which reflect the material coefficients.

In the following, we wish to model a metallic air-filled cavity, $\Omega = [-1, 1]^2$. In this case, we can simplify the equations since $\mu = \mu_0$ and $\varepsilon = \varepsilon_0$ are the constant vacuum values. In fact, if we introduce the vacuum speed of light defined as

$$c_0 = \frac{1}{\sqrt{\varepsilon_0 \mu_0}} \simeq 3 \times 10^8 \text{ m/s},$$

we can consider the normalized system of equations on the form

$$\frac{\partial H^x}{\partial t} = -\frac{\partial E^z}{\partial y}, \qquad (6.9)$$

$$\frac{\partial H^y}{\partial t} = \frac{\partial E^z}{\partial x},$$

$$\frac{\partial E^z}{\partial t} = \frac{\partial H^y}{\partial x} - \frac{\partial H^x}{\partial y},$$

where the unit-free variables are obtained as

$$t = \frac{c_0 \tilde{t}}{L}, \quad \boldsymbol{x} = \frac{\tilde{\boldsymbol{x}}}{L}, \quad \boldsymbol{H} = \frac{\tilde{\boldsymbol{H}}}{H_0}, \quad E^z = (Z_0)^{-1} \frac{\tilde{E}^z}{H_0}.$$

Here, H_0 is a unit magnetic field strength, $Z_0 = \sqrt{\mu_0/\varepsilon_0} \simeq 120\pi$ ohms is the vacuum impedance, and L is some reference length, typically the wavelength of the phenomena of interest.

For the boundary conditions, we assume that the walls of the cavity are perfectly electrically conducting such that the tangential component of the electric field, E^z, vanishes at the wall (i.e., $E^z = 0$ at the wall).

To complete the formulation of the scheme, we need only derive a numerical flux. As for the one-dimensional case, discussed in depth in Section 2.4, the numerical flux for the linear case can be obtained by the use of the Rankine-Hugoniot conditions along a normal, $\hat{\boldsymbol{n}}$. The three-dimensional Maxwell's equations take the form

$$\varepsilon \frac{\partial \boldsymbol{E}}{\partial t} - \nabla \times \boldsymbol{H} = 0, \quad \mu \frac{\partial \boldsymbol{H}}{\partial t} + \nabla \times \boldsymbol{E} = 0,$$

where $\boldsymbol{E} = (E^x, E^y, E^z)$ and $\boldsymbol{H} = (H^x, H^y, H^z)$ are the field components. In this case, one obtains the numerical fluxes

$$-[\hat{\boldsymbol{n}} \times \boldsymbol{H} - (\hat{\boldsymbol{n}} \times \boldsymbol{H})^*] = -\frac{1}{2\{\{Z\}\}}\hat{\boldsymbol{n}} \times [Z^+(\boldsymbol{H}^- - \boldsymbol{H}^+) - \alpha\hat{\boldsymbol{n}} \times (\boldsymbol{E}^- - \boldsymbol{E}^+)],$$

and

$$[\hat{\boldsymbol{n}} \times \boldsymbol{E} - (\hat{\boldsymbol{n}} \times \boldsymbol{E})^*] = \frac{1}{2\{\{Y\}\}}\hat{\boldsymbol{n}} \times [Y^+(\boldsymbol{E}^- - \boldsymbol{E}^+) + \alpha\hat{\boldsymbol{n}} \times (\boldsymbol{H}^- - \boldsymbol{H}^+)],$$

for the equations for the electric and magnetic fields, respectively. In both cases, we have the possibility of the piecewise constant material coefficients, represented by

$$Z^\pm = \frac{1}{Y^\pm} = \sqrt{\frac{\mu^\pm}{\varepsilon^\pm}},$$

as the local impedance and conductance, respectively.

The parameter α in the numerical flux can be used to control dissipation; for example, taking $\alpha = 0$ yields a nondissipative central flux and $\alpha = 1$ results in the classic upwind flux. One is free, however, to take α to be any value in between.

In the simpler case we consider here, we have only three field components and $Z = Y = 1$ due to the normalization used. This yields the numerical flux for the TM form

$$\hat{\boldsymbol{n}} \cdot (\boldsymbol{F} - \boldsymbol{F}^*) = \frac{1}{2}\left\{\begin{array}{l} \hat{n}_y[E^z] + \alpha(\hat{n}_x[\![\boldsymbol{H}]\!] - [H^x]) \\ -\hat{n}_x[E^z] + \alpha(\hat{n}_y[\![\boldsymbol{H}]\!] - [H^y]) \\ \hat{n}_y[H^x] - \hat{n}_x[H^y] - \alpha[E^z], \end{array}\right.$$

where $\boldsymbol{H} = (H^x, H^y)$, and we use the notation

$$[q] = q^- - q^+ = \hat{\boldsymbol{n}} \cdot [\![q]\!].$$

This yields the local semidiscrete scheme

$$\frac{dH_h^x}{dt} = -\mathcal{D}_y E_h^z + \frac{1}{2}(J\mathcal{M})^{-1}\int_{\partial\mathsf{D}^k}(\hat{n}_y[E_h^z] + \alpha(\hat{n}_x[\![\boldsymbol{H}_h]\!] - [H_h^x]))\,\boldsymbol{\ell}(\boldsymbol{x})\,d\boldsymbol{x},$$

$$\frac{dH_h^y}{dt} = \mathcal{D}_x E_h^z + \frac{1}{2}(J\mathcal{M})^{-1}\int_{\partial\mathsf{D}^k}(-\hat{n}_x[E_h^z] + \alpha(\hat{n}_y[\![\boldsymbol{H}_h]\!] - [H_h^y]))\,\boldsymbol{\ell}(\boldsymbol{x})\,d\boldsymbol{x},$$

$$\frac{dE_h^z}{dt} = \mathcal{D}_x H_h^y - \mathcal{D}_y H_h^x + \frac{1}{2}(J\mathcal{M})^{-1}\int_{\partial\mathsf{D}^k}(\hat{n}_y[H_h^x] - \hat{n}_x[H_h^y] - \alpha[E_h^z])\,\boldsymbol{\ell}(\boldsymbol{x})\,d\boldsymbol{x},$$

for the approximation solution (H_h^x, H_h^y, E_h^z). We have suppressed the element index, k, for simplicity. This scheme is implemented in MaxwellRHS2D.m.

—————————— ⏐ MaxwellRHS2D.m ⏐ ——————————

```
function [rhsHx, rhsHy, rhsEz] = MaxwellRHS2D(Hx,Hy,Ez)

% function [rhsHx, rhsHy, rhsEz] = MaxwellRHS2D(Hx,Hy,Ez)
% Purpose  : Evaluate RHS flux in 2D Maxwell TM form

Globals2D;

% Define field differences at faces
dHx = zeros(Nfp*Nfaces,K); dHx(:) = Hx(vmapM)-Hx(vmapP);
dHy = zeros(Nfp*Nfaces,K); dHy(:) = Hy(vmapM)-Hy(vmapP);
dEz = zeros(Nfp*Nfaces,K); dEz(:) = Ez(vmapM)-Ez(vmapP);

% Impose reflective boundary conditions (Ez+ = -Ez-)
dHx(mapB) = 0; dHy(mapB) = 0; dEz(mapB) = 2*Ez(vmapB);

% evaluate upwind fluxes
alpha = 1.0;
ndotdH =  nx.*dHx+ny.*dHy;
fluxHx =  ny.*dEz + alpha*(ndotdH.*nx-dHx);
fluxHy = -nx.*dEz + alpha*(ndotdH.*ny-dHy);
fluxEz = -nx.*dHy + ny.*dHx - alpha*dEz;

% local derivatives of fields
[Ezx,Ezy] = Grad2D(Ez); [CuHx,CuHy,CuHz] = Curl2D(Hx,Hy,[]);

% compute right hand sides of the PDE's
rhsHx = -Ezy  + LIFT*(Fscale.*fluxHx)/2.0;
rhsHy =  Ezx  + LIFT*(Fscale.*fluxHy)/2.0;
rhsEz =  CuHz + LIFT*(Fscale.*fluxEz)/2.0;
return;
```

The boundary condition for E^z is

$$E^z = 0$$

and we do not need to specify any conditions on $\boldsymbol{H}$. This is implemented by setting $\boldsymbol{H}^* = \boldsymbol{H}^-$ at the boundary. The homogeneous condition of E^z can be implemented in different ways. We use a mirror principle, based on assigning $(E^z)^* = -(E^z)^-$ such that $(E^z)^- + (E^z)^* = 0$. This is enforced as follows

$$[E^z] = \hat{\boldsymbol{n}} \cdot [\![E^z]\!] = 2(E^z)^-,$$

at all boundary points, identified by $\boldsymbol{vmapB}$.

With the semidiscrete formulation completed, we use an explicit low-storage Runge-Kutta method to integrate in time, exactly as in the one-dimensional cases discussed in Chapter 3. This is done in Maxwell2D.m, where we also set the timestep as discussed previously.

————————————————————— | Maxwell2D.m | —————————————————————

```
function [Hx,Hy,Ez,time] = Maxwell2D(Hx, Hy, Ez, FinalTime)

% function [Hx,Hy,Ez] = Maxwell2D(Hx, Hy, Ez, FinalTime)
% Purpose :Integrate TM-mode Maxwell's until FinalTime starting
%          with initial conditions Hx,Hy,Ez

Globals2D;
time = 0;

% Runge-Kutta residual storage
resHx = zeros(Np,K); resHy = zeros(Np,K); resEz = zeros(Np,K);

% compute time step size
rLGL = JacobiGQ(0,0,N); rmin = abs(rLGL(1)-rLGL(2));
dtscale = dtscale2D; dt = min(dtscale)*rmin*2/3

% outer time step loop
while (time<FinalTime)

  if(time+dt>FinalTime), dt = FinalTime-time; end

    for INTRK = 1:5
       % compute right hand side of TM-mode Maxwell's equations
       [rhsHx, rhsHy, rhsEz] = MaxwellRHS2D(Hx,Hy,Ez);

       % initiate and increment Runge-Kutta residuals
       resHx = rk4a(INTRK)*resHx + dt*rhsHx;
       resHy = rk4a(INTRK)*resHy + dt*rhsHy;
       resEz = rk4a(INTRK)*resEz + dt*rhsEz;

       % update fields
       Hx = Hx+rk4b(INTRK)*resHx; Hy = Hy+rk4b(INTRK)*resHy;
       Ez = Ez+rk4b(INTRK)*resEz;
    end;
    % Increment time
    time = time+dt;
end
return
```

The driver routine for solving the Maxwell equations is given in MaxwellDriver2D.m. This is where the initial conditions are set and the order, N, of the approximation is specified.

─────────────────────── | MaxwellDriver2D.m | ───────────────────────

```
% Driver script for solving the 2D vacuum Maxwell's equations
%                on TM form
Globals2D;

% Polynomial order used for approximation
N = 10;

% Read in Mesh
[Nv, VX, VY, K, EToV] = MeshReaderGambit2D('Maxwell025.neu');

% Initialize solver and construct grid and metric
StartUp2D;

% Set initial conditions
mmode = 1; nmode = 1;
Ez = sin(mmode*pi*x).*sin(nmode*pi*y); Hx = zeros(Np, K);
Hy = zeros(Np, K);

% Solve Problem
FinalTime = 1;
[Hx,Hy,Ez,time] = Maxwell2D(Hx,Hy,Ez,FinalTime);
```

To validate the code, we evolve the exact cavity solution given as

$$H^x(x,y,t) = -\frac{\pi n}{\omega} \sin(m\pi x) \cos(n\pi y) \sin(\omega t),$$
$$H^y(x,y,t) = \frac{\pi m}{\omega} \cos(m\pi x) \sin(n\pi y) \sin(\omega t),$$
$$E^z(x,y,t) = \sin(m\pi x) \sin(n\pi y) \cos(\omega t),$$

where the resonance frequencies, ω, are given as

$$\omega = \pi\sqrt{m^2 + n^2}, \quad (m,n) \geq 0.$$

For simplicity, we take $m = n = 1$.

In Fig. 6.9 we show the convergence of the electric field, E^z, under both element and order refinement. We show the results computed using either a central flux ($\alpha = 0$) or an upwind flux ($\alpha = 1$). Similar convergence behavior can be observed for the other field components.

When comparing the results, we see that for the scheme based on the upwind flux, we have optimal convergence rates (i.e., $\mathcal{O}(h^{N+1})$ where $h = \sqrt{K}$). For the scheme based on the purely central scheme, the situation is slightly less clear, as there are indications of an even-odd pattern with the accuracy being $\mathcal{O}(h^{N+1})$ for N even and $\mathcal{O}(h^N)$ for N odd. As we will discuss in Chapter 7, such behavior is often observed when central fluxes are used.

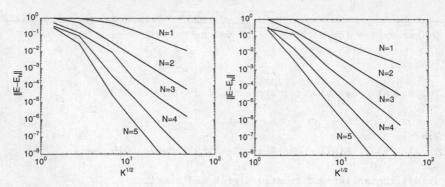

Fig. 6.9. On the left is shown the discrete L^2-error for E_h^z at $T = 10$, obtained using a DG-FEM with central fluxes, for different values of N and K. As a measure of average cell size, we use $\sqrt{K}$. On the right we show similar results obtained using an upwind flux.

6.6 Compressible gas dynamics

Note: Due to the complexity of the problem discussed in the following, we will not list the complete Matlab codes but just those lines essential to the discussion. Parts of the code not displayed are similar to those already introduced.

The Euler equations of inviscid gas dynamics in conservative form are

$$\frac{\partial \rho}{\partial t} + \frac{\partial \rho u}{\partial x} + \frac{\partial \rho v}{\partial y} = 0,$$

$$\frac{\partial \rho u}{\partial t} + \frac{\partial \rho u^2 + p}{\partial x} + \frac{\partial \rho u v}{\partial y} = 0,$$

$$\frac{\partial \rho v}{\partial t} + \frac{\partial \rho u v}{\partial x} + \frac{\partial \rho v^2 + p}{\partial y} = 0,$$

$$\frac{\partial E}{\partial t} + \frac{\partial u\,(E + p)}{\partial x} + \frac{\partial v\,(E + p)}{\partial y} = 0,$$

where ρ is the density of the gas, $(\rho u, \rho v)$ are the x- and y-components of the momentum, p is the internal pressure of the gas, and E is the total energy of the gas. The total energy of the gas is the sum of the potential energy due to the internal pressure and the kinetic energy due to its momentum, given by

$$E = \frac{p}{\gamma - 1} + \frac{\rho}{2}\left(u^2 + v^2\right),$$

where γ is a constant dependent on the type of gas. For this example, we will consider a monoatomic gas with $\gamma = 1.4$. The above equations neglect the effects of viscosity and thermal diffusion, which will be incorporated into a later example that concerns the Navier-Stokes equations for a compressible fluid (see Section 7.5).

To provide some insight into a relatively general numerical discretization for a wide class of nonlinear conservation laws, with Euler's equations as the guiding example, we rewrite the equations in vector form:

$$\frac{\partial \mathbf{q}}{\partial t} + \frac{\partial \mathbf{F}}{\partial x} + \frac{\partial \mathbf{G}}{\partial y} = 0,$$

where

$$\mathbf{q} = \begin{pmatrix} \rho \\ \rho u \\ \rho v \\ E \end{pmatrix}, \mathbf{F} = \begin{pmatrix} \rho u \\ \rho u^2 + p \\ \rho u v \\ u(E + p) \end{pmatrix}, \mathbf{G} = \begin{pmatrix} \rho v \\ \rho u v \\ \rho v^2 + p \\ v(E + p) \end{pmatrix},$$

represent the state vector and the two nonlinear fluxes, respectively. These vector flux functions are implemented in a straightforward manner by first extracting the primitive variables (ρ, u, v, p) from the conserved variable state vector $\mathbf{q}$ at each volume element node in EulerFluxes2D.m. We have chosen to store the state vector as a three-dimensional array, Q, of size $N_p \times K \times 4$, with the last dimension being the number of conserved variables.

───────────── EulerFluxes2D.m ─────────────

```
6   % extract conserved variables
7   rho = Q(:,:,1); rhou = Q(:,:,2); rhov = Q(:,:,3); Ener = Q(:,:,4);
8
9   % compute primitive variables
10  u = rhou./rho; v = rhov./rho;
11  p = (gamma-1)*(Ener - 0.5*(rhou.*u + rhov.*v));
```

Combinations of the primitive and conserved variables are then used to evaluate the vector flux functions $\mathbf{F}$ and $\mathbf{G}$.

───────────── EulerFluxes2D.m ─────────────

```
13  % compute flux functions
14  F = zeros(size(Q));
15  F(:,:,1) = rhou; F(:,:,2) = rhou.*u + p; F(:,:,3) = rhov.*u;
16  F(:,:,4) = u.*(Ener+p);
17
18  G = zeros(size(Q));
19  G(:,:,1) = rhov; G(:,:,2) = rhou.*v; G(:,:,3) = rhov.*v + p;
20  G(:,:,4) = v.*(Ener+p);
```

Following the previous discussions, we represent the state vector as a piecewise N-th-order polynomial, $\boldsymbol{q}_h$, and require it to satisfy a DG statement on weak form for all test functions $\phi_h \in \mathsf{V}_h$, as

$$\int_{\mathrm{D}^k} \left(\frac{\partial \mathbf{q}_h}{\partial t} \phi_h - \mathbf{F}_h \frac{\partial \phi_h}{\partial x} - \mathbf{G}_h \frac{\partial \phi_h}{\partial y} \right) d\boldsymbol{x} + \int_{\partial \mathrm{D}^k} (\hat{n}_x \mathbf{F}_h + \hat{n}_y \mathbf{G}_h)^* \phi_h d\boldsymbol{x} = 0.$$

For the numerical flux, we use the local Lax-Friedrichs flux

$$(\hat{n}_x \mathbf{F}_h + \hat{n}_y \mathbf{G}_h)^* = \hat{n}_x \{\{\mathbf{F}_h\}\} + \hat{n}_y \{\{\mathbf{G}_h\}\} + \frac{\lambda}{2} \cdot [\![\mathbf{q}_h]\!]$$

The dissipative nature of this flux will smear shocks in strongly supersonic and transitional flows but will serve adequately for most subsonic and weakly supersonic flows. To complete the computation of the fluxes, we recover an approximate local maximum linearized acoustic wave speed

$$\lambda = \max_{s \in [q_h^-, q_h^+]} \left(|\mathbf{u}(s)| + \sqrt{\left| \frac{\gamma p(s)}{\rho(s)} \right|} \right).$$

EulerRHS2D.m contains an implementation of the right-hand-side terms. First, we compute the volume terms for each component of the state vector, taking advantage of the local differentiation matrices common to all elements and applying the chain rule to compute the Cartesian derivatives of the fields.

EulerRHS2D.m

```
11  % 1. Compute volume contributions (NOW INDEPENDENT OF
12  %     SURFACE TERMS)
13  gamma = 1.4;
14  [F,G,rho,u,v,p] = EulerFluxes2D(Q, gamma);
15
16  % Compute weak derivatives
17  for n=1:4
18    dFdr = Drw*F(:,:,n); dFds = Dsw*F(:,:,n);
19    dGdr = Drw*G(:,:,n); dGds = Dsw*G(:,:,n);
20    rhsQ(:,:,n) = (rx.*dFdr + sx.*dFds) + (ry.*dGdr + sy.*dGds);
21  end
```

Surface terms are computed by extracting the positive and negative traces of the state vector at each of the nodes along the faces of each element. The positive traces at boundary faces are modified to incorporate boundary conditions. The nodal modified positive and negative traces of the state vectors are passed to EulerFluxes2D.m to evaluate the vector flux functions for both traces at each of the surface nodes.

EulerRHS2D.m

```
23  % 2. Compute surface contributions
24  % 2.1 evaluate '-' and '+' traces of conservative variables
25  for n=1:4
26    Qn = Q(:,:,n);
27    QM(:,:,n) = Qn(vmapM); QP(:,:,n) = Qn(vmapP);
28  end
29
30  % 2.2 set boundary conditions by modifying positive traces
```

```
31  if(~isempty(ExactSolutionBC))
32    QP = feval(ExactSolutionBC, Fx, Fy, nx, ny, ...
33                mapI, mapO, mapW, mapC, QP, time);
34  end
35
36  % 2.3 evaluate primitive variables & flux functions at '-' and '+'
37  % traces
38  [fM,gM,rhoM,uM,vM,pM] = EulerFluxes2D(QM, gamma);
39  [fP,gP,rhoP,uP,vP,pP] = EulerFluxes2D(QP, gamma);
40
41  % 2.4 Compute local Lax-Friedrichs/Rusonov numerical fluxes
42  lambda = max( sqrt(uM.^2+vM.^2) + sqrt(abs(gamma*pM./rhoM)),  ...
43                sqrt(uP.^2+vP.^2) + sqrt(abs(gamma*pP./rhoP)));
44  lambda = reshape(lambda, Nfp, Nfaces*K);
45  lambda = ones(Nfp, 1)*max(lambda, [], 1);
46  lambda = reshape(lambda, Nfp*Nfaces, K);
47
48  % 2.5 Lift fluxes
49  for n=1:4
50    nflux = nx.*(fP(:,:,n) + fM(:,:,n)) ...
51            + ny.*(gP(:,:,n) + gM(:,:,n)) ...
52            + lambda.*(QM(:,:,n) - QP(:,:,n));
53    rhsQ(:,:,n) = rhsQ(:,:,n) - LIFT*(Fscale.*nflux/2);
54  end
```

As with all previous cases, once the right-hand-side is assembled, very few additional changes are needed. However, one does need to choose a timestepping scheme. In the following examples, we continue to use the low-storage RK scheme introduced in Section 3.4, but the strongly stable RK methods introduced in Section 5.7 are equally suitable.

Example 6.1. As a first evaluation of this algorithm, we consider an isentropic vortex test case with an exact solution given by

$$u = 1 - \beta e^{(1-r^2)} \frac{y - y_0}{2\pi},$$

$$v = \beta e^{(1-r^2)} \frac{x - x_0}{2\pi},$$

$$\rho = \left(1 - \left(\frac{\gamma - 1}{16\gamma\pi^2}\right) \beta^2 e^{2(1-r^2)}\right)^{\frac{1}{\gamma-1}},$$

$$p = \rho^\gamma,$$

where $r = \sqrt{(x - t - x_0)^2 + (y - y_0)^2}$, $x_0 = 5$, $y_0 = 0$, $\beta = 5$, and $\gamma = 1.4$ (see IsentropicVortex2D.m for the full implementation). The exact solution is used as the initial condition and to supply exact boundary conditions. The rate of convergence is computed using the sequence of nested refined meshes, shown in Figure 6.10.

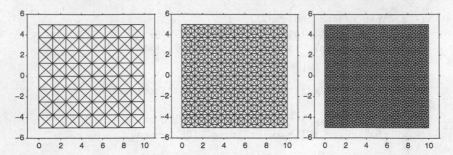

Fig. 6.10. Sequence of three meshes used to perform convergence analysis for isentropic vortex flow solver.

Table 6.2. L^2-errors in ρ and ρu for the isentropic vortex example on a sequence of meshes shown in Fig. 6.10 using different order approximations.

N	\multicolumn{4}{c}{Error in ρ}				\multicolumn{4}{c}{Error in ρu}			
	h	$h/2$	$h/4$	Rate	h	$h/2$	$h/4$	Rate
1	1.28E-02	5.41E-03	1.45E-03	1.57	3.35E-02	1.18E-02	3.14E-03	1.71
2	4.51E-03	6.09E-04	1.12E-04	2.66	8.73E-03	1.23E-03	1.96E-04	2.74
3	9.10E-04	1.09E-04	1.03E-05	3.23	2.05E-03	2.30E-04	1.62E-05	3.49
4	3.44E-04	2.12E-05	1.23E-06	4.06	9.03E-04	3.68E-05	1.54E-06	4.60
5	1.11E-04	5.04E-06	1.41E-07	4.81	2.17E-04	6.89E-06	1.65E-07	5.18
6	7.53E-05	1.59E-06	1.48E-08	6.16	1.28E-04	1.59E-06	1.61E-08	6.48
7	2.21E-05	2.12E-07	2.04E-09	6.70	3.01E-05	2.43E-07	1.91E-09	6.97
8	1.60E-05	7.93E-08	1.89E-10	8.18	1.73E-05	6.08E-08	1.69E-10	8.32

In Table 6.2 we illustrate the accuracy of the scheme through the L^2-errors of the density and one of the momentum components. We note a convergence rate very close to $\mathcal{O}(h^{N+1/2})$, which is the optimal rate of convergence when the Lax-Friedrichs flux is used for a system, as discussed in Section 5.5. As we will also discuss in Section 6.7 this is generally optimal for the two-dimensional case. The small deviations from the optimal order shown in Table 6.2 are caused by the unstructured nature of the grid.

6.6.1 Variational crimes, aliasing, filtering, and cubature integration

In the implementation above, we commit some minor variational crimes in evaluating the spatial terms in the variational statement. To be precise, we consider

$$\int_{\mathsf{D}^k} \left(\frac{\partial \mathbf{q}_h}{\partial t} \phi_h - (\mathcal{I}_N \mathbf{F}_h) \frac{\partial \phi_h}{\partial x} - (\mathcal{I}_N \mathbf{G}_h) \frac{\partial \phi_h}{\partial y} \right) dx$$

$$= -\int_{\partial \mathsf{D}^k} (\mathcal{I}_N (\hat{n}_x \mathbf{F}_h + \hat{n}_y \mathbf{G}_h)))^* \phi_h \, dx,$$

Table 6.3. L^2-errors in ρ and ρu for the isentropic vortex example on a sequence of meshes shown in Fig. 6.10 with different order approximations with an added filter.

N	Error in ρ				Error in ρu			
	h	$h/2$	$h/4$	Rate	h	$h/2$	$h/4$	Rate
1	1.30E-02	5.52E-03	1.50E-03	1.56	3.38E-02	1.20E-02	3.21E-03	1.70
2	4.49E-03	5.90E-04	1.07E-04	2.70	8.69E-03	1.20E-03	1.89E-04	2.76
3	8.96E-04	1.07E-04	1.03E-05	3.22	2.03E-03	2.29E-04	1.63E-05	3.48
4	3.36E-04	2.06E-05	1.21E-06	4.06	8.92E-04	3.63E-05	1.52E-06	4.60
5	1.08E-04	4.97E-06	1.39E-07	4.80	2.16E-04	6.88E-06	1.62E-07	5.19
6	7.33E-05	1.54E-06	1.46E-08	6.15	1.27E-04	1.56E-06	1.58E-08	6.48
7	2.16E-05	2.10E-07	2.01E-09	6.70	3.01E-05	2.40E-07	1.88E-09	6.98
8	1.56E-05	7.80E-08	1.87E-10	8.18	1.71E-05	5.99E-08	1.67E-10	8.32

for all piecewise polynomial test functions, $\phi_h \in V_h$. Here, $\mathcal{I}_N$ is the N-th-order interpolation operator associated with the nodes on the element. As we have discussed in depth in Section 5.3 this interpolation process induces aliasing errors for marginally resolved computations. In practical simulations, with marginally resolved flow features, these errors may drive the simulations unstable. However, as we discussed in Section 5.3, a mild filter applied to the top modes of the polynomial approximation of the right-hand-side residual can be surprisingly effective in controlling stability without a dramatic loss of accuracy. In some cases, scaling the top mode coefficients by 0.95 suffices to maintain stability without greatly affecting accuracy when compared with stable unfiltered simulation results.

As an illustration of this, we show in Table 6.3 the results of the same computation as in Example 6.1, Table 6.2, although with a filter that removes 5% of the top mode. A direct comparison confirms that the impact on the accuracy is marginal.

However, aliasing-driven instabilities can occur for the nodal DG method when used on problems that are more severely underresolved. As discussed in Section 5.3 we can also revert to a more traditional, and more expensive, formulation using exact integration to address this problem. Recall that in Section 5.2 we proved a nonlinear stability result in the absence of aliasing. To benefit from this strong result we must approximate the inner products in the DG variational equations with a cubature rule for the volume integrals [84, 85, 294] and a Gauss quadrature rule for the surface flux integrals. As we will discuss in more detail shortly, one can adjust the accuracy of the integration by choosing sufficiently many points in these integration rules to reduce the aliasing errors. However, we should note that the primitive variables required to compute the nonlinear terms are rational functions of the conserved variables and, thus, the flux functions involved in the inner products are themselves rational polynomials. This implies that we are bound to induce some aliasing errors even with an integration rule approach. One approach to overcome this problem, discussed in [14], is to analytically expand the rational functions in polynomials and then integrate accurately. This approach is, however, quite expensive in computational cost.

The sequence of operations required when using an integration-based DG formulation is slightly different than the very straightforward nodal DG. Before we outline these steps, we will briefly discuss how the cubature rules can be used to evaluate volume integrals and how the Gauss quadrature rules are used for computing surface integrals.

Cubature-Based Integration

A cubature rule approximates the general volume integral

$$\int_I f(\boldsymbol{r})\, d\boldsymbol{r} \approx \sum_{i=1}^{N_c} f(\boldsymbol{r}_i^c) w_i^c,$$

and consists of a set of N_c nodes, $\boldsymbol{r}_i^c$, and weights, w_i^c. The order of accuracy of the rule is determined by assuming that f is the maximum order polynomial that can be integrated exactly. This again determines the number of nodes and weights needed. In the current implementation, cubature node and weight sets are available for exact integration of polynomials up to the 28th order and can be extracted through the function Cubature2D.m (not shown). This routine uses a lookup table to return a vector for the horizontal and vertical coordinates of the nodes and the associated weights, assuming that the domain of integration is the biunit right-angled triangle. For integration rules suitable for exact integration of polynomials with degree greater than 28, the function creates a tensor product of Gauss and Gauss-Radau quadrature rules [89, 197].

To compute an inner product of two polynomial functions, given by their values at the nodes of the triangle, we interpolate both functions to the cubature nodes and compute the weighted discrete inner product of the two vectors of interpolated functions using the cubature integration weights. Suppose we are given the two vectors $\mathbf{f}_h$ and $\mathbf{g}_h$ and we compute $\mathbf{f}^c = \mathcal{V}^c \mathcal{V}^{-1} \mathbf{f}_h$ and $\mathbf{g}^c = \mathcal{V}^c \mathcal{V}^{-1} \mathbf{g}_h$. Here, $\mathcal{V}$ is the regular $N_p \times N_p$ Vandermonde matrix for the standard nodes and triangle and $\mathcal{V}^c$ is the $N_c \times N_p$ matrix with the entries $\mathcal{V}_{ij}^c = \psi_j(\boldsymbol{r}_i^c)$. In the case that $\boldsymbol{f}_h$ and $\boldsymbol{g}_h$ are nonlinear functions, one needs to perform these computations with more care to ensure proper treatment of the high-order polynomial terms. The inner product on the triangle is then approximated by

$$(f_h, g_h)_{\mathsf{D}^k} \approx \sum_{i=1}^{N_c} J_i^{c,k} w_i^c f_i^c g_i^c,$$

where $J_i^{c,k}$ is the Jacobian of the local to physical coordinate map for the k-th triangle. $J_i^{c,k}$ will be independent of N_c for a straight-sided triangle, but it will vary by cubature node for a curvilinear triangle (see Chapter 9).

To ensure a compact implementation of the cubature method, we store the information required to use the cubature rule in a structure, named cub.

This contains the nodal-to-cubature interpolation matrix in cub.V as well as the cubature weights combined with the coordinate map Jacobian in cub.W, the geometric factors in cub.rx, cub.sx, cub.ry, cub.sy, and the local derivatives of the Lagrange basis evaluated at the cubature points

$$\mathcal{D}^c_{r,(i,j)} = \sum_{n=1}^{N_p} \left. \frac{\partial \psi_n}{\partial r} \right|_{\mathbf{r}^c_i} \mathcal{V}^{-1}_{nj},$$

$$\mathcal{D}^c_{s,(i,j)} = \sum_{n=1}^{N_p} \left. \frac{\partial \psi_n}{\partial s} \right|_{\mathbf{r}^c_i} \mathcal{V}^{-1}_{nj}$$

for $1 \leq i \leq N_c$ and $1 \leq j \leq N_p$, in cub.Dr and cub.Ds with respect to the reference triangle, I. The basis ψ_n is the orthonormal basis in Eq. (6.6).

To compute the term

$$\left(\frac{\partial \phi_h}{\partial x}, F_h \right)_{\mathsf{D}^k} + \left(\frac{\partial \phi_h}{\partial y}, G_h \right)_{\mathsf{D}^k},$$

we first interpolate the conservative variables from the triangular nodal points to the cubature nodes, and then compute the vector flux functions at each cubature node as in the following script:

```
                         ┌──────────────────────┐
─────────────────────────│ CurvedEulerRHS2D.m   │─────────────────────────
                         └──────────────────────┘
 9   % 1.1 Interpolate solution to cubature nodes
10   cQ = zeros(cub.Ncub, K, 4);
11   for n=1:4, cQ(:,:,n) = cub.V*Q(:,:,n); end;
12
13   % 1.2 Evaluate flux function at cubature nodes
14   gamma = 1.4;
15   [F,G,rho,u,v,p] = EulerFluxes2D(cQ, gamma);
16
17   % 1.3 Compute volume terms (dphidx, F) + (dphidy, G)
18   rhsQ = zeros(Np, K, 4);
19   for n=1:4
20     ddr = (cub.Dr')*(cub.W.*(cub.rx.*F(:,:,n) + cub.ry.*G(:,:,n)));
21     dds = (cub.Ds')*(cub.W.*(cub.sx.*F(:,:,n) + cub.sy.*G(:,:,n)));
22     rhsQ(:,:,n) = ddr + dds;
23   end
```

Unlike the nodal DG volume flux term, which is exact when F_h and G_h are polynomials of order N on straight-sided elements, this formulation is exact for polynomial fluxes of order $M+1-N$, where M is the order of the cubature scheme. Thus, we can increase the integration order M to reduce the effects of aliasing of the rational vector flux functions.

For curvilinear elements, which we shall discuss in more detail in Chapter 9, we must also carefully treat the geometric factors, as these are also rational polynomials. However, as noted in Section 6.1, the denominator of $\frac{\partial r}{\partial x}$, $\frac{\partial r}{\partial y}$, $\frac{\partial s}{\partial x}$, and $\frac{\partial s}{\partial y}$ is the Jacobian J, which conveniently cancels out the Jacobian in the local-to-physical coordinate mapping. In the case of an isoparametric mapping, the cubature rule will typically need to be exact for polynomials of order $M + 2 - 2N$; that is, we should expect to increase the number of points in the cubature rule used for curvilinear triangles to also account for the deformation of the triangle.

A careful treatment of the surface terms in the variational formulation of DG for Euler's equations requires an accurate evaluation of univariate integrals along each face of the triangle. To be precise we need to accurately approximate

$$\int_{\partial \mathsf{D}^k} (\hat{n}_x \mathbf{F}_h + \hat{n}_y \mathbf{G}_h)^* \phi_h \, d\boldsymbol{x},$$

where $\phi_h(x)$ is a polynomial test function. As with the volume integrals, we must allow for the rational nature of the flux vector functions, and the possibility of nonconstant normals and surface coordinate Jacobians in the case of curvilinear triangles. A judicious application of a univariate quadrature rule, Section 3.1, suits this task admirably. The sequence of operations is similar to that required for the volumetric integrals; that is, we interpolate the conserved variables to Gauss quadrature nodes located on each of the faces of each of the elements. To evaluate F^* and G^*, we need to compute the average of the negative and positive traces of the vector flux functions at each of the Gauss nodes and the jump in the traces of the conserved variables at the Gauss nodes. Next, we evaluate the local Lax-Friedrichs fluxes at the Gauss nodes and integrate against the test functions using the Gauss quadrature rule.

This is achieved in the following section of economical code in CurvedEulerRHS2D.m, which also resorts to the EulerLF2D.m function to evaluate the local Lax-Friedrichs fluxes as per the nodal DG EulerRHS2D.m code.

CurvedEulerRHS2D.m

```
25  % 2.1 SURFACE TERMS (using Gauss nodes on element faces)
26  nx = gauss.nx; ny = gauss.ny;
27  mapW = gauss.mapW; mapI = gauss.mapI; mapO = gauss.mapO;
28  mapB = gauss.mapB; mapC = gauss.mapC;
29
30  % 2.2 Interpolate solution to Gauss surface nodes
31  gQM = zeros(gauss.NGauss*Nfaces, K, 4);
32  gQP = zeros(gauss.NGauss*Nfaces, K, 4);
33  for n=1:4
34    gQ = gauss.interp*Q(:,:,n);
35    gQM(:,:,n) = gQ(gauss.mapM);   gQP(:,:,n) = gQ(gauss.mapP);
36  end
37
38  % 2.3 Apply boundary conditions to '+' traces
39  if(~isempty(SolutionBC))
```

```
40      gQP = feval(SolutionBC, gauss.x, gauss.y, gauss.nx, gauss.ny, ...
41                  gauss.mapI,gauss.mapO, gauss.mapW, gauss.mapC, ...
42                  gQP, time);
43      end
44
45      % 2.4 Evaluate surface flux functions with stabilization
46      switch fluxtype
47        case {'LF'}
48          [flux] = EulerLF2D (nx, ny, gQM, gQP, gamma);
49        case {'Roe'}
50          [flux] = EulerRoe2D(nx, ny, gQM, gQP, gamma);
51        case {'HLL'}
52          [flux] = EulerHLL2D(nx, ny, gQM, gQP, gamma);
53        case {'HLLC'}
54          [flux] = EulerHLLC2D(nx, ny, gQM, gQP, gamma);
55      end
```

We typically prefer a distinct treatment of straight-sided and curvilinear elements. For the straight sided elements, when we multiply by the inverse mass matrix we use the fact that the Jacobians are constant and that the mass matrix inverse on the reference triangle is given by $\mathcal{V}\mathcal{V}^T$. For the curvilinear elements, the mass matrix is different in each element and we use a predetermined Cholesky factorization of the element-specific mass matrices to treat these elements individually. One might hope that the curvilinear elements are in the minority of elements, and this is typically true because elements are only typically curved on or near domain boundaries. Thus, the storage overhead required to store these custom Cholesky factors may not be a significant limiting factor for a simulation.

—————————————————— CurvedEulerRHS2D.m ——————————————————

```
57      % 2.5 Compute surface integral terms
58      for n=1:4
59        rhsQ(:,:,n) = rhsQ(:,:,n) - gauss.interp'*(gauss.W.*flux(:,:,n));
60      end
61
62      % 3.1 Multiply by inverse mass matrix
63      for n=1:4
64        % 3.1.a Multiply straight sided elements by inverse mass matrix
65        rhsQ(:,straight, n) = V*V'*(rhsQ(:,straight, n)./J(:,straight));
66
67        % 3.1.b Multiply curvilinear elements by custom inverse
68        %       mass matrices
69        Ncurved = length(curved);
70        for m=1:Ncurved
71          k = curved(m);
72          mmCHOL = cub.mmCHOL(:,:,k);
```

```
73        rhsQ(:,k,n) = mmCHOL\(mmCHOL'\rhsQ(:,k,n));
74      end
75    end
```

This concludes the evaluation of the right-hand-side residual for the Euler equations. However, the code is not particularly specific to the Euler equations. In fact, to handle a different set of hyperbolic equations, we would need to replace the number of fields being treated (currently four) and the function that evaluates the stabilized normal fluxes into the elements. Otherwise, the code is generic and is easily extendable to alternative hyperbolic systems with smooth solutions.

Example 6.2. As a second example, we consider the flow, often called a Couette flow, between two concentric cylinders spinning at a differential velocity. The exact solution is given as

$$\rho = 1,$$
$$\rho u = -\sin(\theta)\, u_\theta,$$
$$\rho v = \cos(\theta)\, u_\theta,$$
$$E = \frac{p}{\gamma - 1} + \frac{\rho}{2} u_\theta^2,$$

where the azimuthal velocity is given by $u_\theta = \frac{1}{75}\left(-r + \frac{16}{r}\right)$ and the pressure is given by $p = 1 + \frac{1}{75^2}\left(\frac{r^2}{2} - 32\ln(r) - \frac{128}{r^2}\right)$. The polar coordinates (r, θ) are given by $\theta = \tan^{-1}\left(\frac{y}{x}\right)$ and $r = \sqrt{x^2 + y^2}$. The exact solution is used as the initial condition and to supply exact boundary conditions. The rate of convergence is estimated using the sequence of meshes shown in Fig. 6.11 with a full curvilinear representation of the circular boundaries.

In Table 6.4 we illustrate the accuracy of the scheme through the L^2-errors of the density and one of the momentum components. We again note a convergence order close to the expected $\mathcal{O}(h^{N+1/2})$, as discussed in Section 5.5.

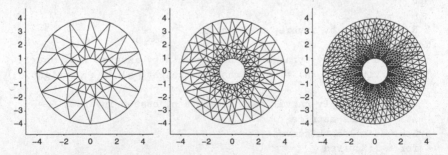

Fig. 6.11. Sequence of three meshes used to perform convergence analysis of the rotating Couette flow between concentric cylinders.

Table 6.4. L^2-errors in ρ and ρu for the rotating Couette flow between concentric cylinders example on a sequence of meshes shown in Fig. 6.11 with different order approximations and full order integration of fluxes and curvilinear representations of the circular boundary. An $*$ indicates that the error level is affected by round-off errors.

	Error in ρ				Error in ρu			
N	h	$h/2$	$h/4$	Rate	h	$h/2$	$h/4$	Rate
1	9.89E-04	3.73E-04	1.02E-04	1.64	3.81E-03	1.04E-03	2.68E-04	1.92
2	2.55E-04	4.62E-05	5.53E-06	2.76	5.87E-04	9.16E-05	1.28E-05	2.76
3	4.23E-05	4.09E-06	2.52E-07	3.70	9.43E-05	8.44E-06	6.60E-07	3.58
4	7.85E-06	3.64E-07	1.05E-08	4.77	1.55E-05	8.41E-07	3.58E-08	4.38
5	1.34E-06	3.53E-08	5.47E-10	5.63	2.69E-06	8.68E-08	2.00E-09	5.20
6	2.31E-07	3.43E-09	3.02E-11	6.45	4.78E-07	9.14E-09	1.14E-10	6.02
7	4.24E-08	3.29E-10	1.76E-12*	7.28*	8.62E-08	9.63E-10	6.37E-12*	6.86*
8	7.52E-09	3.39E-11	2.03E-13*	7.59*	1.54E-08	1.01E-10	8.74E-13*	7.05*

The small deviations from the optimal order shown in Table 6.4 are caused by the unstructured nature of the grid.

6.6.2 Numerical fluxes revisited

The results discussed in the above reflect the behavior of the scheme for problems with smooth solutions. However, as we discussed at length in Section 5.6, the situation becomes considerably more complex when we allow ourselves to consider problems with discontinuous solutions.

In such cases, we are inevitably driven to use a filtering or limiting process to control the Gibbs phenomenon that are responsible for solution spikes. In the case of the Euler equations, dramatic oscillations in the solution near shocks can drive the numerical pressure and densities to be small or even negative. This has a devastating effect, as the physical nature of the equations is destroyed close to the spurious oscillations.

As in the one-dimensional case, we have the choice between filtering, possibly combined with postprocessing, and limiting of the solution. The former approach is exactly as in the one-dimensional case and we refer to Section 5.6 for a discussion of this. However, to entirely remove the oscillations, we must apply a limiting procedure and we will discuss this in some detail shortly. Before doing so, it is worthwhile to revisit the subject of the numerical flux. This is particularly important in this context as the limiters will restrict the DG formulation to behave like a finite volume method in the vicinity of shocks and discontinuities. Thus, it is important to use an accurate numerical flux function, as that is now the main component of the solution process.

The significance of this problem is reflected in the considerable literature devoted to the development of accurate and efficient numerical flux schemes – in particular, in the context of finite volume methods. For a thorough and rigorous discussion and many details on this topic we refer to the excellent texts [218, 301]. A comparative study of a number of different fluxes in the context of a DG-FEM solution of the Euler equations is presented in [263].

Before discussing some specific numerical fluxes for the Euler equations, let us revisit the general problem of the flux construction. In Section 2.4 we discussed in detail the construction of a proper numerical upwind flux for the linear hyperbolic problem

$$\frac{\partial u}{\partial t} + \mathcal{A}\frac{\partial u}{\partial x} = 0.$$

The extension to the two-dimensional case

$$\frac{\partial u}{\partial t} + \mathcal{A}_x \frac{\partial u}{\partial x} + \mathcal{A}_y \frac{\partial u}{\partial y} = 0,$$

is straightforward by considering the Rankine-Hugoniot wave problems associated with the direction matrix,

$$\hat{\mathcal{A}} = \hat{n}_x \mathcal{A}_x + \hat{n}_y \mathcal{A}_y.$$

Examples and further discussions of this for the linear problem are found in Section 6.5.

For the nonlinear problem, however, the situation is more complicated. In Section 2.4 we could, for the linear problem, extract exactly how the waves of the system would interact and, hence, recover the numerical flux from the left and right states of the system. The analytic solution of this basic problem, known as the Riemann problem for general conservation laws, is not possible for a general conservation law and its numerical solution is often prohibitively expensive.

It is therefore natural to seek an approximate solution to this Riemann problem, leading to what is known as approximate Riemann solvers. If we consider the conservation law

$$\frac{\partial u}{\partial t} + \frac{\partial f(u)}{\partial x} = 0,$$

it seems appropriate to approximate the numerical flux by linearization as

$$f^* = \hat{\mathcal{A}}u^*, \tag{6.10}$$

where, clearly, both $\hat{\mathcal{A}}$ and u^* depend on u^- and u^+ (i.e., the left and right states).

If we now assume that $\hat{\mathcal{A}}$ is diagonizable

$$\hat{\mathcal{A}}r_i = \lambda_i r_i,$$

where the eigenvalues, λ_i, are purely real due to the wavelike nature of the problem, we can follow the ideas in Section 2.4 and write

$$u^+ - u^- = \sum_i \alpha_i r_i.$$

Based on this, one easily shows that the solution to the linearized Riemann problem is given as

$$u^* = u^- + \sum_{\lambda_i \leq 0} \alpha_i r_i = u^+ - \sum_{\lambda_i \geq 0} \alpha_i r_i.$$

Multiply this with $\hat{A}$ and take the average of the two expressions to recover

$$\hat{A} u^* = \hat{A} \{\!\{u\}\!\} + \frac{1}{2} |\hat{A}| [\![u]\!], \tag{6.11}$$

where

$$|\hat{A}| = \mathcal{S} |\Lambda| \mathcal{S}^{-1},$$

where $\hat{A} = \mathcal{S} \Lambda \mathcal{S}^{-1}$.

For the nonlinear case, we must place certain conditions on the definition of $\hat{A}$. It is reasonable to require consistency in the sense that

$$\hat{A}(u^-, u^+) \to \frac{\partial f(u)}{\partial u} \quad \text{when} \quad (u^-, u^+) \to u.$$

Furthermore, we must require that $\hat{A}(u^-, u^+)$ is diagonizable and has purely real eigenvalues. Consider the jump in flux as

$$f(u^+) - f(u^-) = \int_0^1 \frac{df(u(\xi))}{d\xi} \, d\xi = \int_0^1 \frac{df(u(\xi))}{du} \frac{du}{d\xi} \, d\xi.$$

We now assume a linear dependence, often known as Roe linearization after [273], as

$$u(\xi) = u^- + (u^+ - u^-)\xi,$$

which, after insertion in the above expression, yields the Roe condition

$$f(u^+) - f(u^-) = \hat{A} \left(u^+ - u^- \right), \tag{6.12}$$

where

$$\hat{A} = \int_0^1 \frac{df(u(\xi)}{du} \, d\xi. \tag{6.13}$$

This suggests that the most natural extension of Eq. (6.11) to the nonlinear conservation law follows directly as

$$f^* = \{\!\{f\}\!\} + \frac{1}{2} |\hat{A}| [\![u]\!].$$

Note in particular that for the scalar problem, this is the local Lax-Friedrichs flux. Unfortunately, the exact evaluation of $\hat{A}$ using Eq. (6.13) is only possible in special cases, some of which we will discuss shortly. However, there are several suitable approximate choices, such as

$$\hat{A} = f_u(\{\{u\}\}),$$

or

$$\hat{A} = \{\{f_u\}\}.$$

In this latter case, one has to be careful to ensure that the flux Jacobian $\hat{A}$ has purely real eigenvalues.

Before we continue, it is illustrative to derive the Roe condition, Eq. (6.12), in a slightly different way. If we assume that the Riemann problem is dominated by one strong wave, propagating with speed s, then the Rankine-Hugoniot condition (see Section 2.4) would require that

$$f(u^+) - f(u^-) = s\left(u^+ - u^-\right).$$

If the solution should also be a solution to the linearized Riemann problem, we recover a similar statement from Eq. (6.10)

$$\hat{A}\left(u^+ - u^-\right) = s\left(u^+ - u^-\right),$$

from which the Roe condition follows directly. Hence, the condition reflects an assumption that the solution is dominated by one strong wave. This is a reasonable condition except in those rare cases where strong shocks interact. In such rare cases, a Riemann solver based on the above principle will be less effective.

Let us now return to specifics of the Euler equations and consider a couple of more advanced choices. In [273], a very elegant approach was introduced to enable the integration of the flux Jacobian and from this, the accurate realization of $\hat{A}$. For the two-dimensional case, the Euler system is linearized along a normal direction. The key issue is to identify the locally linearized mean state, used to define $\hat{A}$.

We use a rotation matrix to compute the normal and tangential components of the momentum and then compute the two flux vector functions in EulerRoe2D.m.

─────────────────────────── | EulerRoe2D.m | ───────────────────────────

```
9   % Rotate "-" trace momentum to face normal-tangent coordinates
10  rhouM = QM(:,:,2); rhovM = QM(:,:,3); EnerM = QM(:,:,4);
11  QM(:,:,2) =  nx.*rhouM + ny.*rhovM;
12  QM(:,:,3) = -ny.*rhouM + nx.*rhovM;
13
14  % Rotate "+" trace momentum to face normal-tangent coordinates
15  rhouP = QP(:,:,2); rhovP = QP(:,:,3); EnerP = QP(:,:,4);
16  QP(:,:,2) = nx.*rhouP + ny.*rhovP;
17  QP(:,:,3) =-ny.*rhouP + nx.*rhovP;
18
```

```
19   % Compute fluxes and primitive variables in rotated coordinates
20   [fxQM,fyQM,rhoM,uM,vM,pM] = EulerFluxes2D(QM, gamma);
21   [fxQP,fyQP,rhoP,uP,vP,pP] = EulerFluxes2D(QP, gamma);
```

The appropriate local mean states are based on weighted averages of left and right states along each face node, yielding

$$\rho^* = \sqrt{\rho^- \rho^+},$$

$$u^* = \frac{\sqrt{\rho^-}\,u^- + \sqrt{\rho^+}\,u^+}{\sqrt{\rho^-} + \sqrt{\rho^+}},$$

$$v^* = \frac{\sqrt{\rho^-}\,v^- + \sqrt{\rho^+}\,v^+}{\sqrt{\rho^-} + \sqrt{\rho^+}},$$

$$H^* = \frac{\sqrt{\rho^-}\,H^- + \sqrt{\rho^+}\,H^+}{\sqrt{\rho^-} + \sqrt{\rho^+}},$$

where the enthalpy is defined by $H = \frac{E+p}{\rho}$. This state is used to evaluate a local speed of sound by

$$c^2 = (\gamma - 1)\left(H^* - \frac{(u^*)^2 + (v^*)^2}{2}\right).$$

────────────────────── EulerRoe2D.m ──────────────────────

```
23   % Compute enthalpy
24   HM = (EnerM+pM)./rhoM; HP = (EnerP+pP)./rhoP;
25
26   % Compute Roe average variables
27   rhoMs = sqrt(rhoM); rhoPs = sqrt(rhoP);
28
29   rho = rhoMs.*rhoPs;
30   u   = (rhoMs.*uM + rhoPs.*uP)./(rhoMs + rhoPs);
31   v   = (rhoMs.*vM + rhoPs.*vP)./(rhoMs + rhoPs);
32   H   = (rhoMs.*HM + rhoPs.*HP)./(rhoMs + rhoPs);
33
34   c2  = (gamma-1)*(H - 0.5*(u.^2 + v.^2)); c = sqrt(c2);
35
36   % Riemann fluxes
37   dW1 = -0.5*rho.*(uP-uM)./c + 0.5*(pP-pM)./c2;
38   dW2 = (rhoP-rhoM) - (pP-pM)./c2;
39   dW3 = rho.*(vP-vM);
40   dW4 = 0.5*rho .*(uP-uM)./c + 0.5*(pP-pM)./c2;
41
42   dW1 = abs(u-c).*dW1; dW2 = abs(u).*dW2;
43   dW3 = abs(u).*dW3; dW4 = abs(u+c).*dW4;
44
```

```
45    % Form Roe fluxes
46    fx = (fxQP+fxQM)/2;
47
48    fx(:,:,1) =fx(:,:,1) ...
49        -(dW1.*1      +dW2.*1 +dW3.*0+dW4.*1        )/2;
50    fx(:,:,2) =fx(:,:,2) ...
51        -(dW1.*(u-c) +dW2.*u  +dW3.*0+dW4.*(u+c)   )/2;
52    fx(:,:,3) =fx(:,:,3) ...
53        -(dW1.*v      +dW2.*v   +dW3.*1+dW4.*v       )/2;
54    fx(:,:,4) =fx(:,:,4) ...
55        -(dW1.*(H-u.*c)+dW2.*(u.^2+v.^2)/2 ...
56        +dW3.*v+dW4.*(H+u.*c))/2;
57
58    % rotate back to Cartesian
59    flux = fx;
60    flux(:,:,2) = nx.*fx(:,:,2) - ny.*fx(:,:,3);
61    flux(:,:,3) = ny.*fx(:,:,2) + nx.*fx(:,:,3);
```

This linearized Roe flux has been used extensively for finite volume methods and we will later show examples confirming its very good properties also in the context of discontinuous Galerkin methods.

As mentioned above, one of the basic assumptions underlying the Roe condition and the associated Roe flux is that the Riemann problem is dominated by one strong wave. For many problems, this is appropriate, but for some more complex situations, the Roe flux may be too simple, giving rise to various spurious phenomena.

The most immediate approach to overcome this restriction is to assume that the Riemann problem consists of three states, separated by two waves, s^- and s^+, with $s^+ > s^-$. This is the approach introduced in [145] which leads to numerical fluxes known as HLL (Harten-Lax-van Leer) fluxes.

Following this idea yields the numerical flux defined as

$$
f^*(u^-, u^+) = \begin{cases} f(u^-), & s^- \geq 0 \\ \frac{s^+ f(u^-) - s^- f(u^+) + s^+ s^- (u^+ - u^-)}{s^+ - s^-} & s^- \leq 0 \leq s^+, \\ f(u^+), & s^+ \leq 0, \end{cases}
$$

where s^- and s^+ must be estimates of the slowest and fastest wave speed, respectively, in the system. Clearly, if $s^- \geq 0$, all information propagates to the right and simple upwinding is appropriate; similarly for $s^+ \leq 0$.

The remaining challenge is to determine the estimates of the wave speeds in the system and there are several options for this. The original approach in [145] is to estimate the two speeds as

$$
s^- = \min\left(\lambda(\hat{A}(u^-)), \lambda(\hat{A}(u^*))\right), \quad s^+ = \max\left(\lambda(\hat{A}(u^+)), \lambda(\hat{A}(u^*))\right),
$$

where the maximum is taken over the eigenvalues of the linearized flux Jacobian. The intermediate state $\boldsymbol{u}^*$ is taken to be the Roe average. An example of this is shown in EulerHLL2D.m.

```
                         ┌─────────────┐
─────────────────────────│ EulerHLL2D.m │─────────────────────────
                         └─────────────┘
24  % Compute Roe average variables
25  rhoMs = sqrt(rhoM); rhoPs = sqrt(rhoP);
26
27  rho = rhoMs.*rhoPs;
28  u   = (rhoMs.*uM + rhoPs.*uP)./(rhoMs + rhoPs);
29  v   = (rhoMs.*vM + rhoPs.*vP)./(rhoMs + rhoPs);
30  H   = (rhoMs.*HM + rhoPs.*HP)./(rhoMs + rhoPs);
31
32  c2  = (gamma-1)*(H - 0.5*(u.^2 + v.^2)); c = sqrt(c2);
33
34  % Compute estimate of waves speeds
35  SL = min(uM-cM, u-c); SR = max(uP+cP, u+c);
36
37  % Compute HLL flux
38  t1 = (min(SR,0)-min(0,SL))./(SR-SL);
39  t2 = 1-t1;
40  t3 = (SR.*abs(SL)-SL.*abs(SR))./(2*(SR-SL));
41
42  for n=1:4
43    fx(:,:,n) = t1.*fxQP(:,:,n) + t2.*fxQM(:,:,n) ...
44                 - t3.*(QP(:,:,n)-QM(:,:,n));
45  end
46
47  % rotate flux back into Cartesian coordinates
48  flux(:,:,1) = fx(:,:,1);
49  flux(:,:,2) = nx.*fx(:,:,2) - ny.*fx(:,:,3);
50  flux(:,:,3) = ny.*fx(:,:,2) + nx.*fx(:,:,3);
51  flux(:,:,4) = fx(:,:,4);;
```

Another useful definition of the wave speeds is suggested in [304] as

$$s^{\mp} = \hat{\boldsymbol{n}} \cdot \boldsymbol{u}^{\mp} \mp c^{\mp} q^{\mp},$$

where

$$q^{\mp} = \begin{cases} 1, & p^* \le p^{\mp} \\ (1 + \frac{\gamma+1}{2\gamma}(p^*/p^{\mp} - 1))^{1/2}, & p^* > p^{\mp}. \end{cases}$$

Here,

$$p^* = \{\{p\}\} + \frac{1}{2}\left(\hat{\boldsymbol{n}} \cdot \boldsymbol{u}^- - \hat{\boldsymbol{n}} \cdot \boldsymbol{u}^+\right)\{\{\rho\}\}\{\{c\}\},$$

with the sound speed being

$$c = \sqrt{\frac{\gamma p}{\rho}}.$$

In relation to the Euler equations, the HLL formulation has the problem that only the dominating sound waves are represented, whereas contact waves are not included in the numerical flux. As already pointed out in the original work on HLL fluxes [145], the most obvious resolution to this shortcoming is to consider an approximate Riemann solver consisting of three or more waves. Fluxes based on such two-state Riemann solvers, often referred to as HLLC fluxes [304], follow the HLL flux schemes very closely, with the numerical flux being

$$
\boldsymbol{f}^*(\boldsymbol{u}^-, \boldsymbol{u}^+) = \begin{cases} \boldsymbol{f}(\boldsymbol{u}^-), & s^- \geq 0 \\ \boldsymbol{f}(\boldsymbol{u}^-) + s^-(\boldsymbol{u}^{*,-} - \boldsymbol{u}^-), & s^- \leq 0 \leq s^* \\ \boldsymbol{f}(\boldsymbol{u}^+) + s^+(\boldsymbol{u}^{*,+} - \boldsymbol{u}^+), & s^* \leq 0 \leq s^+ \\ \boldsymbol{f}(\boldsymbol{u}^+), & s^+ \leq 0, \end{cases}
$$

where the intermediate wave speed, s^*, is given as

$$
s^* = \hat{n} \cdot \{\!\{u\}\!\} + \frac{p^- - p^+}{\{\!\{\rho\}\!\}\{\!\{c\}\!\}}.
$$

The intermediate velocity is given as

$$
\boldsymbol{u}^{*,\mp} = \rho^{\mp} \frac{s^{\mp} - \hat{n} \cdot \boldsymbol{u}^{\mp}}{s^{\mp} - s^*} \begin{bmatrix} 1 \\ s^* \\ \frac{E^{\mp}}{\rho^{\mp}} + (s^* - \hat{n} \cdot \boldsymbol{u}^{\mp})(s^* + \frac{p^{\mp}}{\rho^{\mp}(s^{\mp} - \hat{n} \cdot \boldsymbol{u}^{\mp})}) \end{bmatrix}.
$$

A general-purpose two-state Riemann solver for hyperbolic conservation laws is discussed in [223].

We emphasize that there is a vast literature on this topic with many specialized developments and the reader is encouraged to explore these developments in [218, 263, 301, 303] and the many references therein.

6.6.3 Limiters in two dimensions

At the end of each Runge-Kutta stage we have computed an updated solution vector for each node, of each triangle. Even in the case of linear elements, these solutions may have spuriously large gradients unless the grid is very carefully chosen to align shocks and contact discontinuities in the solution with interelement boundaries. The slope limiters were introduced in finite volume methods to address this issue and we discussed it at length for the one-dimensional case in Section 5.6.

The extension of such techniques to the multidimensional case has been pursued by a number authors, beginning with [64], where a limiter ensuring a maximum principle is introduced. A generalization of this limiter is discussed further in [76]. Adding a fix for small densities/pressures make the methods very robust. Alternatives to such slope limiters have been discussed in [36] and also in [209, 212] to limit not only solution slopes but also higher-order solution moments. An entirely different approach is proposed in [253] based on the modification of the equation with a second order dissipative operator with a locally adapted coefficient. A significant obstacle to keeping an Euler simulation running is the tendency of the generic DG-FEM to generate solutions with locally negative density and/or pressure solution values. This most often happens at shocks where the Gibbs phenomenon creates solution oscillations and in naturally low-pressure/low-density wake regions of obstacles in impulsively started flows. Such negative density or pressure solutions change the nature of the partial differential equation locally, and hyperbolicity is lost. It is critical to minimize the occurrence of negative density and pressures, and, thus, it is sensible to consider limiting of the primitive variables (ρ, u, v, p). Using these variables allows us to apply positivity checks after variable gradients are limited.

Before we continue this discussion, it is worth mentioning that some of the other techniques discussed in Section 5.6 (e.g., filtering and postprocessing) applies directly to the multidimensional case also with the only caveat that most known postprocessing techniques remain essentially one-dimensional. The use of the filter to control aliasing-driven instabilities is equally efficient in multiple dimensions as we already saw in Section 6.6.1.

In the following we discuss an approach to limit solution gradients, but customized for the DG algorithms. We describe in particular the gradient limiting algorithm of [310], which is based on earlier work [186].

The limiting algorithm has a natural place in an $N = 1$ DG simulation, typically called after each stage of the Runge-Kutta timestepping scheme to control the solution gradients of the intermediate solutions as soon as they are computed. Extensions to $N > 1$ can be pursued following the same path as the generalized slope limiter discussed in Section 5.6.

The weighted limiter we consider here consists of six steps:

1. For each element, identify a patch as the element and its three neighbors, creating a ghost element as a neighbor at boundary faces if necessary.
2. Compute the cell averages of the primitive variables and associate these with the element centers of the elements in each patch.
3. For each face of each element, find the average of the primitive variables at each vertex of the face. Averaging is based on the primitive variables at the face vertices of the element and its neighbor (or ghost neighbor).
4. For each face of each element, use the cell average of the primitive variables in the element and its neighbor along that face as well as the averaged values of the primitive variables on the shared face to reconstruct an approximation of the gradient of the primitive variables at the face center. A weighted average of the reconstructed gradients is used to reconstruct a limited gradient at the center of the element for each primitive variable.
5. Once the limited gradients of the primitive variables are computed, we use the product rule to reconstruct the gradients of the conserved variables and then evaluate the first order accurate three-term Taylor expansion of

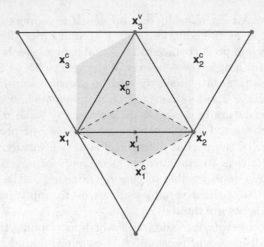

Fig. 6.12. Patch neighborhood used for the limiting algorithm.

the conserved variables while preserving the cell averages of the unlimited conserved variables.

6. The reconstructed conserved variables are tested to see if they will cause negative density or pressure and are moderated until this is avoided.

The limiter itself is implemented in EulerLimiter2D.m and we will describe the purpose of each significant part of the code. The algorithm requires geometric information about the patches of elements formed by grouping each element with its immediate neighbors. In the case of an element with a face on the domain boundary, we create an extra element by reflecting the element about its boundary face.

An example of a four-element patch is shown in Fig. 6.12. The base element is centered at $\mathbf{x}_0^c$ with three neighbors centered at $\mathbf{x}_1^c$, $\mathbf{x}_2^c$, and $\mathbf{x}_3^c$. We will use the cell averages, assumed to be collocated at the element centers (although this is not technically true for higher-order elements). In addition, we use the values of the conserved variables collocated at the element vertices $\mathbf{x}_1^v$, $\mathbf{x}_2^v$, and $\mathbf{x}_3^v$. Finally, we will create intermediate gradient values considered to collocate at the centers of the faces $\mathbf{x}_1^f$, $\mathbf{x}_2^f$, and $\mathbf{x}_3^f$.

We begin by evaluating the various coordinates for all elements and compute the areas of the quadrilaterals formed at each face by the two face vertices and two element centers of the two neighbors. This is encoded in the following, including an adjustment to find the centers of the ghost elements at domain boundaries.

─────────────────── EulerLimiter2D.m ───────────────────

```
14   % 1. compute geometric information for 4 element patch containing
15   %    each element
16   % Build average matrix
```

```
17   AVE = sum(MassMatrix)/2;
18
19   % Compute displacements from center of nodes for Taylor expansion
20   % of limited fields
21   dropAVE = eye(Np)-ones(Np,1)*AVE;
22   dx  = dropAVE*x; dy = dropAVE*y;
23
24   % Find neighbors in patch
25   E1 = EToE(:,1)'; E2 = EToE(:,2)'; E3 = EToE(:,3)';
26
27   % Extract coordinates of vertices and centers of elements
28   v1 = EToV(:,1); xv1 = VX(v1); yv1 = VY(v1);
29   v2 = EToV(:,2); xv2 = VX(v2); yv2 = VY(v2);
30   v3 = EToV(:,3); xv3 = VX(v3); yv3 = VY(v3);
31
32   % compute face unit normals and lengths
33   fnx = [yv2-yv1;yv3-yv2;yv1-yv3]; fny = -[xv2-xv1;xv3-xv2;xv1-xv3];
34   fL = sqrt(fnx.^2 + fny.^2); fnx = fnx./fL; fny = fny./fL;
35
36   % compute element centers
37   xc0 = AVE*x; xc1 = xc0(E1); xc2 = xc0(E2); xc3 = xc0(E3);
38   yc0 = AVE*y; yc1 = yc0(E1); yc2 = yc0(E2); yc3 = yc0(E3);
39
40   % Compute weights for face gradients
41   A0 = AVE*J*2/3; A1 = A0+A0(E1); A2 = A0+A0(E2); A3 = A0+A0(E3);
42
43   % Find boundary faces for each face
44   id1 = find(BCType(:,1)); id2 = find(BCType(:,2));
45   id3 = find(BCType(:,3));
46
47   % Compute location of centers of reflected ghost elements
48   % at boundary faces
49   H1 = 2*(A0(id1)./fL(1,id1));
50   xc1(id1) = xc1(id1) + 2*fnx(1,id1).*H1;
51   yc1(id1) = yc1(id1) + 2*fny(1,id1).*H1;
52
53   H2 = 2*(A0(id2)./fL(2,id2));
54   xc2(id2) = xc2(id2) + 2*fnx(2,id2).*H2;
55   yc2(id2) = yc2(id2) + 2*fny(2,id2).*H2;
56
57   H3 = 2*(A0(id3)./fL(3,id3));
58   xc3(id3) = xc3(id3) + 2*fnx(3,id3).*H3;
59   yc3(id3) = yc3(id3) + 2*fny(3,id3).*H3;
```

We transform the conserved variables to the primitive variables using the definition of the primitive variables:

$$u = \frac{\rho u}{\rho}, \quad v = \frac{\rho v}{\rho},$$

$$p = (\gamma - 1)\left(E - \frac{1}{2}\left(\rho u^2 + \rho v^2\right)\right).$$

For ghost elements at domain boundaries, we compute constrained cell averages of the conserved variables with the usual boundary condition specifications and then evaluate the primitive variables at the ghost cells. This requires one to manipulate the cell averages into a form compatible with the boundary conditions.

```
EulerLimiter2D.m
61  % 2. Find cell averages of conserved & primitive variables in each
62  % 4 element patch
63  % extract fields from Q
64  rho = Q(:,:,1); rhou = Q(:,:,2);
65  rhov = Q(:,:,3); Ener = Q(:,:,4);
66
67  % Compute cell averages of conserved variables
68  rhoC = AVE*rho; rhouC = AVE*rhou; rhovC = AVE*rhov;
69  EnerC = AVE*Ener;
70  averhou = ones(Np,1)*rhouC; averhov = ones(Np,1)*rhovC;
71  averho  = ones(Np,1)*rhoC;  aveEner = ones(Np,1)*EnerC;
72
73  % Compute primitive variables from cell averages of
74  % conserved variables
75  PCO(1,:,1)=rhoC; PCO(1,:,2)= rhouC./rhoC; PCO(1,:,3)=rhovC./rhoC;
76  PCO(1,:,4)=(gamma-1)*(EnerC - 0.5*(rhouC.^2 + rhovC.^2)./rhoC);
77
78  % Find neighbor values of conserved variables
79  PC(:,:,1)=rhoC(EToE'); PC(:,:,2)=rhouC(EToE');
80  PC(:,:,3)=rhovC(EToE'); PC(:,:,4)=EnerC(EToE');
81
82  % Find boundary faces
83  idW = find(BCType'==Wall); idI = find(BCType'==In);
84  idO = find(BCType'==Out); idC = find(BCType'==Cyl);
85
86  % Apply boundary conditions to cell averages of ghost cells
87  % at boundary faces
88  PC = feval(SolutionBC, [xc1;xc2;xc3], [yc1;yc2;yc3], ...
89                  fnx, fny, idI, idO, idW, idC, PC, time);
90  PC(:,:,2) = PC(:,:,2)./PC(:,:,1); PC(:,:,3) = PC(:,:,3)./PC(:,:,1);
91  PC(:,:,4) = (gamma-1)*(PC(:,:,4)-0.5*PC(:,:,1).* ...
92                  (PC(:,:,2).^2 + PC(:,:,3).^2));
```

Next, the average of the conserved variables across inter-element boundaries is computed. This requires a slightly nonstandard access pattern to the face nodes, restricted to the end nodes on each face rather than all the nodes on a face as we have previously required. Thus, we create reduced versions of the face node trace indexing arrays **vmapM** and **vmapP** that only reference the face vertices. Using these restricted arrays, we compute the face-averaged values for the conserved variables at each vertex of each face and, finally, transform from conserved to primitive variables.

```
                                ┌─────────────────┐
                                │ EulerLimiter2D.m │
                                └─────────────────┘
94  % 3. Compute average of primitive variables at face nodes
95  ids = [1;Nfp;Nfp+1;2*Nfp;3*Nfp;2*Nfp+1];
96  vmapP1 = reshape(vmapP, Nfp*Nfaces, K); vmapP1 = vmapP1(ids, :);
97  vmapM1 = reshape(vmapM, Nfp*Nfaces, K); vmapM1 = vmapM1(ids, :);
98
99  rhoA  = ( rho(vmapP1)+ rho(vmapM1))/2;
100 EnerA = (Ener(vmapP1)+Ener(vmapM1))/2;
101 rhouA = (rhou(vmapP1)+rhou(vmapM1))/2;
102 rhovA = (rhov(vmapP1)+rhov(vmapM1))/2;
103
104 uA = rhouA./rhoA; vA = rhovA./rhoA;
105 pA = (gamma-1)*(EnerA - 0.5*rhoA.*(uA.^2 + vA.^2));
106
107 PVA(:,:,1) = rhoA; PVA(:,:,2) = uA; PVA(:,:,3) = vA;
108 PVA(:,:,4) = pA;
```

The next stage of the limiting strategy is to use the cell averages and face vertex values for the primitive variables to reconstruct a gradient for each primitive variable at each face center. Consider each primitive variable independently and generically refer to the current field under consideration as V (not to be confused with the Vandermonde matrix, $\mathcal{V}$). Consider face 1 first and seek gradients at the face centers that satisfy the constraints:

$$V_1^c - V_0^c = (x_1^c - x_0^c) \left.\frac{\partial V}{\partial x}\right|_{(x_1^f, y_1^f)} + (y_1^c - y_0^c) \left.\frac{\partial V}{\partial y}\right|_{(x_1^f, y_1^f)},$$

$$V_2^v - V_1^v = (x_2^v - x_1^v) \left.\frac{\partial V}{\partial x}\right|_{(x_1^f, y_1^f)} + (y_2^v - y_1^v) \left.\frac{\partial V}{\partial y}\right|_{(x_1^f, y_1^f)},$$

where c, v, and f super indices of the x- and y-coordinates imply evaluation at the cell centers, vertices and face centers, respectively, as shown in Fig. 6.12. This can be solved explicitly for $\left.\frac{\partial V}{\partial x}\right|_{(x_1^f, y_1^f)}$, $\left.\frac{\partial V}{\partial y}\right|_{(x_1^f, y_1^f)}$ to give the following equations:

$$\left.\frac{\partial V}{\partial x}\right|_{(x_1^f, y_1^f)} = \frac{1}{2A^1} \left((V_1^c - V_0^c)(y_2^v - y_1^v) + (V_1^v - V_2^v)(y_1^c - y_0^c)\right),$$

$$\left.\frac{\partial V}{\partial y}\right|_{(x_1^f, y_1^f)} = \frac{-1}{2A^1} \left((V_1^c - V_0^c)(x_2^v - x_1^v) + (V_1^v - V_2^v)(x_1^c - x_0^c)\right),$$

with A^1 being the area of the quadrilateral spanned by the two face vertices and the two centers of the elements sharing face 1. Analogous equations hold for the other two faces.

```
                        ┌─────────────────┐
─────────────────────── │ EulerLimiter2D.m │ ───────────────────────
                        └─────────────────┘
115  % Loop over primitive variables
116  for n=1:4
117
118      % find value of primitive variables in patches
119      VC0 = PC0(1,:,n); VC1 = PC(1,:,n); VC2 = PC(2,:,n);
120      VC3 = PC(3,:,n); VA = PVA(:,:,n);
121
122      % Compute face gradients
123      dVdxE1 =  0.5.*( (VC1-VC0).*(yv2-yv1) ...
124                + (VA(1,:)-VA(2,:)).*(yc1 - yc0) )./A1;
125      dVdyE1 = -0.5.*( (VC1-VC0).*(xv2-xv1) ...
126                + (VA(1,:)-VA(2,:)).*(xc1 - xc0) )./A1;
127      dVdxE2 =  0.5.*( (VC2-VC0).*(yv3-yv2) ...
128                + (VA(3,:)-VA(4,:)).*(yc2 - yc0) )./A2;
129      dVdyE2 = -0.5.*( (VC2-VC0).*(xv3-xv2) ...
130                + (VA(3,:)-VA(4,:)).*(xc2 - xc0) )./A2;
131      dVdxE3 =  0.5.*( (VC3-VC0).*(yv1-yv3) ...
132                + (VA(5,:)-VA(6,:)).*(yc3 - yc0) )./A3;
133      dVdyE3 = -0.5.*( (VC3-VC0).*(xv1-xv3) ...
134                + (VA(5,:)-VA(6,:)).*(xc3 - xc0) )./A3;
```

The face gradients are then combined in an area weighted average as

$$
\left. \frac{\partial V}{\partial x} \right|_{(x_0^c, y_0^c)} \approx \frac{A^1 \left. \frac{\partial V}{\partial x} \right|_{(x_1^f, y_1^f)} + A^2 \left. \frac{\partial V}{\partial x} \right|_{(x_2^f, y_2^f)} + A^3 \left. \frac{\partial V}{\partial x} \right|_{(x_3^f, y_3^f)}}{A^1 + A^2 + A^3}.
$$

Following [310], we use the face gradient in lieu of the center gradient for ghost elements. The corresponding implementation is as follows:

```
                        ┌─────────────────┐
─────────────────────── │ EulerLimiter2D.m │ ───────────────────────
                        └─────────────────┘
136      dVdxC0 = (A1.*dVdxE1 + A2.*dVdxE2 + A3.*dVdxE3)./(A1+A2+A3);
137      dVdyC0 = (A1.*dVdyE1 + A2.*dVdyE2 + A3.*dVdyE3)./(A1+A2+A3);
138
139      dVdxC1 = dVdxC0(E1); dVdxC2 = dVdxC0(E2); dVdxC3 = dVdxC0(E3);
140      dVdyC1 = dVdyC0(E1); dVdyC2 = dVdyC0(E2); dVdyC3 = dVdyC0(E3);
141
142      % Use face gradients at ghost elements
143      dVdxC1(id1) = dVdxE1(id1); dVdxC2(id2) = dVdxE2(id2);
144      dVdxC3(id3) = dVdxE3(id3);
145      dVdyC1(id1) = dVdyE1(id1); dVdyC2(id2) = dVdyE2(id2);
146      dVdyC3(id3) = dVdyE3(id3);
```

The square of the magnitude of the maximum gradient is used to form a geometric weighted mean of the neighboring elements of each element according to the following equations:

$$g_1 = \left| \nabla V|_{(x_1^c, y_1^c)} \right|^2, \quad g_2 = \left| \nabla V|_{(x_2^c, y_2^c)} \right|^2, \quad g_3 = \left| \nabla V|_{(x_3^c, y_3^c)} \right|^2.$$

These gradients are subsequently used to form weights to recover a weighted sum of the average gradients in the neighborhood of each triangle; that is, we use a geometric mean to computed a limited gradient, as defined by

$$\nabla V|_{(x_0^c, y_0^c)} \approx w_1 \, \nabla V|_{(x_1^c, y_1^c)} + w_2 \, \nabla V|_{(x_2^c, y_2^c)} + w_3 \, \nabla V|_{(x_3^c, y_3^c)}, \tag{6.14}$$

where

$$w_1 = \frac{g_2 g_3 + \epsilon}{\left(g_1\right)^2 + \left(g_2\right)^2 + \left(g_3\right)^2 + 3\epsilon},$$

$$w_2 = \frac{g_1 g_3 + \epsilon}{\left(g_1\right)^2 + \left(g_2\right)^2 + \left(g_3\right)^2 + 3\epsilon},$$

$$w_3 = \frac{g_1 g_2 + \epsilon}{\left(g_1\right)^2 + \left(g_2\right)^2 + \left(g_3\right)^2 + 3\epsilon},$$

with the ϵ factor set to 10^{-10} used to avoid division by zero. In the case of a solution with constant gradients in the three neighboring elements, the weights will be very close to one-third and the original gradient will be reproduced. The weighted averages are computed as follows:

```
──────────────────────────── EulerLimiter2D.m ────────────────────────────
148   % Build weights used in limiting
149   g1 = (dVdxC1.^2 + dVdyC1.^2); g2 = (dVdxC2.^2 + dVdyC2.^2);
150   g3 = (dVdxC3.^2 + dVdyC3.^2);
151
152   epse = 1e-10; fac = g1.^2 + g2.^2 + g3.^2;
153   w1 = (g2.*g3+epse)./(fac+3*epse);
154   w2 = (g1.*g3+epse)./(fac+3*epse);
155   w3 = (g1.*g2+epse)./(fac+3*epse);
156
157   % Limit gradients
158   LdVdxC0 = w1.*dVdxC1 + w2.*dVdxC2 + w3.*dVdxC3;
159   LdVdyC0 = w1.*dVdyC1 + w2.*dVdyC2 + w3.*dVdyC3;
```

These limited gradients are used in a Taylor expansion about the barycenter of each triangle to reconstruct a gradient limited solution with

$$LV := \bar{V} + \begin{pmatrix} dx \\ dy \end{pmatrix} \cdot L \, \nabla V|_{(x_0^c, y_0^c)},$$

where for the k-th element,

$$dx = x - \frac{\int_{\mathsf{D}^k} x \, d\boldsymbol{x}}{|\mathsf{D}^k|}, \quad dy = y - \frac{\int_{\mathsf{D}^k} y \, d\boldsymbol{x}}{|\mathsf{D}^k|}, \quad \bar{V} = \frac{\int_{\mathsf{D}^k} V \, d\boldsymbol{x}}{|\mathsf{D}^k|},$$

and the limited gradient is given in Eq. (6.14). This increment for the Taylor expansion is implemented in the following as:

──────────────────────── EulerLimiter2D.m ────────────────────────

```
161   % Evaluate limited gradient and cell averages at all nodes of
162   % each element
163   dV(:,:,n) = dx.*(ones(Np,1)*LdVdxCO) + dy.*(ones(Np,1)*LdVdyCO);
164   aV(:,:,n) = ones(Np,1)*VCO;
```

Once the limited gradients of all the primitive variables are computed, the product rule is applied to evaluate the gradients of the related limited conserved variables. However, as noted above, the limited gradient may still yield negative density or pressure values after reconstruction. In the case of negative densities, we repeatedly halve the problematic gradients until all reconstructed densities are above a tolerance. For elements with negative pressures, evident after transforming the limited energy to a pressure, we limit the energy gradient to zero, thus preserving the mean.

──────────────────────── EulerLimiter2D.m ────────────────────────

```
167   % 4. Reconstruct conserved variables using cell averages and
168   % limited gradients
169   averho  = aV(:,:,1); aveu = aV(:,:,2); avev = aV(:,:,3);
170   avep = aV(:,:,4);
171   drho    = dV(:,:,1);  du = dV(:,:,2);   dv = dV(:,:,3);
172   dp = dV(:,:,4);
173
174   % Reconstruct and check for small densities and/or pressures
175   tol = 1e-2;
176
177   Lrho = averho + drho; ids = find(min(Lrho,[],1) < tol);
178   while(~isempty(ids))
179      disp('warning: correcting negative density')
180      drho(:,ids) = .5*drho(:,ids);
181      Lrho = averho + drho; ids = find(min(Lrho,[],1) < tol);
182   end
183
184   % Reconstruct momentum
185   Lrhou = averhou + averho.*du + drho.*aveu;
186   Lrhov = averhov + averho.*dv + drho.*avev;
187
188   % Reconstruct energy
189   dEner = (1/(gamma-1))*dp+0.5*drho.*(aveu.^2+avev.^2) ...
190                   +averho.*(aveu.*du+avev.*dv);
191   LEner = aveEner + dEner;
192
193   % Check for negative pressures => zero gradient
194   Lp   = (gamma-1)*(LEner - 0.5*(Lrhou.^2 + Lrhov.^2)./Lrho);
```

```
195    ids = find(min(Lp,[],1) < tol);
196    if(~isempty(ids))
197      disp('warning: correcting negative pressure');
198      Lp(:,ids) = aveEner(:,ids);
199    end
200
201    % Replace limited gradients with face gradient at boundary faces
202    LQ = zeros(Np, K, 4);
203    LQ(:,:,1) = Lrho; LQ(:,:,2) = Lrhou; LQ(:,:,3) = Lrhov;
204    LQ(:,:,4) = LEner;
```

In the following, we will test the limiter for a number of different problems. To prepare for the presence of strong shocks in the computations, we also switch the timestepping to the second-order, strong stability-preserving RK time integrator, in line with the discussion in Section 5.7.

Example 6.3. To test the impact of the limiter on the accuracy of a smooth solution, we revisit the Couette flow test case, introduced in Example 6.2. We repeat the computations on a sequence of grids, as in the original case. Shown in Table 6.5 are the L^2-errors for all four fields. $N = 1$ is used as we cannot expect higher than $\mathcal{O}(h^2)$ for the general case due to the limiter.

Table 6.5. L^2-errors for the rotating Couette example with limiting and a Roe-averaged linearized Riemann solver and $N = 1$.

Field	h	$h/2$	$h/4$	$h/8$	$h/16$	$h/32$	Rate
ρ	4.58E-03	2.14E-03	7.18E-04	1.75E-04	3.74E-05	8.01E-06	1.87
ρu	5.94E-02	2.03E-02	5.10E-03	1.00E-03	1.79E-04	3.41E-05	2.19
ρv	5.95E-02	2.01E-02	4.97E-03	9.62E-04	1.73E-04	3.32E-05	2.20
E	2.05E-02	9.80E-03	3.50E-03	7.84E-04	1.49E-04	2.95E-05	1.93

Forward Facing Step

Our first, more challenging test case involves supersonic uniform flow encountering a forward facing step. This was originally studied in detail in [325] and later in [76] in the context of DG-FEM. This is an interesting and challenging unsteady test case because the step corner generates a solution singularity and the impulsively started initial condition generates a pressure singularity in time at the front of the step.

The domain is shown in Fig. 6.13. The initial condition is a uniform flow given by

$$\rho = \gamma,$$
$$\rho u = 3\gamma,$$

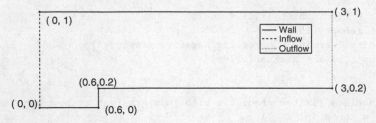

Fig. 6.13. Sketch of the domain and the boundary conditions used in forward facing step case.

$$\rho v = 0,$$
$$E = \frac{1}{\gamma - 1} + \frac{9\gamma}{2}.$$

Boundary conditions are reflective at the wall boundaries as

$$\rho^+ = \rho^-,$$
$$\rho u^+ = \rho u^- - 2\left(n_x \rho u + n_y \rho v\right) n_x,$$
$$\rho v^+ = \rho v^- - 2\left(n_x \rho u + n_y \rho v\right) n_y,$$
$$E^+ = E^-.$$

The inflow boundary condition is set to be the uniform Mach 3 flow, and no boundary conditions are applied at the outflow as it is assumed to be supersonic.

We consider this case with four successively, uniformly, refined meshes and have plotted Mach contours of the solution at time $T = 4$ in Fig. 6.14. In the first few steps of the simulation, the limiter detected and corrected for negative pressures. A Roe flux is used in these computations.

This problem has no analytic solution but a direct comparison with the results in [76, 325] gives confidence in the results.

Scramjet Inlet Flow Model

Our next test is a model for the flow into a scramjet engine. The domain geometry consists of a converging channel containing an obstacle, as shown in Fig. 6.15. The boundary conditions and initial conditions are the same as for the forward facing step problem. However, the flow geometry is quite different, as this simulation reaches steady state. We again compute the solution on a series of three meshes, but this time we apply an error estimate to determine which elements to refine at each level. The following indicator is used to mark elements for refinement:

$$\epsilon^k = \max\left(\frac{|m^+ - m^-| + \delta^2}{|m^-| + \delta}\right),$$

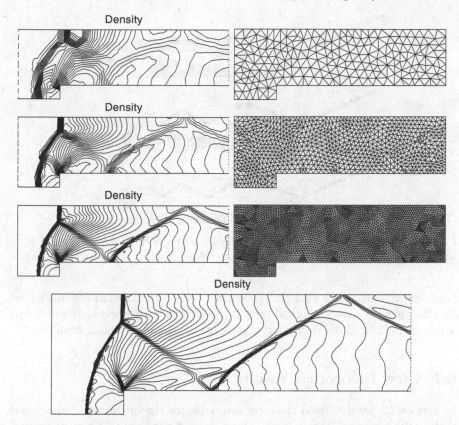

Fig. 6.14. Sequence of solutions to the Mach 3 forward facing step test case (with uniformly refined meshes of size $K = 381$, 1524, 6096, and 24384) using the Tu and Aliabadi limiter with an $N = 1$ DG method. Thirty equally spaced Mach contours are plotted in the range $[0.090388, 6.2365]$.

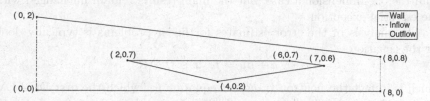

Fig. 6.15. Sketch of the domain and the boundary conditions used in scramjet inlet model case.

where the maximum is taken over all face nodes on element k and δ is a small number used to avoid dividing by zero. Here, m can be any of the conserved variables and we use the density ρ.

To achieve steady state, the simulation is run until a norm of the solution change between time levels drops below a set level of 10^{-3}. The solutions at

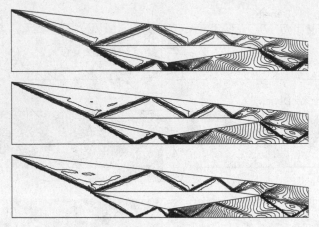

Fig. 6.16. Sequence of solutions to the Mach 3 scramjet inlet flow case (with $K = 9118, 15796, 30253$) using the Tu and Aliabadi limiter with an $N = 1$ DG method.

this nominal steady state for the three mesh levels are shown in Fig. 6.16. The Roe flux is used as a numerical flux. Comparisons with previously published solutions in [244] show very good agreement for the basic shock structure.

6.7 A few theoretical results

In Section 4.5 we discussed the error estimates for the linear one-dimensional scalar problems and in Section 5.5 we reviewed the results for systems and nonlinear problems with smooth solutions. We found in both cases that one can expect optimal convergence properties like $\mathcal{O}(h^{N+1})$. The extension of this result to the multidimensional case is, as it turns out, technically more complex than the one-dimensional case, but the main results remain unchanged with one essential exception.

The analysis of the error estimates for linear problems is typically done for the equation

$$\boldsymbol{\alpha} \cdot \nabla u + \beta u = f, \ \boldsymbol{x} \in \Omega \subset \mathsf{R}^2,$$

which is in fact the neutron transport equation for which the first DG-FEM was introduced in [269]. In early work [217], it was shown that if u is smooth and Ω is tiled with rectangular elements, each using an N-th-order tensor basis, the scheme exhibits optimal convergence; that is,

$$\|u - u_h\|_{\Omega,h} \le Ch^{N+1} \|u\|_{\Omega,N+2,h}.$$

In light of the one-dimensional results derived in Chap. 4.5 this is no surprise. However, it was also shown that for general grids, one should not expect more than $\mathcal{O}(h^N)$, again in agreement with very basic results using the Lax-Richmyer equivalence theorem.

In later work [190], this result was improved to

$$\|u - u_h\|_{\Omega,h} \le Ch^{N+1/2}\|u\|_{\Omega,N+1,h},$$

provided the triangulation is quasi-uniform in the sense that the angles in all triangles are bounded from below by a constant independent of the element size, h. In [190], much broader results were obtained also, including error estimates in general L^p-norms.

It remained unclear, however, whether this result in L^2 was sharp or simply a result of the analysis. In [272], this result was further improved to the optimal result

$$\|u - u_h\|_{\Omega,h} \le Ch^{N+1}\|u\|_{\Omega,N+2,h},$$

provided the grid is quasi-uniform and that all edges are bounded away from the characteristic direction (i.e., $|\boldsymbol{\alpha} \cdot \hat{\boldsymbol{n}}| > 0$ for all outward pointing normals on the element edges). In other words, this analysis does not capture the special case where edges are aligned with the characteristic direction.

To understand whether there are potential problems in this special case, let us consider an example, taken from [257].

Example 6.4. Consider the special case of the neutron equation as

$$\frac{\partial u}{\partial y} = 0, \quad \boldsymbol{x} \in [0,1]^2,$$

and with

$$u(x,0) = x.$$

It is easy to see that the exact solution is $u(x,y) = x$.

We now assume that $\Omega = [0,1]^2$ is triangulated using a simple grid where $h_x = h = 1/I$ and $h_y = h/2$ with I being the number of cells along x. In other words, the edge length along x is twice that along y, but otherwise the grid is entirely regular.

To solve the problem, we will use the simplest possible formulation, based on an $N = 0$ basis; that is, it is a finite volume method given on the cellwise form

$$\int_{\partial D^k} \hat{n}_y u^* \, d\boldsymbol{x} = 0.$$

Using upwinding and exploiting the geometry of the grid, it is easy to see that the local scheme is

$$u_{i,j+1} = \frac{1}{2} \left[u_{i-1/2,j} + u_{i+1/2,j} \right],$$

where we have defined the grid function

$$u_{i,j} = u_h \left(ih, \frac{jh}{2} \right) \quad \text{for} \quad \begin{cases} i = 1/2, 3/2, 5/2, \ldots, I - 1/2, & j = 0, 2, 4, \ldots, J, \\ i = 0, 1, 2, \ldots, I, & j = 1, 3, 5, \ldots, J - 1. \end{cases}$$

Induction now allows one to write down the exact numerical solution as

$$u_{i,j} = \frac{1}{2^j} \sum_{q=0}^{j} \binom{j}{q} u_{j/2-q+i,0} = \frac{1}{2^j} \sum_{q=0}^{j} \binom{j}{q} |h(j/2 - q + i)|,$$

since $u(x,0) = x$.

Note that along $x = 0$ and $x = 1$, the grid is, by necessity, aligned with the characteristic direction. Let us therefore consider the solution $u_{0,1-h/2}$ given as

$$u_{0,J-1} = \frac{1}{2^j} \sum_{q=0}^{j} \binom{j}{q} |h(j/2 - q)| = \sum_{j=1,3,5,\dots}^{J+1} \frac{h}{2^j} \binom{j-1}{(j-1)/2},$$

where we refer to [257] for a derivation of the last reduction. Using Stirling's formula, we have

$$\binom{a}{a/2} = \frac{a!}{(a/2)!(a/2)!} \geq c2^{a+1} a^{-1/2},$$

such that

$$u_{0,J-1} \geq ch \sum_{j=1,3,5,\dots}^{J+1} \frac{1}{\sqrt{j-1}} \geq ch\sqrt{J} \geq ch^{1/2},$$

as $J \propto h^{-1}$. Since the exact solution is $u(0,y) = 0$, this result also reflects the pointwise error and indicates the possibility of losing optimality for certain special grids.

The above example reflects that if $\hat{n} \cdot \boldsymbol{\alpha} = 0$, there may be a loss of optimal convergence rate, at least for $N = 0$. In [257], this was used to construct a problem, based on the above example with a grid with many vertical grid lines, which demonstrates the loss of optimality in L^2 and confirms that the original result in [190] of

$$\|u - u_h\|_{\Omega,h} \leq Ch^{N+1/2}\|u\|_{\Omega,N+1,h},$$

is in fact sharp. It should be emphasized, though, that the suboptimal behavior should be expected for very specific grids only and one can generally expect optimal order for DG-FEM when solving problems with smooth solutions on general unstructured grids. Furthermore, for linear problems one can naturally guarantee that this special case does not happen by constructing the grid appropriately. This conforms well with what we saw in the example of Maxwell's equations in Section 6.5. The above discussion does not cover the central flux and, as we will see in Chapter 7, a loss of optimality in convergence rate is often associated with this choice of flux.

The results in Section 6.6 show that these results, as in the one-dimensional case, can be expected to carry over to nonlinear problems with smooth solutions as long as monotone fluxes are used. The proof of this for the multidimensional case is given in [337, 338], again assuming the use of upwind fluxes.

For the general nonlinear case with nonsmooth solutions, the main complication lies in the incompatibility of the solution being high-order accurate and total variation diminishing (TVD) at the same time. In [132] it is shown that a scheme with the TVD property for solving conservation laws in two dimensions can at most be first-order accurate. The discussion in Section 5.6.2 likewise indicates a reduction of errors for TVD schemes due to the behavior of the limiter around local maxima, even if the solution is smooth.

The best one can therefore hope for when applying the general DG schemes to a nonlinear problem is for it to be total variation bounded (TVB) and the development of such results are presented in [64, 76], including both the multidimensional scalar and system cases. They are similar, in spirit, to those in Section 5.6.2, with a main component being the development of a suitable limiter to enforce the TVB property. The proofs themselves focus on the stability in L^∞ rather than in the TVB norm, which appears too complex. The convergence estimates are similar to those stated in Section 5.8, confirming also optimal error estimates for smooth problems.

6.8 Exercises

1. Consider a smooth function $u(x) = \cos(\pi x)\cos(\pi y)$ for $x \in \mathsf{I}$ (i.e., the reference triangle).
   a) Confirm spectral convergence, exponential decay of the error, of the nodal representation for increasing N. [Note: It is not sufficient to measure the error at the nodal points. Why?]
   b) Repeat the test by showing spectral convergence for the gradient of u using the differentiation matrices.

2. Consider a continuous function $u(x) = |x + 0.5| * |y + 0.5|$ for $x \in \mathsf{I}$ (i.e., the reference triangle). Compute the convergence rate for increasing N – it is as expected. Explain. [Note: It is not sufficient to measure the error at the nodal points. Why?]

3. Consider a continuous function $u(x) = |x + 0.5| * |y + 0.5|$ for $x \in \mathsf{I}$ (i.e., the reference triangle).
   a) Compare the pointwise error for different values of N with that obtained by filtering the approximation to $u(x)$. Do you observe any advantages in the use of the filter?
   b) Experiment with the filter parameters to optimize the quality of the approximation.

4. In this chapter we introduced the upwind DG for Maxwell's equations in the time-domain. All domain boundaries were assumed to be perfectly electrically conducting (i.e. reflective). However, this precludes a whole

class of exterior problems where the goal is to simulate electromagnetic scattering from an object (e.g., radar bouncing off an aircraft) for which a natural condition is the Sommerfeld condition requiring that the scattered field decays with distance from the scatterer. Unfortunately, it is not obvious how to apply this boundary condition since it applies in the limit $|\mathbf{x}| \to \infty$. A popular approach is to create a finite, convex, polygonal, domain containing the scattering object and to pad this domain with a buffer region. Damping terms are added to the original partial differential equations designed to dissipate the solution as it propagates through the extended part of the domain. The perfectly matched layer (PML) originally proposed in [25, 26] enhanced this idea by designing the material properties of the damping regions so that traveling waves enter them without generating spurious reflections at the start of the layer.

A particularly simple PML designed by Abarbanel and Gottlieb [2] is formed by adding extra terms and equations to Maxwell's equations that do not involve spatial derivatives. The modified equations for the transverse mode Maxwell equations are

$$\frac{\partial H^x}{\partial t} = -\frac{\partial E^z}{\partial y} - \sigma^y \left(2H^x + P^y\right),$$

$$\frac{\partial H^y}{\partial t} = \frac{\partial E^z}{\partial x} - \sigma^x \left(2H^y + P^x\right),$$

$$\frac{\partial E^z}{\partial t} = -\frac{\partial H^x}{\partial y} + \frac{\partial H^y}{\partial x} - \frac{d\sigma^x}{dx}Q^x + \frac{d\sigma^y}{dy}Q^y,$$

$$\frac{\partial P^x}{\partial t} = \sigma^x H^y,$$

$$\frac{\partial P^y}{\partial t} = \sigma^y H^x,$$

$$\frac{\partial Q^x}{\partial t} = -\sigma^x Q^x - H^y,$$

$$\frac{\partial Q^y}{\partial t} = -\sigma^y Q^y - H^x,$$

where $\sigma^x = \sigma^x\left(x\right)$ and $\sigma^y = \sigma^y\left(y\right)$ define the material property of rectangular absorbing regions. For instance,

$$\sigma^x := \begin{cases} 0 & |x| < 1 \\ \sigma_0^x\left(x-1\right)^p & x \geq 1 \\ \sigma_0^x\left(x+1\right)^p & x \leq -1, \end{cases}$$

$$\sigma^y := \begin{cases} 0 & |y| < 1 \\ \sigma_0^y\left(y-1\right)^p & y \geq 1 \\ \sigma_0^y\left(y+1\right)^p & y \leq -1, \end{cases}$$

defines a PML that will damp the solution in the region exterior to $|x|, |y| \leq 1$. The solution will decay exponentially fast as it propagates through the layer, with the rate determined by the reference parameters σ_0^x and σ_0^y. p is a measure of the smoothness of the absorption profile. Take $p = 2$ unless otherwise stated.

Modify MaxwellRHS2D.m and Maxwell2D.m to include these extra terms. A simple way to achieve this is to evaluate the σ^x and σ^y fields (and derivatives) at every point in the entire domain and solve the full PML equations everywhere despite the observation that these terms are only nonzero in the damping region. The extra terms also involve products of the damping factors and the solution fields. One approach to handling this is to evaluate these products at every interpolation node point – for instance, evaluating $\mathcal{I}_N \left(-\sigma^y \left(2H^x + P^y \right) \right)$ and adding it to the right-hand-side term for H^x and similarly for the other additional terms. Finally, in practice you should align the boundaries of the inner region with element domain boundaries, as the solution may exhibit low regularity exactly at these interfaces.

To test your PML enhanced Maxwell's solver, create a mesh for a square domain $|x|, |y| \leq 1$ and extend it to $|x|, |y| \leq 2$ in order to accommodate the PML buffer region. Introduce a Gaussian pulse in the electric field, E^z, and solve to FinalTime $= 10$. Experiment with the values used for the reference damping factors σ_0^x and σ_0^y to determine how they affect the solution decay rate in the PML. Do you notice backscatter from the outer boundary domain? You may well need to reduce the timestep and increase resolution in the PML layer, if you choose large reference damping factors, in order to accurately resolve the fast decay of the solution as it enters the layers.

One way to quantify the quality of the PML method is to solve the problem in a large domain, $|x|, |y| < 4$. For the time $t < 6$, reflections from the outer boundary cannot impact the physical domain, $|x|, |y| < 1$, where we can assume to have an exact solution. Use this approach to evaluate the quality of the PML as a function of p and the width of the PML layer.

7

Higher-order equations

So far, we have only considered problems with first order spatial derivatives (e.g., as in conservation laws), and shown the methods to perform well and in agreement with the strong theoretical foundation. It is natural to ask whether one can extend the formulation to include more general problem types. This is the topic of this chapter and, as we will see shortly, the generalization to deal with higher-order spatial operators is less direct than one would expect. Let us consider the following simple example, taken from [288].

Example 7.1. Consider the linear heat equation

$$\frac{\partial u}{\partial t} = \frac{\partial^2 u}{\partial x^2}, \quad x \in [0, 2\pi],$$

with periodic boundary conditions and $u(x, 0) = \sin(x)$. The exact solution is easily found as

$$u(x, t) = e^{-t} \sin(x).$$

Based on the previous discussions of the discontinuous Galerkin methods, it is tempting to simply write the heat equation as

$$\frac{\partial u}{\partial t} - \frac{\partial}{\partial x} u_x = 0,$$

and then identify u_x as the flux in the first order equation. The resulting scheme becomes

$$v_h^k = \mathcal{D}_r u_h^k, \quad \mathcal{M}^k \frac{d u_h^k}{dt} - \mathcal{S} v_h^k = -\int_{\partial \mathsf{D}^k} \hat{n} \cdot \left(v_h^k - v^* \right) \ell^k(x) \, dx,$$

in each element, k. A reasonable choice for the flux could be a simple central flux (i.e., $v^* = \{\{v_h\}\}$) since the heat equation has no preferred direction of propagation.

Table 7.1. Global L^2-errors for solving the heat equation using K elements, each with a local order of approximation, N, using the scheme in Example 7.1. A '–' marks that the algorithm is unstable.

$N \backslash K$	10	20	40	80	160
1	4.27E-1	4.34E-1	4.37E-1	4.38E-1	4.39E-1
2	5.00E-1	4.58E-1	4.46E-1	4.43E-1	4.42E-1
4	1.68E-1	1.37E-1	1.28E-1	1.26E-1	–
8	7.46E-3	8.60E-3	–	–	–

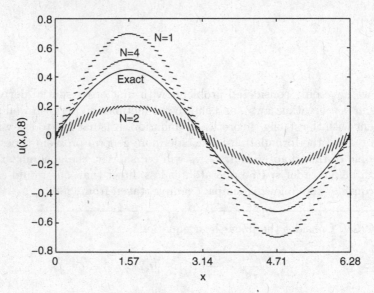

Fig. 7.1. Computed solutions to the heat equation using the scheme in Example 7.1 with $K = 80$ elements and different orders of approximation.

In Table 7.1 we show the global L^2-error (i.e., the error measured in $\| \cdot \|_{\Omega,h}$), under refinement both in K and N. The results are disappointing, displaying both lack of convergence and/or regular instability at high resolutions.

To further illustrate the problem, we show the computed solutions in Fig. 7.1 for fixed K and increasing values of N. Increasing K yields similar results as expected from Table 7.1.

A careful analysis [335] reveals that the scheme is both inconsistent and weakly unstable, consistent with the results in Table 7.1. The instability is driven by roundoff errors, which, in combination with the weak nature of the instability, explains why one does not observe the instability for some low-resolution cases in Table 7.1.

7.1 Higher-order time-dependent problems

From the above example it is clear that a new idea is required to use the discontinuous Galerkin method for problems with higher spatial derivatives. This new idea, first proposed in [22], is to rewrite the high spatial derivative as a system of first-order equations; for example, if we need to solve the (well-posed) problem

$$\frac{\partial u}{\partial t} = \sum_{p=1}^{P} \frac{\partial}{\partial x}\left(a_p \frac{\partial^{p-1}u}{\partial x^{p-1}}\right) + a_0 u,$$

subject to suitable boundary conditions, we discretizing the system

$$\frac{\partial u}{\partial t} = \sum_{p=1}^{P} \frac{\partial a_p q_{p-1}}{\partial x} + a_0 u,$$

with

$$\forall p = 1, \ldots, P-1: \; q_p = \frac{\partial^p q}{\partial x^p}, \;\; q_0 = u.$$

At first, this seems to introduce several disadvantages (e.g., the memory usage increases significantly with the use of the system rather than the scalar formulation). Furthermore, it also appears to be more expensive to evaluate the spatial derivative using a first-order derivative twice rather than a directly defined second-order operator.

In the following, we nevertheless pursue this idea in some detail for different type of problems to illustrate that this overhead is not as bad as it appears if we choose the numerical fluxes using certain guidelines. Subsequently, we also briefly revisit the issue of computational cost and consider a few special ways of addressing this.

7.1.1 The heat equation

Let us first return to the heat equation discussed in Example 7.1 and follow the idea introduced above. We consider

$$\frac{\partial u}{\partial t} = \frac{\partial}{\partial x} a(x) \frac{\partial u}{\partial x},$$

with $u = 0$ at the outer boundaries. It is easily shown that $a(x) > 0$ suffices to guarantee well-posedness. We rewrite this problem as

$$\frac{\partial u}{\partial t} = \frac{\partial}{\partial x}\sqrt{a}q, \;\; q = \sqrt{a}\frac{\partial u}{\partial x},$$

to recover a system of first order equations. This we can discretize using the techniques developed for the conservation laws; that is, we assume that (u, q) can be approximated as

$$\begin{bmatrix} u(x,t) \\ q(x,t) \end{bmatrix} \simeq \begin{bmatrix} u_h(x,t) \\ q_h(x,t) \end{bmatrix} = \bigoplus_{k=1}^{K} \begin{bmatrix} u_h^k(x,t) \\ q_h^k(x,t) \end{bmatrix} = \bigoplus_{k=1}^{K} \sum_{i=1}^{N_p} \begin{bmatrix} u_h^k(x_i,t) \\ q_h^k(x_i,t) \end{bmatrix} \ell_i^k(x),$$

where, as usual, we represent (u,q) by N-th-order piecewise polynomials on K elements. We recover the strong form

$$\mathcal{M}^k \frac{d\boldsymbol{u}_h^k}{dt} = \tilde{\mathcal{S}}^{\sqrt{a}} \boldsymbol{q}_h^k - \int_{\partial \mathsf{D}^k} \hat{\boldsymbol{n}} \cdot \left((\sqrt{a} q_h^k) - (\sqrt{a} q_h^k)^* \right) \boldsymbol{\ell}^k(x)\, dx,$$

$$\mathcal{M}^k \boldsymbol{q}_h^k = \mathcal{S}^{\sqrt{a}} \boldsymbol{u}_h^k - \int_{\partial \mathsf{D}^k} \hat{\boldsymbol{n}} \cdot \left(\sqrt{a} u_h^k - (\sqrt{a} u_h^k)^* \right) \boldsymbol{\ell}^k(x)\, dx,$$

and the weak form

$$\mathcal{M}^k \frac{d\boldsymbol{u}_h^k}{dt} = -(\mathcal{S}^{\sqrt{a}})^T \boldsymbol{q}_h^k + \int_{\partial \mathsf{D}^k} \hat{\boldsymbol{n}} \cdot (\sqrt{a} q_h^k)^* \boldsymbol{\ell}^k(x)\, dx, \qquad (7.1)$$

$$\mathcal{M}^k \boldsymbol{q}_h^k = -(\tilde{\mathcal{S}}^{\sqrt{a}})^T \boldsymbol{u}_h^k + \int_{\partial \mathsf{D}^k} \hat{\boldsymbol{n}} \cdot (\sqrt{a} u_h^k)^* \boldsymbol{\ell}(x)\, dx.$$

As in Section 5.2, we have introduced the two special operators

$$\tilde{\mathcal{S}}_{ij}^{\sqrt{a}} = \int_{\mathsf{D}^k} \ell_i^k(x) \frac{d\sqrt{a(x)}\ell_j^k(x)}{dx}\, dx, \quad \mathcal{S}_{ij}^{\sqrt{a}} = \int_{\mathsf{D}^k} \sqrt{a(x)}\ell_i^k(x) \frac{d\ell_j^k(x)}{dx}\, dx.$$

We note that these operators are closely connected as

$$\tilde{\mathcal{S}}_{ij}^{\sqrt{a}} + \mathcal{S}_{ji}^{\sqrt{a}} = \left[\sqrt{a(x)}\ell_i(x)\ell_j(x) \right]_{x_l}^{x_r}.$$

Before defining the numerical flux, it is worth making a few observations. In general, it is reasonable that the numerical fluxes can have dependencies as

$$(\sqrt{a}q_h)^* = f((\sqrt{a}q_h)^-, (\sqrt{a}q_h)^+, (\sqrt{a}u_h)^-, (\sqrt{a}u_h)^+),$$

$$(\sqrt{a}u_h)^* = g((\sqrt{a}q_h)^-, (\sqrt{a}q_h)^+, (\sqrt{a}u_h)^-, (\sqrt{a}u_h)^+).$$

The problem with this generic form is that the two first-order equations are tightly coupled through the numerical flux and, hence, must be solved simultaneously as a globally coupled system. If, however, we restrict the generality of the numerical flux as

$$(\sqrt{a}q_h)^* = f((\sqrt{a}q_h)^-, (\sqrt{a}q_h)^+, (\sqrt{a}u_h)^-, (\sqrt{a}u_h)^+),$$

$$(\sqrt{a}u_h)^* = g((\sqrt{a}u_h)^-, (\sqrt{a}u_h)^+),$$

we see that $\boldsymbol{q}_h^k$ can be recovered through a local operation. Hence, the auxiliary function, $q(x,t)$, is a truly local variable, used only on each element to compute the derivatives and impose boundary conditions.

Keeping in mind the inherent properties of the heat equation (i.e., there is no preferred direction of propagation), it is natural to consider the simple central flux

$$(\sqrt{a}q_h)^* = \{\{\sqrt{a}q_h\}\}, \quad (\sqrt{a}u_h)^* = \{\{\sqrt{a}u_h\}\}.$$

Semidiscrete stability of this scheme is established in the following.

Theorem 7.2. *The discontinuous Galerkin scheme with central fluxes for the heat equation is stable.*

Proof. Let us first consider the situation in a single interval, k, bounded by $[x_l^k, x_r^k]$. We form the local elementwise operator

$$\mathcal{B}_h(u_h, q_h; \phi_h, \pi_h) = \phi_h^T \mathcal{M} \frac{d}{dt} u_h - \phi_h^T \tilde{\mathcal{S}}^{\sqrt{a}} q_h + \left[\phi_h(\sqrt{a}q_h) - (\sqrt{a}q_h)^*\right]_{x_l}^{x_r}$$
$$+ \pi_h^T \mathcal{M} q_h - \pi_h^T \mathcal{S}^{\sqrt{a}} u_h + \left[\pi_h(\sqrt{a}u_h) - (\sqrt{a}u_h)^*\right]_{x_l}^{x_r},$$

where we have left out the k superscript for the local element for simplicity. We first note that if (u_h, q_h) satisfies the numerical scheme, then

$$\mathcal{B}_h(u_h, q_h; \phi_h, \pi_h) = 0, \quad \forall(\phi_h, \pi_h) \in V_h,$$

where V_h is the space of N-th-order polynomials with support on the element, D. If we now choose the test functions as

$$\phi_h = u_h, \quad \pi_h = q_h,$$

and use that

$$u_h^T \tilde{\mathcal{S}}^{\sqrt{a}} q_h + q_h^T \mathcal{S}^{\sqrt{a}} u_h = u_h^T \tilde{\mathcal{S}}^{\sqrt{a}} q_h + u_h^T (\mathcal{S}^{\sqrt{a}})^T q_h = \left[\sqrt{a} u_h q_h\right]_{x_l}^{x_r},$$

we recover

$$\frac{1}{2} \frac{d}{dt} \|u_h\|_D^2 + \|q_h\|_D^2 + \Theta_r - \Theta_l = 0,$$

where

$$\Theta = \sqrt{a} u_h q_h - (\sqrt{a}q_h)^* u_h - (\sqrt{a}u_h)^* q_h.$$

First, assume that $a(x)$ is continuous and use the central fluxes

$$(\sqrt{a}q_h)^* = \sqrt{a}\{\!\{q_h\}\!\}, \quad (\sqrt{a}u_h)^* = \sqrt{a}\{\!\{u_h\}\!\}.$$

Consider x_r, where we have the term

$$\Theta_r = -\frac{\sqrt{a}}{2} \left(u_h^- q_h^+ + u_h^+ q_h^-\right).$$

Summering over all elements, we immediately recover

$$\frac{1}{2} \frac{d}{dt} \|u_h\|_{\Omega,h}^2 + \|q_h\|_{\Omega,h}^2 = 0,$$

and, thus, stability.

For the slightly more general case of $a(x)$ being only piecewise smooth, we consider the numerical flux

$$(\sqrt{a}q_h)^* = \{\!\{\sqrt{a}q_h\}\!\} + \frac{1}{2}[\![\sqrt{a}]\!] q_h^+,$$

$$(\sqrt{a}u_h)^* = \{\{\sqrt{a}u_h\}\} + \frac{1}{2}[\![\sqrt{a}]\!]u_h^+,$$

which reduces to the central flux for the continuous case above. This yields

$$\Theta_r = -\frac{1}{2}\{\{\sqrt{a}\}\}\left(u_h^- q_h^+ + u_h^+ q_h^-\right)$$

and, hence, stability by summation over all elements.   □

Based on this, there is good reason to believe that the proposed scheme is functional. To confirm this, let us consider the implementation of the scheme outlined in the above.

──────────────── | HeatCRHS1D.m | ────────────────

```
function [rhsu] = HeatCRHS1D(u,time)

% function [rhsu] = HeatCRHS1D(u,time)
% Purpose  : Evaluate RHS flux in 1D heat equation
%      using central flux

Globals1D;

% Define field differences at faces
du = zeros(Nfp*Nfaces,K); du(:)  = (u(vmapM)-u(vmapP))/2.0;

% impose boundary condition -- Dirichlet BC's
uin  = -u(vmapI); du(mapI) = (u(vmapI)-uin)/2.0;
uout = -u(vmapO); du(mapO)=(u(vmapO) - uout)/2.0;

% Compute q and form differences at faces
q = rx.*(Dr*u) - LIFT*(Fscale.*(nx.*du));
dq = zeros(Nfp*Nfaces,K); dq(:)  = (q(vmapM)-q(vmapP))/2.0;

% impose boundary condition -- Neumann BC's
qin  = q(vmapI); dq(mapI) = (q(vmapI)- qin )/2.0;
qout = q(vmapO); dq(mapO) = (q(vmapO)-qout)/2.0;

% compute right hand sides of the semi-discrete PDE
rhsu = rx.*(Dr*q) - LIFT*(Fscale.*(nx.*dq));
return
```

In HeatCRHS1D.m, we show the routine to compute the right-hand side for the semidiscrete discretization of the simplest problem with $a(x) = 1$. We impose homogenous Dirichlet boundary conditions by defining the exterior ghost states (u_h^+, q_h^+),

$$u_h^+ = -u_h^-, \quad q_h^+ = q_h^- \quad \Rightarrow \quad \begin{cases} \{\{u_h\}\} = 0, \quad [\![u_h]\!] = 2\hat{n}^- u_h^- \\ \{\{q_h\}\} = q_h^-, \quad [\![q_h]\!] = 0. \end{cases}$$

In a similar fashion, Neumann conditions are imposed as

$$u_h^+ = u_h^-, \quad q_h^+ = -q_h^- \quad \Rightarrow \quad \begin{cases} \{\{u_h\}\} = u_h^-, \ [\![u_h]\!] = 0 \\ \{\{q_h\}\} = 0, \quad [\![q_h]\!] = 2\hat{n}^- q_h^-. \end{cases}$$

If the boundary conditions are inhomogeneous, this is straightforwardly modified; that is,

$$u_h^+ = -u_h^- + 2f(t), \quad q_h^+ = q_h^-,$$

for a Dirichlet boundary condition $u(x,t) = f(t)$.

Heat1D.m

```
function [u,time] = Heat1D(u,FinalTime)

% function [u] = Heat1D(u,FinalTime)
% Purpose  : Integrate 1D heat equation until
%            FinalTime starting with initial condition, u.

Globals1D;
time = 0;

% Runge-Kutta residual storage
resu = zeros(Np, K);

% compute time step size
xmin = min(abs(x(1,:)-x(2,:)));
CFL=0.25;dt   = CFL*(xmin)^2;
Nsteps = ceil(FinalTime/dt); dt = FinalTime/Nsteps;

% outer time step loop
for tstep=1:Nsteps
  for INTRK = 1:5
    timelocal = time + rk4c(INTRK)*dt;

    % compute right hand side of 1D advection equations
    [rhsu] = HeatCRHS1D(u,timelocal);

    % initiate and increment Runge-Kutta residuals
    resu = rk4a(INTRK)*resu + dt*rhsu;

    % update fields
    u = u+rk4b(INTRK)*resu;
  end;
  % Increment time
  time = time+dt;
end
return
```

As in previous cases, we use an explicit Runge-Kutta method to advance the solution in time. This is implemented in Heat1D.m. Selecting the timestep follows the discussion for the first-order operator and we choose the timestep as

$$\Delta t \leq C \min_{i,k} \frac{1}{a(x_i^k)} (\Delta x_i^k)^2,$$

where, again, C is of order 1. The driver for initializing and solving the heat equation is shown in HeatDriver1D.m, following the well-known pattern.

HeatDriver1D.m

```
% Driver script for solving the 1D advection equations
% with variable coefficient

Globals1D;

% Polynomial order used for approximation
N = 8;

% Read in Mesh
[Nv, VX, K, EToV] = MeshGen1D(0,2*pi,20);

% Initialize solver and construct grid and metric
StartUp1D;

% Set initial conditions
u = sin(x);

% Solve Problem
FinalTime = 0.8;
[u,time] = Heat1D(u,FinalTime);
```

With confidence in the stability of the scheme, let us investigate the quality of the solution as measured through the convergence rate. For this, we return to the problem in Example 7.1:

$$\frac{\partial u}{\partial t} = \frac{\partial^2 u}{\partial x^2}, \quad x \in [0, 2\pi],$$

with homogeneous Dirichlet boundary conditions and $u(x,0) = \sin(x)$. In Fig. 7.2 we show the computed global error as a function of both order, N, and grid refinement through increasing K (i.e., decreasing grid size h).

A couple of things are worth emphasizing. First, we note with content that the scheme, in contrast to the obvious one in Example 7.1, converges. However, the rates of convergence are nonoptimal since we observe that for N odd, the rate appears to be $\mathcal{O}(h^N)$, while for N even the rate is indeed optimal (i.e., $\mathcal{O}(h^{N+1})$). Recall that a similar phenomenon was observed in

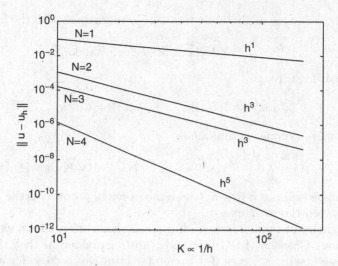

Fig. 7.2. Convergence of the scheme for solving the heat equation using a central flux.

Section 6.5 when solving Maxwell's equations using a central flux. For the heat problem, this behavior is confirmed in the following theorem [78]:

Theorem 7.3. *Let $\varepsilon_u = u_h - u$ and $\varepsilon_q = q_h - q$ signify the pointwise errors for the heat equation with periodic boundaries and a constant coefficient $a(x)$, computed with Eq. (7.1) and central fluxes. Then*

$$\|\varepsilon_u(T)\|_{\Omega,h}^2 + \int_0^T \|\varepsilon_q(s)\|_{\Omega,h}^2 \, ds \leq Ch^{2N},$$

where C depends on the regularity of u, T, and N. For N even, C is $\mathcal{O}(h^2)$.

The proof is technical and can be found in [78]. The computational results confirm that the theorem is sharp and we can only in special cases expect optimal convergence. However, we also note that the approximation error for u and q is of the same order.

While this loss of optimal convergence rate may appear as a minor thing, it suggests that one should consider alternative formulations. It is worth keeping in mind that we have considerable freedom in choosing the numerical flux and we can use the stability considerations to guide this choice. Additional inspiration can be gained by recalling that upwind fluxes most often leads to the schemes with optimal convergence rates.

From the proof of Theorem 7.2 we obtain at each interface a term like

$$\Theta = u_h q_h \sqrt{a} - (\sqrt{a} q_h)^* u_h - (\sqrt{a} u_h)^* q_h.$$

To guarantee stability, we must ensure

$$\Theta_r^- - \Theta_l^+ \geq 0$$

at each interface. One easily shows that this is guaranteed with the flux choice

$$(\sqrt{a}u_h)^* = \sqrt{a^-}u_h^-, \quad (\sqrt{a}q_h)^* = \{\!\{\sqrt{a}\}\!\}q_h^+,$$

or, alternatively, through

$$(\sqrt{a}u_h)^* = \{\!\{\sqrt{a}\}\!\}u_h^+, \quad (\sqrt{a}q_h)^* = \sqrt{a^-}q_h^-.$$

A slightly more symmetric solution is

$$(\sqrt{a}u_h)^* = \{\!\{\sqrt{a}u_h\}\!\} + \hat{\boldsymbol{\beta}} \cdot [\![\sqrt{a}u_h]\!], \quad (\sqrt{a}q_h)^* = \{\!\{\sqrt{a}q_h\}\!\} - \hat{\boldsymbol{\beta}} \cdot [\![\sqrt{a}q_h]\!],$$

Here, $\hat{\boldsymbol{\beta}}$ can be taken as $\hat{\boldsymbol{n}}$ or $-\hat{\boldsymbol{n}}$, where the essential property is the difference in sign between the two fluxes.

These choice of these fluxes, leading to methods often known as the local discontinuous Galerkin (LDG) methods, are remarkable in that they effectively utilize upwinding, even if it is counter-intuitive to do so for a problem like the heat equation. A careful inspection of the approach reveals, however, that the upwinding is done in a very careful way, always doing upwinding for u_h and q_h in opposite directions. This is essential for the stability. In HeatLD-GRHS1D.m, we illustrate an implementation of this for solving the constant coefficient heat equation.

───────────────────────── **HeatLDGRHS1D.m** ─────────────────────────

```
function [rhsu] = HeatLDGRHS1D(u,time)

% function [rhsu] = HeatLDGRHS1D(u,time,a,ax)
% Purpose: Evaluate RHS flux in 1D heat equation using an LDG flux

Globals1D;

% Define field differences at faces
du = zeros(Nfp*Nfaces,K);
du(:) = (1.0+nx(:)).*(u(vmapM)-u(vmapP))/2.0;

% impose boundary condition -- Dirichlet BC's
uin  = -u(vmapI); du(mapI) = (1.0+nx(mapI)).*(u(vmapI)- uin)/2.0;
uout = -u(vmapO); du(mapO) = (1.0+nx(mapO)).*(u(vmapO)-uout)/2.0;

% Compute q
q = rx.*(Dr*u)- LIFT*(Fscale.*(nx.*du));
dq = zeros(Nfp*Nfaces,K);
dq(:) = (1.0-nx(:)).*(q(vmapM)-q(vmapP))/2.0;

% impose boundary condition -- Neumann BC's
qin  = q(vmapI); dq(mapI) = (1.0-nx(mapI)).*(q(vmapI)- qin)/2.0;
qout = q(vmapO); dq(mapO) = (1.0-nx(mapO)).*(q(vmapO)-qout)/2.0;
```

```
% compute right hand sides of the semi-discrete PDE
rhsu = rx.*(Dr*q) - LIFT*(Fscale.*(nx.*dq));
return
```

As elegant as this construction appears, the motivation for considering it is whether one can improve the convergence rate for N being odd. In Fig. 7.3 we show results for solving the heat equation, and these appear to show that optimal order of accuracy is restored.

This observation is confirmed by the following theorem, the proof of which can be found in [78].

Theorem 7.4. *Let $\varepsilon_u = u - u_h$ and $\varepsilon_q = q - q_h$ signify the pointwise errors for the heat equation with periodic boundaries and a constant coefficient $a(x)$, computed with Eq. (7.1) and LDG fluxes. Then*

$$\|\varepsilon_u(T)\|^2_{\Omega,h} + \int_0^T \|\varepsilon_q(s)\|^2_{\Omega,h}\, ds \leq Ch^{2N+2},$$

where C depends on the regularity of u, T, and N.

Figure 7.3 also shows that one has to be careful to claim similar results for nonperiodic cases. Indeed, extending the LDG fluxes directly to impose Dirichlet boundary conditions on both ends of the domain appears to destroy the optimal convergence rate. We are not aware of an analysis confirming this or suggesting a way to overcome the loss of optimality for the general case with nontrivial boundary conditions.

The analysis of the schemes (i.e., stability and error estimates) assumes that all inner products are done exactly. As discussed at length in Chapter 5, it is often computationally advantageous to avoid this and compute with an aliased solution and stabilization if needed. However, in contrast to pure

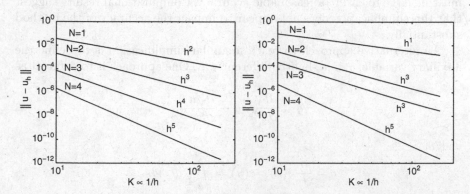

Fig. 7.3. On the left, we show convergence of scheme for solving the periodic heat equation using an LDG flux. On the right is shown the result of the same scheme used for solving the problem with homogeneous boundary conditions.

conservation laws, the inherent dissipative nature of the heat equations makes stabilization much less of a concern here.

In the nodal approach, the simpler scheme becomes

$$\mathcal{M}^k \frac{d\boldsymbol{u}_h^k}{dt} = \mathcal{S}^k \mathcal{A}^k \boldsymbol{q}_h^k - \int_{\partial \mathsf{D}^k} \hat{\boldsymbol{n}} \cdot ((\sqrt{a}q_h^k) - (\sqrt{a}q_h^k)^*) \, \boldsymbol{\ell}^k(x) \, dx,$$

$$\mathcal{M}^k \boldsymbol{q}_h^k = \mathcal{A}^k \mathcal{S}^k \boldsymbol{u}_h^k - \int_{\partial \mathsf{D}^k} \hat{\boldsymbol{n}} \cdot (\sqrt{a}u_h^k - (\sqrt{a}u_h^k)^*) \, \boldsymbol{\ell}^k(x) \, dx,$$

and the weak form becomes

$$\mathcal{M}^k \frac{d\boldsymbol{u}_h^k}{dt} = -(\mathcal{S}^k)^T \mathcal{A}^k \boldsymbol{q}_h^k + \int_{\partial \mathsf{D}^k} \hat{\boldsymbol{n}} \cdot (\sqrt{a}q_h^k)^* \boldsymbol{\ell}^k(x) \, dx,$$

$$\mathcal{M}^k \boldsymbol{q}_h^k = -\mathcal{A}^k (\mathcal{S}^k)^T \boldsymbol{u}_h^k + \int_{\partial \mathsf{D}^k} \hat{\boldsymbol{n}} \cdot (\sqrt{a}u_h^k)^* \boldsymbol{\ell}^k(x) \, dx,$$

where $\mathcal{A}_{ii}^k = \sqrt{a(x_i^k)}$.

A further interesting simplification would be

$$\mathcal{M}^{-1} \mathcal{A} \mathcal{S} \simeq \mathcal{A} \mathcal{D}_r,$$

in which case only the standard differentiation matrix $\mathcal{D}_r$ is required. The error made originates the commutation error between $\mathcal{A}$ and $\mathcal{M}$, much like the errors associated with the aliasing discussed in Section 5.3 A local and very conservative bound on this is

$$\left| (\mathcal{A}\mathcal{M} - \mathcal{M}\mathcal{A})_{ij} \right| \leq \max |a_x||x_i - x_j|N^{-1}.$$

Clearly, if there is limited local variation of $a(x)$, this term is very small. If, on the other hand, there is substantial variation (i.e., $|a_x|$ is large), then also N must be large to ensure a reasonable accuracy. Computational results suggest that this simplification does not appear to impact the accuracy of the method substantially.

An alternative approach to reach a similar simplification is to define the auxiliary variable, q, in a slightly different way. One approach is to define it as

$$q = \frac{\partial}{\partial x} \sqrt{a}u - \frac{\partial \sqrt{a}}{\partial x} u,$$

or [78]

$$q = \frac{\partial}{\partial x} g(u), \quad g(u) = \int_u \sqrt{a(s)} ds.$$

In both cases, $a(x)$, is inside the differentiation and, thus, only the standard discrete operator $\mathcal{D}_r$ is needed.

7.1.2 Extensions to mixed and higher-order problems

With the approach for the heat equation in place, it is natural to consider further extensions. For instance, a very important class of problems, known as convection-diffusion problems, is of the type

$$\frac{\partial u}{\partial t} + \frac{\partial}{\partial x} f(u) = \frac{\partial}{\partial x} a(x) \frac{\partial u}{\partial x},$$

subject to appropriate boundary and initial conditions.

To develop a suitable discretization for such problems, we write it in first-order form as

$$\frac{\partial u}{\partial t} + \frac{\partial}{\partial x} \left(f(u) - \sqrt{a} q \right) = 0, \tag{7.2}$$

$$q = \sqrt{a} \frac{\partial u}{\partial x},$$

by combining the knowledge we have from solving both conservation laws and heat equations. We can now discretize this exactly as above. The only problem that requires attention is the choice of the numerical fluxes, $(f(u) - \sqrt{a} q)^*$ and $(\sqrt{a} u)^*$.

However, based on our past discussions, it is natural to use a monotone flux for f^*; for example, a Lax-Friedrichs flux like

$$f(u)^* = \{\{f(u)\}\} + \frac{C}{2}[\![u]\!], \quad C \geq \max|f'(u)|.$$

Furthermore, for the parts corresponding to the dissipative operator, $(\sqrt{a} q)^*$ and $(\sqrt{a} u)^*$, we rely on the results of the previous section and choose either a central flux or the LDG flux, which, at least for the purely diffusive problem, yields optimal convergence rates.

To illustrate the behavior of this mixed approach, let us consider the following example:

Example 7.5. We consider the solution of the viscous Burgers' equation

$$\frac{\partial u}{\partial t} + \frac{\partial}{\partial x} \left(\frac{u^2}{2} \right) = \varepsilon \frac{\partial^2 u}{\partial x^2}, \quad x \in [-1, 1],$$

which has an exact traveling wave solution

$$u(x, t) = -\tanh\left(\frac{x + 0.5 - t}{2\varepsilon} \right) + 1.$$

To solve this, we use a standard DG method with a Lax-Friedrichs flux for the nonlinear flux, $f(u) = u^2/2$, and central fluxes for the dissipative operator. In BurgersRHS1D.m, we illustrate the implementation of this approach for the evaluation of the right-hand side, with the exact solution being used to impose boundary conditions.

_____ BurgersRHS1D.m _____

```
function [rhsu] = BurgersRHS1D(u,epsilon,xL,xR,time)

% function [rhsu] = BurgersRHS1D(u,epsilon,xL, xR, time)
% Purpose  : Evaluate RHS flux in 1D viscous Burgers equation

Globals1D;

% Define field differences at faces
du = zeros(Nfp*Nfaces,K); du(:) = u(vmapM)-u(vmapP);

% impose boundary condition at x=0
uin =-tanh((xL+0.5-time)/(2*epsilon))+1.0;
du(mapI) = 2.0*(u(vmapI)-uin);
uout=-tanh((xR+0.5-time)/(2*epsilon))+1.0;
du(mapO) = 2.0*(u(vmapO)-uout);

% Compute q and jumps
q = sqrt(epsilon)*(rx.*(Dr*u) - LIFT*(Fscale.*(nx.*du/2.0)));
dq = zeros(Nfaces,K); dq(:) = (q(vmapM)-q(vmapP))/2.0;

% impose boundary condition - Dirichlet conditions
dq(mapI) = 0.0; dq(mapO) = 0.0;

% Evaluate nonlinear flux
du2 = zeros(Nfp*Nfaces,K); du2(:) = (u(vmapM).^2-u(vmapP).^2)/2.0;

% impose boundary condition
du2(mapI)=(u(vmapI).^2-uin.^2); du2(mapO)=(u(vmapO).^2-uout.^2);

% Compute flux
maxvel = max(max(abs(u)));

% penalty scaling -- See Chapter 7.2
%tau = .25*reshape(N*N./max(2*J(vmapP),2*J(vmapM)), Nfp*Nfaces, K);
tau=0;

% flux term
flux = nx.*(du2/2.0 - sqrt(epsilon)*dq) - maxvel/2.0.*du ...
     - sqrt(epsilon)*tau.*du;

% local derivatives of field
dfdr = Dr*(u.^2/2 - sqrt(epsilon)*q);

% compute right hand sides of the semi-discrete PDE
rhsu = -(rx.*dfdr - LIFT*(Fscale.*flux));
return
```

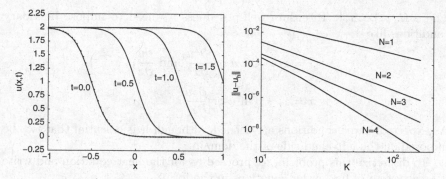

Fig. 7.4. On the left, we illustrate the moving front solution to the viscous Burgers' equation. On the right, we illustrate the convergence of the scheme, showing an $\mathcal{O}(h^{N+1/2})$ convergence rate for all orders.

In Fig. 7.4 we show both a few snapshots of the solution with $\varepsilon = 0.1$ and the convergence of the error in the maximum norm. The convergence rate generally behaves as $\mathcal{O}(h^{N+1/2})$, although there appears to be an optimal convergence rate [i.e., $\mathcal{O}(h^{N+1})$], for N even. Note that the convergence rate does not degrade to $\mathcal{O}(h^N)$, as for the pure heat problem with central fluxes.

The analysis of the stability of the discretization for such mixed problems follows, to a large extent, from the results for the individual operators (i.e., the nonlinear stability through entropy conditions combined with the bounds of the individual terms for the linear operators). The nonlinear stability for the convection-diffusion problem is shown rigorously in [78].

The analysis of the error (i.e., a quantitative understanding of the convergence rate of the scheme discussed in detail in Chapter 5), would suggest an expected convergence rate of $\mathcal{O}(h^{N+1/2})$ from the use of the Lax-Friedrichs flux, in agreement with what we found in the above example. The possible loss of accuracy from the linear operator does not appear to impact this, although this is based on experimental evidence only.

Let us briefly also consider general higher-order spatial operators and how to discretize these. As an example, consider the problem

$$\frac{\partial u}{\partial t} = \frac{\partial^3 u}{\partial x^3}.$$

First, we need to understand what kind of boundary conditions are required to ensure well-posedness in a finite domain. We can understand this through the energy method

$$\frac{1}{2}\frac{d}{dt}\|u\|_\Omega^2 = \left[u\frac{\partial^2 u}{\partial x^2} - \frac{1}{2}\left(\frac{\partial u}{\partial x}\right)^2 \right]_{x_l}^{x_r},$$

where $\Omega = [x_l, x_r]$. To ensure well-posedness we need to impose boundary conditions like

$$x = x_l : \text{ On } u \text{ or } \frac{\partial^2 u}{\partial x^2} \text{ and } \frac{\partial u}{\partial x},$$

$$x = x_r : \text{ On } u \text{ or } \frac{\partial^2 u}{\partial x^2}.$$

As expected, three conditions are needed, although it is essential that two be imposed on the left-hand side of the domain.

To discretize this problem, we proceed as for the heat equation and write it as a system of first-order equations in the form

$$\frac{\partial u}{\partial t} = \frac{\partial q}{\partial x}, \quad q = \frac{\partial p}{\partial x}, \quad p = \frac{\partial u}{\partial x}.$$

This can now be discretized in a straightforward manner. Proving stability can be done in exactly the same way as for the heat equation; that is, one forms the elementwise operator, $\mathcal{B}_h(u_h, q_h, p_h; \phi_h, \pi_h, \theta_h)$, as the sum of the three first order terms. Since $\mathcal{B}_h(u_h, q_h, p_h; \phi_h, \pi_h, \theta_h) = 0$ for any N-th-order polynomial $(\phi_h, \pi_h, \theta_h) \in V_h$ when (u_h, q_h, p_h) is a solution, we can choose $(\phi_h, \pi_h, \theta_h) = (u_h, -p_h, q_h)$. A bit of manipulation yields a local elementwise balance as

$$\frac{1}{2}\frac{d}{dt}\|u_h\|_{\mathsf{D}^k}^2 = \Theta_r - \Theta_l,$$

$$\Theta = \frac{p_h^2}{2} - u_h q_h + u_h(q_h)^* + q_h(u_h)^* - p_h(p_h)^*.$$

We can use this to guide the choice of fluxes leading to stable schemes; for example, the use of central fluxes on all variables yields

$$\Theta = \frac{1}{2}\left(u_h^+ q_h^- + u_h^- q_h^+ - p_h^- p_h^+ \right),$$

which, after summation over all edges, vanishes. This establishes energy conservation in the periodic case, as for the continuous equation.

Similar to the heat equation, we can also use an LDG-type upwinding by choosing

$$(u_h)^* = u_h^-, \quad (q_h)^* = q_h^+, \quad (p_h)^* = p_h^-,$$

or

$$(u_h)^* = u_h^+, \quad (q_h)^* = q_h^-, \quad (p_h)^* = p_h^-.$$

A slightly different way of writing this

$$(u_h)^* = \{\{u_h\}\} + \hat{\beta} \cdot [\![u_h]\!], \quad (q_h)^* = \{\{q_h\}\} - \hat{\beta} \cdot [\![q_h]\!], \quad (p_h)^* = p_h^-,$$

where, again, $\hat{\beta} = \pm\hat{n}$. The key point here is that $(p_h)^*$ has to be taken from the left, in agreement with the discussion of the well-posedness of the continuous problem. In DispersiveLDGRHS1D.m, we illustrate how to implement this for the dispersive problem, subject to periodic boundary conditions.

────────────────── DispersiveLDGRHS1D.m ──────────────────

```
function [rhsu] = DispersiveLDGRHS1D(u,time)

% function [rhsu] = DispersiveLDGRHS1D(u,time)
% Purpose  : Evaluate RHS flux in 1D u_xxx using LDG fluxes and
%            periodic BC's

Globals1D;

% Define field differences at faces, incl BC
du   = zeros(Nfp*Nfaces,K); du(:)     = u(vmapM)-u(vmapP);
uin  = u(vmap0);            du (mapI) = u(vmapI) - uin;
uout = u(vmapI);           du (mapO) = u(vmap0) - uout;
fluxu = nx.*(1.0+nx).*du/2.0;

% Compute local variable p, define differences, incl BC
p = rx.*(Dr*u) - LIFT*(Fscale.*fluxu);
dp   = zeros(Nfp*Nfaces,K); dp(:)     = p(vmapM)-p(vmapP);
pin  = p(vmap0);           dp(mapI) = p(vmapI) - pin;
pout = p(vmapI);           dp(mapO) = p(vmap0) - pout;
fluxp = nx.*(1.0-nx).*dp/2.0;

% Compute local variable q, define differences, incl BC
q = rx.*(Dr*p) - LIFT*(Fscale.*fluxp);
dq   = zeros(Nfp*Nfaces,K); dq(:)     = q(vmapM)-q(vmapP);
qin  = q(vmap0);           dq (mapI) = q(vmapI) - qin;
qout = q(vmapI);           dq (mapO) = q(vmap0) - qout;
fluxq = nx.*(1.0-nx).*dq/2.0;

% compute right hand sides of the semi-discrete PDE
rhsu = rx.*(Dr*q) - LIFT*(Fscale.*fluxq);
return
```

──

Let us conclude with an example.

──

Example 7.6. We consider the solution of the dispersive equation

$$\frac{\partial u}{\partial t} = \frac{\partial^3 u}{\partial x^3}, \quad x \in [-1, 1],$$

with periodic boundary conditions and $u(x,0) = \cos(\pi x)$. One easily shows that the exact solution is given as

$$u(x,t) = \cos(\pi^3 t + \pi x).$$

This problem is solved using both the central fluxes and the upwind-style LDG fluxes, with the convergence results shown in Fig. 7.5.

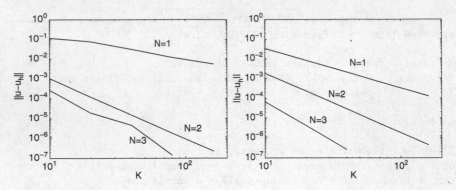

Fig. 7.5. On the left, we show the convergence rate for the linear third-order dispersive equation, solved using a DG method with central fluxes, while the right shows the results obtained using the LDG-style upwind fluxes.

As expected, we have suboptimal convergence for certain orders when using central fluxes, while the LDG flux delivers optimal rates of convergence. However, it comes at a cost as the time-step has to be reduced by a factor of 2 for the LDG fluxes. We shall return to this caveat shortly.

The extension of the above techniques to general higher-order equations and mixed equations is straightforward. Analysis and details of such extension can be found in [219, 329, 330, 331, 332, 333], covering a variety of linear and nonlinear problems. One should generally expect convergence rates of $\mathcal{O}(h^{N+1/2})$ when combining a Lax-Freidrichs flux for the nonlinear flux and an LDG flux for the linear operators.

While the careful attention to the design of the fluxes suffices to ensure that all the auxiliary variables are local, one can still voice concern about the computational cost of this approach. In particular, the computation of a derivative of order s requires the computation of s first derivatives in contrast to more standard methods where an s-th-order operator is defined and applied once.

This concern has been addressed in recent work [124, 311] for the second order diffusion operator and in [57] for more general operators. We illustrate the basic idea of this alternative approach by the heat equation

$$\frac{\partial u}{\partial t} = \frac{\partial^2 u}{\partial x^2},$$

with periodic conditions. On each element, D^k, we require that

$$\left(\phi_h, \frac{\partial u_h}{\partial t} \right)_{D^k} = \left(\phi_h, \frac{\partial^2 u_h}{\partial x^2} \right)_{D^k},$$

for all polynomial test functions, $\phi_h \in V_h$. Here, u_h represent the usual local polynomial approximation to u. Integration by parts twice now results in

$$\left(\phi_h, \frac{\partial u_h}{\partial t}\right)_{\mathsf{D}^k} = \left(\frac{d^2\phi_h}{dx^2}, u_h\right)_{\mathsf{D}^k} + \left(\hat{\boldsymbol{n}} \cdot \frac{\partial u_h}{\partial x}, \phi_h\right)_{\partial\mathsf{D}^k} - \left(\hat{\boldsymbol{n}} \cdot u_h, \frac{d\phi_h}{dx}\right)_{\partial\mathsf{D}^k}.$$

To complete the scheme, we introduce the numerical fluxes for u_h and $(u_h)_x$ to obtain

$$\left(\phi_h, \frac{\partial u_h}{\partial t}\right)_{\mathsf{D}^k} = \left(\frac{\partial^2\phi_h}{\partial x^2}, u_h\right)_{\mathsf{D}^k} + \left(\hat{\boldsymbol{n}} \cdot (u_h)_x^*, \phi_h\right)_{\partial\mathsf{D}^k} - \left(\hat{\boldsymbol{n}} \cdot u_h^*, \frac{\partial\phi_h}{\partial x}\right)_{\partial\mathsf{D}^k}.$$

The flux choices leading to stable schemes can then be found by energy methods or entropy inequalities, resulting in both central fluxes and LDG-type fluxes being acceptable [57]. For even-order operators a stabilizing penalty term is required, much as we will see in more detail in the next chapter. The error analysis for this and more general higher-order operators indicates a suboptimal convergence rate for many cases, but computational results confirms optimality. See [57] for the details.

The main advantage of this formulation is the reduction of the computational work as well as a more compact formulation since the fluxes only connect to the neighboring elements. The impact of this on the spectral radius and, thus, the timestep is not understood at this point. Another constraint of these methods is their less systematic nature, requiring analysis for each new type of equation.

7.2 Elliptic problems

With the ability to solve problems with higher spatial derivatives we can consider problems of the type

$$\frac{\partial u}{\partial t} = \frac{\partial^2 u}{\partial x^2} - f(x),$$

with appropriate boundary and initial conditions. For this problem, any of the methods discussed in the previous section can be applied.

If, however, we are interested in a steady-state solution (i.e., $u_t = 0$), it is more natural to consider the elliptic problem

$$\frac{\partial^2 u}{\partial x^2} = f(x),$$

and invert the discrete operator to recover the approximate solution, u_h. We recognize this as the classic Poisson problem. It is well known that under mild conditions on $f(x)$, this problem has a unique solution [59].

As an example, let us consider the following problem

$$\frac{\partial^2 u}{\partial x^2} = -\sin(x), \quad x \in [0, 2\pi],$$

subject to $u(0) = u(2\pi) = 0$ and with the exact solution $u(x) = \sin(x)$. We use HeatCRHS1D.m to build the discrete approximation

$$\mathcal{A}u_h = f_h,$$

to the second order problem by simply calling HeatCRHS1D.m with the $N_p \times K$ size identity matrix, one unit-vector at a time.

When attempting to invert $\mathcal{A}$, one quickly realizes, however, that the matrix is singular. This is in contrast to the continuous problem, which is uniquely solvable. The question arises of what is causing this sudden breakdown and why we did not experience problems when solving the heat equation.

The singularity of the discrete operator indicates that it has at least one zero eigenvalue – in contrast to the continuous case in which there are none once the boundary conditions are enforced. To understand the severity of the problem, we show in Fig. 7.6 the computed eigenvalues of $\mathcal{A}$ close to the real axis, confirming that all eigenvalues are negative and real but also that there is at least one zero eigenvalue. A closer inspection reveals that there is exactly one such zero eigenvalue and in Fig. 7.6 we also show the corresponding eigenvector, v, for $N = 1, 2, 4$ and $K = 6$.

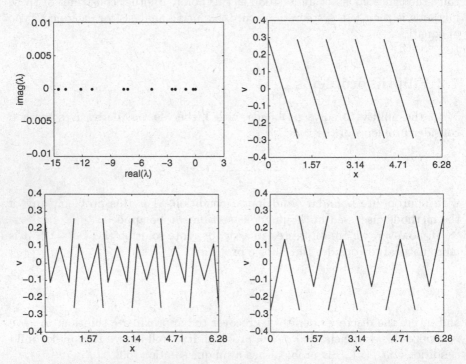

Fig. 7.6. Top left shows the eigenvalues of the discrete Poisson operator, confirming that it is singular. In the remaining three figures is shown the corresponding singular eigenmode for $K = 6$ and $N = 1$, $N = 2$, and $N = 4$, respectively in a clockwise direction with $N = 1$ in the top right corner.

The eigenvectors for the three different orders of approximation share some distinct features. First, we notice that they are all piecewise linear inside the elements, as required to ensure that their second derivative reduces to zero. However, we also note that the eigenvectors v all share the property that

$$v^- = -v^+, \tag{7.3}$$

at any internal interface. Since we are using a central flux, which vanishes exactly for this choice of values, all the elements effectively decouple. On the other hand, this condition is the same condition that we use to impose homogeneous Dirichlet boundary conditions so that at the interfaces it appears as if we have the correct zero solution. It is the added freedom in the discontinuous basis that comes back to bite us by allowing too much flexibility, which we cannot control using the simple central flux.

This also explains why this particular spurious mode caused no problems in the time-dependent problem as it is characterized by being constant in time and very rapidly varying in space. As long as the solutions are well resolved, this mode will not cause any problems. However, for marginally resolved problems, the situation may be a little more delicate and the spurious mode may impact the accuracy. In such cases, the stabilization methods developed for the elliptic case can also be useful to improve the accuracy for time-dependent problems.

With this added understanding, we can ask ourselves what to do about it. An obvious solution would be, at least asymptotically in N and/or h, to disallow the eigenmode with the properties in Eq. (7.3) without it also vanishing inside the element; that is, to remove the artificial null-space.

The discontinuous Galerkin (DG) method now shows one of its strengths by allowing us to modify the choice of the numerical flux. In particular, let us consider a set of numerical fluxes as

$$q^* = \{\{q\}\} - \tau[[u]], \quad u^* = \{\{u\}\}.$$

For $\tau = 0$ this reduces to the central flux. The role of the added term is to penalize the solution to disallow large jumps in u; clearly if we introduce one of the null vectors with the property in Eq. (7.3), the added penalty term would highlight the jump. The parameter τ is chosen to control this jump.

The minor change of the flux required is illustrated in HeatCstabRHS1D.m, which should be compared with HeatCRHS1D.m discussed Section 7.1. Indeed, as expected, taking $\tau > 0$ makes $\mathcal{A}$, the discrete approximation to the Poisson equation, invertible by pushing the zero eigenmode out of the kernel.

──────────────────────── | HeatCstabRHS1D.m | ────────────────────────

```
function [rhsu] = HeatCstabRHS1D(u,time)

% function [rhsu] = HeatCstabRHS1D(u,time)
% Purpose : Evaluate RHS flux in 1D heat equation using stabilized
% central flux
```

```
Globals1D;

% Define field differences at faces
du    = zeros(Nfp*Nfaces,K); du(:) = (u(vmapM)-u(vmapP))/2.0;

% impose boundary condition -- Dirichlet BC's
uin  = -u(vmapI); du (mapI) = (u(vmapI) -  uin )/2.0;
uout = -u(vmapO); du (mapO) = (u(vmapO) - uout)/2.0;

% Compute q
q = rx.*(Dr*u) - LIFT*(Fscale.*(nx.*du));
dq = zeros(Nfp*Nfaces,K);dq(:) = q(vmapM)-q(vmapP);

% impose boundary condition -- Neumann BC's
qin = q(vmapI);  dq (mapI) = q(vmapI)-  qin;
qout = q(vmapO); dq (mapO) = q(vmapO)- qout;

% evaluate fluxes
tau = 1.0;
fluxq = nx.*(dq/2.0+tau*nx.*du);

% compute right hand sides of the semi-discrete PDE
rhsu = rx.*(Dr*q) - LIFT*(Fscale.*fluxq);
return
```

Before we finalize matters and consider some more examples, let us understand whether $\mathcal{A}$ is symmetric. There are several reasons why this is both desired and expected. First, if $\mathcal{A}$ is symmetric, we can take advantage of this when solving the linear system, hence reducing the storage need and computational cost (see Section 7.3). Furthermore, for a classic finite element or finite difference method, $\mathcal{A}$ is well-known to be symmetric.

If we draw inspiration from a classic finite element method, we recall that the resulting linear system takes the form

$$\mathcal{S}u_h = \mathcal{M}f_h,$$

where we have the symmetric, global matrices, $\mathcal{S}$ and $\mathcal{M}$.

Now, recall the simple formulation we have considered so far where we have the local statement

$$\mathcal{D}_r q_h - \mathcal{M}^{-1} \int_{\partial \mathsf{D}^k} \hat{n} \cdot (q_h - q^*) \ell(x)\, dx = f_h,$$

$$q_h = \mathcal{D}_r u_h - \mathcal{M}^{-1} \int_{\partial \mathsf{D}^k} \hat{n} \cdot (u_h - u^*) \ell(x)\, dx,$$

where we have dropped the element index k for simplicity. The equivalent to this, when comparing to the finite element scheme above, would be

$$\mathcal{M}^{-1}\mathcal{S}u_h = f_h.$$

In this case, the matrix to invert, $\mathcal{M}^{-1}\mathcal{S}$, is no longer symmetric, even in the finite element case (where there are other good reasons never to consider this – e.g., it is expensive to invert $\mathcal{M}$).

If we wish to mimic the original finite element scheme, we should consider the scheme

$$\mathcal{S}\boldsymbol{q}_h - \int_{\partial D^k} \hat{n} \cdot (\boldsymbol{q}_h - \boldsymbol{q}^*) \, \ell(x) \, dx = \mathcal{M}\boldsymbol{f}_h,$$

$$\boldsymbol{q}_h = \mathcal{D}_r \boldsymbol{u}_h - \mathcal{M}^{-1} \int_{\partial D^k} \hat{n} \cdot (\boldsymbol{u}_h - \boldsymbol{u}^*) \, \ell(x) \, dx.$$

As we will prove shortly (Section 7.2.2), this is indeed symmetric on general grids under very light conditions on the choice of fluxes.

In PoissonCstabRHS1D.m, we show the minimal change required to modify the setup of the discrete operator to ensure that it is fully symmetric.

PoissonCstabRHS1D.m

```
function [rhsu] = PoissonCstabRHS1D(u,time)

% function [rhsu] = PoissonCstabRHS1D(u,time)
% Purpose  : Evaluate RHS in 1D Poisson equation on
%                symmetric form using stabilized central flux

Globals1D;

% Define field differences at faces
du = zeros(Nfp*Nfaces,K); du(:) = (u(vmapM)-u(vmapP))/2.0;

% impose boundary condition -- Dirichlet BC's
uin  = -u(vmapI); du (mapI) = (u(vmapI) -  uin )/2.0;
uout = -u(vmapO); du (mapO) = (u(vmapO) - uout)/2.0;

% Compute q
q = rx.*(Dr*u) - LIFT*(Fscale.*(nx.*du));
dq = zeros(Nfp*Nfaces,K); dq(:) = q(vmapM)-q(vmapP);

% impose boundary condition -- Neumann BC's
qin  = q(vmapI); dq (mapI) = q(vmapI)- qin;
qout = q(vmapO); dq (mapO) = q(vmapO)-qout;

% evaluate fluxes
tau = 1.0;
fluxq = nx.*(dq/2.0+tau*nx.*du);

% compute right hand sides of the semi-discrete PDE
rhsu = J.*((invV'*invV)*(rx.*(Dr*q) - LIFT*(Fscale.*fluxq)));
return
```

Let us consider a simple example to illustrate the performance when solving the one-dimensional Poisson equation.

Example 7.7. Consider the problem

$$\frac{d^2u}{dx^2} = -\sin(x), \ x \in [0, 2\pi],$$

with $u(0) = u(2\pi) = 0$.

The routine for setting up the discrete Poisson operator is shown in PoissonCstabRHS1D.m, and in PoissonCstabDriver1D.m, we illustrate the driver routine needed to solve the problem.

─────────────────────── | PoissonCstab1D.m | ───────────────────────

```
function [A] = PoissonCstab1D();

% function [A] = PoissonCstab1D();
% Purpose: Set up symmetric Poisson matrix with estabilized
%          central fluxes

Globals1D;
A = zeros(K*Np); g = zeros(K*Np,1);

% Build matrix -- one column at a time
for i=1:K*Np
    g(i) = 1.0;
    gmat = reshape(g,Np,K);
    Avec = PoissonCstabRHS1D(gmat);
    A(:,i) = reshape(Avec,K*Np,1);
    g(i)=0.0;
end
return
```

─────────────────────── | PoissonCDriver1D.m | ───────────────────────

```
% Driver script for solving the 1D Poisson equation
Globals1D;

% Polynomial order used for approximation
N =4;

% Read in Mesh
[Nv, VX, K, EToV] = MeshGen1D(0,2*pi,10);

% Initialize solver and construct grid and metric
StartUp1D;
```

```
% Set RHS
f = -J.*((invV'*invV)*sin(x));

% Set up operator
[A] = PoissonCstab1D();

% Solve Problem
solvec = A\f(:);
u = reshape(solvec,Np,K);
```

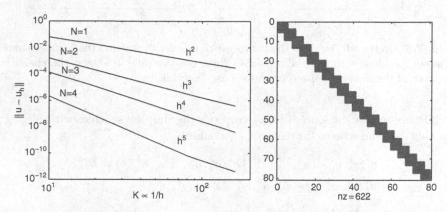

Fig. 7.7. On the left, we show the convergence rate for the solving the homogeneous Poisson problem with a stabilized central flux. On the right is shown the sparsity pattern for $K = 20$ and $N = 3$ of the discrete Poisson operator based on this flux.

The problem is solved with different number of elements and orders of approximation, and we show the results in Fig. 7.7, obtained with $\tau = 1$ for the stabilization parameter. The results indicate optimal rates of convergence, $\mathcal{O}(h^{N+1})$.

One can naturally question whether the LDG flux, introduced to recover optimal convergence in Section 7.1 for the heat equation, can also be used to discretize the Poisson problem. This is naturally possible and we show in Fig. 7.8 the results of solving the problem in Example 7.5 using a local stabilized flux of the kind

$$q_h^* = \{\{q_h\}\} + \hat{\beta} \cdot [\![q_h]\!] - \tau[\![u_h]\!], \quad u_h^* = \{\{u_h\}\} - \hat{\beta} \cdot [\![u_h]\!],$$

with $\tau = 1$ for all cases and $\hat{\beta} = \pm\hat{n}$. As in Fig. 7.7, we recover an optimal convergence rate. The code used to solve the problem is a direct combination of PoissonCstabRHS1D.m and HeatLDGRHS1D.m.

If we now compare the two plots of the nonzero elements in Fig. 7.7 and Fig. 7.8, we note that the latter, based on the LDG fluxes, is considerably sparser that the former, based on the central fluxes.

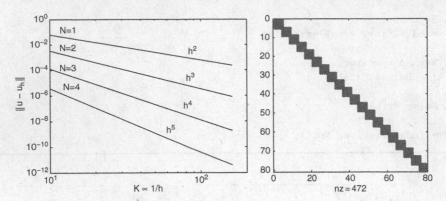

Fig. 7.8. On the left, we show the convergence rate for the solving the homogeneous Poisson problem with a stabilized LDG flux. On the right is shown the sparsity pattern of the discrete Poisson operator based on this flux.

To appreciate the source of this, consider the simplest schemes with $N = 0$. In this case the scheme for element k becomes

$$q_h^*(q_h^k, q_h^{k+1}, u_h^k, u_h^{k+1}) - q_h^*(q_h^k, q_h^{k-1}, u_h^k, u_h^{k-1}) = hf_h^k,$$
$$u_h^*(u_h^k, u_h^{k+1}) - u_h^*(u_h^k, u_h^{k-1}) = hg_h^k.$$

If we first consider the central flux with

$$q_h^*(q_h^-, q_h^+, u_h^-, u_h^+) = \{\!\{q_h\}\!\} - \tau[\![u_h]\!], \quad u_h^*(u_h^-, u_h^+) = \{\!\{u_h\}\!\},$$

we recover

$$\frac{u_h^{k+2} - 2u_h^k + u_h^{k-2}}{(2h)^2} + \tau\frac{u_h^{k+1} - u_h^{k-1}}{h} = f_h^k. \tag{7.4}$$

One notices that the stencil is wider than would be expected; that is, the standard second-order finite difference method uses only the nearest neighbors in contrast to the central stencil in Eq. (7.4), which engages two levels of elements.

If we instead consider the LDG flux

$$q_h^*(q_h^-, q_h^+, u_h^-, u_h^+) = q_h^- - \tau[\![u_h]\!], \quad u_h^*(u_h^-, u_h^+) = u_h^+,$$

we recover the stencil

$$\frac{u_h^{k+1} - 2u_h^k + u_h^{k-1}}{h^2} + \tau\frac{u_h^{k+1} - u_h^{k-1}}{h} = f_h^k. \tag{7.5}$$

This leads to a more compact stencil by relying only on the nearest neighbors. The sparsity patterns in Figs. 7.7 and 7.8 reflect how this carries over to the general high-order case where the choice of the LDG flux generally leads to a much sparser operator. The extension to the multidimensional case is less dramatic in terms of sparsity as LDG no longer guarantees nearest neighbor connections on general grids, although it will remain sparser than the multidimensional version of the central fluxes [252, 283].

The more compact discrete operator/stencil does, however, come at a price. If we compute the condition number, κ, one observes that

$$\kappa(\mathcal{A}_{LDG}) \simeq 2\kappa(\mathcal{A}_C);$$

that is, the condition number of the LDG-based discrete operators is about twice that of the operator based on central fluxes. The impact of this is a slightly worse accuracy of direct solvers and increased iteration counts when an iterative solver is used to solve the linear system.

Trying to balance sparsity and the conditioning, a possible compromise is the internal penalty flux, defined as

$$q_h^* = \{\{(u_h)_x\}\} - \tau[\![u_h]\!], \quad u_h^* = \{\{u_h\}\}.$$

We note that for $N = 0$, the flux q_h^* depends solely on the penalty term. This indicates, as we will discuss shortly, that the value of τ plays a larger role for the internal penalty flux than for the previous two cases.

Nevertheless, to first convince ourselves that this approach works, we repeat the problem in Example 7.5 using the internal penalty flux, and we show the results in Fig. 7.9. As one could hope, we observe optimal rates of convergence and, when comparing with Figs. 7.7 and 7.8, a sparsity pattern in the discrete operator in between those obtained with the central flux and the LDG flux. Furthermore, the condition number behaves as

$$\kappa(\mathcal{A}_C) \simeq \kappa(\mathcal{A}_{IP});$$

that is, the internal penalty method appears to offer a suitable compromise between central fluxes and LDG fluxes. An example of the implementation of internal penalty flux is shown in PoissonIPstabRHS1D.m.

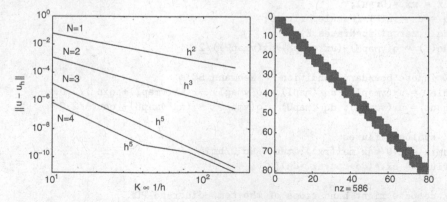

Fig. 7.9. On the left, we show the convergence rate for the solving the homogeneous Poisson problem with a stabilized internal penalty flux. On the right is shown the sparsity pattern of the discrete Poisson operator based on this flux.

Let us now return to the question of how to choose the penalty parameter, τ. As we will show rigorously in Section 7.2.2, for schemes based on central or LDG fluxes, it suffices to take $\tau > 0$ to ensure invertibility and a purely negative real spectrum [13, 48, 49]. However, for the internal penalty-based scheme, we must require that

$$\tau \geq C\frac{(N+1)^2}{h}, \quad C \geq 1,$$

to guarantee similar properties. The bound on C is derived in [281].

Whereas these results ensure invertibility and definiteness, the actual value of τ may well affect both the conditioning and the accuracy of the scheme. In fact, as we will see, the actual value of τ has quite a different impact on the schemes.

PoissonIPstabRHS1D.m

```
function [rhsu] = PoissonIPstabRHS1D(u)

% function [rhsu] = PoissonIPstabRHS1D(u)
% Purpose  : Evaluate RHS in 1D Poisson equation on symmetric form
%            using stabilized internal penalty flux

Globals1D;

% Define field differences at faces
du  = zeros(Nfp*Nfaces,K); du(:)  = u(vmapM)-u(vmapP);

% impose boundary condition -- Dirichlet BC's
uin  = -u(vmapI); du(mapI) = u(vmapI)-uin;
uout = -u(vmapO); du(mapO) = u(vmapO)-uout;

% Compute q
fluxu = nx.*du/2.0;
ux = rx.*(Dr*u);
q = ux - LIFT*(Fscale.*fluxu);
dq = zeros(Nfp*Nfaces,K);
dq(:) = q(vmapM)-(ux(vmapM)+ux(vmapP))/2.0;

% impose boundary condition -- Neumann BC's
qin  = ux(vmapI); dq (mapI) = q(vmapI) - (ux(vmapI)+ qin )/2.0;
qout = ux(vmapO); dq (mapO) = q(vmapO) - (ux(vmapO)+ qout)/2.0;

% evaluate fluxes
hmin = 2.0/max(max(rx)); tau = Np^2/hmin;
fluxq = nx.*(dq+tau*nx.*du);

% compute right hand sides of the semi-discrete PDE
rhsu = J.*((invV'*invV)*(rx.*(Dr*q) - LIFT*(Fscale.*fluxq)));
return
```

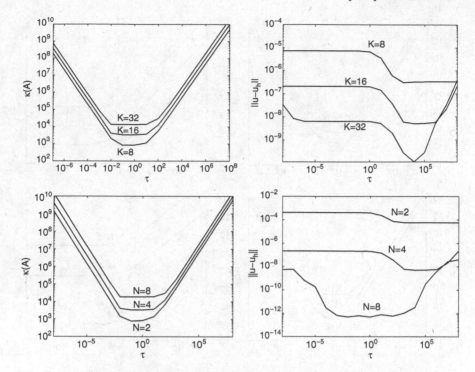

Fig. 7.10. Sensitivity to the penalty parameter in the central flux-based scheme. In the left column, we show the condition number, $\kappa(\mathcal{A})$, for variations in K, with $N = 4$ fixed in the top row, and n, with $K = 16$ fixed in the bottom row, as a function of the penalty parameter, τ. In the right column is shown the L^2-error as a function of τ for a similar variation of the parameters.

Let us first consider the case of the central flux. We consider again the simple problem in Example 7.5 and solve it for various combinations of order, N, and number of elements, K, while varying τ over a large range of values. In Fig. 7.10 we show the impact of the variation of τ on the condition number, $\kappa(\mathcal{A})$, and accuracy measured in the broken L^2-norm.

We note in particular that outside a region of τ centered around 1, the condition number of $\kappa(\mathcal{A})$ basically scales as

$$\kappa(\mathcal{A}) \propto \begin{cases} \tau^{-1}, & \tau \ll 1 \\ \tau, & \tau \gg 1, \end{cases}$$

while inside the region of $\tau \simeq \mathcal{O}(1)$, there is limited impact of τ on the condition number. For the accuracy, there appears to be some advantages of having τ significantly larger than 1, although this should be balanced against solution cost if an iterative solver is used. We do note, however, that taking extreme values of τ generally has limited impact on the accuracy, except when the conditioning becomes extremely poor, as is indicated for the $N = 8$, $K = 16$ case.

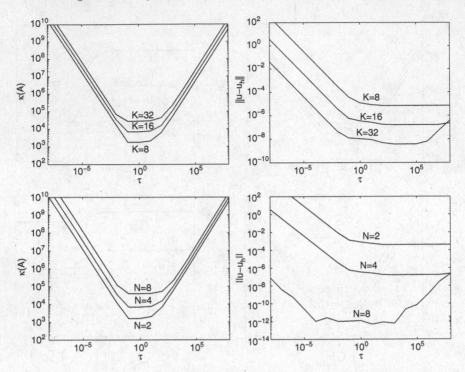

Fig. 7.11. Sensitivity to the penalty parameter in the LDG based scheme. In the left column, we show the condition number, $\kappa(\mathcal{A})$, for variations in K, with $N = 4$ fixed in the top row, and N, with $K = 16$ fixed in the bottom row, as a function of the penalty parameter, τ. In the right column is shown the L^2-error as a function of τ for a similar variation of the parameters.

If we now repeat the example for the LDG-based flux, the situation is very different. We show the results in Fig. 7.11. While the behavior of the conditioning of the discrete operator is similar to the previous case, changes in τ have a dramatic impact on the error, at least when $\tau \ll 1$, in which case we observe an error proportional to τ^{-1}. However, again taking $\tau \simeq \mathcal{O}(1)$ appears to be a safe choice both in terms of conditioning and accuracy.

For the internal penalty-based scheme, it only makes sense to consider the value of the constant in front of the correct scaling. In Fig. 7.12 we show the results of the same computations as above. For the condition number, we observe a direct proportionality to the value of constant, C, in the penalty term. This is similar to what we observed for the two previous formulations. In terms of the accuracy, the internal penalty method seems to behave like the two other schemes when the penalty parameter becomes large. However, since the actual value of the penalty parameter is $C(N+1)^2/h$, it is, in effect, significantly larger, impacting the conditioning and, thus, the accuracy of the solution process more severely.

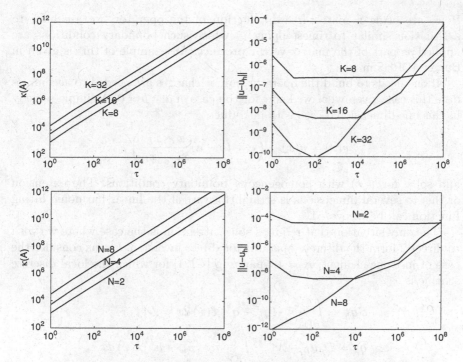

Fig. 7.12. Sensitivity to the penalty parameter in the internal penalty-based scheme; in this case, τ refers to the value of the constant in front of the correct scaling. In the left column, we show the condition number, $\kappa(\mathcal{A})$, for variations in K, with $N = 4$ fixed in the top row, and N, with $K = 16$ fixed in the bottom row, as a function of the penalty parameter, τ. In the right column is shown the L^2-error as a function of τ for a similar variation of the parameters.

As we have seen, the discontinuous Galerkin method works well also for elliptic problems such as the ones considered here. Which one of the three fluxes one prefers to use may well be problem dependent. Indeed, if a direct solver is used, sparsity may be more important than conditioning, possibly favoring an LDG-based approach. On the other hand, for iterative methods, a central or internal penalty flux would seem most appropriate.

Extension to the inhomogenous problem

Before we continue with more general aspects, let us consider the closely related problem

$$\frac{\partial^2 u}{\partial x^2} = f(x), \quad x \in \Omega,$$

subject to inhomogeneous boundary conditions as

$$u(x_L) = a, \quad u(x_R) = b.$$

If we are using a matrix-free application of the operator, we are back to a situation similar to timestepping in which such boundary conditions are applied as part of the matrix-vector product. An example of this is given in Burgers1DRHS.m.

If one needs to build the operator, minor changes are needed to accommodate this case given what we know for the case of $a = b = 0$. A simple remedy in this one-dimensional case is to introduce

$$v(x) = u(x) - \left(a + (b - a)\frac{x - x_L}{x_R - x_L} \right),$$

and solve for $v(x)$ with homogeneous boundary conditions. The extension of this to several dimensions is straightforward if the linear harmonic lifting function can be recovered.

The only situation that requires some attention is the case where we wish to directly form the discrete operator for direct inversion. Let us consider the case of just one element, $K = 1$, spanning $[-1, 1]$ for which the local discrete scheme is

$$\mathcal{S}q_h - \int_{\partial \mathsf{D}^k} \hat{n} \cdot (q_h - q^*) \ell(x)\, dx = \mathcal{M}f_h,$$

$$q_h = \mathcal{D}_r u_h - \mathcal{M}^{-1} \int_{\partial \mathsf{D}^k} \hat{n} \cdot (u_h - u^*) \ell(x)\, dx.$$

For the Dirichlet boundary conditions, we use

$$x = -1: \ u^*(-1) = a, \ q^*(-1) = q_h(-1),$$

$$x = 1: \ u^*(1) = b, \ q^*(1) = q_h(1),$$

resulting in the following scheme:

$$\mathcal{S}q_h = \mathcal{M}f_h,$$
$$q_h = \mathcal{D}_r u_h - \mathcal{M}^{-1} \left(e_{N_p}(u_h(1) - b) - e_1(u_h(-1) - a) \right),$$

where e_i is an N_p zero vector with element i being 1. If we now define

$$q_h = \tilde{q}_h + \mathcal{M}^{-1} \left(e_{N_p}b - e_1 a \right) = \tilde{q}_h + \mathcal{M}^{-1} \int_{\partial \mathsf{D}^k} (\hat{n} \cdot g(x)) \ell(x)\, dx,$$

where $\tilde{q}_h$ is the solution to the homogeneous problem and $u(x) = g(x)$ for $x \in \partial \Omega$, we recover the modified scheme as

$$\mathcal{A}u_h = \mathcal{M}\tilde{f}_h = \mathcal{M}f_h - \mathcal{S}\mathcal{M}^{-1} \int_{\partial \mathsf{D}^k} (\hat{n} \cdot g(x)) \ell(x)\, dx.$$

In other words, the system to solve remains unchanged from the homogenous case, but the right hand side changes to reflect the inhomogenous boundary conditions.

In the case of Neumann conditions

$$\frac{\partial u}{\partial x} = h(x), \ \ x \in \partial\Omega,$$

a similar line of argument leads to

$$\mathcal{A}u_h = \mathcal{M}\tilde{f}_h = \mathcal{M}f_h - \int_{\partial \mathrm{D}^k} (\hat{n} \cdot h(x)) \, \ell(x) \, dx.$$

Clearly, all internal boundaries are treated as for the homogeneous case and no changes are needed.

7.2.1 Two-dimensional Poisson and Helmholtz equations

Let us continue by extending the development of DG methods for problems with higher spatial operators beyond one dimension. As we will see, this extension is entirely straightforward and the behavior is as expected.

Let us first consider the solution of the two-dimensional Poisson equation,

$$\nabla^2 u = f, \ \ \boldsymbol{x} \in \Omega,$$

with

$$u(\boldsymbol{x}) = 0, \ \ \boldsymbol{x} \in \partial\Omega,$$

and $f(\boldsymbol{x})$ is a given function defined in Ω.

To follow the development of the one-dimensional methods, we introduce a new vector function, $\boldsymbol{q}$, to recover the first-order system

$$\nabla \cdot \boldsymbol{q} = f, \ \ \nabla u = \boldsymbol{q}.$$

If we now discretize this in the usual way, we assume that $(u, \boldsymbol{q}) = (u, q^x, q^y)$ are approximated by piecewise N-th-order polynomials, (u_h, q_h^x, q_h^y), on each triangle and recover the local formulation

$$\mathcal{S}_x q_h^x + \mathcal{S}_y q_h^y - \int_{\partial \mathrm{D}^k} \hat{n} \cdot ((q_h^x, q_h^y) - q_h^*) \, \ell(\boldsymbol{x}) \, d\boldsymbol{x} = \mathcal{M}^k f_h,$$

$$\mathcal{M}^k q_h^x = \mathcal{S}_x u_h - \int_{\partial \mathrm{D}^k} \hat{n}_x (u_h - u_h^*) \, \ell(\boldsymbol{x}) \, d\boldsymbol{x},$$

$$\mathcal{M}^k q_h^y = \mathcal{S}_x u_h - \int_{\partial \mathrm{D}^k} \hat{n}_y (u_h - u_h^*) \, \ell(\boldsymbol{x}) \, d\boldsymbol{x}.$$

The internal penalty fluxes are given as

$$q_h^* = \{\!\{\nabla u_h\}\!\} - \tau[\![u]\!], \ \ u_h^* = \{\!\{u_h\}\!\}.$$

An example of implementing this in a matrix-free form is given in Poisson-RHS2D.m.

──────────────────── PoissonRHS2D.m ────────────────

```
function [rhsu,Mu] = PoissonRHS2D(u)

% function [rhsu] = PoissonRHS2D(u)
% Purpose  : Evaluate RHS flux in 2D Heat/Poisson equation using a
%            stabilized internal penalty flux

Globals2D;

% Define field differences at faces and impose Dirichlet BCs
du = zeros(Nfp*Nfaces,K);
du(:)=u(vmapM)-u(vmapP); du(mapD) = 2*u(vmapD);

% Compute qx and qy, define differences and impose Neumann BC's
[dudx,dudy] = Grad2D(u);

% Compute DG gradient with central fluxes
fluxxu = nx.*du/2.0; qx = dudx - LIFT*(Fscale.*fluxxu);
fluxyu = ny.*du/2.0; qy = dudy - LIFT*(Fscale.*fluxyu);

% Compute minimum height of elements either side of each edge
hmin = min(2*J(vmapP)./sJ(mapP), 2*J(vmapM)./sJ(mapM));
tau = reshape(Np./hmin, Nfp*Nfaces, K);

% Evaluate jumps in components of q at element interfaces
dqx=zeros(Nfp*Nfaces,K); dqx(:)=qx(vmapM)-qx(vmapP);
dqx(mapN) = 2*qx(vmapN);
dqy=zeros(Nfp*Nfaces,K); dqy(:)=qy(vmapM)-qy(vmapP);
dqy(mapN) = 2*qy(vmapN);

% Evaluate flux function
fluxq = (nx.*dqx + ny.*dqy + tau.*du)/2;

% Compute right hand side
divq = Div2D(qx,qy);

% compute right hand side residual
rhsu = J.*((invV'*invV)*(divq - LIFT*(Fscale.*fluxq)));
Mu = J.*((invV'*invV)*u);
return;
```

──

   While the matrix-free form is ideally suited for iterative solvers, it is illustrative to form the matrix to inspect symmetries and sparsity patterns. This
is done in Poisson2D.m in a way similar to the one-dimensional case.

| Poisson2D.m |

```
function [A,M] = Poisson2D();

% function [A,M] = Poisson2D()
% Purpose: Set up matrix for 2D Poisson equation based on
%          stabilized internal fluxes on symmetric form

Globals2D;
g = zeros(K*Np,1);
A = spalloc(K*Np, K*Np, 3*Np);   M = spalloc(K*Np, K*Np, 3*Np);

% Build matrix -- one column at a time
for i=1:K*Np
    g(i) = 1.0;
    gmat = reshape(g,Np,K);
    [Avec,Mvec] = PoissonRHS2D(gmat);

    ids = find(Avec); A(ids,i) = Avec(ids);
    ids = find(Mvec); M(ids,i) = Mvec(ids);
    g(i)=0.0;
end
return
```

Finally, we have in PoissonDriver2D.m the driver routine that is used to solve the Poisson problem with a specific right-hand side.

| PoissonDriver2D.m |

```
% Driver script for solving the 2D Poisson equation
Globals2D;

% Polynomial order used for approximation
N = 5;

% Read in Mesh
[Nv, VX, VY, K, EToV] = MeshReaderGambitBC2D('circA01.neu');

% Initialize solver and construct grid and metric
StartUp2D;

% set up boundary conditions
BuildBCMaps2D;

% set up right hand side for homogeneous Poisson
%[A,M] = Poisson2D(); % Setup using PoissonRHS2D.m
[A,M] = PoissonIPDG2D(); % Setup using PoissonIPDG2D.m

% set up Dirichlet boundary conditions
uD = zeros(Nfp*Nfaces, K);
```

```
uD(mapD) = sin(pi*Fx(mapD)).*sin(pi*Fy(mapD));

% set up Neumann boundary conditions
qN = zeros(Nfp*Nfaces, K);
qN(mapN) = nx(mapN).*(pi*cos(pi*Fx(mapN)).*sin(pi*Fy(mapN))) + ...
           ny(mapN).*(pi*sin(pi*Fx(mapN)).*cos(pi*Fy(mapN))) ;

% evaluate boundary condition contribution to rhs
Aqbc = PoissonIPDGbc2D(uD, qN);

% set up right hand side forcing
rhs = -2*(pi^2)*sin(pi*x).*sin(pi*y);
rhs = -MassMatrix*(J.*rhs) + Aqbc;

% solve system
u = A\rhs(:);
u = reshape(u, Np, K);
```

The way the discrete operator is set up is clearly not ideal, as it requires storage of the full matrix, even if a very large fraction of the entries are zero; that is, the matrix is very sparse. This problem can be overcome in one of two different ways. One can either set up the matrix directly in sparse format, hence only storing entries different from zero. This is exemplified in PoissonIPDG2D, where the discrete Poisson operator is set up, using a weak from and a stabilized central flux.

───────────────────────── PoissonIPDG2D.m ─────────────────────────

```
function [OP,MM] = PoissonIPDG2D()

% Purpose: Set up the discrete Poisson matrix directly
%          using LDG. The operator is set up in the weak form
Globals2D;

% build local face matrices
massEdge = zeros(Np,Np,Nfaces);
Fm = Fmask(:,1); faceR = r(Fm);
V1D = Vandermonde1D(N, faceR);  massEdge(Fm,Fm,1) = inv(V1D*V1D');
Fm = Fmask(:,2); faceR = r(Fm);
V1D = Vandermonde1D(N, faceR);  massEdge(Fm,Fm,2) = inv(V1D*V1D');
Fm = Fmask(:,3); faceS = s(Fm);
V1D = Vandermonde1D(N, faceS);  massEdge(Fm,Fm,3) = inv(V1D*V1D');

% build local volume mass matrix
MassMatrix = invV'*invV;

% build DG derivative matrices
MM  = zeros(K*Np*Np, 3);  OP = zeros(K*Np*Np*(1+Nfaces), 3);
```

```
% global node numbering
entries = (1:Np*Np)'; entriesMM = (1:Np*Np)';
for k1=1:K
  if(~mod(k1,1000)) k1, end;
  rows1 = ((k1-1)*Np+1:k1*Np)'*ones(1,Np); cols1 = rows1';

  % Build local operators
  Dx = rx(1,k1)*Dr + sx(1,k1)*Ds;  Dy = ry(1,k1)*Dr + sy(1,k1)*Ds;

  OP11 = J(1,k1)*(Dx'*MassMatrix*Dx + Dy'*MassMatrix*Dy);

  % Build element-to-element parts of operator
  for f1=1:Nfaces
    k2 = EToE(k1,f1); f2 = EToF(k1,f1);

    rows2 = ((k2-1)*Np+1:k2*Np)'*ones(1,Np); cols2 = rows2';

    fidM  = (k1-1)*Nfp*Nfaces + (f1-1)*Nfp + (1:Nfp);
    vidM = vmapM(fidM); Fm1 = mod(vidM-1,Np)+1;
    vidP = vmapP(fidM); Fm2 = mod(vidP-1,Np)+1;

    id = 1+(f1-1)*Nfp + (k1-1)*Nfp*Nfaces;
    lnx = nx(id);  lny = ny(id); lsJ = sJ(id);
    hinv = max(Fscale(id), Fscale(1+(f2-1)*Nfp, k2));

    Dx2 = rx(1,k2)*Dr + sx(1,k2)*Ds;
    Dy2 = ry(1,k2)*Dr + sy(1,k2)*Ds;

    Dn1 = lnx*Dx  + lny*Dy ;
    Dn2 = lnx*Dx2 + lny*Dy2;

    mmE = lsJ*massEdge(:,:,f1);

    gtau = 100*2*(N+1)*(N+1)*hinv; % set penalty scaling
    switch(BCType(k1,f1))
      case {Dirichlet}
    OP11 = OP11 + ( gtau*mmE - mmE*Dn1 - Dn1'*mmE ); % ok
      case {Neuman}
        % nada
      otherwise
    % interior face variational terms
    OP11        = OP11 + 0.5*( gtau*mmE - mmE*Dn1 - Dn1'*mmE );

    OP12 = zeros(Np);
    OP12(:,Fm2) =                 - 0.5*( gtau*mmE(:,Fm1) );
    OP12(Fm1,:) = OP12(Fm1,:) ...
            - 0.5*(        mmE(Fm1,Fm1)*Dn2(Fm2,:) );
    OP12(:,Fm2) = OP12(:,Fm2) - 0.5*(-Dn1'*mmE(:, Fm1) );
    OP(entries(:), :) = [rows1(:), cols2(:), OP12(:)];
```

```
    entries = entries + Np*Np;

    end
  end
  OP(entries(:),:)    = [rows1(:), cols1(:), OP11(:)];
  MM(entriesMM(:),:) = [rows1(:), cols1(:), J(1,k1)*MassMatrix(:)];
  entries = entries + Np*Np; entriesMM = entriesMM + Np*Np;
end

OP    =    OP(1:max(entries)  -Np*Np,:);
OP    = myspconvert(OP, Np*K, Np*K, 1e-15);
MM    =    MM(1:max(entriesMM)-Np*Np,:);
MM    = myspconvert(MM, Np*K, Np*K, 1e-15);
return
```

A well-known alternative to this approach is to use a iterative solver where all that is needed is the ability to perform a matrix-vector product – exactly as PoissonRHS2D.m does. The only minor change needed is that PoissonRHS2D.m returns an $N_p \times K$ matrix whereas an iterative method would require a vector. With this minor change, however, one can solve the problem using an appropriate iterative solver such as a preconditioned conjugate gradient method. A possible preconditioner could be an incomplete Cholesky factorization, as we discuss in Section 7.3.

In the following example, we will explore these options in a little more detail.

Example 7.8. Consider the two-dimensional Helmholtz equation

$$u - \nabla^2 u = f, \ \boldsymbol{x} \in [-1,1]^2,$$

with $u = 0$ at the boundary. This is a classic problem that appears in many areas of the sciences. Furthermore, it is an example of a problem one often needs to solve when using implicit time-integration methods for the heat equation.

It is, naturally, closely related to the Poisson equation discussed in the above, with the linear operator being

$$\mathcal{L}u = f, \ \mathcal{L} = 1 - \nabla^2.$$

Thus, the modification from the Poisson equation is minimal and we can use the schemes for this directly.

We solve the problem with f such that it has the exact solution

$$u(x,y) = \sin(\pi x)\sin(\pi y).$$

The aim here is to understand both the convergence rate and to provide an example of how a preconditioner can be useful.

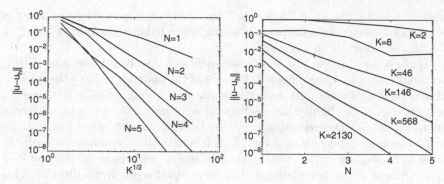

Fig. 7.13. On the left, we show the $h \simeq \sqrt{K}$ convergence for different orders of local approximation and the right side shows the effect of order increases.

Table 7.2. Number of preconditioned conjugate gradient iterations to reach a tolerance of 10^{-8}, using different preconditioners on the discrete Helmholtz problem. In this two-dimensional case, h is the approximate average length of the simplices in the grid and N is the order of the local approximation.

$h \backslash N$	No preconditioning					Jacobi precon.					Block-Jacobi precon.				
	1	2	3	4	5	1	2	3	4	5	1	2	3	4	5
2.0	2	4	6	9	12	2	4	6	9	12	2	4	5	6	7
1.0	13	27	41	60	80	13	26	38	56	72	12	22	33	46	56
1/2	60	95	137	191	245	58	89	117	151	194	49	85	112	138	162
1/4	121	195	274	396	480	117	170	228	294	380	105	160	210	260	310
1/8	228	393	514	710	921	212	321	428	551	721	212	313	411	505	599
1/16	465	751	1084	1476	1915	449	666	883	1142	1487	445	644	847	1048	1246

To study the convergence rate, we solve the problem on a series of progressively refined grids, both in grid size and order of approximation. In Fig. 7.13 we show the convergence results, confirming the expected convergence when refining in order N or in elementsize h. To better understand the challenges associated with solving the linear systems, we briefly consider the effect of preconditioning for an iterative solver. Since the linear system resulting from the discretization of the Helmholtz equation is symmetric positive definite, it is most natural to use a preconditioned conjugate gradient solver for solving the problem.

In Table 7.2 we show the iteration counts as a function of both $h \simeq 1/\sqrt{K}$ and N. The results are obtained with a conjugate gradient method with a small tolerance of 10^{-8}. We observe that the iteration count approximately scales as N/h (i.e., higher order and/or smaller grids), results in more iterations. To improve on this, we can seek a preconditioner, $\mathcal{C}$, in the hope that the system

$$\mathcal{C}^{-1}\mathcal{A}u_h = \mathcal{C}^{-1}f_h$$

is easier to solve than the original problem. Finding a good preconditioner is often an art more than a science where one has to balance the cost of inverting $\mathcal{C}$ with the reduction in iterations.

Using a very simple preconditioner such as a Jacobi preconditioner (i.e., preconditioning with the diagonal elements of the matrix), reduces the number of iterations somewhat, although the qualitative scaling remains the same, in particular in h. A slightly more advanced approach is to use a block-Jacobi approach, where the preconditioner consists of diagonal blocks of $\mathcal{A}$, making it slightly more expensive to invert. This is a natural preconditioner for high-order methods, as it will address the local elements and one should expect it to be superior at larger order compared to the simple Jacobi preconditioner. This is confirmed in Table 7.2 where we observe a significant drop in the number of iterations for N being large. However, the scaling in h remains essentially unchanged.

Generally, it has been found that incomplete factorization methods, (e.g., incomplete LU or Cholesky methods), work well for the class of methods discussed here as we will also see in Section 7.3. However, these method, have the disadvantage that one generally needs to assemble the full matrix to compute the incomplete factorization. For problems of even medium size, this is often impractical.

The design and analysis of efficient preconditioning techniques remains an area of significant activity and excellent starting points for such further studies are [275, 305].

A Poisson solver for curved elements

We recall the Poisson equation, given here in strong form

$$-\nabla^2 u = f$$
$$u = u^D \text{ on } \partial\Omega^D,$$
$$\frac{\partial u}{\partial n} = g \text{ on } \partial\Omega^N.$$

where Ω^D is the subset of the boundary with Dirichlet boundary conditions and Ω^N is the remainder of the boundary where Neumann boundary conditions are specified. The Interior Penalty DG (IPDG) [12, 13] variational statement requires us to find $u_h \in V_h$ such that

$$(\nabla\phi_h, \nabla u_h)_{\mathsf{D}^k} - \frac{1}{2}(\hat{\boldsymbol{n}} \cdot \nabla\phi_h, \hat{\boldsymbol{n}} \cdot [\![u_h]\!])_{\partial\mathsf{D}^k\backslash\partial\Omega} - (\phi_h, \hat{\boldsymbol{n}} \cdot \{\!\{\nabla u_h\}\!\})_{\partial\mathsf{D}^k\backslash\partial\Omega}$$
$$- (\phi_h, \hat{\boldsymbol{n}} \cdot \nabla u_h^-)_{\partial\mathsf{D}^k\cap\partial\Omega^N} + (\tau^k\phi_h, \hat{\boldsymbol{n}} \cdot [\![u_h]\!])_{\partial\mathsf{D}^k\backslash\partial\Omega} + (\tau^k\phi_h, u_h^-)_{\partial\mathsf{D}^k\cap\partial\Omega^D}$$
$$= (\phi_h, f_h)_{\mathsf{D}^k} - (\hat{\boldsymbol{n}} \cdot \nabla\phi_h, p^O)_{\partial\mathsf{D}^k\cap\partial\Omega^D} + (\phi_h, g)_{\partial\mathsf{D}^k\cap\partial\Omega^N}$$
$$+ (\tau^k\phi_h, u^D)_{\partial\mathsf{D}^k\cap\partial\Omega^D} \tag{7.6}$$

is true for all $\phi_h \in V_h = \bigoplus_{k=1}^{K} P_N(D^k)$ – a piecewise polynomial basis. We recall that τ^k depends on the mesh and the approximation order to ensure solvability.

We create a discrete representation of this system stored in a matrix $\mathcal{A}$ as demonstrated in CurvedPoissonIPDG2D.m and then solve

$$\mathcal{A}u_h = f_h.$$

Before discussing the implementation and evaluation of the discrete operator in detail, we recall the discussion of variational crimes in Section 6.6.1. In the context of a time-dependent computation, we could effectively manage anomalous behavior caused by an inaccurate evaluation of the inner-products required in the DG discretizations. Specifically, for hyperbolic partial differential equations we could filter the solution to control effects of aliasing and still maintain nearly optimal convergence rates. In the current context it is possible that a mesh contains some curvilinear elements with nonpolynomial mapping Jacobians. This complicates the evaluation of the inner products on the left-hand side of Eq. (7.6) and may cause a lack of convergence or reduced order of convergence if not handled carefully. Even a few curvilinear elements can cause large global errors because of the elliptic nature of this problem, as we will discuss in more detail in Section 9.1.

We resolve this by using the integration-based approach discussed in Section 6.6.1. All volume integrals are evaluated using a cubature rule, designed to exactly integrate polynomial functions up to a desired order. The global stiffness matrices are stored in a sparse-coordinate format [i.e., an array of size (number of nonzeros)$\times 3$, where we estimate the number of nonzeros based on the number of elements and order of approximation being used]. The CurvedPoissonIPDG2D.m function assumes that the cubature and Gauss rules have been set up previously and stored in the cub and gauss structures by CubatureVolumeMesh2D.m and GaussFaceMesh2D.m, respectively.

The storage for the sparse stiffness and mass matrices are initiated by the following

────────────────────── CurvedPoissonIPDG2D.m ──────────────────────

```
1   function [OP,MM] = CurvedPoissonIPDG2D()
2
3   % function [OP,MM] = CurvedPoissonIPDG2D()
4   % Purpose: Set up a discrete Poisson matrix and mass matrix
5   %          using Interior Penalty Discontinuous Galerkin (IPDG).
6
7   Globals2D;
8
9   NGauss = gauss.NGauss;
10
11  % build DG derivative matrices
12  MM  = zeros(K*Np*Np, 3);   OP = zeros(K*Np*Np*(1+Nfaces), 3);
```

We then loop through each element and evaluate the local stiffness matrix term

$$(\nabla \ell_n^k, \nabla \ell_m^k)_{\mathsf{D}^k} = \sum_{i=1}^{N_c} w_i^c J_i^{c,k} \left(\nabla \ell_n^k (r_i^c, s_i^c) \cdot \nabla \ell_m^k (r_i^c, s_i^c) \right)$$

for the Lagrange basis functions $\ell_n^k(\boldsymbol{x})$ and $\ell_m^k(\boldsymbol{x})$ on the k-th element. Here (r_i^c, s_i^c) represents the N_c cubature points. This construction is achieved in a matrix approach as:

———————— CurvedPoissonIPDG2D.m ————————

```
14    % global node numbering
15    entries = (1:Np*Np)'; entriesMM = (1:Np*Np)';
16    for k1=1:K
17      if(~mod(k1,200)) k1, end;
18
19      % Location of k1'th diagonal block entries in OP matrix
20      rows1 = ((k1-1)*Np+1:k1*Np)'*ones(1,Np); cols1 = rows1';
21
22      % Extract local mass matrix and cubature weights
23      locmm = cub.mm(:,:,k1);
24      cw  = spdiags(cub.W (:,k1), 0, cub.Ncub, cub.Ncub);
25
26      % Evaluate derivatives of Lagrange basis functions at
27      %          cubature nodes
28      [cDx, cDy] = PhysDmatrices2D(x(:,k1), y(:,k1), cub.V);
29
30      % Evaluate local stiffness matrix
31      OP11 = cDx'*cw*cDx + cDy'*cw*cDy;
```

The remaining terms on the left-hand side of Eq. (7.6) all involve surface integrals of the Lagrange polynomials and normal derivatives at both internal and external traces of each face. These functions are evaluate at the Gauss nodes on each face:

———————— CurvedPoissonIPDG2D.m ————————

```
34      for f1=1:Nfaces
35
36        % Find neighbor
37        k2 = EToE(k1,f1); f2 = EToF(k1,f1);
38
39        idsM = (f1-1)*NGauss+1:f1*NGauss;
40
41        % Extract Lagrange basis function
42        %   -> Gauss node interpolation matrix
43        gVM = gauss.finterp(:,:,f1);
44        gVP = gauss.finterp(:,:,f2);      gVP = gVP(NGauss:-1:1,:);
45
```

```
46      % Evaluate spatial derivatives of  Lagrange basis function
47      % at Gauss nodes
48      [gDxM, gDyM] = PhysDmatrices2D(x(:,k1), y(:,k1),gVM);
49      [gDxP, gDyP] = PhysDmatrices2D(x(:,k2), y(:,k2),gVP);
50
51      % Evaluate normals at Gauss nodes on face
52      gnx = spdiags(gauss.nx(idsM, k1), 0, NGauss, NGauss);
53      gny = spdiags(gauss.ny(idsM, k1), 0, NGauss, NGauss);
54      gw  = spdiags(gauss.W(idsM, k1),  0, NGauss, NGauss);
55
56      % Compute normal derivatives of Lagrange basis functions
57      % at Gauss nodes
58      gDnM = gnx*gDxM + gny*gDyM;
59      gDnP = gnx*gDxP + gny*gDyP;
60
61      % Locate global numbers of Lagrange nodes in neighboring
62      % element
63      cols2 = ones(Np,1)*((k2-1)*Np+1:k2*Np);
64
65      % Find minimum height of two elements sharing this face
66      hinv = max(Fscale( 1 + (f1-1)*Nfp, k1), ...
67                 Fscale( 1 + (f2-1)*Nfp, k2));
68
69      % Set penalty scaling
70      gtau = 20*(N+1)*(N+1)*hinv;
```

This supplies enough information to evaluate all the surface integral terms to build the discrete Poisson operator, and what remains is to determine whether the face in question is an internal element-element interface, a Dirichlet boundary face, or a face on a Neumann boundary. Each type of face requires a slightly different treatment. An internal face contributes two terms to the matrix, one on the diagonal block of the matrix and one to an off-diagonal block. A Dirichlet face contributes a term just to the diagonal block, and a Neumann face contributes no additional terms to the matrix. This choice is made in the following and surface integrals are handled in a blocked approach. As each coupling term is computed they are added to the list of nonzero entries. The diagonal block for each element is only added after all face terms for the element have been computed.

—————————— CurvedPoissonIPDG2D.m ——————————

```
72      % Determine type of face
73      switch(BCType(k1,f1))
74        case {Dirichlet}
75          % Dirichlet boundary face variational terms
76      OP11 = OP11 + (gVM'*gw*gtau*gVM - gVM'*gw*gDnM - gDnM'*gw*gVM);
77
78        case {Neuman}
```

```
79              % Do nothing
80           otherwise
81         % Interior face variational terms for stiffness matrix
82         OP11 = OP11 + 0.5*(gVM'*gw*gtau*gVM ...
83                       - gVM'*gw*gDnM - gDnM'*gw*gVM);
84
85         OP12 = - 0.5*(gVM'*gw*gtau*gVP + gVM'*gw*gDnP - gDnM'*gw*gVP);
86
87            % Store self-neighbor interaction term in global stiffness
88            %   matrix
89            OP(entries(:), :) = [rows1(:), cols2(:), OP12(:)];
90         entries = entries + Np*Np;
91         end
92     end
93
94     % Store k1'th self-self interaction term in global stiffness
95     % matrix
96     OP(entries(:), :)   = [rows1(:), cols1(:), OP11(:)];
97     MM(entriesMM(:), :) = [rows1(:), cols1(:), locmm(:)];
98     entries = entries + Np*Np; entriesMM = entriesMM + Np*Np;
99 end
```

The last step in building the sparse stiffness and mass matrices is to convert from coordinate sparse form into Matlab's intrinsic sparse matrix format.

—————————————————————— | CurvedPoissonIPDG2D.m | ——————————————————————

```
101 % Convert OP and MM from coordinate storage format to Matlab's
102 %    intrinsic sparse matrix format
103 OP = OP(1:max(entries)-Np*Np,:);
104 OP = myspconvert(OP, Np*K, Np*K, 1e-15);
105 MM = myspconvert(MM, Np*K, Np*K, 1e-15);
```

A companion function CurvedPoissonIPDGbc2D.m creates a second matrix $\mathcal{A}_{bc}$ that allows one to introduce the inhomogeneous boundary condition data that occurs on the right-hand side of Eq. (7.6). For an example of how this works in practice we include a piece of CurvedPoissonIPDGDriver2D.m where we invoke these functions, form the right-hand-side data from the forcing function and the boundary condition data, and solve the problem.

—————————————————————— | CurvedPoissonIPDGDriver2D.m | ——————————————————————

```
30 % build weak Poisson operator matrices
31 [A, M] = CurvedPoissonIPDG2D();
32 Abc = CurvedPoissonIPDGbc2D();
33
34 % set up right hand side
35 f = (-2*pi^2-1)*sin(pi*x).*sin(pi*y);
36
```

```
37   % set up boundary condition
38   ubc = zeros(gauss.NGauss*Nfaces*K,1);
39   xbc = gauss.x(gauss.mapD); ybc = gauss.y(gauss.mapD);
40   ubc(gauss.mapD) = sin(pi*xbc).*sin(pi*ybc);
41   xbc = gauss.x(gauss.mapN); ybc = gauss.y(gauss.mapN);
42   ubc(gauss.mapN) = ...
43       gauss.nx(gauss.mapN).*(pi*cos(pi*xbc).*sin(pi*ybc)) + ...
44       gauss.ny(gauss.mapN).*(pi*sin(pi*xbc).*cos(pi*ybc));
45
46   ubc = Abc*ubc;
47
48   % solve linear system
49   solvec = (A+M)\(M*(-f(:)) + ubc);
50   u = reshape(solvec, Np, K);
```

In this particular example, we take a minor liberty with the evaluation of the right-hand-side portion of the load vector by interpolating f at the elemental Lagrange interpolation nodes and then evaluating the inner products. Alternatively, one can evaluate f at cubature nodes in each element and then evaluate the inner products accordingly.

7.2.2 A look at basic theoretical properties

To gain a better understanding of the behavior we have observed in the last sections, let us develop some of the basic theory for DG approximations of elliptic problems.

To keep things simple, we primarily discuss the different schemes in the context of Poisson's equation but conclude with a few remarks and references for more general problems.

Recall the Poisson equation

$$-\nabla^2 u(\boldsymbol{x}) = f(\boldsymbol{x}), \ \ \boldsymbol{x} \in \Omega,$$

subject to homogeneous Dirichlet boundary conditions as

$$u(\boldsymbol{x}) = 0, \ \ \boldsymbol{x} \in \partial\Omega.$$

Existence and uniqueness of solutions to this problem is well understood [59] under light conditions on f and Ω. The variational formulation for finding $u \in H_0^1(\Omega)$ is

$$a(u, \phi) = \int_\Omega \nabla u \cdot \nabla \phi \, d\boldsymbol{x} = (f, v)_\Omega, \ \ \forall \phi \in H_0^1(\Omega).$$

Here, $H_0^1(\Omega)$ is the space of functions $u \in H^1(\Omega)$ for which $u = 0$ on $\partial\Omega$. This is the essential statement of coercivity.

Table 7.3. Overview of numerical flux choices

	u_h^*	q_h^*
Central flux	$\{\!\{u_h\}\!\}$	$\{\!\{q_h\}\!\} - \tau[\![u_h]\!]$
Local DG flux (LDG)	$\{\!\{u_h\}\!\} + \boldsymbol{\beta} \cdot [\![u_h]\!]$	$\{\!\{q_h\}\!\} - \boldsymbol{\beta}[\![q_h]\!] - \tau[\![u_h]\!]$
Internal penalty flux (IP)	$\{\!\{u_h\}\!\}$	$\{\!\{\nabla u_h\}\!\} - \tau[\![u_h]\!]$

We follow the approach outlined previously and rewrite the problem as a first-order system

$$-\nabla \cdot \boldsymbol{q} = f, \quad \boldsymbol{q} = \nabla u. \tag{7.7}$$

Let us now seek polynomial solutions, $u_h \in V_h = \oplus_{k=1}^K P_N(D^k)$, where we recall that $P_N(D^k)$ is the space of N-th-order polynomials defined on D^k and $\boldsymbol{q}_h \in U_h = V_h \times V_h := V_h^2$. Following the standard DG approach, we seek $(u_h, \boldsymbol{q}_h)$ for all test functions $(\phi_h, \boldsymbol{\pi}_h) \in V_h \times U_h$ such that

$$(\boldsymbol{q}_h, \nabla \phi_h)_{\Omega,h} - \sum_{k=1}^K (\hat{\boldsymbol{n}} \cdot \boldsymbol{q}_h^*, \phi_h)_{\partial D^k} = (f, \phi_h)_{\Omega,h}, \tag{7.8}$$

$$(\boldsymbol{q}_h, \boldsymbol{\pi}_h)_{\Omega,h} = \sum_{k=1}^K (u_h^*, \hat{\boldsymbol{n}} \cdot \boldsymbol{\pi}_h)_{\partial D^k} - (u_h, \nabla \cdot \boldsymbol{\pi}_h)_{\Omega,h}. \tag{7.9}$$

We have introduced the numerical flux pair $(u_h^*, \boldsymbol{q}_h^*)$ to connect elements and we summarize in Table 7.3 the choices discussed previously. Recall the standard notation that

$$\{\!\{u\}\!\} = \frac{u^- + u^+}{2},$$

where u can be both a scalar or a vector. Similarly, we define the jumps along a normal, $\hat{\boldsymbol{n}}$, in the standard notation as

$$[\![u]\!] = \hat{\boldsymbol{n}}^- u^- + \hat{\boldsymbol{n}}^+ u^+, \quad [\![\boldsymbol{u}]\!] = \hat{\boldsymbol{n}}^- \cdot \boldsymbol{u}^- + \hat{\boldsymbol{n}}^+ \cdot \boldsymbol{u}^+.$$

We note that there are several lesser used alternatives to these three options for the numerical fluxes and we refer to [13] for a complete discussion of these.

Primal forms and consistency

To get a better understanding of the different schemes, let us eliminate the auxiliary variable, $\boldsymbol{q}_h$, to recover the primal form of the discretization. To achieve this we need the following result:

Lemma 7.9. *Assume that Ω has been triangulated into K elements, D^k. Then*

$$\sum_{k=1}^K (\hat{\boldsymbol{n}} \cdot \boldsymbol{u}, v)_{\partial D^k} = \oint_\Gamma \{\!\{\boldsymbol{u}\}\!\} \cdot [\![v]\!] \, d\boldsymbol{x} + \oint_{\Gamma_i} \{\!\{v\}\!\}[\![\boldsymbol{u}]\!] \, d\boldsymbol{x},$$

where Γ represents the set of unique edges and Γ_i the set of unique purely internal edges.

The result is easy to obtain by considering the contribution along one edge from two elements and rewrite the terms.

Combine this with Eq. (7.9) to obtain

$$(q_h, \pi_h)_{\Omega,h} = \oint_\Gamma u_h^* [\![\pi_h]\!]\, dx - (u_h, \nabla \cdot \pi_h)_{\Omega,h}.$$

Here, we have used that the numerical flux is single valued (i.e., $\{\!\{u_h^*\}\!\} = u_h^*$ and $[\![u_h^*]\!] = 0$). Integration by parts and the application of Lemma 7.9 yields

$$(q_h, \pi_h)_{\Omega,h} = (\nabla u_h, \pi_h)_{\Omega,h} + \oint_\Gamma u_h^* [\![\pi_h]\!]\, dx \qquad (7.10)$$

$$- \oint_\Gamma \{\!\{\pi_h\}\!\} \cdot [\![u_h]\!]\, dx - \oint_{\Gamma_i} \{\!\{u_h\}\!\} [\![\pi_h]\!]\, dx.$$

Homogeneous boundary conditions on u_h along $\Gamma_b = \Gamma/\Gamma_i$ are imposed in the usual way (i.e., $u_h^+ = -u_h^-$) and note that all numerical fluxes u_h^* in Table 7.3 can be written as

$$u_h^* = \{\!\{u_h\}\!\} + \hat{\beta} \cdot [\![u_h]\!],$$

to obtain

$$(q_h, \pi_h)_{\Omega,h} = (\nabla u_h, \pi_h)_{\Omega,h} - \oint_{\Gamma_b} u_h \hat{n} \cdot \pi_h\, dx - \oint_{\Gamma_i} (\{\!\{\pi_h\}\!\} - \beta [\![\pi_h]\!]) \cdot [\![u_h]\!]\, dx. \qquad (7.11)$$

We now define a lifting operator $\mathcal{L}(\phi_h) \in U_h$ for $\phi_h \in V_h$ as

$$(\mathcal{L}(\phi_h), \pi_h)_{\Omega,h} = \oint_{\Gamma_i} (\{\!\{\pi_h\}\!\} - \beta [\![\pi_h]\!]) \cdot [\![\phi_h]\!]\, dx + \oint_{\Gamma_b} \phi_h \hat{n} \cdot \pi_h\, dx, \quad (7.12)$$

and obtain from Eq. (7.11),

$$(q_h, \pi_h)_{\Omega,h} = (\nabla u_h - \mathcal{L}(u_h), \pi_h)_{\Omega,h}$$

or

$$q_h = \nabla u_h - \mathcal{L}(u_h). \qquad (7.13)$$

Insert this into Eq. (7.8) to recover the primal problem of finding u_h as

$$\mathcal{A}(u_h, \phi_h) + \mathcal{I}(u_h, \phi_h) = \mathcal{F}(\phi_h), \quad \forall \phi_h \in V_h,$$

where we have

$$\mathcal{A}(u_h, \phi_h) = (\nabla u_h - \mathcal{L}(u_h), \nabla \phi_h - \mathcal{L}(\phi_h))_{\Omega,h},$$

$$\mathcal{I}(u_h, \phi_h) = \int_{\Gamma_i} \tau [\![u_h]\!] \cdot [\![\phi_h]\!]\, dx + \int_{\Gamma_b} \tau u_h \phi_h\, dx, \qquad (7.14)$$

$$\mathcal{F}(\phi_h) = (f, \phi_h)_{\Omega,h}.$$

We recognize $\mathcal{A}$ as the discrete approximation to the Laplacian, $\mathcal{I}$ is the stabilization needed to ensure invertibility, and $\mathcal{F}$ is the projection of f onto V_h.

To fit the internal penalty form into this general form, the numerical flux requires that the lifting operator $\mathcal{L}(\phi_h) \in U_h$ for $\phi_h \in V_h$ be defined as

$$(\mathcal{L}(\phi_h), \boldsymbol{\pi}_h)_{\Omega,h} = \oint_{\Gamma_i} \{\!\{\boldsymbol{\pi}_h\}\!\} \cdot [\![\phi_h]\!]\, d\boldsymbol{x} + \oint_{\Gamma_b} \phi_h \hat{\boldsymbol{n}} \cdot \boldsymbol{\pi}_h \, d\boldsymbol{x},$$

with

$$\mathcal{A}(u_h, \phi_h) = (\nabla u_h - \mathcal{L}(u_h), \nabla \phi_h - \mathcal{L}(\phi_h))_{\Omega,h} - (\mathcal{L}(u_h), \mathcal{L}(\phi_h))_{\Omega,h}. \quad (7.15)$$

An immediate consequence of the general expression in Eq. (7.14) is that all operators are symmetric, hence proving this previously made important observation.

The extension of the above discussion to the strong form relies on the identity

$$(\nabla \cdot \boldsymbol{\pi}, \phi)_{\Omega,h} = -(\boldsymbol{\pi}, \nabla \phi)_{\Omega,h} + \oint_{\Gamma} \{\!\{\boldsymbol{\pi}\}\!\} \cdot [\![\phi]\!]\, d\boldsymbol{x} + \oint_{\Gamma_i} \{\!\{\phi\}\!\}[\![\boldsymbol{\pi}]\!]\, d\boldsymbol{x}, \quad (7.16)$$

as discussed in [283]. The proof of this identity is straightforward by considering one face, and the details can be found in [165].

Let us express the general discrete problem as

$$\mathcal{B}_h(u_h, \phi_h) = \mathcal{A}(u_h, \phi_h) + \mathcal{I}(u_h, \phi_h) = \mathcal{F}(\phi_h), \quad \forall \phi_h \in V_h,$$

and rewrite Eq. (7.10) as

$$(\boldsymbol{q}_h, \boldsymbol{\pi}_h)_{\Omega,h} = (\nabla u_h, \boldsymbol{\pi}_h)_{\Omega,h} - \oint_{\Gamma} [\![u_h - u_h^*]\!] \cdot \{\!\{\boldsymbol{\pi}_h\}\!\} d\boldsymbol{x} - \oint_{\Gamma_i} \{\!\{\boldsymbol{\pi}_h\}\!\} [\![u_h - u_h^*]\!]\, d\boldsymbol{x}.$$

Taking $\boldsymbol{\pi}_h = \nabla \phi_h$, we can now combine this with Eq. (7.8) to obtain a slightly different expression for $\mathcal{B}_h$ as

$$\mathcal{B}_h(u_h, \phi_h) = (\nabla u_h, \nabla \phi_h)_{\Omega,h} - \oint_{\Gamma} ([\![u_h - u_h^*]\!] \cdot \{\!\{\nabla \phi_h\}\!\} + \{\!\{\boldsymbol{q}_h^*\}\!\} \cdot [\![\phi_h]\!])\, d\boldsymbol{x}$$

$$- \oint_{\Gamma_i} ([\![\nabla \phi_h]\!]\{\!\{u_h - u_h^*\}\!\} + [\![\boldsymbol{q}_h^*]\!]\{\!\{\phi_h\}\!\})\, d\boldsymbol{x}. \quad (7.17)$$

From Eq. (7.16) with $\boldsymbol{r} = \nabla u$ we recover the identity

$$(\nabla u, \nabla \phi)_{\Omega,h} = -(\nabla \cdot \nabla u, \phi)_{\Omega,h} + \oint_{\Gamma} \{\!\{\nabla u\}\!\} \cdot [\![\phi]\!]\, d\boldsymbol{x} + \oint_{\Gamma_i} \{\!\{\phi\}\!\}[\![\nabla u]\!]\, d\boldsymbol{x},$$

which inserted into Eq.(7.17) yields

$$\mathcal{B}_h(u_h, \phi_h) = -(\nabla \cdot \nabla u_h, \phi_h)_{\Omega,h}$$

$$- \oint_\Gamma (\llbracket u_h - u_h^* \rrbracket \cdot \{\{\nabla \phi_h\}\} + \{\{q_h^* - \nabla u_h\}\} \cdot \llbracket \phi_h \rrbracket) \, d\boldsymbol{x}$$

$$- \oint_{\Gamma_i} (\llbracket \nabla \phi_h \rrbracket \{\{u_h - u_h^*\}\} + \llbracket q_h^* - \nabla u_h \rrbracket \{\{\phi_h\}\}) \, d\boldsymbol{x}.$$

To test consistency, we insert a smooth solution, u, into the bilinear form to obtain

$$\mathcal{B}_h(u, \phi_h) = -(\nabla \cdot \nabla u, \phi_h)_{\Omega,h} + \oint_\Gamma (\llbracket u^* \rrbracket \cdot \{\{\nabla \phi_h\}\} + (\nabla u - \{\{q^*\}\}) \cdot \llbracket \phi_h \rrbracket) \, d\boldsymbol{x}$$

$$- \oint_{\Gamma_i} (\llbracket \nabla \phi_h \rrbracket (u - \{\{u^*\}\}) + \llbracket q^* \rrbracket \{\{\phi_h\}\}) \, d\boldsymbol{x},$$

since $\{\{u\}\} = u$, $\{\{\nabla u\}\} = \nabla u$, and $\llbracket u \rrbracket = \llbracket \nabla u \rrbracket = 0$. If we assume that the numerical flux is consistent (i.e., single valued such that $\{\{u^*\}\} = u$ and $\llbracket u^* \rrbracket = 0$ on the trace of the elements), this further simplifies as

$$\mathcal{B}_h(u, \phi_h) = -(\nabla \cdot \nabla u, \phi_h)_{\Omega,h} + \oint_\Gamma ((\nabla u - \{\{q^*\}\}) \cdot \llbracket \phi_h \rrbracket) \, d\boldsymbol{x} - \oint_{\Gamma_i} \llbracket q^* \rrbracket \{\{\phi_h\}\} \, d\boldsymbol{x}.$$

Now recall from Eq. (7.13) that

$$q = \nabla u - \mathcal{L}(u) = \nabla u,$$

where $\llbracket u \rrbracket = 0$ ensures that $\mathcal{L}(u) = 0$ by the definition of the lifting operator, Eq. (7.12). Considering the numerical fluxes in Table 7.3, all of which are consistent, we recover that $\llbracket q^* \rrbracket = 0$ and $\{\{q^*\}\} = \nabla u$, from which

$$\mathcal{B}_h(u, \phi_h) = \mathcal{F}(\phi_h), \quad \forall \phi_h \in V_h.$$

This establishes consistency of the discontinuous Galerkin approximation provided only that the numerical fluxes are consistent which, as we have already discussed, is the case for those considered here. Furthermore, an immediate consequence of this is Galerkin orthogonality

$$\mathcal{B}_h(u - u_h, \phi_h) = 0, \quad \forall \phi_h \in V_h.$$

If we briefly consider the adjoint problem

$$\mathcal{B}_h(v, w) = (g, v)_{\Omega,h}, \quad v \in H_0^2,$$

where w solves

$$-\nabla \cdot \nabla w = g, \quad w = 0, \quad \boldsymbol{x} \in \partial \Omega,$$

a similar line of arguments implies that

$$\mathcal{B}_h(\phi_h, w) = (g, \phi_h)_{\Omega,h} + \oint_\Gamma \llbracket u^*(\phi_h) \rrbracket \cdot \nabla w \, d\boldsymbol{x} - \oint_{\Gamma_i} \llbracket q^*(\phi_h) \rrbracket w \, d\boldsymbol{x}, \quad \forall \phi_h \in V_h.$$

Thus, if $[\![u^*(\phi_h)]\!] = [\![q^*(\phi_h)]\!] = 0$, the solution to the adjoint problem is likewise consistent. Fluxes with this property are known as conservative and this is guaranteed for single-valued fluxes; that is, this property is shared by all fluxes in Table 7.3. The analysis of several other flux choices is discussed in [13].

Stability and penalty terms

As we have seen in the computational experiments, there is a need to penalize the fluxes to ensure invertibility and the existence of unique discrete solutions. To gain a better understanding of this, we will discuss the coercivity of $\mathcal{B}_h$ to establish the conditions under which

$$\mathcal{B}_h(\phi_h, \phi_h) \geq C_c \|\phi_h\|_{\mathrm{DG}}^2, \quad \forall \phi_h \in \mathsf{V}_h.$$

Showing this suffices to guarantee the existence of a unique solution.

Here, we have defined the natural DG norm as

$$\|\phi\|_{\mathrm{DG}}^2 = \|\nabla\phi\|_{\Omega,h}^2 + \left\| h^{-1/2}[\![\phi]\!] \right\|_{\Gamma_i}^2 + \left\| h^{-1/2}\phi \right\|_{\Gamma_b}^2, \tag{7.18}$$

where the boundary norms are defined as

$$\|\phi\|_{\Gamma_i}^2 = \oint_{\Gamma_i} \phi^2 \, d\boldsymbol{x}, \quad \|\phi\|_{\Gamma_b}^2 = \oint_{\Gamma_b} \phi^2 \, d\boldsymbol{x}.$$

In Eq. (7.18) we define the local h as $h = \min((h^k)^-, (h^k)^+)$ (i.e., as the minimum cell size between two neighbor elements) and the factor $h^{-1/2}$ arises naturally from balancing the terms and using the trace inequality [320] for $u_h \in P_N(\mathsf{D})$:

$$\|u_h\|_{\partial \mathsf{D}} \leq \sqrt{\frac{(N+1)(N+d)}{d} \frac{|\partial \mathsf{D}|}{|\mathsf{D}|}} \, \|u_h\|_{\mathsf{D}}, \tag{7.19}$$

where d is the dimension of the simplex, D. Note that

$$\frac{|\partial \mathsf{D}|}{|\mathsf{D}|} \propto h^{-1},$$

giving rise to the classic result [59]

$$\|u_h\|_{\partial \mathsf{D}} \leq C_i h^{-1/2} \|u_h\|_{\mathsf{D}},$$

where the inverse inequality constant C_i depends on N and the shape of the simplex.

To control C_i, we assume that the grid remains reasonable under refinement. Such grid regularity statements can be cast in many ways; for example,

$$h_e \leq h^k \leq C_e h_e,$$

where h_e is the cell size from a uniform refinement. An alternative condition can be obtained through the angle condition, bounding the maximum angle [19].

The three schemes discussed so far can all be formulated as

$$\mathcal{B}_h(\phi_h, \phi_h) = \|\nabla\phi_h - \mathcal{L}(\phi_h)\|_{\Omega,h}^2 - \gamma\|\mathcal{L}(\phi_h)\|_{\Omega,h}^2 + \mathcal{I}(\phi_h, \phi_h),$$

where $\gamma = 1$ for the internal penalty method and $\gamma = 0$ otherwise.

First, recall that

$$\mathcal{I}(\phi_h, \phi_h) = \oint_{\Gamma_i} \tau[\![\phi_h]\!] \cdot [\![\phi_h]\!]\, d\boldsymbol{x} + \oint_{\Gamma_b} \tau\phi_h\phi_h\, d\boldsymbol{x}.$$

For reasons that will become apparent shortly, let us take the local stabilization factor as

$$\tau = \frac{\tilde{\tau}^k}{h},$$

where we again recall that $h = \min((h^k)^-, (h^k)^+)$. Here, $\tilde{\tau}^k$ is a local constant that may depend on the local order of approximation. With this, we recover

$$\mathcal{I}(\phi_h, \phi_h) = \|\tau^{1/2}[\![\phi_h]\!]\|_{\Gamma_i}^2 + \|\tau^{1/2}\phi_h\|_{\Gamma_b}^2$$
$$= \|(\tilde{\tau}^k)^{1/2}h^{-1/2}[\![\phi_h]\!]\|_{\Gamma_i}^2 + \|(\tilde{\tau}^k)^{1/2}h^{-1/2}\phi_h\|_{\Gamma_b}^2.$$

Exploring the inverse inequality, Eq. (7.19), one can furthermore show that [13]

$$\|\mathcal{L}(\phi_h)\|_{\Omega,h}^2 \leq C_l^2 \left(\|h^{-1/2}[\![\phi_h]\!]\|_{\Gamma_i}^2 + \|h^{-1/2}\phi_h\|_{\Gamma_b}^2\right). \qquad (7.20)$$

Finally, we consider

$$\|\nabla\phi_h - \mathcal{L}(\phi_h)\|_{\Omega,h}^2 = \|\nabla\phi_h\|_{\Omega,h}^2 + \|\mathcal{L}(\phi_h)\|_{\Omega,h}^2 - 2\sum_{k=1}^{K} (\nabla\phi_h, \mathcal{L}(\phi_h))_{\mathrm{D}^k}.$$

Using the arithmetic-geometric inequality $2ab \leq \varepsilon a^2 + \varepsilon^{-1}b^2$ for all $\varepsilon > 0$, we have

$$\|\nabla\phi_h - \mathcal{L}(\phi_h)\|_{\Omega,h}^2 \geq (1-\varepsilon)\|\nabla\phi_h\|_{\Omega,h}^2 + (1-\varepsilon^{-1})\|\mathcal{L}(\phi_h)\|_{\Omega,h}^2.$$

If we now assume that $\varepsilon < 1$, we see that $(1-\varepsilon) > 0$ and, hence, $(1-\varepsilon^{-1}) < 0$. Comparing Eq. (7.18) and Eq. (7.20), it is clear that

$$C_l\|\phi_h\|_{\mathrm{DG}} \geq \|\mathcal{L}(\phi_h)\|_{\Omega,h} \;\Rightarrow\; (1-\varepsilon^{-1})C_l^2\|\phi_h\|_{\mathrm{DG}}^2 \leq \|\mathcal{L}(\phi_h)\|_{\Omega,h}^2.$$

Combining these pieces, we recover

$$\mathcal{B}_h(\phi_h, \phi_h) \geq \left(1 - \varepsilon + C_l^2\left(1 - \gamma - \varepsilon^{-1}\right)\right)\|\nabla\phi_h\|_{\Omega,h}^2$$
$$+ \left(C_l^2\left((1-\gamma) - \varepsilon^{-1}\right) + \tilde{\tau}\right)\|h^{-1/2}[\![\phi_h]\!]\|_{\Gamma_i}^2 + \|h^{-1/2}\phi_h\|_{\Gamma_b}^2,$$

where $\tilde{\tau} \leq \min_k(\tilde{\tau}^k)$, with $\tilde{\tau}^k$ being the local stabilization factor on element k.

Let us first consider the case with $\gamma = 0$ (i.e., the stabilized central fluxes and LDG). To establish coercivity, $\mathcal{B}_h(\phi_h, \phi_h)$ must be strictly positive. Straightforward algebra shows that the two terms are both positive, provided

$$\frac{C_l^2}{C_l^2 + \tilde{\tau}} < \varepsilon < 1,$$

which is clearly guaranteed as long as $\tilde{\tau} > 0$. This confirms that $\mathcal{B}_h$ is invertible as long as $\tilde{\tau} > 0$ in the stabilization.

For the internal penalty case, $\gamma = 1$, and we have the condition

$$\tilde{\tau} - \frac{C_l^2}{\varepsilon} > 0.$$

The situation is different in this case and we need to require $\tilde{\tau} > \tau^*$. Here, τ^* depends on C_l, which again depends on N and the regularity of the grid. This analysis is carried through more carefully in [281], where it is shown that

$$\tilde{\tau}^k \geq \frac{(N+1)(N+d)}{d} \frac{|\partial \mathsf{D}^k|}{|\mathsf{D}^k|},$$

suffices to guarantee coercivity.

Combining the definition of $\mathcal{B}_h(u_h, \phi_h)$, $(u_h, \phi_h) \in \mathsf{V}_h$, with Eq. (7.20), the definition of the DG norm, Eq. (7.18), and the Cauchy-Schwarz inequality, one easily shows continuity of B_h as

$$\mathcal{B}_h(u_h, \phi_h) \leq C_k \, \|u_h\|_{\mathrm{DG}} \, \|\phi_h\|_{\mathrm{DG}} \, , \quad \forall \phi_h \in \mathsf{V}_h.$$

Error estimates

Having established consistency, coercivity, and continuity of the discrete operator, we know that a unique solution exists and that the scheme converges to the solution of Poisson equation. All that is left is to establish error estimates to understand how these methods perform under grid refinement.

Let us first establish a basic result as

$$\|u - u_h\|_{\mathrm{DG}}^2 = \|\nabla u - \nabla u_h\|_{\Omega,h}^2 + \left\| h^{-1/2} [\![u - u_h]\!] \right\|_{\Gamma_i}^2 + \left\| h^{-1/2} u - u_h \right\|_{\Gamma_b}^2$$
$$\leq \|u\|_{\Omega,1,h}^2 + h^{-2} \|u - u_h\|_{\Omega,h}^2,$$

where the last step follows from the use of Eq. (7.19). Now recall from Section 4.3, Theorem 4.8 that

$$\|u - u_h\|_{\Omega,q,h} \leq C(N, p, q) h^{\sigma - q} |u|_{\Omega,\sigma,h},$$

for $u \in H^p(\Omega)$, $0 \leq q \leq \sigma$ and $\sigma = \min(N+1, p)$. This immediately yields

$$\|u - u_h\|_{\mathrm{DG}} \leq C h^{\sigma - 1} |u|_{\Omega,\sigma,h}.$$

In case u is very smooth (i.e., $p > N + 1$), we recover

$$\|u - u_h\|_{\mathrm{DG}} \leq Ch^N |u|_{\Omega, N+1, h},$$

whereas for $N + 1 > p$ (i.e., a very high resolution), we have

$$\|u - u_h\|_{\mathrm{DG}} \leq Ch^{p-1} |u|_{\Omega, p, h}.$$

Recall the coercivity result as

$$\mathcal{B}_h(u_h, u_h) \geq C_c \|u_h\|_{\mathrm{DG}}^2, \quad u_h \in \mathsf{V}_h.$$

We now define the numerical solution, u_h, and the projection of the exact solution, $\mathcal{P}_N u$, and consider

$$\mathcal{B}_h(u_h - \mathcal{P}_N u, u_h - \mathcal{P}_N u) = \mathcal{B}_h(u - \mathcal{P}_N u, u_h - \mathcal{P}_N u) \geq C_c \|u_h - \mathcal{P}_N u\|_{\mathrm{DG}}^2,$$

where the equality follows from Galerkin orthogonality, which is a consequence of consistency, as we have already discussed. This implies that

$$
\begin{aligned}
C_c \|u_h - \mathcal{P}_N u\|_{\mathrm{DG}}^2 &\leq \mathcal{B}_h(u - \mathcal{P}_N u, u_h - \mathcal{P}_N u) \\
&\leq C_k \|u - \mathcal{P}_N u\|_{\mathrm{DG}} \|u_h - \mathcal{P}_N u\|_{\mathrm{DG}} \\
&\leq Ch^{\sigma-1} |u|_{\Omega, \sigma, h} |u|_{\Omega, \sigma, h} \|u_h - \mathcal{P}_N u\|_{\mathrm{DG}},
\end{aligned}
$$

where we have used the continuity of $\mathcal{B}_h$. This yields

$$\|u_h - \mathcal{P}_N u\|_{\mathrm{DG}} \leq Ch^{\sigma-1} |u|_{\Omega, \sigma, h}.$$

Using the triangle inequality, we obtain

$$\|u - u_h\|_{\mathrm{DG}} \leq \|u - \mathcal{P}_N u\|_{\mathrm{DG}} + \|u_h - \mathcal{P}_N u\|_{\mathrm{DG}} \leq Ch^{\sigma-1} |u|_{\Omega, \sigma, h},$$

for $u \in H_0^p(\Omega)$ and $\sigma = \min(N + 1, p)$. The constant C can depend on N and p, but not on h. Hence in the DG norm we can expect $\mathcal{O}(h^N)$ convergence for sufficiently smooth solutions.

To obtain a result in the L^2-norm, we will need to consider the adjoint problem

$$-\nabla \cdot \nabla \psi = u - u_h, \quad \psi = 0, \quad \boldsymbol{x} \in \partial\Omega.$$

As mentioned previously, all the methods we consider here are also adjoint consistent, such that

$$\mathcal{B}_h(\phi, \psi) = (u - u_h, v)_{\Omega, h}, \quad \forall \phi \in H_0^2(\Omega).$$

Now, take $\phi = u - u_h$ and consider ψ_h as a linear polynomial approximation to ψ to obtain

$$\|u - u_h\|_{\Omega, h}^2 = \mathcal{B}_h(u - u_h, \psi) = \mathcal{B}_h(u - u_h, \psi - \psi_h) \leq C \|u - u_h\|_{\mathrm{DG}} \|\psi - \psi_h\|_{\mathrm{DG}},$$

where we have exploited consistency, Galerkin orthogonality, and continuity of $\mathcal{B}_h$. The first term we discussed above and one easily establishes that

$$\|\psi - \psi_h\|_{\mathrm{DG}} \leq Ch\|\psi\|_{\Omega,2,h} \leq Ch\|u - u_h\|_{\Omega,h},$$

where the latter follows from regularity conditions of the adjoint equation itself. With this, we recover the result

$$\|u - u_h\|_{\Omega,h} \leq Ch^\sigma|u|_{\Omega,\sigma,h},$$

for $u \in H_0^p(\Omega)$ and $\sigma = \min(N + 1, p)$. The constant C depends on N and p but not on h. This confirms the many previous results showing $\mathcal{O}(h^{N+1})$ convergence for sufficiently smooth solutions.

Although we have focused on the three different methods summarized in Table 7.3, most of the results and techniques discussed above apply also to the many alternatives discussed in [13] and elsewhere.

The only difference is for schemes that are not conservative and, thus, not adjoint consistent, in which case the duality argument applied above to obtain optimal L^2-convergence fails. For such cases, one can only prove suboptimal convergence as $\mathcal{O}(h^N)$, although it is often observed that for N odd, optimal convergence is restored in L^2.

Generalizations and extensions

Attempts to extend and generalize the above results can be taken in many directions, with the most straightforward one being to more general boundary conditions, as discussed in [13, 254].

There have also been significant developments toward the understanding of the DG approximations to the harmonic Helmholtz equations and Maxwell's equations. These developments include both the low-frequency case [255, 256] and the high-frequency, indefinite case [166, 167, 168]. A nonstabilized scheme is discussed in [169]. Closely related efforts for saddlepoint problems include the Stokes problem [66, 67, 278, 279] and the equations of linearized elasticity [73, 172]. Most of these different problems require techniques related to the above steps, combined with new ideas to address the more complex nature of the problems.

Analysis of cases involving different order elements and nonconforming elements have been discussed in numerous works; for example [12, 50, 254].

7.3 Intermission of solving linear systems

Before we continue and consider more complex applications, it is useful to review a few standard techniques for solving linear systems. For the one-dimensional case discussed previously, this is less of a concern, as all linear systems are banded, making a direct solver the natural choice. However, for the multidimensional case, the situation is more complex, as the linear systems are sparse but with a structure that depends on the grid connectivity.

This is not intended to be a comprehensive review, but rather a study of solver technology conveniently made available through Matlab. Further information be found in some of the many excellent texts on computational linear algebra [104, 141, 275, 308]. We will include some comments on our experience of their positive and negative aspects. At this time, it is unclear which approach is best for general DG discretizations and we demur on some more aggressive strategies such as the algebraic multigrid [119, 150, 245] and domain decomposition-based multilevel solvers [117, 133, 193, 213]. In addition, we mostly discuss techniques that are suitable for solving the same linear systems with multiple different right-hand sides, as one might encounter when implicitly timestepping partial differential equations.

As an example for experimentation, we consider the linear Poisson equation

$$\nabla^2 u = f(x, y) = \left(\left(16 - n^2 \right) r^2 + \left(n^2 - 36 \right) r^4 \right) \sin(n\theta), \quad x^2 + y^2 \leq 1,$$

where $n = 12$, $r = \sqrt{x^2 + y^2}$, $\theta = \arctan(y, x)$ and the exact solution, $u(x, y)$ is imposed at the boundary of Ω.

Using a DG formulation with internal penalty fluxes and $K = 512$ elements, each of fourth order, we have a total of 7680 degrees of freedom. To compute these, we need to solve $\mathcal{A}u_h = f_h$ for the solution vector u. The discrete operator $\mathcal{A}$, is a square matrix with a total of 284,100 nonzero entries, reflecting a sparsity exceeding 95%.

7.3.1 Direct methods

We first consider the use of an LU-factorization of the matrix such that $\mathcal{A} = \mathcal{L}\mathcal{U}$, where the factor matrices are logically lower and upper triangular to allow their efficient inversion. In Matlab the factorization and solution is achieved with

```
>> [L, U] = lu(A);
>> u = U\(L\f);
```

This requires storage for the sparse matrices $(\mathcal{A}, \mathcal{L}, \mathcal{U})$ which, in our test case, required storage for 8,149,974 extra nonzeros in $(\mathcal{L}, \mathcal{U})$. It is evident that the sparsity pattern of $\mathcal{A}$ is not compactly represented near the diagonal, so the LU-factorization process creates factor matrices with significantly fill in; that is, significantly more storage is required for $\mathcal{L}$ and $\mathcal{U}$ than for $\mathcal{A}$. For smaller problems, this is less of a problem, but as problems increase in size this growth in memory usage becomes a significant concern.

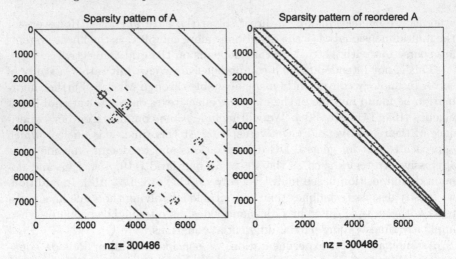

Fig. 7.14. On the left, we show the sparsity pattern of the discrete Poisson matrix before reverse Cuthill-McKee reordering. On the right, we show the sparsity pattern after the reordering.

To reduce the amount of fill in for the LU-factors, one can explore reordering of the degrees of freedom to minimize the bandwidth of the matrix as much a possible. A powerful approach for this is the reverse Cuthill-McKee numbering system [125]. This is a built-in method in Matlab and provides a permutation vector for a given sparse symmetric matrix to significantly reduce the distance of any nonzero entry from the diagonal. To illustrate the effect of this reordering, we contrast in Fig. 7.14 the sparsity patterns of $\mathcal{A}$ before and after reordering.

To take advantage of this in the solution of the linear system, one can use the following approach in which a permutation vector is computed, applied to the row and column swaps to $\mathcal{A}$, permute f, solve the two systems, and permute the solution:

```
>> P = symrcm(A);
>> A = A(P,P);
>> rhs = rhs(P);
>> [L,U] = lu(A);
>> u = U\(L\f);
>> u(P) = u;
```

This simple procedure reduces the number of nonzeros in the LU-factors to 3,802,198 (i.e., a reduction of more than 50%). All subsequent tests use the reordered $\mathcal{A}$ matrix.

We can further reduce the storage requirements by taking advantage of $\mathcal{A}$ being a positive definite symmetric matrix and use a Cholesky factorization; that is, find the matrix $\mathcal{C}$ such that $\mathcal{A} = \mathcal{C}^T \mathcal{C}$ using

```
>> C = chol(A);
>> u = C\(C'\f);
```

In combination with the reordering, the number of extra nonzeros is reduced to 1,901,109.

7.3.2 Iterative methods

For the direct methods discussed above, it is evident that the additional storage requirement for the factors may become excessive. With this in mind, it is reasonable to consider Krylov subspace iterative methods as a suitable alternative. Because matrix $\mathcal{A}$ is symmetric positive-definite, we use the conjugate gradient (CG) method to solve the system to a chosen tolerance. Again, we can use Matlab's suite of iterative solvers to our advantage, allowing a solution to be obtained by:

```
>> ittol = 1e-8; maxit = 1000;
>> u = pcg(A, f, ittol, maxit);
```

This solves the system using the CG method. There are no significant extra storage requirements beyond the system matrix $\mathcal{A}$ and f, which is a definite advantage. However, one also observes that it takes more than 100 times longer to finish the required 818 iterations than when using the reordered LU-factorization. In the iterative approach, each iteration involves one matrix-vector between $\mathcal{A}$ and a residual vector along with a several less expensive operations. Clearly, storage is traded for additional execution time.

We can seek to reduce the amount of solver time taken for the conjugate gradient solver by choosing to solve

$$\mathcal{C}^{-1}\mathcal{A}u_h = \mathcal{C}^{-1}f_h,$$

where we require that $\mathcal{C}^{-1}v$ is easily evaluated for a given vector v and that the preconditioner, $\mathcal{C}$, approximates $\mathcal{A}$ well. A straightforward way to form such a preconditioner is to compute the Cholesky factorization of $\mathcal{A}$ but drop entries that do not appear in the original sparse pattern of $\mathcal{A}$. Such an incomplete Cholesky factorization and its use with the CG methods can be accomplished in Matlab through

```
>> ittol = 1e-8; maxit = 1000;
>> Cinc = cholinc(OP, '0')
>> u = pcg(A, f, ittol, maxit, Cinc', Cinc);
```

For our sample system, the zero-fill incomplete Cholesky factor requires the storage of 154,082 entries, but the timing is improved by about a factor of 2 (i.e., it remains about 50 times slower than the reordered LU-factorization method). However, notably the number of iterations is reduced to 138 from 818 when compared to the nonpreconditioned case.

We finally supply a finite cut-off tolerance to the incomplete factorization that limits the minimum of the entries that are added to the sparsity pattern of the factor. Using a tolerance of 10^{-4} as in the following:

```
>> ittol = 1e-8; maxit = 1000; droptol = 1e-4;
>> Cinc = cholinc(A, droptol);
>> u = pcg(A, b, ittol, maxit, Cinc', Cinc);
```

reduces the number of iterations to 17 and yields a dramatic reduction in the execution time to just four times that of the direct solver. The amount of extra storage increases to 197,927. This again highlights the ability to balance storage and performance by considering different solution techniques and preconditioners.

It is important to appreciate that these results are specific to the problem being considered. Generally, one should consider iterative problems for large sparse problems for which excessive memory requirements would otherwise be a bottleneck. A distinct problem for the straightforward use of iterative methods is their inability to work with many right-hand sides at the same time. This is in contrast to direct solvers where the reordering and factorization is independent of the right-hand side. For further discussion and many more details on these issues, we refer to [104, 141, 308, 275] and references therein.

7.4 The incompressible Navier-Stokes equations

The dynamics of an incompressible fluid flow in two spatial dimensions is described by the Navier-Stokes equations as

$$\frac{\partial \boldsymbol{u}}{\partial t} + (\boldsymbol{u} \cdot \nabla)\,\boldsymbol{u} = -\nabla p + \nu \nabla^2 \boldsymbol{u}, \quad \boldsymbol{x} \in \Omega,$$

$$\nabla \cdot \boldsymbol{u} = 0,$$

where $\boldsymbol{u} = (u, v)$, and p are the x-component of the velocity, the y-component of the velocity, and the scalar pressure field, respectively. In conservative flux form, the equations are

$$\frac{\partial \boldsymbol{u}}{\partial t} + \nabla \cdot \mathcal{F} = -\nabla p + \nu \nabla^2 \boldsymbol{u},$$

$$\nabla \cdot \boldsymbol{u} = 0,$$

with the flux $\mathcal{F}$ being

$$\mathcal{F} = [\boldsymbol{F}_1, \boldsymbol{F}_2] = \begin{bmatrix} u^2 & uv \\ uv & v^2 \end{bmatrix}.$$

The equations are closed with initial conditions on the velocity fields and boundary conditions. In particular, we shall assign different parts of the boundary according to specific boundary conditions; for example, as inflow

$\partial\Omega^I$, outflow $\partial\Omega^O$, or walls $\partial\Omega^W$. The specific boundary conditions applied on each subset of the boundary will be discussed later.

These equations constitute a mixture of a conservation law, diffusion, and constrained evolution. The mixed nature of these equations makes discrete timestepping a more complicated affair than has been case for the equations we have discussed previously. There are a number of different procedures mentioned in the literature and we will make no effort to provide a thorough overview of these techniques. Good starting points for such explorations are [97, 198].

7.4.1 Temporal splitting scheme

We consider here the stiffly stable timestepping method [196, 282] in which each timestep breaks into three stages. The first stage amounts to explicitly integrating the conservation law component of the equations using an Adams-Bashforth second-order scheme. The second stage involves the projection of the updated velocity components onto the space of weakly divergence-free functions and in the final stage, the viscous term is treated implicitly.

First define the nonlinear functions, $\mathcal{N}(\boldsymbol{u}) = (N_x, N_y) = \nabla \cdot \mathcal{F}$, as

$$N_x(\boldsymbol{u}) = \nabla \cdot \boldsymbol{F}_1 = \frac{\partial(u^2)}{\partial x} + \frac{\partial(uv)}{\partial y}, \quad N_y(\boldsymbol{u}) = \nabla \cdot \boldsymbol{F}_2 = \frac{\partial(uv)}{\partial x} + \frac{\partial(v^2)}{\partial y},$$

$$(7.21)$$

and express the first part of the timestep as

$$\frac{\gamma_0\tilde{\boldsymbol{u}} - \alpha_0\boldsymbol{u}^n - \alpha_1\boldsymbol{u}^{n-1}}{\Delta t} = -\beta_0\mathcal{N}(\boldsymbol{u}^n) - \beta_1\mathcal{N}(\boldsymbol{u}^{n-1}). \qquad (7.22)$$

This method is second order in time for the velocity (see [143] for details). However, it is not self-starting, in contrast to the Runge-Kutta schemes we have used thus far. For the first timestep we choose the coefficients $\alpha_0, \alpha_1, \beta_0, \beta_1$ to reduce the scheme to the forward Euler method,

$$\gamma_0 = 0, \ \alpha_0 = 1, \ \alpha_1 = 0, \ \beta_0 = 1, \ \beta_1 = 0,$$

consistent with the assumption that u^0 and v^0 are provided as initial conditions. The subsequent timesteps are done with

$$\gamma_0 = \frac{3}{2}, \ \alpha_0 = 2, \ \alpha_1 = -\frac{1}{2}, \ \beta_0 = 2, \ \beta_1 = -1,$$

reflecting a second-order Adams-Bashforth method.

In the second stage, the pressure projection step, the intermediate velocities $\tilde{\tilde{\boldsymbol{u}}}$ are updated using

$$\gamma_0\frac{\tilde{\tilde{\boldsymbol{u}}} - \tilde{\boldsymbol{u}}}{\Delta t} = -\nabla\bar{p}^{n+1}.$$

To close this system we seek $\bar{p}^{n+1}$ such that the divergence of $\tilde{\tilde{u}}$ is zero by solving for $\bar{p}^{n+1}$ as

$$-\nabla^2 \bar{p}^{n+1} = -\frac{\gamma_0}{\Delta t} \nabla \cdot \tilde{u}. \tag{7.23}$$

This Poisson problem is closed with Neumann boundary conditions at inflow and wall boundaries derived from the equation as

$$\hat{n} \cdot \nabla \bar{p}^{n+1} = \frac{\partial \bar{p}^{n+1}}{\partial \hat{n}} \tag{7.24}$$

$$= -\beta_0 \hat{n} \cdot \left(\frac{Du^n}{Dt} - \nu \nabla^2 u^n \right) - \beta_1 \hat{n} \cdot \left(\frac{Du^{n-1}}{Dt} - \nu \nabla^2 u^{n-1} \right)$$

$$= -\beta_0 \hat{n} \cdot \left(\frac{Du^n}{Dt} + \nu \nabla \times \omega^n \right) - \beta_1 \hat{n} \cdot \left(\frac{Du^{n-1}}{Dt} + \nu \nabla \times \omega^{n-1} \right)$$

where

$$\frac{Du}{Dt} = \frac{\partial u}{\partial t} + \nabla \cdot \mathcal{F}(u),$$

is the material derivative and $\omega^n := \nabla \times u^n$ is the vorticity at time $t = t^n$. At outflow boundaries, Dirichlet pressure boundary conditions can be specified.

We then use $\bar{p}^{n+1}$ to update the intermediate velocity by

$$\tilde{\tilde{u}} = \tilde{u} - \frac{\Delta t}{\gamma_0} \nabla \bar{p}^{n+1}.$$

Finally, the timestep is completed by solving:

$$\gamma_0 \left(\frac{u^{n+1} - \tilde{\tilde{u}}}{\Delta t} \right) = \nu \nabla^2 u^{n+1}, \tag{7.25}$$

which expressed as implicit Helmholtz equations for the velocity components

$$-\nabla^2 u^{n+1} + \frac{\gamma_0}{\nu \Delta t} u^{n+1} = \frac{\gamma_0}{\nu \Delta t} \tilde{\tilde{u}}.$$

Adding the convection equations, Eq. (7.22), the pressure projection equation, Eq. (7.23), and the viscous equations, Eq. (7.25), the effective equation to be integrated is

$$\frac{\gamma_0 u^{n+1} - \alpha_0 u^n - \alpha_1 u^{n-1}}{\Delta t} = -\nabla \bar{p}^{n+1} - \beta_0 \mathcal{N}(u^n) - \beta_1 \mathcal{N}(u^{n-1}) + \nu \nabla^2 u^{n+1}.$$

7.4.2 The spatial discretization

After the temporal discretization, there are a number of possible strategies for handling the spatial derivatives. For the nonlinear advection, Eq. (7.22), we employ an upwind DG treatment to improve stability in advection dominated flows. For the pressure correction step, Eq. (7.23) we use an internal penalty formulation, with sufficient stabilization to avoid spurious solution jumps between elements. A basic treatment of the Neumann pressure condition, Eq. (7.24), is outlined. Likewise, for the viscous correction step, Eq. (7.25), we again consider an internal penalty DG formulation.

The advection step

We first consider the discretization of the nonlinear terms $\mathcal{N}$ in Eq. (7.21). Using the Lax-Friedrichs flux-based DG discretization as in Chapter 6, we find $\mathcal{N}$ such that

$$(\phi_h, \mathcal{N})_{\mathsf{D}^k} = (\phi_h, \nabla \cdot \mathcal{I}_N \mathcal{F}(\boldsymbol{u}_h)))_{\mathsf{D}^k} - \frac{1}{2}(\phi_h, [\![\mathcal{I}_N \mathcal{F}(\boldsymbol{u}_h)]\!])_{\partial \mathsf{D}^k}$$

$$+\frac{\lambda}{2}(\phi_h, \hat{\boldsymbol{n}} \cdot [\![\boldsymbol{u}_h]\!])_{\partial \mathsf{D}^k}, \quad \forall \phi_h \in \mathsf{V}_h, \tag{7.26}$$

with $\mathsf{V}_h = \bigoplus_{k=1}^{K} \mathsf{P}_N(\mathsf{D}^k)$. At inflow and walls, boundary conditions through $\boldsymbol{u}^+$ are specified. We also recall that $\mathcal{I}_N f$, refers to the N-th-order interpolation of f.

The evaluations of $\mathcal{N}$ are performed in INSAdvection2D.m and in the following we discuss how each term in the above variational equation is evaluated in more detail. The flux functions are first evaluated at the triangle nodes by

---
| INSAdvection2D.m |

```
3    % evaluate flux vectors
4    fxUx = Ux.^2;   fyUx = Ux.*Uy;   fxUy = Ux.*Uy; fyUy = Uy.^2;
```

The old values of $\mathcal{N}$ are saved and we begin to evaluate the new versions by computing the divergence of the flux functions by

---
| INSAdvection2D.m |

```
6    % save old nonlinear terms
7    NUxold = NUx; NUyold = NUy;
8
9    % evaluate inner-product of test function gradient and flux
10   % functions
11   NUx = Div2D(fxUx, fyUx); NUy = Div2D(fxUy, fyUy);
```

Note that Div2D.m evaluates the volume inner products and multiplies by the inverse mass matrix simultaneously, as described in Section 6.2. To evaluate the flux terms, the velocities are first subtracted at the internal and external traces of the element boundary nodes on each face.

| INSAdvection2D.m |

```
14   UxM = zeros(Nfp*Nfaces, K); UxP = zeros(Nfp*Nfaces, K);
15   UyM = zeros(Nfp*Nfaces, K); UyP = zeros(Nfp*Nfaces, K);
16   UxM(:) = Ux(vmapM); UyM(:) = Uy(vmapM);
17   UxP(:) = Ux(vmapP); UyP(:) = Uy(vmapP);
```

We then apply the predetermined boundary conditions to the external traces at nodes that are on the faces coinciding with the boundaries as

```
┌─────────────────┐
──────────────────│ INSAdvection2D.m │──────────────────
                  └─────────────────┘
20  UxP(mapI) = bcUx(mapI); UxP(mapW) = bcUx(mapW);
21  UxP(mapC) = bcUx(mapC);
22  UyP(mapI) = bcUy(mapI); UyP(mapW) = bcUy(mapW);
23  UyP(mapC) = bcUy(mapC);
```

and use these values to evaluate the surface flux functions at each trace for the face node by

```
┌─────────────────┐
──────────────────│ INSAdvection2D.m │──────────────────
                  └─────────────────┘
26  fxUxM = UxM.^2; fyUxM = UxM.*UyM; fxUyM = UxM.*UyM; fyUyM = UyM.^2;
27  fxUxP = UxP.^2; fyUxP = UxP.*UyP; fxUyP = UxP.*UyP; fyUyP = UyP.^2;
```

The trace values are used to determine the Lax-Friedrichs penalty multiplier λ for the local Lax-Friedrichs flux at each face and to form the full flux function

```
┌─────────────────┐
──────────────────│ INSAdvection2D.m │──────────────────
                  └─────────────────┘
38  % form local Lax-Friedrichs/Rusonov fluxes
39  fluxUx = 0.5*( -nx.*(fxUxM-fxUxP) - ny.*(fyUxM-fyUxP) ...
40           - maxvel.*(UxP - UxM));
41  fluxUy = 0.5*( -nx.*(fxUyM-fxUyP) - ny.*(fyUyM-fyUyP) ...
42           - maxvel.*(UyP - UyM));
```

and subsequently apply the lifting operator to these surface fluxes to obtain the contribution to the evaluation of the nonlinear terms by

```
┌─────────────────┐
──────────────────│ INSAdvection2D.m │──────────────────
                  └─────────────────┘
44  % put volume and surface terms together
45  NUx = NUx + LIFT*(Fscale.*fluxUx);
46  NUy = NUy + LIFT*(Fscale.*fluxUy);
```

This relatively simple sequence of operations completes the evaluation of the nonlinear terms in our algorithm. Having computed the nonlinear terms, we recover the intermediate velocities $\tilde{u}$ with

```
┌─────────────────┐
──────────────────│ INSAdvection2D.m │──────────────────
                  └─────────────────┘
49  UxT = ((a0*Ux+a1*Uxold) - dt*(b0*NUx + b1*NUxold))/g0;
50  UyT = ((a0*Uy+a1*Uyold) - dt*(b0*NUy + b1*NUyold))/g0;
```

This completes the advection step of the splitting method.

The pressure step

There are three main considerations when computing the pressure update. These involve evaluating the divergence of the intermediate velocity resulting from the advection step, computing the pressure Neumann boundary conditions, Eq. (7.24), and solving the Poisson equation to recover the pressure. An implementation of these tasks is given in INSPressure2D.m.

Recall the Poisson equation for pressure:

$$-\nabla^2 \bar{p}^{n+1} = -\frac{\gamma_0}{\Delta t} \nabla \cdot \tilde{u}.$$

The divergence of the intermediate velocity field $\tilde{u}$ is first evaluated by calling Div2D.m. Next, we recall the formula for the Neumann pressure condition on boundaries in Eq. (7.24). For simplicity we compute the curl of the vorticity at all volume nodes, add in the nonlinear terms, and then extract out the values at the Neumann boundaries as

---
INSPressure2D.m

```
4    DivUT = Div2D(UxT, UyT);
5
6    % dp/dn = -n.(u.grad u + curl curl u)
7    [tmp,tmp,CurlU] = Curl2D(Ux,Uy,[]);
8    [dCurlUdx,dCurlUdy] = Grad2D(CurlU);
9
10   res1 =  -NUx - nu*dCurlUdy; res2 = -NUy + nu*dCurlUdx;
11
12   % save old and compute new dp/dn
13   dpdnold = dpdn; dpdn = zeros(Nfp*Nfaces, K);
14
15   % decide which nodes to apply Neumann boundary conditions for
16   % pressure
17   nbcmapD = [mapI; mapW; mapC]; vbcmapD = [vmapI; vmapW; vmapC];
18   dpdn(nbcmapD) = (nx(nbcmapD).*res1(vbcmapD) ...
19         + ny(nbcmapD).*res2(vbcmapD));
20   dpdn = dpdn - bcdUndt;
```

Note that this assumes that the time derivative of the normal component of the velocity at these boundary nodes has been precomputed. In the tests follow, this quantity is known and computed by the formula in CurvedINS2D.m.

We chose to use an internal penalty DG formulation for the Laplace operator in the Poisson equation as discussed in Section 7.2.1. Hence, we seek a pressure field $p_h \in V_h$ that satisfies

$$(\nabla \phi_h, \nabla p_h)_{\mathsf{D}^k} - \frac{1}{2} \left(\hat{\boldsymbol{n}} \cdot \nabla \phi_h, \hat{\boldsymbol{n}} \cdot [\![p_h]\!] \right)_{\partial \mathsf{D}^k \backslash \partial \Omega} - \left(\phi_h, \hat{\boldsymbol{n}} \cdot \{\!\{ \nabla p_h \}\!\} \right)_{\partial \mathsf{D}^k \backslash \partial \Omega}$$

$$+ \left(\tau^k \phi_h, \hat{\boldsymbol{n}} \cdot [\![p_h]\!] \right)_{\partial \mathsf{D}^k \backslash \partial \Omega} - \left(\hat{\boldsymbol{n}} \cdot \nabla \phi_h, p_h^- \right)_{\partial \mathsf{D}^k \cap \partial \Omega^O}$$

$$- \left(\phi_h, 2 \hat{\boldsymbol{n}} \cdot \nabla p_h^- \right)_{\partial \mathsf{D}^k \cap \partial \Omega^{IWC}} + \left(\tau^k \phi_h, p_h^- \right)_{\partial \mathsf{D}^k \cap \partial \Omega^O}$$

$$= \left(\phi_h, -\frac{\gamma_0}{\Delta t} \nabla \cdot \tilde{\boldsymbol{u}} \right)_{\mathsf{D}^k} - \left(\hat{\boldsymbol{n}} \cdot \nabla \phi_h, p^O \right)_{\partial \mathsf{D}^k \cap \partial \Omega^O}$$

$$+ \left(\phi_h, \beta_0 \frac{\partial p_h^n}{\partial \hat{\boldsymbol{n}}} + \beta_1 \frac{\partial p_h^{n-1}}{\partial \hat{\boldsymbol{n}}} \right)_{\partial \mathsf{D}^k \cap \partial \Omega^{IW}} + \left(\tau^k \phi_h, p^O \right)_{\partial \mathsf{D}^k \cap \partial \Omega^O},$$

for all $\phi_h \in \mathsf{V}_h$. Here,

$$\tau^k = \tau \frac{(N+1)^2}{h^k},$$

on all elements D^k. The last term on the right hand side refers to the Neumann data for the pressure computed above, and not the negative trace of the normal derivative of the computed pressure fields.

Having computed sufficient data to form the right-hand side for the pressure Poisson solve, we can use a template mass matrix to compute the inner product. With the right-hand side of above equation and assuming the discrete operator has been set up, we follow Section 7.3 and permute the right-hand-side vector so it is compatible with the reverse Cuthill-McKee ordering of the pressure system, back-solve using the Cholesky factors, and finally reorder to the elementally numbered nodes with

---
INSPressure2D.m
```
22   % Evaluate right hand side term for Poisson equation for pressure
23   PRrhs = MassMatrix*(J.*(-DivUT*g0/dt) ...
24             + LIFT*(sJ.*(b0*dpdn + b1*dpdnold)));
25
26   % Add Dirichlet boundary condition forcing
27   PRrhs(:) = PRrhs(:) + rhsbcPR(:);
28
29   % Pressure Solve, -laplace PR = +(div UT)/dt
30                                + LIFT*dpdn on boundaries
31   PRrhs(:) = PRrhs(PRperm); tmp =  PRsystemCT\PRrhs(:);
32   PR(PRperm) = PRsystemC\tmp;
```

With the updated pressure field in hand, the gradient is computed with Grad2D.m (see Section 6.2) and we can update $\tilde{\boldsymbol{u}}$ and $\tilde{\tilde{\boldsymbol{u}}}$ with

---
INSPressure2D.m
```
34   % compute  (U~~,V~~) = (U~,V~) - dt*grad PR
35   [dPRdx,dPRdy] = Grad2D(PR);
36
37   % increment (Ux~,Uy~) to (Ux~~,Uy~~)
38   UxTT = UxT - dt*(dPRdx)/g0; UyTT = UyT - dt*(dPRdy)/g0;
```

Viscous step

We recall the Helmholtz equation from Eq. (7.25), which we need to solve to fully update the velocity components. For simplicity, we adopt the same solution strategy as we did for the pressure solve. Considering just the x-component of the velocity we seek $u_h \in V_h$ such that

$$
(\nabla\phi_h, \nabla u_h)_{\mathsf{D}^k} + \frac{\gamma_0}{\nu\Delta t}(\phi_h, u_h)_{\mathsf{D}^k}
$$

$$
-\frac{1}{2}(\hat{\boldsymbol{n}}\cdot\nabla\phi_h, \hat{\boldsymbol{n}}\cdot[\![u_h]\!])_{\partial\mathsf{D}^k\backslash\partial\Omega} - (\hat{\boldsymbol{n}}\cdot\nabla\phi_h, u_h^-)_{\partial\mathsf{D}^k\cap\partial\Omega^{IW}}
$$

$$
-(\phi_h, \hat{\boldsymbol{n}}\cdot\{\!\{\nabla u_h\}\!\})_{\partial\mathsf{D}^k\backslash\partial\Omega} - (\phi_h, \hat{\boldsymbol{n}}\cdot\nabla u_h^-)_{\partial\mathsf{D}^k\cap\partial\Omega^O}
$$

$$
+(\tau^k\phi_h, \hat{\boldsymbol{n}}\cdot[\![u_h]\!])_{\partial\mathsf{D}^k\backslash\partial\Omega} + (\tau^k\phi_h, u_h^-)_{\partial\mathsf{D}^k\cap\partial\Omega^{IW}}
$$

$$
= \left(\phi_h, \frac{\gamma_0}{\nu\Delta t}\tilde{\tilde{u}}_h\right)_{\mathsf{D}^k} - (\hat{\boldsymbol{n}}\cdot\nabla\phi_h, u^{IW})_{\partial\mathsf{D}^k\cap\partial\Omega^{IW}}
$$

$$
+\left(\phi_h, \frac{\partial u_h^{n+1}}{\partial\hat{\boldsymbol{n}}}\right)_{\partial\mathsf{D}^k\cap\partial\Omega^O} + (\tau^k\phi_h, -u^{IW})_{\partial\mathsf{D}^k\cap\partial\Omega^{IWC}}
$$

for all $\phi_h \in V_h$. We solve a similar system for the y-component of the velocity v_h, with the same matrix on the left-hand side. This system is similar to the pressure system, with the addition of the scaled mass matrix term and the reversal of the treatments for the boundary conditions since the velocity requires Dirichlet data at inflow and wall, and boundaries and Neumann data for the outflow boundary. The introduction of the scaled mass matrix also makes the problem considerably easier to solve than the Poisson problem.

The heavy lifting of setting up the viscous system matrix is done using CurvedPoissonIPDG2D.m and the boundary terms on the right-hand side are assumed to have been evaluated with CurvedPoissonIPDGbc2D.m. Thus, the actions required to update the velocity components are succinctly shown in the following piece of CurvedINSViscous2D.m:

CurvedINSViscous2D.m

```
5   % save Ux,Uy
6   Uxold = Ux; Uyold = Uy;
7
8   % viscous solves (RCM, backsolve twice, undo RCM)
9   Uxrhs = Uxrhs(VELperm); tmp = VELsystemCT\Uxrhs(:);
10  Ux(VELperm) = VELsystemC\tmp;
11  Uyrhs = Uyrhs(VELperm); tmp = VELsystemCT\Uyrhs(:);
12  Uy(VELperm) = VELsystemC\tmp;
```

Completing the splitting scheme

The full sequence of advection, pressure correction, and the viscous update is shown in CurvedINS2D.m, where the three stages are called within a simple timestepping loop in the following way:

──────────────────────── CurvedINS2D.m ────────────────────────

```
79    % script to compute pressure PR and intermediate UxTT, UyTT
80    INSPressure2D;
81
82    % script to compute viscous solves and update velocity
83    CurvedINSViscous2D;
84
85    % Increment time
86    time = tstep*dt;
```

The CurvedINS2D.m function involves more setup, and post-processing tasks, but the above listing represents the core solution steps required at each timestep. For further information and insight, we recommend a careful reading of this script. In particular, the transition from first-order to a second-order timestepping scheme is clearly presented within the code. Furthermore, we have restricted these scripts to the use of direct sparse linear algebra solvers as discussed in Section 7.3. In practice, as the number of degrees of freedom in a simulation increases, it is natural to transition to a preconditioned iterative solver to perform the pressure and viscous solves (see Section 7.3).

7.4.3 Benchmarks and validations

The incompressible Navier-Stokes equations are complicated enough that in order to test the implementation of the numerical scheme, we are obliged to be creative. In the following sections we will show a number of computational results to validate the algorithm on some benchmark problems. For the first two cases, we have exact solutions to compare the numerical solution against and we compute maximum nodal errors for each solution field component to estimate the order of accuracy. For the third test case, we rely on estimates of the maximum lift and drag on the cylinder as well as the drop in base pressure across the cylinder from [189].

The Kovasznay flow solution

The first test case is the Kovasznay [207] analytical solution for laminar flow behind a two-dimensional grid. This is a steady-state solution, so it does not provide a rigorous test of the timestepping method. However, it does provide a valuable validation of the treatment of the nonlinear terms, the pressure treatment, and the viscous terms.

The proposed solution is an exact, steady-state solution for the incompressible Navier-Stokes equations given by

$$\lambda := \frac{1}{2\nu} - \sqrt{\frac{1}{4\nu^2} + 4\pi^2},$$
$$u = 1 - e^{\lambda x} \cos(2\pi y),$$

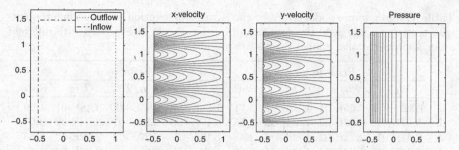

Fig. 7.15. On the left is shown the geometry for the Kovasznay solution; the remaining contours show the exact solution to the incompressible Navier-Stokes equations.

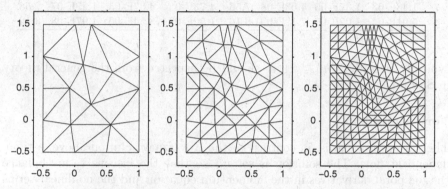

Fig. 7.16. Sequence of three meshes used to perform convergence analysis for the Kovasznay solution.

$$v = \frac{\lambda}{2\pi} e^{\lambda x} \sin(2\pi y),$$

$$p = \frac{1}{2} \left(1 - e^{2\lambda x}\right).$$

The solution is used to apply Dirichlet boundary conditions for the velocity at the inflow boundary shown in Fig. 7.15 as well as Neumann boundary conditions for the velocity at the outflow boundary. Similarly, we use the definition of the pressure for Dirichlet boundary conditions at the outflow. For the Neumann pressure boundary condition at the inflow, we apply the formula in Eq. (7.24).

To test the algorithm, we perform an experimental convergence analysis for polynomials up to eighth order. The simulation is performed for one time unit with the viscosity set to $\nu = \frac{1}{40}$. The base mesh is shown in Fig. 7.16 and two uniform element refinements were applied to the mesh.

In Table 7.4 we show the errors tabulated by polynomial order and mesh size for the horizontal velocity component u and the pressure p. Results for v are similar. For $N > 1$, we see convincing convergence in all cases and, in general, the rate of convergence is between $\mathcal{O}(h^N)$ and $\mathcal{O}(h^{N+1})$ and

Table 7.4. Maximum nodal errors for the Kovasznay solution. The maximum point-wise error for u and p on the sequence of meshes shown in Fig. 7.16 with different order approximations. Similar results are found for v.

	Error in u				Error in p			
N	h	$h/2$	$h/4$	Rate	h	$h/2$	$h/4$	Rate
1	1.32E+00	7.05E-01	1.23E-01	1.71	3.13E+00	1.53E+00	3.48E-01	1.59
2	5.01E-01	9.67E-02	1.45E-02	2.55	1.47E+00	2.54E-01	2.08E-02	3.07
3	2.41E-01	2.74E-02	1.89E-03	3.49	5.02E-01	2.79E-02	2.42E-03	3.85
4	6.40E-02	3.34E-03	1.00E-04	4.66	9.31E-02	8.87E-03	2.02E-04	4.42
5	1.87E-02	6.92E-04	8.96E-06	5.51	6.15E-02	1.02E-03	1.89E-05	5.84
6	9.07E-03	5.41E-05	6.93E-07	6.84	1.23E-02	2.17E-04	2.63E-06	6.09
7	1.43E-03	1.37E-05	4.03E-08	7.56	4.73E-03	4.67E-05	1.29E-07	7.58
8	5.91E-04	1.02E-06	7.17E-09	8.16	1.52E-03	3.54E-06	1.97E-08	8.12

typically close to $\mathcal{O}(h^{N+1/2})$, as should be expected from the general theory in Section 6.7.

Vortex problem

The vortex problem is, in contrast to the Kovasznay solution above, a time-dependent flow. The solution is set up so that the viscous terms balance the temporal derivatives in the momentum equation and the nonlinear terms balance the gradient of the pressure. The solution is known everywhere for all time and is given by

$$u = -\sin(2\pi y)e^{-\nu 4\pi^2 t},$$
$$v = \sin(2\pi x)e^{-\nu 4\pi^2 t},$$
$$p = -\cos(2\pi x)\cos(2\pi y)e^{-\nu 8\pi^2 t}.$$

We use the exact solution to specify the velocity and pressure boundary conditions, excluding the Neumann pressure boundary conditions at the inflows shown in Fig. 7.17, where Eq. (7.24) is applied. A minor difference from the previous test case is that the temporal derivative of the normal component of velocity is required for the pressure boundary equation, and this is computed analytically and applied in the formulas. Since we are using a Lax-Friedrichs flux, it is necessary to determine which parts of the boundary are inflows and outflows. In this particular case there are four inflow and four outflow regions, as shown Fig. 7.17.

In Table 7.5 we show convergence results for the vortex case. To achieve convincing convergence we were obliged to reduce the timestep by a factor of 10. Having done this, we recover nearly optimal order convergence for the velocity and pressure apart from the $N = 8$ case, where the error at the highest resolution is dominated by timestepping errors and the resulting estimate of the convergence rate is suboptimal.

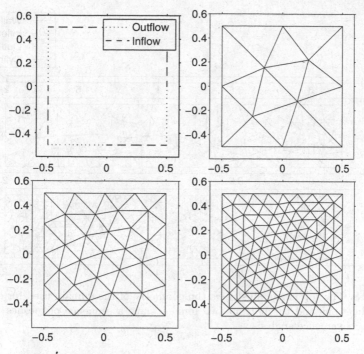

Fig. 7.17. Top left shows the geometry of the domain used for the vortex problem test case. The remaining figures show the sequence of meshes used to perform convergence analysis.

Table 7.5. Maximum nodal errors for the vortex solution. The maximum pointwise error for u and p on a sequence of meshes shown in Fig 7.17 with different order approximations. Similar results are found for v. An $*$ indicates that the error levels are affected by temporal errors.

		Error in u				Error in p		
N	h	$h/2$	$h/4$	Rate	h	$h/2$	$h/4$	Rate
1	8.38E-01	2.85E-01	5.39E-02	1.98	2.09E+00	5.42E-01	9.39E-02	2.24
2	7.42E-02	3.57E-02	1.04E-02	1.42	3.14E-01	8.78E-02	1.78E-02	2.07
3	1.22E-01	4.66E-03	5.61E-04	3.88	2.32E-01	1.56E-02	8.79E-04	4.02
4	1.02E-02	2.14E-03	4.48E-05	3.92	2.75E-02	4.07E-03	1.19E-04	3.93
5	1.83E-02	7.52E-05	3.03E-06	6.28	3.23E-02	3.27E-04	6.84E-06	6.10
6	5.97E-04	4.10E-05	1.70E-07	5.89	9.67E-04	6.73E-05	4.91E-07	5.47
7	1.39E-03	6.32E-07	4.45E-08	7.46	1.85E-03	3.94E-06	4.13E-08	7.73
8	1.80E-05	3.27E-07*	2.73E-08*	4.68*	2.71E-05	5.39E-07*	2.75E-08*	4.97*

Unsteady flow around cylinder in channel

As the last test case, we consider a slightly more complicated flow problem, which is designed to be relatively simple to implement for the incompressible

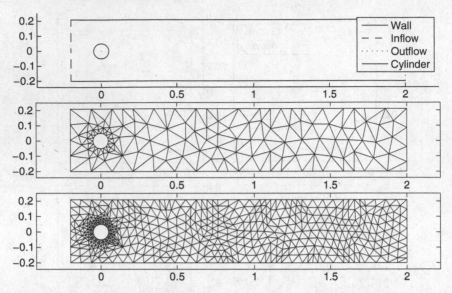

Fig. 7.18. The top image illustrates the geometry of the domain used for cylinder flow simulations and the bottom two images show a sequence of meshes used to perform convergence analysis.

Navier-Stokes equations. The flow configuration is a time-dependent two-dimensional flow through a channel, with a circular obstacle in the channel. The cylinder is placed off the centerline of the channel, so it begins to shed vortices as soon as the parabolic inflow is ramped up from its zero initial condition. This test case was suggested as a benchmark in the mid-1990s [277] and numerous simulation codes have been compared for this case. Peak drag and lift coefficients and base pressure drop over the cylinder have been accurately obtained by high resolution simulations and we can compared the DG based solver for these quantities.

The spatial domain for this problem is shown in Fig. 7.18 and the simulations are run until $T = 8$. At initial time $T = 0$ zero flow is assumed. The inflow boundary conditions for this case are given by a parabolic inflow profile, modulated by a sine function depending on time as

$$u(x, y, t) = 0.41^{-2} \sin\left(\frac{\pi t}{8}\right) 6(y + 0.2)(0.21 - y)),$$

$$v(x, y, t) = 0.$$

No slip boundary conditions are imposed on the cylinder and channel walls. At the outflow, a zero pressure is assumed and natural boundary conditions for velocity. A viscosity of $\nu = 10^{-3}$ is assumed, and a rough estimate yields a Reynolds number for the flow of Re ≈ 100.

A closed-form exact solution is not known for this problem. Instead, we compute the drag and lift coefficients for the cylinder defined as the horizontal and vertical components of the total force on the cylinder. These are given by the following formulas:

$$C_d\left(t\right) = -\oint_{\text{Cylinder}} -p\hat{n}_x + \nu\left(\hat{n}_x 2\frac{\partial u}{\partial x} + \hat{n}_y\left(\frac{\partial v}{\partial x} + \frac{\partial u}{\partial y}\right)\right) ds,$$

$$C_l\left(t\right) = -\oint_{\text{Cylinder}} -p\hat{n}_y + \nu\left(\hat{n}_x\left(\frac{\partial u}{\partial y} + \frac{\partial v}{\partial x}\right) + \hat{n}_y 2\frac{\partial v}{\partial y}\right) ds,$$

where $(\hat{n}_x, \hat{n}_y)$ is the outward facing normals along the cylinder surface. These numbers are computed for every timestep in INSLiftDrag2D.m in the following segment of code:

INSLiftDrag2D.m

```
23  % compute derivatives
24  [dUxdx,dUxdy] = Grad2D(Ux); dUxdx = dUxdx(vmapC);
25  dUxdy = dUxdy(vmapC);
26  [dUydx,dUydy] = Grad2D(Uy); dUydx = dUydx(vmapC);
27  dUydy = dUydy(vmapC);
28
29  PR = PR(vmapC); nxC = nx(mapC); nyC = ny(mapC); sJC = sJ(mapC);
30
31  hforce = -PR.*nxC + nu*(nxC.*(2*dUxdx)      + nyC.*(dUydx + dUxdy));
32  vforce = -PR.*nyC + nu*(nxC.*(dUydx+dUxdy) + nyC.*(2*dUydy)      );
33
34  hforce = reshape(hforce, Nfp, length(mapC)/Nfp);
35  vforce = reshape(vforce, Nfp, length(mapC)/Nfp);
36  sJC    = reshape(   sJC, Nfp, length(mapC)/Nfp);
37
38  % compute weights for integrating (1,hforce) and (1,vforce)
39  V1D = Vandermonde1D(N, r(Fmask(:,1)));
40  MM1D = inv(V1D)'/V1D;
41  w = sum(MM1D, 1);
42
43  % Compute drag coefficient
44  Cd = sum(w*(sJC.*hforce));
45
46  % Compute lift coefficient
47  Cl = sum(w*(sJC.*vforce));
```

This code first computes the gradient of the velocity, extracts out the velocities on the cylinder boundary, and then evaluates the above integrands. Finally, the integrations are performed around the cylinder to obtain the estimates for lift and drag.

In addition, we measure the drop in pressure between the leading point of the cylinder and the trailing point; that is, we compute

$$\Delta p = p\left(-0.05, 0\right) - p\left(0.05, 0\right).$$

Table 7.6. Maximum lift, drag coefficients, and final time pressure drop for the cylinder in the channel test case.

K	N	t_{C_d}	C_d	t_{C_l}	C_l	$\Delta p\,(t=8)$
115	6	3.9394263	2.9529140	5.6742937	0.4966074	-0.1095664
460	6	3.9363751	2.9509030	5.6930431	0.4778757	-0.1116310
236	8	3.9343595	2.9417190	5.6990205	0.4879853	-0.1119122
236	10	3.9370396	2.9545659	5.6927772	0.4789706	-0.1116177
[189]	N/A	3.93625	2.950921575	5.693125	0.47795	-0.1116

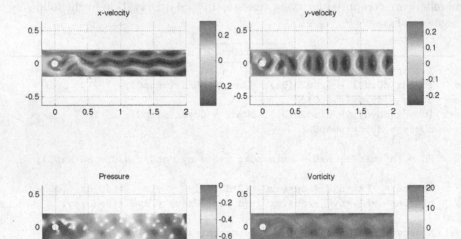

Fig. 7.19. Simulation results for a $K = 236$, $N = 10$ study of the unsteady cylinder test case.

Once the simulation is complete, the maximum lift and drag is identified and the times at which these maxima are achieved were compared with the reference values from [189]. In Table 7.6 we show excellent agreement with the reference values, despite the code only using a few hundred triangles with $N = 6$ compared to the many thousands of degrees of freedom used to compute the reference values with a low-order finite element code.

In Fig. 7.19 we show a snapshot of the computed solution as an illustration of the complexity of the computed flows.

7.5 The compressible Navier-Stokes equations

Let us finally return to the problem of fully compressible gas dynamics, already discussed in Section 6.6 for the inviscid case, and consider the time-dependent compressible Navier-Stokes equations. This serves as an application of DG methods to problems that involve a combination of a hyperbolic system of nonlinear conservation laws with second-order dissipative terms. Both the compressible Euler equations discussed in Section 6.6 and the incompressible Navier-Stokes equations from Section 7.4 are special cases of these equations, obtained by dropping the viscous terms and enforcing the velocity field to be incompressible, respectively.

During the last two decades, a number of techniques, all related to the basic DG formulation, have been considered for effectively handling the combined hyperbolic/parabolic nature of these equations in a high-order accurate scheme. The focus on high-order accuracy has been to enable the efficient modeling of boundary layers, shear layers, and complex vortical solutions typical of these equations. One of the first methods to achieve high-order accuracy for these equations involved the use of a staggered-grid Chebyshev multidomain method [202, 203, 204]. The approach relied heavily on the use of quadrilateral elements and strong imposition of fluxes, but showed the power of high-order elements to resolve solution structures inherent for these equations.

At approximately the same time, an approach, even closer to the spirit of DG methods, was outlined as a constructive penalty formulation that provably achieved a stable coupling between high-order elements treated in a pseudo-spectral manner [152, 153, 157]. This approach is very similar to DG methods with weakly imposed boundary conditions but with the equations being satisfied in a pointwise collocation sense as illustrated in Example 2.2.

The emergence of schemes based on more classic DG formulations happened almost simultaneously with the two previous developments. The first schemes seem to have been proposed independently in [22, 23] and in [228, 230] for triangular meshes. This was extended to meshes of mixed element types in [323] and time-dependent meshes in [229]. These efforts established that the DG method could successfully resolve shock structures in numerical simulations of complex phenomena and applications.

The DG-based approach has met a certain amount of popularity and further practical results have been obtained with these methods in [37, 38], in which a dynamic order adaptive refinement algorithm is explored to resolve shock structures.

There are, however, some significant challenges for large-scale time-dependent simulations. The presence of the second-order differential operators can induce very small timesteps, which we incidentally avoided in the incompressible Navier-Stokes case by employing a semi-implicit time integrator. In the case of the compressible Navier-Stokes equations, the diffusion operator is no longer linear and additional consideration are required. In [116] a similar semi-implicit philosophy is explored and in [191], it was demonstrated how one can address this problem by using an implicit explicit Runge-Kutta time integrator to achieve high-order temporal accuracy as well as increased timestep size.

It is likewise clear that the resolution requirements for direct numerical simulation of turbulent/high Reynolds number flows are beyond current computing facilities [83]. Thus, a number of efforts have been directed at combining DG methods with Reynolds-averaged Navier-Stokes and k-ω turbulence modes [20, 247]. Further investigations of the use of a variational multiscale formulation with DG have also been initiated [31].

In the following sections we describe the original formulation in [21] and outline an example implementation using the integration-based algorithm we discussed for the Euler equations in Section 6.6. We conclude with a few test cases demonstrating the effectiveness of this formulation.

7.5.1 Integration-based gradient, divergence, and jump operators

Before we introduce the formulation for the compressible Navier-Stokes equations, it is appropriate to build a little more infrastructure to make that solution procedure more transparent. We will frequently need to compute gradients and derivatives and thus introduce extensions to the basic Grad2D.m and Div2D.m scripts that allowed us to compute the local element gradients and derivatives of our piecewise polynomial solution. Those scripts work best on straight-sided elements and when the function being differentiated is exactly supported by the nodal basis functions. In the extension of these scripts we generalize and allow the field to be differentiated to be of higher order than the solution and its derivatives. Furthermore, we take into account the contributions to the derivatives from the interelement jumps in the field being differentiated.

We recall a standard variational statement of the discontinuous Galerkin gradient with a central flux as follows: given $u_h \in V_{h,M} = \bigoplus_{k=1}^{K} P_M(D^k)$, we find $\mathcal{G}u_h = (\mathcal{G}^x u_h, \mathcal{G}^y u_h) \in V_{h,N} \times V_{h,N}$ such that

$$(\phi_h, \mathcal{G}^x u_h)_{\Omega,h} = -\left(\frac{\partial \phi_h}{\partial x}, \mathcal{I}_N u_h\right)_{\Omega,h} + \sum_{k=1}^{K} (\hat{n}_x \{\{\mathcal{I}_N u_h\}\}, \phi_h)_{\partial D^k},$$

$$(\phi_h, \mathcal{G}^y u_h)_{\Omega,h} = -\left(\frac{\partial \phi_h}{\partial y}, \mathcal{I}_N u_h\right)_{\Omega,h} + \sum_{k=1}^{K} (\hat{n}_y \{\{\mathcal{I}_N u_h\}\}, \phi_h)_{\partial D^k},$$

for all $\phi_h \in V_{h,N}$. Notice that we have carefully allowed the function being differentiated to be of different order (M) to the approximation of its derivative (N). The combined application of cubature-based integration for the elemental integrals, and Gauss-based quadrature for the surface integrals of Section 6.6.1, used to discretize the Euler equations of gas dynamics is repeated to evaluate this generic gradient. Since the values of the field are different at the elemental cubature/Gauss nodes for each element, the following function computes the DG gradient, using the integration infrastructure stored in the cub and gauss structures as discussed in Section 6.6.1.

─────────────────────── CurvedDGGrad2D.m ───────────────────────

```
1   function [dUdx, dUdy] = CurvedDGGrad2D(cU, gU, gmapD, bcU)
2
3   % function [dUdx, dUdy] = CurvedDGGrad2D(cU, gU, gmapD, bcU)
4   % Purpose: compute DG derivative of field given at cubature volume
5   %          and Gauss surface nodes
6
7   Globals2D;
8
9   % Volume terms: dUdx and dUdy
10  dUdx = (cub.Dr')*(cub.W.*(cub.rx.*cU)) ...
11        + (cub.Ds')*(cub.W.*(cub.sx.*cU));
12  dUdy = (cub.Dr')*(cub.W.*(cub.ry.*cU)) ...
13        + (cub.Ds')*(cub.W.*(cub.sy.*cU));
14
15  % Surface traces at Gauss nodes
16  gUM = gU(gauss.mapM); gUP = gU(gauss.mapP);
17
18  % Apply boundary conditions
19  gUP(gmapD) = bcU(gmapD);
20
21  % Normal flux terms
22  fx = 0.5*(gauss.nx.*(gUM + gUP)); fy = 0.5*(gauss.ny.*(gUM + gUP));
23
24  % Add lifted flux terms to volume terms
25  dUdx = dUdx - gauss.interp'*(gauss.W.*fx);
26  dUdy = dUdy - gauss.interp'*(gauss.W.*fy);
27
28  % Multiply by inverse mass matrix templated for straight
29  % sided triangles
30  dUdx(:,straight) = V*V'*(dUdx(:,straight)./J(:,straight));
31  dUdy(:,straight) = V*V'*(dUdy(:,straight)./J(:,straight));
32
33  % Multiply by custom inverse mass matrix for each curvilinear
34  % triangle
35  Ncurved = length(curved);
36  for m=1:Ncurved
37    k = curved(m);
38    mmCHOL = cub.mmCHOL(:,:,k);
39    dUdx(:,k) = mmCHOL\(mmCHOL'\dUdx(:,k));
40    dUdy(:,k) = mmCHOL\(mmCHOL'\dUdy(:,k));
41  end
42
43  % Correct sign
44  dUdx = -dUdx; dUdy = -dUdy;
45  return;
```

Following the same approach, we discretize a divergence operator $u_h \in V_{h,M} \times V_{h,M}$ by finding $\mathcal{D}u_h \in V_{h,N}$ such that

$$(\phi_h, \mathcal{D}u_h)_{\Omega,h} = -(\nabla\phi_h, \mathcal{I}_N u_h)_{\Omega,h} + \sum_{k=1}^{K}(\hat{n} \cdot \{\{\mathcal{I}_N u_h\}\}, \phi_h)_{\partial \mathsf{D}^k}, \quad \forall \phi_h \in V_{h,N}.$$

The procedure is outlined in CurvedDGDiv2D.m.

CurvedDGDiv2D.m

```
1   function [divU] = CurvedDGDiv2D(cU, cV, gU, gV, gmapN, bcNdotU)
2
3   % function [divU] = CurvedDGDiv2D(cU, cV, gU, gV, gmapN, bcNdotU)
4   % Purpose: compute the divergence of a vectorial function given
5   %          at cubature and surface Gauss nodes
6
7   Globals2D;
8
9   % volume terms: U
10  divU = (cub.Dr')*(cub.W.*(cub.rx.*cU+cub.ry.*cV)) + ...
11         (cub.Ds')*(cub.W.*(cub.sx.*cU+cub.sy.*cV));
12
13  % surface traces at Gauss nodes
14  gUM = gU(gauss.mapM); gUP = gU(gauss.mapP);
15  gVM = gV(gauss.mapM); gVP = gV(gauss.mapP);
16
17  % normal fluxes
18  gFxM = gauss.nx.*gUM + gauss.ny.*gVM;
19  gFxP = gauss.nx.*gUP + gauss.ny.*gVP;
20
21  % Apply boundary conditions
22  gFxP(gmapN) = bcNdotU(gmapN);
23
24  % add flux terms to divergence
25  divU = divU - gauss.interp'*(gauss.W.*(gFxM + gFxP))/2;
26
27  % multiply straight sided triangles by inverse mass matrix
28  divU(:,straight) = V*V'*(divU(:,straight)./J(:,straight));
29
30  % multiply straight sided triangles by custom inverse mass matrices
31  Ncurved = length(curved);
32  for m=1:Ncurved
33    k = curved(m);
34    mmCHOL = cub.mmCHOL(:,:,k);
35    divU(:,k) = mmCHOL\(mmCHOL'\divU(:,k));
36  end
37
38  % Correct sign
39  divU = -divU;
40  return
```

Finally, for this prelude, we will repeatedly need to evaluate terms of the form

$$(\phi_h, \mathcal{J} u_h)_{\partial \mathsf{D}^k} = (\phi_h, \hat{\boldsymbol{n}} \cdot [\![\mathcal{I}_N u_h]\!])_{\partial \mathsf{D}^k}, \quad \phi_h \in \mathsf{V}_{h,N},$$

in the penalization of inter element jumps of the solution. Here, the jump operator $\mathcal{J}$ is not to be confused with the elemental coordinate Jacobian. This time, there are no volume integrals, so the solution is only needed at elemental boundary Gauss nodes as illustrated in CurvedDGJump2D.m.

─────────────────────────── CurvedDGJump2D.m ───────────────────────────

```
1   function [jumpU] = CurvedDGJump2D(gU, gmapD, bcU)
2
3   % function [jumpU] = CurvedDGJump2D(gU, gmapD, bcU)
4   % purpose: compute discontinuous Galerkin jump applied
5   %          to a field given at cubature and Gauss nodes
6
7   Globals2D;
8
9   % surface traces at Gauss nodes
10  gUM = gU(gauss.mapM); gUP = gU(gauss.mapP);
11  gUP(gmapD) = bcU(gmapD);
12
13  % compute jump term and lift to triangle interiors
14  fx = gUM - gUP;
15  jumpU = -gauss.interp'*(gauss.W.*fx);
16
17  % multiply straight sided triangles by inverse mass matrix
18  jumpU(:,straight) = V*V'*(jumpU(:,straight)./J(:,straight));
19
20  % multiply by custom inverse mass matrix for each curvilinear
21  % triangle
22  Ncurved = length(curved);
23  for m=1:Ncurved
24    k = curved(m);
25    mmCHOL = cub.mmCHOL(:,:,k);
26    jumpU(:,k) = mmCHOL\(mmCHOL'\jumpU(:,k));
27  end
28  return
```

───

We will use the notation $\mathcal{G}^x u$, $\mathcal{G}^y u$, $\mathcal{J} u$ to denote the action of the central flux-based gradient and stabilization operators on u and $\mathcal{D} u$ to denote the action of the divergence on the vector $\boldsymbol{u}$.

7.5.2 Solver for the compressible Navier-Stokes equations of gas dynamics

A reduced model of the compressible Navier-Stokes equations of gas dynamics in conservative form is given as

$$\frac{\partial \rho}{\partial t} + \frac{\partial \rho u}{\partial x} + \frac{\partial \rho v}{\partial y} = 0, \tag{7.27}$$

$$\frac{\partial \rho u}{\partial t} + \frac{\partial \rho u^2 + p}{\partial x} + \frac{\partial \rho u v}{\partial y} = \frac{\partial \tau_{11}}{\partial x} + \frac{\partial \tau_{12}}{\partial y},$$

$$\frac{\partial \rho v}{\partial t} + \frac{\partial \rho u v}{\partial x} + \frac{\partial \rho v^2 + p}{\partial y} = \frac{\partial \tau_{21}}{\partial x} + \frac{\partial \tau_{22}}{\partial y},$$

$$\frac{\partial E}{\partial t} + \frac{\partial u\,(E + p)}{\partial x} + \frac{\partial v\,(E + p)}{\partial y} = \frac{\partial \tau_{31}}{\partial x} + \frac{\partial \tau_{32}}{\partial y},$$

where ρ is the density of the gas, ρu and ρv are the x- and y-components of the momentum, p is the internal pressure of the gas, E is the total energy of the gas, and τ is the stress tensor. We have assumed that the kinematic viscosity is independent of density and temperature and is homogeneous in space. For simplicity we assume that the thermal diffusion is negligible.

The stress tensor, τ, is defined by

$$\tau_{11} = \mu \left(2\frac{\partial u}{\partial x} - \frac{2}{3} \left(\frac{\partial u}{\partial x} + \frac{\partial v}{\partial y} \right) \right)$$

$$\tau_{12} = \tau_{21} = \mu \left(\frac{\partial u}{\partial y} + \frac{\partial v}{\partial x} \right)$$

$$\tau_{22} = \mu \left(2\frac{\partial v}{\partial y} - \frac{2}{3} \left(\frac{\partial u}{\partial x} + \frac{\partial v}{\partial y} \right) \right)$$

$$\tau_{31} = u\tau_{11} + v\tau_{12}$$

$$\tau_{32} = u\tau_{21} + v\tau_{22}.$$

As with the Euler equations, we write the problem in a general vector conservation law form

$$\frac{\partial \mathbf{q}}{\partial t} = \frac{\partial}{\partial x} \left(-\mathbf{F} + \tau_{:1} \right) + \frac{\partial}{\partial y} \left(-\mathbf{G} + \tau_{:2} \right),$$

with $\mathbf{q}$, $\mathbf{F}$, and $\mathbf{G}$ defined as for the Euler equations in Section 6.6.

To evaluate the right-hand side we follow the approach of [21] coupled with the use of integration rules to evaluate the stress tensor and the nonlinear fluxes. All spatial derivatives can be evaluated using the gradient and divergence functions of Section 7.5.1. We provide an implementation of the sequence of operations required to evaluate the right-hand side in CurvedC-NSRHS2D.m. There are quite a few steps required, and we have broken the process into several stages. The function takes as input an array q containing nodal data for the density, momentum, and energy fields as well as the current time, viscosity parameter, and a function that evaluates the boundary conditions.

──────────── CurvedCNSRHS2D.m ────────────

```
 1  function [rhsQ] = CurvedCNSRHS2D(Q, mu, time, SolutionBC, fluxtype)
 2
 3  % function [rhsQ] = CurvedCNSRHS2D(Q, mu, time,
 4  %                                   SolutionBC, fluxtype)
 5  % Purpose: evaluate right hand side residual of the
 6  %          compressible Navier-Stokes equations
 7  Globals2D;
 8
 9  % Gas constant
10  gamma = 1.4;
```

The first stage is to interpolate the input density, momentum, and energy fields to cubature and Gauss nodes on each element and element face, respectively.

──────────── CurvedCNSRHS2D.m ────────────

```
14  % Extract fields from three dimensional array
15  rho = Q(:,:,1); rhou = Q(:,:,2); rhov = Q(:,:,3); Ener = Q(:,:,4);
16
17  % Interpolate fields to volume cubature nodes & Gauss quadrature
18  % surface nodes
19  crho  = cub.V*rho;  grho  = gauss.interp*rho;
20  crhou = cub.V*rhou; grhou = gauss.interp*rhou;
21  crhov = cub.V*rhov; grhov = gauss.interp*rhov;
22  cEner = cub.V*Ener; gEner = gauss.interp*Ener;
```

Once we have evaluated the fields at the Gauss face nodes, we use these values at domain boundary faces to evaluate the external boundary values suitable for these fields.

──────────── CurvedCNSRHS2D.m ────────────

```
26  % Compute primitive fields at Gauss quadrature surface nodes
27  gu = grhou./grho; gv = grhov./grho;
28  gPr = (gamma-1)*(gEner-0.5*grho.*(gu.^2+gv.^2));
29
30  if(~isempty(SolutionBC))
31    % create boundary condition variables
32    [brho,brhou,brhov,bEner] = ...
33        feval(SolutionBC, grho, grhou, grhov, gEner, time);
34    gmapB = gauss.mapB;
35
36    % compute primitive variables of boundary data
37    bu = gu; bv = gv; bPr = gPr;
38    bu(gmapB) = brhou(gmapB)./brho(gmapB);
39    bv(gmapB) = brhov(gmapB)./brho(gmapB);
```

```
40     bPr(gmapB) = (gamma-1)*(bEner(gmapB) ...
41        -0.5*brho(gmapB).*(bu(gmapB).^2+bv(gmapB).^2));
42   end
```

To compute the stress tensor, we need to evaluate the derivatives of the components of the velocity vector. The DG central flux derivative operators are used to evaluate these derivatives, and, in particular, we use the product rule to evaluate the spatial derivatives of the conserved momentum and density variables. This approach requires the evaluation of spatial derivatives as follows

$$\nabla u = \frac{1}{\rho}\left(\nabla \rho u - u \nabla \rho\right), \quad \nabla v = \frac{1}{\rho}\left(\nabla \rho v - v \nabla \rho\right).$$

In practice we evaluate the derivatives with the divergence or gradient operator and then evaluate the nonlinear products at the Gauss nodes as:

$$\mathcal{G}u_h = \frac{1}{\rho_h}\left(\mathcal{G}\rho u_h - u_h \mathcal{G}\rho_h\right), \quad \mathcal{G}v_h = \frac{1}{\rho_h}\left(\mathcal{G}\rho v_h - v_h \mathcal{G}\rho_h\right).$$

Using the approach outlined in Section 7.5.1 for the derivatives this is achieved as

```
────────────────── CurvedCNSRHS2D.m ──────────────────
46   % Compute gradients of the conserved variables
47   [drhodx,  drhody] = CurvedDGGrad2D(crho,  grho,  gmapB, brho);
48   [drhoudx, drhoudy] = CurvedDGGrad2D(crhou, grhou, gmapB, brhou);
49   [drhovdx, drhovdy] = CurvedDGGrad2D(crhov, grhov, gmapB, brhov);
```

We then apply the product rule to evaluate the spatial derivatives of the velocity components and finally combine the derivatives using the definition of the stress tensor. This is first done at the cubature nodes

```
────────────────── CurvedCNSRHS2D.m ──────────────────
53   % Compute primitive fields at cubature nodes & Gauss quadrature
54   %    surface nodes
55   cu  = crhou./crho; cv  = crhov./crho;
56   cPr = (gamma-1)*(cEner - 0.5*crho.*(cu.^2+cv.^2));
57
58   % Interpolate derivatives of conserved variables to cubature nodes
59   cdrhodx  = cub.V*drhodx;  cdrhody  = cub.V*drhody;
60   cdrhoudx = cub.V*drhoudx; cdrhoudy = cub.V*drhoudy;
61   cdrhovdx = cub.V*drhovdx; cdrhovdy = cub.V*drhovdy;
62
63   % Use product-rule to evaluate gradients of velocity components
64   % at cubature nodes
65   cdudx = (cdrhoudx  - cdrhodx.*cu)./crho;
66   cdudy = (cdrhoudy  - cdrhody.*cu)./crho;
```

```
67   cdvdx = (cdrhovdx  - cdrhodx.*cv)./crho;
68   cdvdy = (cdrhovdy  - cdrhody.*cv)./crho;
69
70   % Compute viscous stress tensor at cubature nodes
71   ct11 = mu.*(2*cdudx - (2/3)*(cdudx + cdvdy));
72   ct12 = mu.*(cdudy + cdvdx);
73   ct22 = mu.*(2*cdvdy - (2/3)*(cdudx + cdvdy));
74   ct31 = cu.*ct11 + cv.*ct12;
75   ct32 = cu.*ct12 + cv.*ct22;
```

and subsequently repeated at the Gauss face nodes

--------------------------------- CurvedCNSRHS2D.m ---------------------------------

```
79   % Interpolate derivatives of conserved variables to Gauss nodes
80   gdrhodx  = gauss.interp*drhodx;   gdrhody  = gauss.interp*drhody;
81   gdrhoudx = gauss.interp*drhoudx;  gdrhoudy = gauss.interp*drhoudy;
82   gdrhovdx = gauss.interp*drhovdx;  gdrhovdy = gauss.interp*drhovdy;
83
84   % Use product-rule to evaluate gradients of velocity components
85   % at Gauss nodes
86   gdudx = (gdrhoudx  - gdrhodx.*gu)./grho;
87   gdudy = (gdrhoudy  - gdrhody.*gu)./grho;
88   gdvdx = (gdrhovdx  - gdrhodx.*gv)./grho;
89   gdvdy = (gdrhovdy  - gdrhody.*gv)./grho;
90
91   % Compute viscous stress tensor at Gauss nodes
92   gt11 = mu.*(2*gdudx - (2/3)*(gdudx + gdvdy));
93   gt12 = mu.*(gdudy + gdvdx);
94   gt22 = mu.*(2*gdvdy - (2/3)*(gdudx + gdvdy));
95   gt31 = gu.*gt11 + gv.*gt12;
96   gt32 = gu.*gt12 + gv.*gt22;
```

With the approximation of the stress tensor in hand we now return to the task of evaluating the right-hand side for each of the conservation equations. For the discrete density equation the right-hand side is evaluated as

$$\frac{\partial \rho}{\partial t} = -\mathcal{G}^x\left(\rho u\right) - \mathcal{G}^y\left(\rho v\right). \tag{7.28}$$

Note that we require the samples of ρu_h and ρv_h at the Gauss face nodes, cubature volume nodes, and Gauss boundary face nodes to perform the DG differentiation as follows:

--------------------------------- CurvedCNSRHS2D.m ---------------------------------

```
100  % Local copy of normals at Gauss nodes
101  gnx = gauss.nx; gny = gauss.ny;
102
```

```
103   % Add mass conservation terms together and compute divergence
104   cF = -crhou;   cG = -crhov;
105   gF = -grhou;   gG = -grhov;
106   bF = -brhou;   bG = -brhov;
107   rhsQ(:,:,1) = CurvedDGDiv2D(cF, cG, gF, gG, ...
108                   gmapB, gnx.*bF+gny.*bG);
```

For the x-momentum equation we again rely on the generic CurvedDG-Div2D.m to evaluate the derivatives using central fluxes as

$$\frac{\partial \rho u}{\partial t} = \mathcal{G}^x \left(-\left(\rho u^2 + p \right) + \tau_{11} \right) + \mathcal{G}^y \left(-\left(\rho u v \right) + \tau_{12} \right). \qquad (7.29)$$

———————————— CurvedCNSRHS2D.m ————————————

```
110   % Add x-momentum conservation terms together and compute
111   % divergence
112   cF = -(crhou.*cu + cPr) + ct11;  cG = -(crhou.*cv) + ct12;
113   gF = -(grhou.*gu + gPr) + gt11;  gG = -(grhou.*gv) + gt12;
114   bF = -(brhou.*bu + bPr) + gt11;  bG = -(brhou.*bv) + gt12;
115   rhsQ(:,:,2) = CurvedDGDiv2D(cF, cG, gF, gG, ...
116                   gmapB, gnx.*bF+gny.*bG);
```

Similarly for the y-momentum equation we evaluate the right-hand side of

$$\frac{\partial \rho v}{\partial t} = \mathcal{G}^x \left(-\left(\rho u v \right) + \tau_{21} \right) + \mathcal{G}^y \left(-\left(\rho v^2 + p \right) + \tau_{22} \right). \qquad (7.30)$$

———————————— CurvedCNSRHS2D.m ————————————

```
118   % Add y-momentum conservation terms together and compute
119   % divergence
120   cF = -(crhou.*cv) + ct12;  cG = -(crhov.*cv + cPr) + ct22;
121   gF = -(grhou.*gv) + gt12;  gG = -(grhov.*gv + gPr) + gt22;
122   bF = -(brhou.*bv) + gt12;  bG = -(brhov.*bv + bPr) + gt22;
123   rhsQ(:,:,3) = CurvedDGDiv2D(cF, cG, gF, gG, ...
124                   gmapB, gnx.*bF+gny.*bG);
```

Finally, for the energy equation, the right-hand side is evaluated as

$$\frac{\partial E}{\partial t} = \mathcal{G}_x \left(-u \left(E + p \right) + \tau_{31} \right) + \mathcal{G}_y \left(-v \left(E + p \right) + \tau_{32} \right). \qquad (7.31)$$

———————————— CurvedCNSRHS2D.m ————————————

```
126   % Add Energy conservation terms together and compute divergence
127   cF = cu.*(-cEner - cPr) + ct31;  cG = cv.*(-cEner - cPr)  + ct32;
128   gF = gu.*(-gEner - gPr) + gt31;  gG = gv.*(-gEner - gPr)  + gt32;
```

```
129   bF = bu.*(-bEner - bPr) + gt31; bG = bv.*(-bEner - bPr)  + gt32;
130   rhsQ(:,:,4) = CurvedDGDiv2D(cF, cG, gF, gG, ...
131                 gmapB, gnx.*bF+gny.*bG);
```

Thus far we have used central fluxes in the evaluation of all derivatives. To maintain stability for the nonlinear terms, we employ Lax-Friedrichs fluxes as for the Euler equations. Using this approach (see Section 6.6 for details), we achieve the following:

```
                    ┌─────────────────┐
──────────────────── CurvedCNSRHS2D.m ────────────────────
                    └─────────────────┘
135   % Add Lax-Friedrichs jump stabilization
136   glambda = sqrt(gu.^2 + gv.^2) + sqrt(abs(gamma*gPr./grho));
137   glambda = max(glambda(gauss.mapM), glambda(gauss.mapP));
138   glambda = reshape(glambda, gauss.NGauss, Nfaces*K);
139   glambda = ones(gauss.NGauss, 1)*max(glambda, [], 1);
140   glambda = reshape(glambda, gauss.NGauss*Nfaces, K);
141
142   rhsQ(:,:,1) = rhsQ(:,:,1) ...
143       + CurvedDGJump2D(glambda.*grho,  gmapB, glambda.*brho)/2;
144   rhsQ(:,:,2) = rhsQ(:,:,2) ...
145       + CurvedDGJump2D(glambda.*grhou, gmapB, glambda.*brhou)/2;
146   rhsQ(:,:,3) = rhsQ(:,:,3) ...
147       + CurvedDGJump2D(glambda.*grhov, gmapB, glambda.*brhov)/2;
148   rhsQ(:,:,4) = rhsQ(:,:,4) ...
149       + CurvedDGJump2D(glambda.*gEner, gmapB, glambda.*bEner)/2;
```

As this completes the evaluation of the right-hand sides for the four conserved variables, it is an opportune moment to discuss some negative aspects of this formulation. The timestep restriction scales asymptotically as

$$\Delta t \approx \frac{h}{N^2} \frac{C}{(|u| + |c|) + \frac{N^2 \mu}{h}} \propto \frac{C \Delta x}{(|u| + |c|) + \mu (\Delta x)^{-1}}, \qquad (7.32)$$

where c is the local speed of sound and C is a constant depending on the timestepping method being used. The observant reader will notice that it is likely that the maximum allowable timestep will be impractically small for a high-order ($N \gg 1$) computation. One might think that this will not be an issue for a low viscosity case ($\mu \ll 1$) but in that case thin boundary layers or shear layers will form that require small elements and/or high-order approximations. In the DG formulation for the incompressible Navier-Stokes discussed in Section 7.4, this problem was avoided by an implicit treatment of the diffusion operator in the momentum equations. However, in this instance, the diffusion terms are actually nonlinear in the momentum and density variables and so not tractable by the same approach.

7.5.3 A few test cases

In the following we will discuss a few simple test cases to confirm the correct behavior of the proposed algorithm. It should be noted that these tests are not exhaustive due to the cost of solving a problem as complex as the compressible Navier-Stokes equations. However, all results indicate excellent behavior and robust convergence rates for smooth problems.

Test Case I: Steady shear flow

As a very simple steady-state test to evaluate the generic solver strategy we consider the following shear flow which is an exact solution for our model of the compressible Navier-Stokes equations

$$\rho(x, y, t) = 1,$$
$$\rho u(x, y, t) = y^2,$$
$$\rho v(x, y, t) = 0,$$
$$E(x, y, t) = \frac{2\mu x + 10}{\gamma - 1} + \frac{y^4}{2},$$
$$\gamma = \frac{3}{2}, \quad \mu = 0.01.$$

We consider convergence rates for a uniformly refined sequence of meshes run with polynomial order 1 to 4 in Table 7.7, confirming expected convergence rates. Since the solution itself is a global polynomial of maximum order 4 we see that the method exactly resolves the solution with $N = 4$ as expected and achieves optimal order convergence rates for $N < 4$.

Test Case II: A pressure pulse in periodic domain

There are surprisingly few exact, nontrivial, reference solutions for the compressible Navier-Stokes equations to help us determine the performance characteristics and correctness of the solver. We, instead, resort to a classic compromise and compare the results from the DG algorithm with the results from a simple pseudo-spectral Fourier solver for the equations on a periodic domain

Table 7.7. L^2-errors and convergence rates for the x-momentum field u and the energy, E, in the steady shear flow computational experiment with exact solution.

	Error in u				Error in E			
N	h	$h/2$	$h/4$	Rate	h	$h/2$	$h/4$	Rate
1	1.99E-03	5.93E-04	1.60E-04	1.82	5.98E-02	1.77E-02	4.79E-03	1.82
2	4.67E-05	8.35E-06	1.70E-06	2.39	1.03E-03	1.27E-04	1.59E-05	3.01
3	2.00E-06	1.39E-07	9.18E-09	3.88	5.12E-05	3.18E-06	2.02E-07	3.99
4	7.10E-15	2.72E-14	9.85E-14	Converged	9.42E-14	1.54E-13	3.20E-13	Converged

Table 7.8. L^2-errors and convergence rates for the energy field in the pressure pulse computational experiment. The reference data are obtained from 256 modes pseudo-spectral method for compressible Navier-Stokes.

N	h	$h/2$	$h/4$	Rate
1	5.77E-02	2.38E-02	8.98E-03	1.34
2	2.38E-02	4.96E-03	3.55E-04	3.03
3	9.88E-03	5.43E-04	1.78E-05	4.56
4	2.76E-03	5.25E-05	1.56E-06	5.39

[159]. The pseudo-spectral Fourier code is available to the interested reader, starting with SpectralCNSDriver2D.m.

The domain was chosen to be periodic, with dimensions $[0,1) \times [0,1)$ and the fields were initialized with

$$\rho(x,y,t=0) = 1,$$
$$\rho u(x,y,t=0) = 0,$$
$$\rho v(x,y,t=0) = 0,$$
$$E(x,y,t=0) = \frac{12}{\gamma-1} + \frac{\rho}{2}\exp\left(-4\left(\cos(\pi x)^2 + \cos(\pi y)^2\right)\right),$$

which is analogous to an infinite array of pressure pulses in a quiescent fluid. The viscosity parameter is set to $\mu = 0.01$ and the system is integrated until time $t = 0.1$. The pseudo-spectral Fourier solver was used with 256 modes as a reasonable resolution. The results in Table 7.8 show optimal order convergence of the discontinuous Galerkin solution, and these rates were unchanged when comparing against a finer 512 modes pseudo-spectral computation.

In a real sense, this is not a particularly demanding test case. Although it exercises all of the algorithmic components of the solver, excluding the imposition of domain boundary conditions, the simulation does not involve strongly nonlinear effects, with the solution mainly consisting of an acoustic perturbation about a quiescent state.

7.6 Exercises

1. Consider the linear Schrödinger equation

$$\frac{\partial u}{\partial t} = i\frac{\partial^2 u}{\partial x^2}, \quad x = [-1,1].$$

   a) Assume that a unique solution exist and show that the boundary conditions

$$u(\pm 1) = 0,$$

   suffice to guarantee the well-posedness of the problem.

b) Formulate a DG scheme with central fluxes for this problem and prove its stability. Implement the scheme and evaluate it accuracy to estimate the order as h^α. What is α?

c) Formulate a DG scheme with LDG fluxes for this problem and prove its stability. Implement the scheme and evaluate it accuracy to estimate the order as h^α. What is α?

2. Consider the biharmonic problem

$$\frac{\partial u}{\partial t} = -\frac{\partial^4 u}{\partial x^4}, \quad x = [-1, 1].$$

a) Assume that a unique solution exist and show that the boundary conditions

$$u(\pm 1) = \frac{\partial u}{\partial x}(\pm 1) = 0,$$

suffice to guarantee the well-posedness of the problem.

b) Formulate a DG scheme with central fluxes for this problem and prove its stability. Implement the scheme and evaluate its accuracy to estimate the order as h^α. What is α?

c) Formulate a DG scheme with LDG fluxes for this problem and prove its stability. Implement the scheme and evaluate its accuracy to estimate the order as h^α. What is α?

d) If using an explicit timestepping scheme, how does Δt relate to h and N?

3. Consider the dispersive problem

$$\frac{\partial u}{\partial t} = \frac{\partial^5 u}{\partial x^5}, \quad x = [-1, 1].$$

a) Assume that a unique solution exists and derive a set of boundary conditions that guarantees the well-posedness of the problem.

b) Formulate a DG scheme with central fluxes for this problem and prove its stability. Implement the scheme and evaluate its accuracy to estimate the order as h^α. What is α?

c) Formulate a DG scheme with LDG fluxes for this problem and prove its stability. Implement the scheme and evaluate it accuracy to estimate the order as h^α. What is α?

d) If using an explicit time stepping scheme, how does Δt relate to h and N?

4. Consider the Korteveg-de Vries equation

$$\frac{\partial u}{\partial t} + \frac{\partial u^2}{\partial x} + \sigma \frac{\partial^3 u}{\partial x^3} = 0, \quad x \in [-1, 1].$$

a) What boundary conditions do you expect would be needed for this problem and how would this depend on the sign of σ?

b) Formulate and implement a DG method for this problem and evaluate its accuracy for a constructed solution; that is, choose $u(x,t)$ to be a solution and find a forcing function such that $u(x,t)$ satisfies the modified Korteveg de-Vries equation.

5. Consider the two-dimensional wave equation

$$\frac{\partial^2 u}{\partial t^2} - \nabla u = 0, \quad \boldsymbol{x} \in [-1,1]^2,$$

with $u = 0$ at the boundary.

a) Construct and implement a two-dimensional DG scheme for solving this equation with initial conditions $u(\boldsymbol{x},0) = f(\boldsymbol{x})$. Discuss in particular the choice of the numerical flux.

b) Validate the accuracy of the scheme by considering an exact solution of the form

$$u(x,y,t) = \cos(\omega t)\sin(\pi k x)\sin(\pi l y),$$

where the mode number is (k,l). What is ω? Do you see optimal convergence for all values of N, and if not, why?

6. Consider the two-dimensional wave equation

$$\frac{\partial^2 u}{\partial t^2} - \nabla u = f(\boldsymbol{x}), \boldsymbol{x} \in [-1,1]^2,$$

with $u = 0$ at the boundary and look for solutions of the form

$$u(\boldsymbol{x},t) = \exp(i\omega t)\hat{u}(\boldsymbol{x}),$$

leading to the harmonic Helmholtz equation

$$-\omega^2 \hat{u} - \nabla \hat{u} = f.$$

a) Modify the provided codes; for example, PoissonRHS2D.m, to enable the solution of this problem and test the accuracy of the solver for some chosen right-hand side, f, so the exact solution is known.

b) Compute the solution for different values of ω. What do you observe when ω is increased?

c) Rather than using a direct solver, it is tempting to use an iterative solver as was so successful for the Poisson and Helmholtz equation. Experiment with this – do you observe additional problems with simpler schemes such as conjugate gradients when increasing ω - can you explain why?

d) What about preconditioning?

7. For simplicity, the contribution to the energy conservation law from thermal diffusion was omitted from Eq. (7.27). The task here is to modify CurvedCNSRHS2D.m to include a thermal diffusion term given by

$$\frac{\partial}{\partial x}\left(\kappa\frac{\partial T}{\partial x}\right) + \frac{\partial}{\partial y}\left(\kappa\frac{\partial T}{\partial y}\right),$$

in the right-hand side of the energy conservation law in Eq. (7.27). The variable κ is the thermal conductivity of the fluid, and T is the temperature given by $T = \frac{p}{\rho}$ in the current normalization.

a) Use DG derivatives with central fluxes to evaluate each of the spatial derivatives, applying Dirichlet boundary condition at walls and inflows, and zero Neumann boundary conditions at any outflows. For Dirichlet boundary conditions, assume that the external temperature is given by a reference temperature T_0.

b) Assuming a constant thermal conductivity, note that the steady shear flow (see Section 7.5.3) is also an exact solution for this test case, as a consequence of the pressure being linear in space variables and the density being constant. Verify that your implementation computes the correct steady state solution with and without the thermal diffusion term.

8. In CurvedCNSRHS2D.m it was also assumed that the dynamic viscosity (μ) are constant. In practice, however, the dynamic viscosity actually depends on the temperature of the fluid. A common model for this dependence is given by Sutherland's formula

$$\mu = \mu_0 \frac{T_0 + C}{T + C}\left(\frac{T}{T_0}\right)^{\frac{3}{2}},$$

with μ_0 being the dynamic viscosity at temperature T_0 and C being Sutherland's constant for the fluid/gas in question.

a) Assuming $C = 111$, $T_0 = 273$, and $\mu_0 = 1.716\mathrm{E}{-5}$ for air [324], modify CurvedCNSRHS2D.m to include the variable dynamic viscosity, noting that the viscosity is no longer constant and is required to be evaluated at both the cubature and Gauss integration nodes according to this formula.

b) Add an appropriate forcing function to the right hand side of the full compressible Navier-Stokes equations to ensure that the shear layer test case in Section 7.5.3 is still an exact solution to the equations when using Sutherland's formula for the dynamic viscosity. Verify the generalized solver using this solution and forcing function.

8

Spectral properties of discontinuous Galerkin operators

In the preceding chapters, we have developed schemes and discussed basic properties such as stability and convergence. We have not, however, discussed the spectral properties of the operators apart from the discussion of the numerical dispersion relations in Section 4.6. In that discussion it became clear that the eigenvalues of the operators play a key role when determining the properties of the discrete operators and, thus, the behavior of the scheme.

In the previous discussion of dispersion relations and eigenmodes, we already saw some indications of potential problems due to the introduction of additional unphysical modes in the dispersion diagrams. We refer to such nonphysical modes as spurious modes and, as we will see, they can manifest themselves in various ways. These spurious modes can be nonphysical eigenvalues that change in a nonconvergent way, although generally one would expect convergence to all physical eigenvalues. A different, and particularly difficult, situation arises if there is also convergence to nonphysical eigenvalues.

To illustrate some of these issues, let us consider a classic example, introduced in [289].

Example 8.1. We consider the two-dimensional Maxwell's equations, describing a harmonic wave of frequency ω, on the curl-curl form

$$\nabla \times \nabla \times \hat{E} = \omega^2 \hat{E}, \tag{8.1}$$

where $\hat{E}(x, y)$ is the electric field components in the plane (see Section 8.2 for a derivation). For simplicity, we seek the solution inside a square metallic domain, $\Omega \in [-1, 1]^2$, and require that all tangential components of $\hat{E}$ vanish at the boundary. This produces an eigenvalue problem for which we would expect only discrete solutions. In fact, one easily shows that the exact eigenvalues are given as $\omega^2 = n^2 + m^2$, with (n, m) being integers.

By inspecting the above problem, a number of issues may cause some concern. In particular, we notice that any function of the form $\hat{E} = \nabla \psi$ yields a zero eigenvalue; that is, the nullspace of the curl-curl operator is very large.

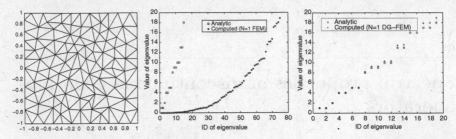

Fig. 8.1. On the left is shown a typical unstructured grid for the perfectly electrically conducting (PEC) metallic $[-1, 1]^2$ square, tiled with 192 elements. In the middle is shown the lower part of the eigenvalue spectrum computed using a continuous second order finite element method. On the right is shown the result of the same computation using a second order accurate discontinuous Galerkin method.

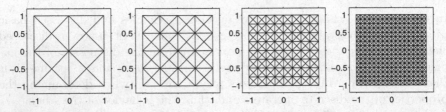

Fig. 8.2. Sequence of cross-hatch grids used to compute the eigenvalues in a PEC square, using h-refinement.

We now solve the eigenvalue problem using a standard second order accurate nodal continuous Galerkin finite element approximation of the curl-curl operator [188, 314]. In Fig. 8.1 we show the unstructured grid as well as the computed eigenvalues, plotted by magnitude. It is evident that the physical eigenvalues are completely hidden in a cloud of spurious small eigenvalues and that there is no easy way to distinguish between physical and nonphysical eigenvalues. This cloud of spurious eigenvalues is associated with the large nullspace of the operator. The approximation perturbs the many zero eigenvalues into a cloud of small nonphysical eigenvalues, as in Fig. 8.1. We also note that Fig. 8.1 indicates no such spurious modes for the discontinuous Galerkin finite element method (DG-FEM), although, as we will see later in this chapter, the general situation is not quite that simple and fortuitous.

The perturbations of the large nullspace in the continuous finite element method can be avoided entirely by using a special, highly structured grid, a sequence of which is illustrated in Fig. 8.2. Unfortunately, this test highlights another type of spurious nonphysical eigenvalue. By inspecting the first eight computed eigenvalues listed in Table 8.1, we notice an eigenvalue apparently converging toward a value of 6.0. As the eigenvalues of this simple test are $n^2 + m^2$, with (n, m) being integers, a value of 6.0 is clearly not possible.

Table 8.1. First eight eigenvalues for the square PEC resonator, computed by solving the curl-curl equation on a sequence of cross-hatch grids (Fig. 8.2) using a second order continuous Galerkin finite element method.

h	1.00	0.500	0.250	0.125	Exact
	1.0661	1.0170	1.0043	1.0011	1.0000
	1.0661	1.0170	1.0043	1.0011	1.0000
	2.2035	2.0678	2.0171	2.0043	2.0000
	4.8634	4.2647	4.0680	4.0171	4.0000
ω^2	4.8634	4.2647	4.0680	4.0171	4.0000
	6.4846	5.3971	5.1063	5.0267	5.0000
	6.4846	5.3971	5.1063	5.0267	5.0000
	6.1338	5.6712	5.9229	5.9807	—
	11.092	8.8141	8.2713	8.0685	8.0000

This kind of spurious eigenvalue is potentially more troublesome than those caused by the perturbation of the nullspace, as it appears to be a convergent eigenvalue; that is, there is no easy way to distinguish this spurious eigenvalue from the remaining physical ones.

An intuitive understanding of the source of this spurious eigenvalue has been offered [29, 30] and can be traced to the nonmonotonic nature of the numerical dispersion properties of the continuous finite element method when this is used to approximate first-order derivatives, as happens in the curl-curl operator.

In light of the above example, it is clear that a number of questions need to be addressed to understand whether a discrete approximation is also spectrally correct. Following [95, 96], the following issues are central to such an inquiry into the problem of spurious modes and spectrally correct approximations:

- *Completeness of the spectrum.* This is to guarantee that all eigenvalues, below a given upper bound, of the continuous operator can be represented by the discrete operator, provided the resolution is sufficient.
- *Isolation of the discrete kernel modes.* This addresses the pollution problem shown in Fig. 8.1 by guaranteeing that there is no overlap between the eigenvalues of the discrete kernel modes and the physical eigenvalues.
- *Nonpollution of the spectrum.* This addresses the example shown in Table 8.1 by disallowing convergent discrete eigenvalues.
- *Nonpollution and completeness of the eigenvectors.* This reflects a similar set of requirements for the eigenvectors as for the eigenvalues.

In this chapter we will dig a little deeper into these issues and discuss the spectral properties of some of the discrete differential operators constructed using the DG-FEM's. We will in particular highlight the issues associated with spurious modes and the impact on the spectral properties of the various choices made when formulating the DG-FEM. To complement the discussion in Section 4.6, we primarily discuss properties of global second order operators (e.g., the Laplacian and the curl-curl operator associated with electromagnetics). We conclude, however, with a return to the time-dependent problem to illustrate the relevance of these considerations for such problems also.

8.1 The Laplace eigenvalue problem

We will begin by discussing how to formulate the Laplace eigenvalue problem using the DG discretization introduced in Section 7.2. The eigenvalue problem can be succinctly stated as finding a eigenpair (u, λ) that satisfies

$$-\nabla^2 u = -\frac{\partial^2 u}{\partial x^2} - \frac{\partial^2 u}{\partial y^2} = \lambda u, \quad \boldsymbol{x} \in \Omega, \tag{8.2}$$

$$u = 0, \quad \boldsymbol{x} \in \partial\Omega.$$

Since the operator is self-adjoint, all eigenvalues are real. Furthermore, the Laplacian is coercive, guaranteeing that all eigenvalues in Eq. (8.2) are positive.

For the square domain $[-1, 1]^2$, the eigenpairs of Eq. (8.2) are found as

$$u(x, y) = \sin\left(\frac{\pi n (x+1)}{2}\right) \sin\left(\frac{\pi m (y+1)}{2}\right), \quad (n, m) = 1, 2, 3, 4, 5 \ldots, \tag{8.3}$$

$$\lambda = \frac{\pi^2 (n^2 + m^2)}{4}.$$

Clearly, we should expect that the eigenvalues and eigenvectors of the discrete DG operator should approximate the above continuous eigenpairs. Furthermore, with a sequence of refined meshes we also expect that the lower part of the discrete spectrum should only contain eigenvalues close to the exact ones and they should have the correct multiplicity. However, as we have seen in the previous example, these properties are not guaranteed.

Following the notation of Section 7.2, we write the operator as

$$-\nabla \cdot \boldsymbol{q} = \lambda u, \quad \nabla u = \boldsymbol{q}.$$

We now discretize this by assuming that $(u, \boldsymbol{q}) = (u, q^x, q^y)$ are approximated by piecewise N-th-order polynomials, $(u_h, q_h^x, q_h^y) \in V_h$, on each triangle and seek a solution to the variational statement

$$-(\nabla \cdot \boldsymbol{q}, \phi_h)_{\Omega,h} + \sum_{k=1}^{K} (\hat{\boldsymbol{n}} \cdot (\boldsymbol{q}_h - \boldsymbol{q}_h^*), \phi_h)_{\partial \mathsf{D}^k} = -\lambda (u_h, \phi_h)_{\Omega,h},$$

$$(\boldsymbol{q}_h, \boldsymbol{\pi}_h)_{\Omega,h} = (\nabla u_h, \boldsymbol{\pi}_h)_{\Omega,h} - \sum_{k=1}^{K} (\hat{\boldsymbol{n}} (u_h - u_h^*), \boldsymbol{\pi}_h)_{\partial \mathsf{D}^k},$$

for all $\phi_h \in V_h$ and $\boldsymbol{\pi}_h \in V_h \times V_h$. As usual, we define V_h as the space of piecewise polynomials of order N defined on Ω_h. From this, we recover the elementwise scheme

$$-\mathcal{S}_x q_h^x - \mathcal{S}_y q_h^y + \int_{\partial D^k} \hat{\boldsymbol{n}} \cdot [(q_h^x, q_h^y) - q_h^*] \, \boldsymbol{\ell}(\boldsymbol{x}) \, d\boldsymbol{x} = -\lambda \mathcal{M} u_h,$$

$$\mathcal{M} q_h^x = \mathcal{S}_x u_h - \int_{\partial D^k} \hat{n}_x \, [u_h - u_h^*] \, \boldsymbol{\ell}(\boldsymbol{x}) \, d\boldsymbol{x},$$

$$\mathcal{M} q_h^y = \mathcal{S}_x u_h - \int_{\partial D^k} \hat{n}_y \, [u_h - u_h^*] \, \boldsymbol{\ell}(\boldsymbol{x}) \, d\boldsymbol{x}.$$

For simplicity, we have removed the reference to each local element k. We first consider the spectral behavior of the scheme based on the stabilized central flux already discussed in Section 7.2. In this case, the fluxes are given as

$$q^* = \{\!\{q\}\!\} - \sigma[\![u]\!], \quad u^* = \{\!\{u\}\!\},$$

where we must take $\sigma > 0$ to guarantee stability. In general, we use

$$\sigma = \tau \frac{(N+1)^2}{h}.$$

where $\tau > 0$.

We express the discrete eigenvalue problem as

$$\mathcal{L} u_h = (\mathcal{A} - \sigma \mathcal{I}) \, u_h = \lambda \mathcal{M} u_h.$$

Here, $\mathcal{M}$ is the standard block-diagonal mass matrix, while $\mathcal{A}$ and $\mathcal{I}$ represent the nonstabilized discrete operator and the stabilization operator, respectively. To get a sense of the sparsity patterns with the stabilized central flux, we show these patterns for the two operators in Fig. 8.3 for the case of $K = 16$ and $N = 4$ on a hash grid as in Fig. 8.2.

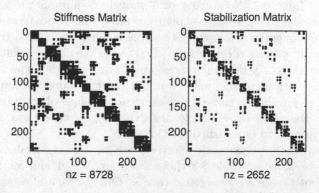

Fig. 8.3. Sparsity patterns for the discrete Laplace stiffness matrix (left) and stabilization matrix (right) using a central flux.

Table 8.2. 10 smallest eigenvalues of the stabilized central flux discrete Laplace operator on the hash mesh in Fig. 8.2.

(n,m)	$\frac{4\lambda_{n,m}}{\pi^2}$	Approximation	Error
$(1, 1)$	2	2.0000E+00	1.58E-06
$(2, 1)$	5	5.0002E+00	1.89E-04
$(1, 2)$	5	5.0002E+00	1.89E-04
$(2, 2)$	8	8.0031E+00	3.09E-03
$(1, 3)$	10	1.0010E+01	1.01E-02
$(3, 1)$	10	1.0010E+01	1.01E-02
$(2, 3)$	13	1.3017E+01	1.71E-02
$(3, 2)$	13	1.3017E+01	1.71E-02
$(1, 4)$	17	1.7186E+01	1.86E-01
$(4, 1)$	17	1.7186E+01	1.86E-01

Table 8.3. h-Convergence study for the 10 smallest discrete eigenvalues of the Laplace problem using a stabilized central flux. The domain is the biunit square with Dirichlet data.

Mode	$N = 1$					$N = 3$				
	h	$h/2$	$h/4$	$h/8$	Rate	h	$h/2$	$h/4$	$h/8$	Rate
(1,1)	2.78E-01	7.96E-02	2.12E-02	5.46E-03	1.94	1.66E-04	3.27E-06	5.59E-08	9.07E-10	5.94
(1,2)	2.92E+00	6.95E-01	1.78E-01	4.52E-02	1.97	1.15E-02	2.06E-04	3.46E-06	5.56E-08	5.96
(2,1)	2.92E+00	6.95E-01	1.78E-01	4.52E-02	1.97	1.15E-02	2.06E-04	3.46E-06	5.56E-08	5.96
(2,2)	1.72E+00	1.21E+00	3.34E-01	8.69E-02	1.92	4.45E-03	7.65E-04	1.40E-05	2.31E-07	5.92
(1,3)	5.16E+00	2.26E+00	6.16E-01	1.60E-01	1.93	1.32E-01	3.21E-03	5.74E-05	9.37E-07	5.94
(3,1)	1.50E+01	2.34E+00	6.27E-01	1.61E-01	1.95	1.36E-01	3.26E-03	5.77E-05	9.39E-07	5.94
(2,3)	1.55E+01	2.83E+00	7.96E-01	2.07E-01	1.93	2.04E-01	5.51E-03	1.05E-04	1.75E-06	5.90
(3,2)	1.55E+01	2.83E+00	7.96E-01	2.07E-01	1.93	2.04E-01	5.51E-03	1.05E-04	1.75E-06	5.90
(1,4)	2.39E+01	6.18E+00	1.82E+00	4.78E-01	1.90	4.86E-01	2.70E-02	5.10E-04	8.43E-06	5.92
(4,1)	2.40E+01	6.18E+00	1.82E+00	4.78E-01	1.90	4.860E-01	2.70E-02	5.10E-04	8.43E-06	5.92

We compute the 10 smallest eigenvalues on the first mesh in Fig. 8.2 and list them in Table 8.2. No sorting or removal of spurious modes has been done; that is, the 10 eigenvalues shown are really the 10 smallest eigenvalues of the discrete operator. Compared to these eigenvalues, scaled by $4/\pi^2$, are the analytically predicted eigenvalues, Eq. (8.3). We observe that the lowest mode is correct to six decimal places and, as one might expect, the error grows quickly with increasing mode number. Furthermore, the multiplicity of the discrete eigenvalues is correct and there are no signs of additional discrete eigenvalues. Based on these simple observations, all conditions for spectral correctness appear to be satisfied for this case.

To consider the asymptotic convergence of the eigenvalues under h-refinement, we show in Table 8.3 the results for $N = 1$ and for $N = 3$. We observe in both cases a convergence rate of $\mathcal{O}(h^{2N})$ for the eigenvalues. For the higher values of (n, m) we see a slow deterioration of the convergence rate due to the low accuracy of the approximation; that is, this is simply due the

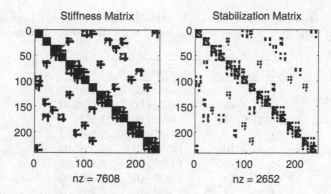

Fig. 8.4. Sparsity patterns for the discrete Laplace stiffness matrix (left) and stabilization matrix (right) using an internal penalty flux.

eigensolutions not being sufficiently well resolved to exhibit the asymptotic convergence rates.

To appreciate the possible impact of changing the flux, let us consider the same problem but with an interior penalty DG (IPDG) formulation. As discussed in Section 7.2 this amounts to choosing a numerical flux of the form

$$q^* = \{\{\nabla u\}\} - \sigma[\![u]\!], \ \ u^* = \{\{u\}\},$$

and we must take $\sigma > 0$ to guarantee stability. In general, we again take it to be

$$\sigma = \tau \frac{(N+1)^2}{h}.$$

In contrast to the central flux case, τ now has a lower limit, as we also discussed in Section 7.2. In Fig. 8.4 we show the sparsity pattern for $K = 16$, $N = 4$. This should be contrasted with Fig. 8.3. As expected, the stiffness matrix, $\mathcal{A}$, is sparser for the IPDG flux than for the central flux.

Repeating the computation of the 10 smallest eigenvalues yields results very similar to those of the stabilized central flux, as shown in Table 8.4. Again, there are no signs of spurious eigenvalues and all the physical eigenvalues appear with correct multiplicity and appear to converge. Computations of the actual convergence rates for the eigenvalues closely mimics those of the stabilized central flux scheme and are shown in Table 8.3.

The results above indicate rapid convergence of all eigenvalues at order $\mathcal{O}(h^{2N})$ with no signs of spurious eigenvalues of the kind discussed previously. This has been confirmed rigorously in [11] through the following theorem:

Theorem 8.2. *Consider a discontinuous Galerkin approximation of the Laplacian where the grid size is h and the order of approximation is N. Then for h being small enough, there exists an upper bound L such that*

$$\sup_{1 \leq i \leq L} |\lambda^i - \lambda_h^i| \leq C h^\sigma,$$

Table 8.4. 10 smallest eigenvalues of the stabilized IPDG discrete Laplace operator on the 16 triangle hash mesh in Fig. 8.2 using a fourth order polynomial representation.

(n,m)	$\frac{4\lambda_{n,m}}{\pi^2}$	Approximation	Error
$(1, 1)$	2	2.00E+00	1.46E-06
$(2, 1)$	5	5.00E+00	1.75E-04
$(1, 2)$	5	5.00E+00	1.75E-04
$(2, 2)$	8	8.00E+00	2.99E-03
$(1, 3)$	10	1.00E+01	9.57E-03
$(3, 1)$	10	1.00E+01	9.92E-03
$(2, 3)$	13	1.30E+01	1.73E-02
$(3, 2)$	13	1.30E+01	1.73E-02
$(1, 4)$	17	1.72E+01	1.86E-01
$(4, 1)$	17	1.72E+01	1.86E-01

where $\sigma = \min(N,p)$ and the eigenvector associated to λ^i is $u^i \in H^{p+1}(\Omega)$.

If the discrete discontinuous Galerkin approximation is Hermitian this improves as

$$\sup_{1 \leq i \leq L} |\lambda^i - \lambda_h^i| \leq Ch^{2\sigma}.$$

In [11] it is furthermore shown that there are no spurious eigenvalues and that the kernel is isolated. All of the schemes we have considered, both here and in Section 7.2, are symmetric/Hermitian. However, the above result is more general in that it also includes a number of other formulations, leading to nonsymmetric operators. The analysis confirms that the Hermitian form is a preferred choice.

A similar result can be obtained for the eigenfunctions [11]:

Theorem 8.3. *Consider a discontinuous Galerkin approximation of the Laplacian where the grid size of h and the order of approximation is N. Then for h being small enough, there exists an upper bound L such that there is convergence for the eigenvectors, u^i, as*

$$\sup_{1 \leq i \leq L} |u^i - u_h^i| \leq Ch^{\sigma},$$

where $\sigma = \min(N,p)$ and $u^i \in H^{p+1}(\Omega)$.

Numerical evidence in [11] indicates that these results hold also for nonconforming grids.

8.1.1 Impact of the penalty parameter on the spectrum

The above experiments and the associated theoretical experiments have all been done under the assumption that the stabilization parameter, τ, is sufficiently large. However, similar to what we did in Section 7.2, it is worthwhile

to consider the impact of the penalty parameter on the spectral properties of
the scheme.

For the Poisson problem discussed in Section 7.2, the role of this parameter
is clear. The stabilization term is used to ensure that the bilinear form is
coercive and the linear system is uniquely solvable with the given boundary
and forcing data. However, for the eigenvalue problem, we are seeking a set
of correct eigenmodes and the role of the penalty is less well defined. This is
because an eigenfunction obtained for this problem may not necessarily live
in the correct energy space for this problem [i.e $H^1(\Omega)$], since the eigenspaces
for the DG eigenfunctions are represented in a broken polynomial space.

To draw comparisons with a traditional $H_0^1(\Omega)$ conforming finite element
method, we partition the approximation space, V_h, into

$$\mathsf{X}_{h,0} = \left\{ u_h \in H_0^1(\Omega) \middle|\ u_h \in \bigoplus_{k=1}^{K} \mathsf{P}_N\left(\mathsf{D}^k\right) \right\}, \quad N^C = \dim \mathsf{X}_{h,0},$$

recognized as the space of functions that would be appropriate for a continuous
finite element method and

$$\mathsf{Y}_h = \left\{ u_h \in \bigoplus_{k=1}^{K} \mathsf{P}_N\left(\mathsf{D}^k\right) \middle|\ (u_h, \phi_h)_{\Omega,h} = 0, \ \forall \phi_h \in \mathsf{X}_{h,0} \right\}, \quad N^D = \dim \mathsf{Y}_h,$$

that contains the additional freedom associated with the DG formulation. We
recall that $\mathsf{P}_N(\mathsf{D}^k)$ represents the space of polynomials up to order N defined
on D^k.

We decompose the approximation of the solution, by orthogonal projection, into two components

$$u_h = u_h^C + u_h^D,$$

where $u_h^C \in \mathsf{X}_{h,0}$ and $u_h^D \in \mathsf{Y}_h$. The superscripts C and D denote the continuous and discontinuous parts of the vector obtained by L^2-projection. For con-
venience, we choose suitable L^2-orthonormal bases for $\mathsf{X}_{h,0} = \{\psi_i^C\}_{1\leq i \leq N^C}$
and the complement $\mathsf{Y}_h = \{\psi_i^D\}_{1\leq i \leq N^D}$.

Now write the generic eigenvalue problem as

$$\begin{pmatrix} \mathcal{A} & \mathcal{B} \\ \mathcal{B}^T & \mathcal{C} + \sigma\mathcal{I} \end{pmatrix} \begin{pmatrix} \hat{u}^C \\ \hat{u}^D \end{pmatrix} = \lambda \begin{pmatrix} \hat{u}^C \\ \hat{u}^D \end{pmatrix}, \tag{8.4}$$

The entries of the vector $\hat{u}^C$ are coefficients of the basis vectors for $\mathsf{X}_{h,0}$,
while those of $\hat{u}^D$ are coefficients of a basis for the complement Y_h. Here,
$\mathcal{A}$ reflects the coupling between the continuous modes, $\mathcal{B}$ the coupling between the continuous and discontinuous modes, and, finally, $\mathcal{C}$ and $\mathcal{I}$, the self-
coupling between the discontinuous modes and the stabilization. The matrix
$\mathcal{A}$ is symmetric positive-semidefinite and the matrices $\mathcal{C}$ and $\mathcal{I}$ are symmetric
positive-definite.

We denote by $\mathcal{A}(\sigma)$ the matrix in the eigenvalue problem, Eq. (8.4), and
use the notation

$$\lambda_1 \leq \cdots \leq \lambda_{N^C + N^D} : \text{eigenvalues of } \mathcal{A}(\sigma);$$
$$\alpha_1 \leq \cdots \leq \alpha_{N^C} : \text{eigenvalues of } \mathcal{A};$$
$$\delta_1 \leq \cdots \leq \delta_{N^D} : \text{eigenvalues of } \mathcal{I}.$$

Since $\mathcal{A}$ and $\mathcal{I}$ are symmetric, they admit diagonalizations $\mathcal{U}^T \mathcal{A} \mathcal{U} = \Sigma = \text{diag}(\alpha_1, \ldots, \alpha_{N^C})$ and $\mathcal{Q}^T \mathcal{I} \mathcal{Q} = \Delta = \text{diag}(\delta_1, \ldots, \delta_{N^D})$, where $\mathcal{U}$ and $\mathcal{Q}$ are unitary matrices. The eigenvalues of $\mathcal{A}(\sigma)$ are preserved by the similarity transformation

$$\begin{pmatrix} \mathcal{U}^T & 0 \\ 0 & \mathcal{Q}^T \end{pmatrix} \begin{pmatrix} \mathcal{A} & \mathcal{B} \\ \mathcal{B}^t & \mathcal{C} + \sigma \mathcal{I} \end{pmatrix} \begin{pmatrix} \mathcal{U} & 0 \\ 0 & \mathcal{Q} \end{pmatrix} = \begin{pmatrix} \Sigma & \mathcal{U}^T \mathcal{B} \mathcal{Q} \\ \mathcal{Q}^T \mathcal{B}^T \mathcal{U} & \mathcal{Q}^T \mathcal{C} \mathcal{Q} + \sigma \Delta \end{pmatrix}. \tag{8.5}$$

Because σ only appears in the diagonal entries of Eq. (8.5), we can apply the Gerschgorin theorem from linear algebra to realize that the first N^C rows Eq. (8.5) yield N^C intervals whose centers, the eigenvalues of $\mathcal{A}$, are independent of σ, while the remaining N^D intervals have centers of the form $\gamma_j + \sigma \delta_j$ for $j = 1, \ldots, N^D$. Since $\mathcal{I}$ is positive-definite, $\delta_j > 0$ for all j, and thus these intervals diverge with σ. Thus, the spectrum of $\mathcal{A}(\sigma)$ splits into a group of N^C eigenvalues bounded independent of σ, and N^D that diverge as $\sigma \to \infty$.

Consider the partial diagonalization of $\mathcal{A}(\sigma)$ corresponding to the smallest N^C eigenvalues:

$$\begin{pmatrix} \mathcal{A} & \mathcal{B} \\ \mathcal{B}^T & \mathcal{C} + \sigma \mathcal{I} \end{pmatrix} \begin{pmatrix} \mathcal{V} \\ \mathcal{W} \end{pmatrix} = \begin{pmatrix} \mathcal{V} \\ \mathcal{W} \end{pmatrix} (\Lambda_1), \tag{8.6}$$

where $\Lambda_1 = \text{diag}(\lambda_1, \ldots, \lambda_{N^C})$, $\mathcal{V}$ is an $N^C \times N^C$-matrix, and $\mathcal{W}$ is an $N^D \times N^C$-matrix. These satisfy the orthogonality constraint $\mathcal{V}^T \mathcal{V} + \mathcal{W}^T \mathcal{W} = \mathcal{I}$. The three matrices Λ_1, $\mathcal{V}$, and $\mathcal{W}$ all vary with σ. The second row of the diagonalization, Eq. (8.6), implies that $\sigma \mathcal{I} \mathcal{W} = \mathcal{W} \Lambda_1 - \mathcal{C} \mathcal{W} - \mathcal{B}^T \mathcal{V}$. The norm of this quantity can be bounded independent of σ: $\sigma \|\mathcal{I} \mathcal{W}\| \leq \|\Lambda_1\| + \|\mathcal{C}\| + \|\mathcal{B}^T\|$, since the orthogonality condition ensures that $\|\mathcal{V}\|, \|\mathcal{W}\| \leq 1$ and Gerschgorin's theorem implies the boundedness of $\|\Lambda_1\|$. Thus as $\sigma \to \infty$, $\|\mathcal{I} \mathcal{W}\| \to 0$. Since $\mathcal{I}$ is symmetric positive-definite, $\delta_1 \|\mathcal{W}\| \leq \|\mathcal{I} \mathcal{W}\| \leq \delta_{N^D} \|\mathcal{W}\|$, we conclude that $\|\mathcal{W}\| = O(1/\sigma)$. Hence, $\|\mathcal{V}^T \mathcal{V} - \mathcal{I}\| = O(1/\sigma^2)$ as $\sigma \to \infty$, and from the first row of the diagonalization, Eq. (8.6), we observe that

$$\|\mathcal{A} \mathcal{V} - \mathcal{V} \Lambda_1\| = \|\mathcal{B} \mathcal{W}\| = O(1/\sigma).$$

If $\mathbf{v}_j$ denotes the j-th column of $\mathcal{V}$, then $\|\mathcal{A} \mathbf{v}_j - \lambda_j \mathbf{v}_j\| = O(1/\sigma)$. Thus, the bounded eigenvalues of $\mathcal{A}(\sigma)$ must converge to those of $\mathcal{A}$, and since $\mathcal{A}$ is symmetric, we recall the error estimate $|\alpha_j - \lambda_j| \leq \|\mathcal{A} \mathbf{v}_j - \lambda_j \mathbf{v}_j\| / \|\mathbf{v}_j\| = O(1/\sigma)$ for $j = 1, \ldots, N^C$. These findings are summarized in the following theorem.

Theorem 8.4. *As $\sigma \to \infty$, the spectrum of the symmetric discontinuous Galerkin discretization operator $\mathcal{A}(\sigma)$ decouples into a set of N^C eigenvalues that are bounded independent of σ, and N^D remaining eigenvalues that* *diverge as $\sigma \to \infty$. The bounded eigenvalues $\{\lambda_1, \ldots, \lambda_{NC}\}$ converge to the eigenvalues $\{\alpha_1, \ldots, \alpha_{NC}\}$ of $\mathcal{A}$ with error $|\alpha_j - \lambda_j| = O(1/\sigma)$ as $\sigma \to \infty$.*

This theorem indicates that we will always be able to choose a σ large enough to clean the lower spectrum. It also highlights that if the underlying continuous component of the discretization has spectral problems, so will the richer DG formulation. It is essential to appreciate that this is not special to the Laplacian problem but is a general property.

However, the result does not directly reveal how large the parameter must be to guarantee the desired decoupling. From Gerschgorin's theorem we expect that the stabilization term will elevate the spurious spectrum by approximately $\sigma \delta_1$, and hence the smallest eigenvalue of the stabilization matrix will dictate our choice of σ.

To demonstrate the impact of a range of choices for the penalty constant, τ, we show in Fig. 8.5 a parameter study. The spectrum of the stabilized central flux approximation of the Laplacian is shown with $N = 2$ and $K = 16$ elements for increasing values of τ. The results confirm the separation of the two groups of eigenmodes, as discussed above. However, the experiment also shows that the condition for invertibility (i.e., $\tau > 0$), does not appear to suffice to guarantee an unpolluted spectrum. It also highlights that the results in Theorem 8.2 are of an asymptotic nature and for a finite resolution, the situation may be considerably more complex. For further discussions of this matter, we refer to [319].

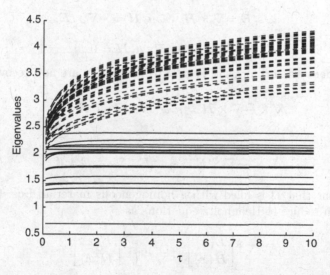

Fig. 8.5. Dependence of the eigenvalue spectrum of the discrete stabilized central flux Laplace operator on the penalty parameter τ. The solid lines represent the physical eigenvalues and the dashed lines mark the unphysical eigenvalues.

8.2 The Maxwell eigenvalue problem

Consider first the normalized time-domain form of Maxwell's equations:

$$\frac{\partial D}{\partial t} = \nabla \times H, \quad \frac{\partial B}{\partial t} = -\nabla \times E, \tag{8.7}$$

$$\nabla \cdot D = 0, \quad \nabla \cdot B = 0, \tag{8.8}$$

within the general charge-free three-dimensional domain, Ω. The electric field, $E(x,t)$, and the electric flux density, $D(x,t)$, as well as the magnetic field, $H(x,t)$, and the magnetic flux density, $B(x,t)$, are related through the simplified constitutive relations

$$D = \varepsilon_r E, \quad B = \mu_r H.$$

Here, $\varepsilon_r(x)$ and $\mu_r(x)$ refer to the relative permittivity and permeability, respectively, of the materials. Most of the results discussed later can be expected to generalize to more complex materials with few changes.

An assumed purely harmonic time variation as

$$\begin{bmatrix} E(x,t) \\ H(x,t) \end{bmatrix} = \exp(i\omega t) \begin{bmatrix} \hat{E}(x) \\ \hat{H}(x) \end{bmatrix}$$

yields the frequency domain form of Eqs. (8.7) and (8.8)

$$i\omega\varepsilon_r \hat{E} = \nabla \times \hat{H}, \quad i\omega\mu_r \hat{H} = -\nabla \times \hat{E}, \tag{8.9}$$

$$\nabla \cdot \varepsilon_r \hat{E} = 0, \quad \nabla \cdot \mu_r \hat{H} = 0. \tag{8.10}$$

Second order curl-curl forms of Eqs. (8.9) and (8.10) are now recovered as

$$\nabla \times \frac{1}{\varepsilon_r} \nabla \times \hat{H} = \mu_r \omega^2 \hat{H}, \quad \nabla \cdot \mu_r \hat{H} = 0, \tag{8.11}$$

and its dual

$$\nabla \times \frac{1}{\mu_r} \nabla \times \hat{E} = \varepsilon_r \omega^2 \hat{E}, \quad \nabla \cdot \hat{\varepsilon}_r \hat{E} = 0. \tag{8.12}$$

In the event that Ω is filled with a homogeneous material, Eqs. (8.11) and (8.12) both reduce to Helmholtz equations as

$$-\nabla^2 \begin{bmatrix} \hat{E}(x) \\ \hat{H}(x) \end{bmatrix} = \omega^2 \mu_r \varepsilon_r \begin{bmatrix} \hat{E}(x) \\ \hat{H}(x) \end{bmatrix}. \tag{8.13}$$

Note that this is identical to the Laplace eigenvalue problem we discussed in Section 8.1 and for which we have a good understanding of the spectral properties. In the following we refer to Eqs. (8.7) and (8.8) as the time-domain form of Maxwell's equations, Eqs. (8.9) and (8.10) as the first order form of

Maxwell's equations in the frequency domain, Eqs. (8.11) and (8.12) as the curl-curl form and Eq. (8.13) as the Helmholtz form.

Regardless of the particular form of the equations, the boundary conditions remain the same. At a point of a perfectly electrically conducting (PEC) wall, endowed with a unique outward pointing vector, $\hat{n}$, the conditions are

$$\hat{n} \times \boldsymbol{E} = 0, \quad \hat{n} \cdot \boldsymbol{H} = 0. \tag{8.14}$$

Alternatively, using Eq. (8.7) in combination with the first boundary condition yields the equivalent condition

$$(\hat{n} \cdot \nabla) \, \boldsymbol{H} = 0. \tag{8.15}$$

We will refer to the former as the Dirichlet condition and the latter as the Neumann boundary condition.

The relations between the field components in adjacent materials are

$$\hat{n} \times [\boldsymbol{E}] = 0, \quad \hat{n} \times [\boldsymbol{H}] = 0, \tag{8.16}$$

where

$$[\boldsymbol{u}] = \boldsymbol{u}^- - \boldsymbol{u}^+ \tag{8.17}$$

represents the jump in field value across the interface, with $\boldsymbol{u}^+$ representing the neighboring field value and $\boldsymbol{u}^-$ representing the local field value.

Let us define the two weak curl operators $(\forall \phi \in \bigoplus_k L^2(\mathrm{D}^k))$

$$\left(\tilde{\nabla}_D \times \boldsymbol{\psi}, \boldsymbol{\phi} \right)_{\Omega,h}^a = (\nabla \times \boldsymbol{\psi}, \boldsymbol{\phi})_{\Omega,h} - \frac{1}{2} \sum_{k=1}^{K} \left[\left(\frac{a^+}{\overline{a}} \hat{n} \times [\boldsymbol{\psi}], \boldsymbol{\phi} \right)_{\partial \mathrm{D}^k \backslash \partial\Omega} \right.$$
$$\left. - (\hat{n} \times [\boldsymbol{\psi}], \boldsymbol{\phi})_{\partial \mathrm{D}^k \cup \partial\Omega} \right] \tag{8.18}$$

and

$$\left(\tilde{\nabla}_N \times \boldsymbol{\psi}, \boldsymbol{\phi} \right)_{\Omega,h}^a = (\nabla \times \boldsymbol{\psi}, \boldsymbol{\phi})_{\Omega,h} - \frac{1}{2} \sum_{k=1}^{K} \left(\frac{a^+}{\overline{a}} \hat{n} \times [\boldsymbol{\psi}], \boldsymbol{\phi} \right)_{\partial \mathrm{D}^k \backslash \partial\Omega}, \tag{8.19}$$

reflecting the Dirichlet and Neumann PEC conditions, respectively. The constant a is a place holder, used to specify the materials, as will become clear shortly. In the simplest case of homogeneous materials, we will omit the superscript a, as it takes the value of 1.

Using these, the semidiscrete scheme for solving the time-domain solution, $(\boldsymbol{E}_h, \boldsymbol{H}_h) \in V_h^3$ to Maxwell's equations is given as

$$\frac{d}{dt} (\varepsilon_r \boldsymbol{E}_h, \boldsymbol{\phi}_h)_{\Omega,h} = \left(\tilde{\nabla}_N \times \boldsymbol{H}_h, \boldsymbol{\phi}_h \right)_{\Omega,h}^Z$$
$$+ \frac{\alpha}{2} \sum_{k=1}^{K} \left(\frac{1}{\overline{Z}} \hat{n} \times \hat{n} \times [\boldsymbol{E}_h], \boldsymbol{\phi}_h \right)_{\partial \mathrm{D}^k \backslash \partial\Omega}$$
$$+ \alpha \sum_{k=1}^{K} (Y \hat{n} \times \hat{n} \times [\boldsymbol{E}_h], \boldsymbol{\phi}_h)_{\partial \mathrm{D}^k \cup \partial\Omega}, \tag{8.20}$$

$$\frac{d}{dt}\left(\mu_r \boldsymbol{H}_h, \phi_h\right)_{\Omega,h} = -\left(\tilde{\nabla}_D \times \boldsymbol{E}_h, \phi_h\right)_{\Omega,h}^{Y}$$

$$+\frac{\alpha}{2}\sum_{k=1}^{K}\left(\frac{1}{\bar{Y}}\hat{n}\times\hat{n}\times[\boldsymbol{H}_h],\phi_h\right)_{\partial D^k\backslash\partial\Omega} \qquad (8.21)$$

for all $\phi_h \in \mathsf{V}_h^3$. Here, V_h is defined in the usual fashion as

$$\mathsf{V}_h = \bigoplus_{k=1}^{K} \mathsf{P}_N(\mathsf{D}^k), \quad \mathsf{V}_h^3 := \mathsf{V}_h \times \mathsf{V}_h \times \mathsf{V}_h.$$

Following the discussion in Section 6.5 we use a general flux, with $\alpha = 0$ corresponding to central fluxes and $\alpha = 1$ to upwind fluxes. Let us consider the question of spurious solutions appearing through perturbations of the large nullspace of the curl-curl operator.

As a first step, we show that the discrete approximation of the first-order equations, Eqs. (8.9) and (8.10), and the discrete curl-curl approximation, Eqs. (8.11) and (8.12), have eigensolutions that are closely connected.

Lemma 8.5. *Assume that (ε_r, μ_r) are constant throughout Ω. Then there exists an eigensolution, $[(\hat{\boldsymbol{E}}_h, \hat{\boldsymbol{H}}_h), \omega]$, $\omega \neq 0$, to the first order problem*

$$\left(\tilde{\nabla}_N \times \hat{\boldsymbol{H}}_h, \phi_h\right)_{\Omega,h} = i\omega\left(\varepsilon_r \hat{\boldsymbol{E}}_h, \phi_h\right)_{\Omega,h} \qquad (8.22)$$

$$\left(\tilde{\nabla}_D \times \hat{\boldsymbol{E}}_h, \phi_h\right)_{\Omega,h} = -i\omega\left(\mu_r \hat{\boldsymbol{H}}_h, \phi_h\right)_{\Omega,h}, \qquad (8.23)$$

for any $\phi_h \in \mathsf{V}_h^3$ if and only if $[\hat{\boldsymbol{H}}_h, \omega^2]$, $\omega \neq 0$, is an eigensolution to

$$\left(\frac{1}{\varepsilon_r}\tilde{\nabla}_N \times \hat{\boldsymbol{H}}_h, \phi_h\right)_{\Omega,h} = \omega^2\left(\mu_r \hat{\boldsymbol{H}}_h, \psi_h\right)_{\Omega,h}, \qquad (8.24)$$

$$(\phi_h, \boldsymbol{q}_h)_{\Omega,h} = \left(\tilde{\nabla}_N \times \psi_h, \boldsymbol{q}_h\right)_{\Omega,h} \qquad (8.25)$$

for all $\psi_h, \boldsymbol{q}_h \in \mathsf{V}_h^3$.
Equivalently, $[\hat{\boldsymbol{E}}_h, \omega^2]$, $\omega \neq 0$, is an eigensolution to

$$\left(\frac{1}{\mu_r}\tilde{\nabla}_D \times \hat{\boldsymbol{E}}_h, \phi_h\right)_{\Omega,h} = \omega^2\left(\varepsilon_r \hat{\boldsymbol{E}}_h, \psi_h\right)_{\Omega,h}, \qquad (8.26)$$

$$(\phi_h, \boldsymbol{q}_h)_{\Omega,h} = \left(\tilde{\nabla}_D \times \psi_h, \boldsymbol{q}_h\right)_{\Omega,h} \qquad (8.27)$$

for all $\psi_h, \boldsymbol{q}_h \in \mathsf{V}_h^3$.

Before we prove this, we need the following result.

Lemma 8.6. *For $i=1,\dots,\ \dim(\Omega)$ and $(\psi,\phi)\in\bigoplus_k L^2(\mathsf{D}^k)$, we have*

$$
(\partial_i\psi,\phi)_{\Omega,h}+\sum_k\left[\left(\frac{a^+}{a^++a^-}\hat n_i[\psi],\phi\right)_{\partial\mathsf{D}^k\backslash\partial\Omega}-(\hat n_i\psi,\phi)_{\partial\mathsf{D}^k\cup\partial\Omega}\right]
$$

$$
=-(\psi,\partial_i\phi)_{\Omega,h}-\sum_k\left[\left(\frac{a^-}{a^++a^-}\hat n_i[\phi],\psi\right)_{\partial\mathsf{D}^k\backslash\partial\Omega}\right],
$$

where a is assumed piecewise smooth.

Proof. The result follows directly by writing out the terms; that is,

$$
\sum_k\left[(\partial_i\psi,\phi)_{\mathsf{D}^k}+\left(\frac{a^+}{a^++a^-}\hat n_i[\psi],\phi\right)_{\partial\mathsf{D}^k\backslash\partial\Omega}-(\hat n_i\psi,\phi)_{\partial\mathsf{D}^k\cup\partial\Omega}\right]
$$

$$
=\sum_k\left[-(\psi,\partial_i\phi)_{\mathsf{D}^k}+\left(\frac{a^+}{a^++a^-}\hat n_i[\psi],\phi\right)_{\partial\mathsf{D}^k\backslash\partial\Omega}+(\hat n_i\psi,\phi)_{\partial\mathsf{D}^k\backslash\partial\Omega}\right]
$$

$$
=\sum_k-(\psi,\partial_i\phi)_{\mathsf{D}^k}+\sum_{\text{Faces}}\left(\frac{a^+}{a^++a^-}\hat n_i\psi^+,\phi^-\right)_{\partial\mathsf{D}^k\backslash\partial\Omega}
$$

$$
-\left(\frac{a^+}{a^++a^-}\hat n_i\psi^-,\phi^-\right)_{\partial\mathsf{D}^k\backslash\partial\Omega}+(\hat n_i\psi^-,\phi^-)_{\partial\mathsf{D}^k\backslash\partial\Omega}
$$

$$
-\left(\frac{a^-}{a^++a^-}\hat n_i\psi^-,\phi^+\right)_{\partial\mathsf{D}^k\backslash\partial\Omega}+\left(\frac{a^-}{a^++a^-}\hat n_i\psi^+,\phi^+\right)_{\partial\mathsf{D}^k\backslash\partial\Omega}
$$

$$
-(\hat n_i\psi^+,\phi^+)_{\partial\mathsf{D}^k\backslash\partial\Omega}
$$

$$
=\sum_k\left[-(\psi,\partial_i\phi)_{\mathsf{D}^k}-\left(\frac{a^+}{a^++a^-}\hat n_i\phi^-+\hat n_i\phi^--\frac{a^-}{a^++a^-}\hat n_i\phi^+,\phi^-\right)_{\partial\mathsf{D}^k\backslash\partial\Omega}\right]
$$

$$
=\sum_k\left[-(\psi,\partial_i\phi)_{\mathsf{D}^k}-\left(\frac{a^-}{a^++a^-}\hat n_i[\phi],\psi\right)_{\partial\mathsf{D}^k\backslash\partial\Omega}\right].
$$

$\square$

We can now return to the proof of the previous result

Proof. Let us define a test function, ϕ_h

$$
(\phi_h,q_h)_{\mathsf{D}^k}=\left(\frac{1}{\varepsilon_r}\tilde\nabla_N\times\psi_h,q_h\right)_{\mathsf{D}^k},
$$

for all $\psi_h,q_h\in\mathsf{V}_h^3$. Abusing the notation slightly since $\tilde\nabla_N\times\psi_N$ is defined in a weak sense only, we obtain from Eq. (8.22)

$$
\left(\frac{1}{\varepsilon_r}\tilde\nabla_N\times\psi_h,\tilde\nabla_N\times\hat H_h\right)_{\Omega,h}=i\omega\left(\frac{1}{\varepsilon_r}\tilde\nabla_N\times\psi_h,\varepsilon_r\hat E_h\right)_{\Omega,h}
$$

$$
=i\omega\left(\psi_h,\tilde\nabla_D\times\hat E_h\right)_{\Omega,h}=\omega^2\left(\psi_h,\mu_r\hat H_h\right)_{\Omega,h}.
$$

The key step here is the nearly self-adjoint nature of the global operator, which is a result of Lemma 8.6 and used in the second step. The reverse follows by assuming that $[\hat{\boldsymbol{H}}_h, \omega]$ satisfy

$$
\left(\tilde{\nabla}_N \times \boldsymbol{\psi}_h, \frac{1}{\varepsilon_r} \tilde{\nabla}_N \times \hat{\boldsymbol{H}}_h \right)_{\Omega,h} = \omega^2 \left(\boldsymbol{\psi}_h, \mu_r \hat{\boldsymbol{H}}_h \right)_{\Omega,h},
$$

for all $\boldsymbol{\psi}_h \in \mathsf{V}_h^3$. We will also define $\boldsymbol{E}_h \in \mathsf{V}_h^3$ as satisfying

$$
i\omega \left(\boldsymbol{\phi}_h, \varepsilon_r \hat{\boldsymbol{E}}_h \right)_{\Omega,h} = \left(\boldsymbol{\phi}_h, \tilde{\nabla}_N \times \hat{\boldsymbol{H}}_h \right)_{\Omega,h}.
$$

Combining these two results yields

$$
\omega^2 \left(\boldsymbol{\psi}_h, \mu_r \hat{\boldsymbol{H}}_h \right)_{\Omega,h} = \left(\tilde{\nabla}_N \times \boldsymbol{\psi}_h, \frac{1}{\varepsilon_r} \tilde{\nabla}_N \times \hat{\boldsymbol{H}}_h \right)_{\Omega,h}
$$

$$
= i\omega \left(\tilde{\nabla}_N \times \boldsymbol{\psi}_h, \hat{\boldsymbol{E}}_h \right)_{\Omega,h} = i\omega \left(\boldsymbol{\psi}_h, \tilde{\nabla}_D \times \hat{\boldsymbol{E}}_h \right)_{\Omega,h}.
$$

Since $\omega \neq 0$, this recovers Eq. (8.23).

An equivalent approach can be used to show that Eqs. (8.22), (8.23), and (8.26) have related eigensolutions for $\omega \neq 0$, thus completing the proof. □

This result has a number of consequences, with the immediate one being that the eigensolutions for homogeneous problems with nonzero eigenvalues are essentially the same, regardless of whether one solves the first-order form, Eqs. (8.9) and (8.10), or the corresponding curl-curl form, Eqs. (8.11) and (8.12), in terms of the spectral properties. This remains true for both the two-dimensional and three-dimensional cases. For the two-dimensional case, however, things can be reduced further.

Lemma 8.7. *There exists an eigensolution, $[(\hat{E}_h^x, \hat{E}_h^y, \hat{H}_h^z), \omega], \omega \neq 0$, to the first order transverse electric(TE)-polarized problem*

$$
i\omega \left(\varepsilon_r \hat{E}_h^x, \phi_h \right)_{\Omega,h} = \left(\frac{\partial \hat{H}_h^z}{\partial y}, \phi_h \right)_{\Omega,h} - \sum_k \left(\hat{n}_y [\hat{H}_h^z], \phi_h \right)_{\partial \mathsf{D}^k},
$$

$$
i\omega \left(\varepsilon_r \hat{E}_h^y, \phi_h \right)_{\Omega,h} = -\left(\frac{\partial \hat{H}_h^z}{\partial x}, \phi_h \right)_{\Omega,h} + \sum_k \left(\hat{n}_x [\hat{H}_h^z], \phi_h \right)_{\partial \mathsf{D}^k},
$$

$$
i\omega \left(\mu_r \hat{H}_h^z, \phi_h \right)_{\Omega,h} = \left(\frac{\partial \hat{E}_h^x}{\partial y} - \frac{\partial \hat{E}_h^y}{\partial x}, \phi_h \right)_{\Omega,h} - \sum_k \left(\hat{n}_y [\hat{E}_h^x] - \hat{n}_x [\hat{E}_h^y], \phi_h \right)_{\partial \mathsf{D}^k},
$$

for all $\phi_h \in \mathsf{V}_h$ if and only if $(\omega, \hat{H}_h^z)$ satisfies the local discontinuous Galerkin form of the Helmholtz problem in the form

$$
\omega^2 \left(\mu_r \hat{H}_h^z, \phi_h \right)_{\Omega,h} = \left(\tilde{\nabla} \cdot \boldsymbol{q}_h, \phi_h \right)_{\Omega,h},
$$

$$
\left(\varepsilon_r \boldsymbol{q}_h, \phi_h \right)_{\Omega,h} = \left(\tilde{\nabla} \hat{H}_h^z, \phi_h \right)_{\Omega,h},
$$

for all $\phi_h \in V_h$. The variable, $q_h \in P_N(D^k)^2$, is defined locally only.

Similarly, there exists an eigensolution, $[(\hat{H}_h^x, \hat{H}_h^y, \hat{E}_h^z), \omega]$, $\omega \neq 0$ to the first-order transverse magnetic(TM)-polarized problem

$$i\omega \left(\mu_r \hat{H}_h^x, \phi_h \right)_{\Omega,h} = -\left(\frac{\partial \hat{E}_h^z}{\partial y}, \phi_h \right)_{\Omega,h} + \sum_k \left(\hat{n}_y [\hat{E}_h^z], \phi_h \right)_{\partial D^k},$$

$$i\omega \left(\mu_r \hat{H}_h^y, \phi_h \right)_{\Omega,h} = \left(\frac{\partial \hat{E}_N^z}{\partial x}, \phi_h \right)_{\Omega,h} - \sum_k \left(\hat{n}_x [\hat{E}_h^z], \phi_h \right)_{\partial D^k},$$

$$i\omega \left(\varepsilon_r \hat{E}_h^z, \phi_h \right)_{\Omega,h} = \left(\frac{\partial \hat{H}_h^y}{\partial x} - \frac{\partial \hat{H}_h^x}{\partial y}, \phi_h \right)_{\Omega,h} - \sum_k \left(\hat{n}_x [\hat{E}_h^y] - \hat{n}_y [\hat{E}_h^x], \phi_h \right)_{\partial D^k},$$

for all $\phi_h \in V_h$ if and only if $(\omega, \hat{E}_h^z)$ satisfies the local discontinuous Galerkin form of the Helmholtz problem on the form

$$\omega^2 \left(\varepsilon_r \hat{E}_h^z, \phi_h \right)_{\Omega,h} = \left(\tilde{\nabla} \cdot q_h, \phi_h \right)_{\Omega,h},$$

$$(\mu_r q_h, \phi_h)_{\Omega,h} = \left(\tilde{\nabla} \hat{E}_h^z, \phi_h \right)_{\Omega,h},$$

for all $\phi_h \in V_h$.

Proof. The result follows directly by realizing that since $\omega \neq 0$, we can write the TE-polarized first-order equations as

$$i\omega \varepsilon_r \tilde{q} = \nabla \hat{H}^z,$$
$$i\omega \mu_r \hat{H}^z = \nabla \cdot \tilde{q},$$

with $(\tilde{q}^x, \tilde{q}^y) = (-\hat{E}^y, \hat{E}^x)$. The local discrete form

$$i\omega (\varepsilon_r \tilde{q}_h, \phi_h)_{D^k} = \left(\tilde{\nabla} \hat{H}_h^z, \phi_h \right)_{D^k},$$

$$i\omega \left(\mu_r \hat{H}_h^z, \phi_h \right)_{D^k} = \left(\phi, \tilde{\nabla} \cdot \tilde{q}_h \right)_{D^k},$$

is identical to the discrete first-order form.

Multiplying the additional equation by $i\omega$ and redefining

$$q = i\omega \tilde{q},$$

yields the auxiliary variable introduced in the discontinuous Galerkin formulation of the Helmholtz equation.

Conversely one recovers the first order form directly from the discrete Helmholtz form by dividing with $i\omega$ and renaming the variables. Clearly, the same approach remains valid for the transverse magnetic formulation. □

This two-dimensional result establishes a close connection among the nonzero eigensolutions to the first-order problem, the curl-curl equations, and the Helmholtz problem for the case of homogeneous materials.

8.2.1 The two-dimensional eigenvalue problem

In the following, we return to the problem discussed in Example 8.1 and discuss DG-FEM's for solving the TM-mode Maxwell's equations, already introduced in Section 6.5. We follow the approach taken above and consider the harmonic form

$$i\omega\hat{H}^x = -\frac{\partial\hat{E}^z}{\partial y},$$

$$i\omega\hat{H}^y = \frac{\partial\hat{E}^z}{\partial x},$$

$$i\omega\hat{E}^z = -\frac{\partial\hat{H}^x}{\partial y} + \frac{\partial\hat{H}^y}{\partial x}.$$

If we further assume that all boundaries of our domain are perfectly electrically conducting, then this set of partial differential equations is closed with the following conditions:

$$\hat{\boldsymbol{n}} \cdot \hat{\boldsymbol{H}} = 0, \quad \hat{E}^z = 0.$$

Let us now consider the slightly generalized local scheme, adopted from Section 6.5, as

$$i\omega H_h^x = -\mathcal{D}_y\hat{E}_h^z + \frac{1}{2J}\mathcal{M}^{-1}\int_{\partial\mathsf{D}^k}\left(\hat{n}_y\left[\hat{E}_h^z\right] - \alpha\left[\hat{H}_h^n\right]\hat{n}_x - \beta\left[\hat{H}_h^\tau\right]\hat{n}_y\right)\ell(\boldsymbol{x})\,d\boldsymbol{x}, \quad (8.28)$$

$$i\omega\hat{H}_h^y = \mathcal{D}_x\hat{E}_h^z + \frac{1}{2J}\mathcal{M}^{-1}\int_{\partial\mathsf{D}^k}\left(-\hat{n}_x\left[\hat{E}_h^z\right] - \alpha\left[\hat{H}_h^n\right]\hat{n}_y + \beta\left[\hat{H}_h^\tau\right]\hat{n}_x\right)\ell(\boldsymbol{x})d\boldsymbol{x},$$

$$i\omega\hat{E}_h^z = -\mathcal{D}_y\hat{H}_h^x + \mathcal{D}_x\hat{H}_h^y + \frac{1}{2J}\mathcal{M}^{-1}\int_{\partial\mathsf{D}^k}\left(-\gamma\left[\hat{E}_h^z\right] + \left[\hat{H}_h^x\right]\hat{n}_y - \left[\hat{H}_h^y\right]\hat{n}_x\right)\ell(\boldsymbol{x})\,d\boldsymbol{x}.$$

Here, we have neglected the indication of the local elements for simplicity (i.e., the above statement holds on all elements). Furthermore, we have introduced

$$\hat{H}^n = \hat{H}^x\hat{n}_x + \hat{H}^y\hat{n}_y,$$

$$\hat{H}^\tau = \hat{H}^x\hat{n}_y - \hat{H}^y\hat{n}_x,$$

representing the normal and tangential components, respectively, of the magnetic field.

The formulation in Eq. (8.28) is general enough to recover all widely used DG-FEM schemes for solving Maxwell's equations (e.g., the central flux formulation [118, 258, 260], the upwind formulations [164, 205, 318], the non-conforming penalty formulation [34, 165], and also the classic Lax-Friedrichs flux formula). The various parameter choices for each method are shown in Table 8.5. Note in particular the complementary nature of the upwind flux, penalizing the tangential component of the magnetic field, and the penalty scheme, penalizing the normal component of the magnetic field. In the Lax-Friedrichs flux, both components are included in the flux.

Table 8.5. Penalty parameters to recover specific schemes methods from the local general statement, Eq. (8.28).

Method	α	β	γ
Central fluxes	0	0	0
Upwind fluxes	0	1	1
Penalty fluxes	1	0	1
Lax-Friedrichs fluxes	1	1	1

Setting the test function equal to the trial function, we recover

$$\text{Re}(i\omega) = -\frac{\sum_k \oint_{\partial \mathsf{D}^k} \left(\gamma \left[\hat{E}_h^z \right]^2 + \alpha \left[\hat{H}_h^n \right]^2 + \beta \left[\hat{H}_h^\tau \right]^2 \right) dx}{\left\| \hat{H}_h^x \right\|_{\Omega,h}^2 + \left\| \hat{H}_h^y \right\|_{\Omega,h}^2 + \left\| \hat{E}_h^z \right\|_{\Omega,h}^2}. \tag{8.29}$$

For simplicity, we have neglected special attention to the boundary terms, as these will not impact the subsequent discussion in a substantial way.

An immediate consequence of Eq. (8.29) is that any non-negative choice of α, β, γ will yield a semidiscrete L^2-stable scheme. This result also indicates that asymptotically the choice of nonzero γ will require eigenfunctions with bounded real part to have a continuous electric field component E^z. Similarly, choosing a positive α will asymptotically force normal continuity of the magnetic field for bounded eigenvalues. Finally, a positive β will asymptotically require tangential continuity of the magnetic field for bounded eigenvalues.

Let us relate this to the following natural Sobolev spaces for solutions to Maxwell's equations:

$$H^1(\Omega) = \left\{ v \in \left(L^2(\Omega) \right)^2 | \nabla v \in \left(L^2(\Omega) \right)^2 \right\},$$

$$H(\text{curl}, \Omega) = \left\{ v \in \left(L^2(\Omega) \right)^2 | \nabla \times v \in \left(L^2(\Omega) \right)^2 \right\},$$

$$H(\text{div}, \Omega) = \left\{ v \in \left(L^2(\Omega) \right)^2 | \nabla \cdot v \in L^2(\Omega) \right\}.$$

and subspaces with boundary conditions built in

$$H_0^1(\Omega) = \left\{ v \in H^1(\Omega) | v = 0 \text{ on } \partial\Omega \right\},$$

$$H_0(\text{curl}, \Omega) = \left\{ v \in H(\text{curl}, \Omega) | \hat{n} \times v = 0 \text{ on } \partial\Omega \right\},$$

$$H_0(\text{div}, \Omega) = \left\{ v \in H(\text{div}, \Omega) | \hat{n} \cdot v = 0 \text{ on } \partial\Omega \right\}.$$

In Table 8.6 we relate these observations in terms of these Sobolev spaces.

In the following, we discuss some results obtained when forming the matrix systems associated with each of the central, upwind, penalty, and Lax-Friedrichs flux formulations. The grid is the hash mesh shown in Fig. 8.2. This particular grid is chosen because the spectra of the associated discrete operators exhibit a number of desirable and undesirable properties. We will consider

Table 8.6. Flux parameters to recover specific named DG methods from the general variational statement, Eq. (8.28).

Method	Target space for H	Target space for E_z
Central fluxes	?	?
Upwind fluxes	$H(\mathrm{curl}, \Omega)$	$H_0^1(\Omega)$
Penalty fluxes	$H_0(\mathrm{div}, \Omega))$	$H_0^1(\Omega)$
Lax-Friedrichs fluxes	$H_0(\mathrm{div}, \Omega) \cap H(\mathrm{curl}, \Omega)$	$H_0^1(\Omega)$

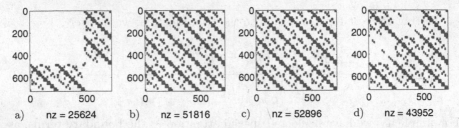

a)   nz = 25624   b)   nz = 51816   c)   nz = 52896   d)   nz = 43952

Fig. 8.6. DG-FEM system for the TM equations for a $K = 16$ hash mesh grid with fourth order elements. (a) central fluxes; (b) upwind fluxes; (c) penalty fluxes; (d) Lax-Friedrichs fluxes.

two different local basis functions. In the first set of studies, we use a complete N-th-order polynomial basis while in the second set of experiments we consider the impact of combining the different fluxes with a locally divergence free basis.

Spectral properties for the complete polynomial basis

In Fig. 8.6 we show the block structure of the discrete operators based on a complete polynomial basis. This highlights the impact on the operator of the choice of terms in the system when the stabilization terms are activated.

We consider the test in Example 8.1 and estimate h-convergence on the sequence of meshes shown in Fig. 8.2. For each combination of flux and basis we plot the eigenvalues of the TM mode Maxwell's operator in the complex plane and focus on the area near the origin. In addition we include tables demonstrating the rate of convergence of the 10 nonzero eigenvalues nearest to $4i$ for a PEC waveguide and compare with the expected set of eigenvalues

$$i \left[\sqrt{2}, \sqrt{5}, \sqrt{5}, \sqrt{8}, \sqrt{10}, \sqrt{10}, \sqrt{13}, \sqrt{13}, \sqrt{17}, \sqrt{17} \right] \frac{\pi}{2} .$$

Based on the results in Section 4.6, the analysis discussed in relation to Theorem 8.2 and the close connection between the discrete Helmholtz problems and the two-dimensional curl-curl equations discussed in Lemma 8.7, we generally expect asymptotic rates of convergence for the eigenvalues of DG discretizations for the second-order Maxwell's equations of $\mathcal{O}(h^{2N})$ and no significant problems with nullspace pollution.

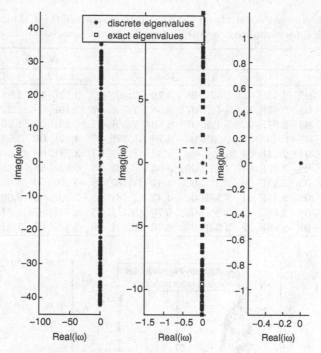

Fig. 8.7. Left: Full spectrum of a central flux discretization of a $K = 16$ hash mesh with an fourth order polynomial basis. Middle: Close up of region near the imaginary axis and the origin. Right: Closer view of the origin.

Central flux formulation

In Fig. 8.7 we show the eigenvalues computed for this test. As we expect from the skew-Hermitian nature of the global discrete operator, all the eigenvalues are on the imaginary axis up to machine precision. Because we have not constrained the eigenfunctions to be divergence-free we see that the discrete kernel is in fact not empty, similar to what we discussed in Example 8.1. However, it appears to be stable, in agreement with the discussion above. The first eight nonzero eigenvalues are visually well resolved.

The rates of convergence of the first 10 eigenvalues are shown in Table 8.7 for $N = 3$ and $N = 5$. While the spurious kernel modes appear stable, the convergence rate for the continuous eigenvalues is erratic and far from the optimal rate. This behavior can only be attributed to the lack of stabilization, the inclusion of which leads to the scheme analyzed in detail in [35] and for which optimal convergence rates have been shown. We will return to a discussion of this scheme in Section 8.2.2.

Upwind flux formulation

In Fig. 8.8 we show the eigenvalues for the discrete DG operator with $N = 4$ for the coarsest mesh in Fig. 8.1. As we expect from the L^2-stability result for the global discrete operator, all the eigenvalues are in the left half-plane.

Table 8.7. h-Convergence study for the first 10 eigenvalues of the TM mode Maxwell operator with an unstabilized central flux on the biunit square.

Mode	$N = 3$				$N = 5$			
	h	$h/2$	$h/4$	Rate	h	$h/2$	$h/4$	Rate
(1,1)	4.53E-05	9.13E-07	1.52E-08	5.78	2.49E-09	3.37E-12	2.98E-13	6.52
(1,2)	1.20E-03	2.97E-05	5.42E-07	5.56	7.59E-07	1.10E-09	1.22E-12	9.63
(2,1)	1.20E-03	2.97E-05	5.42E-07	5.56	7.59E-07	1.10E-09	1.26E-12	9.60
(2,2)	1.24E-03	8.49E-05	1.63E-06	4.78	5.87E-07	4.42E-09	5.53E-12	8.35
(1,3)	2.91E-04	2.13E-04	5.02E-06	2.93	8.23E-07	3.74E-08	5.69E-11	6.91
(3,1)	1.29E-02	3.16E-04	5.93E-06	5.54	5.83E-05	6.98E-08	7.36E-11	9.80
(2,3)	3.44E-03	3.25E-04	7.55E-06	4.42	1.18E-08	5.20E-08	8.09E-11	3.60
(3,2)	3.44E-03	3.25E-04	7.55E-06	4.42	1.18E-08	5.20E-08	8.09E-11	3.59
(1,4)	3.97E-01	2.98E-02	3.36E-05	6.76	9.11E-05	8.51E-07	1.31E-09	8.04
(4,1)	3.97E-01	2.98E-02	3.36E-05	6.76	9.11E-05	8.51E-07	1.31E-09	8.04

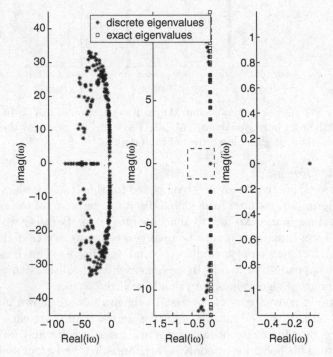

Fig. 8.8. Left: Full spectrum of an upwind flux DG discretization of a $K = 16$ triangle hash mesh with a fourth order polynomial basis. Middle: Close up of region near the imaginary axis and the origin. Right: Closer view of the origin.

The first eight eigenvalues located near the imaginary axis represent visually accurate eigenvalues. Following the eigenvalues to higher frequencies, we see that their real parts become increasingly negative. This is a result of the corresponding eigenfunctions being successively further away from H (curl, Ω)

Table 8.8. h-Convergence study for the first 10 eigenvalues of the TM mode Maxwell operator with an upwind flux on the biunit square.

| | $N = 3$ | | | | $N = 5$ | | | |
Mode	h	$h/2$	$h/4$	Rate	h	$h/2$	$h/4$	Rate
(1,1)	2.92E-06	1.11E-08	4.18E-11	8.05	8.15E-11	2.13E-14	5.86E-14	5.22
(1,2)	2.95E-04	1.23E-06	4.89E-09	7.94	9.43E-08	2.40E-11	3.20E-14	10.75
(2,1)	2.95E-04	1.23E-06	4.89E-09	7.94	9.43E-08	2.40E-11	7.68E-14	10.11
(2,2)	1.44E-04	4.65E-06	1.84E-08	6.47	5.95E-08	1.41E-10	9.68E-14	9.62
(1,3)	5.32E-03	3.17E-05	1.35E-07	7.63	9.84E-06	3.19E-09	8.25E-13	11.75
(3,1)	5.88E-03	3.19E-05	1.36E-07	7.70	1.01E-05	3.20E-09	8.63E-13	11.74
(2,3)	6.29E-03	4.39E-05	1.87E-07	7.52	1.13E-05	4.64E-09	1.24E-12	11.55
(3,2)	6.29E-03	4.39E-05	1.87E-07	7.52	1.13E-05	4.64E-09	1.30E-12	11.53
(1,4)	3.20E-02	3.53E-04	1.63E-06	7.13	1.95E-04	1.15E-07	3.18E-11	11.28
(4,1)	3.20E-02	3.53E-04	1.63E-06	7.13	1.95E-04	1.15E-07	3.18E-11	11.27

which increases the tangential jumps in the formula for the real part of the eigenvalue given in Eq. (8.29), hence adding dissipation to these modes. In a time-domain simulation the components of the initial condition projected onto these eigenfunctions will decay very rapidly in time. As for the central flux, we see that there is a spuriously large kernel, corresponding to static modes, which have magnetic fields not belonging to H_0 (div$_0$, Ω) (i.e. the space of divergence-free magnetic fields that satisfy the PEC boundary condition). This is to be expected because neither the divergence free condition nor the boundary condition are being enforced explicitly or via a penalty term. However, these kernel modes do not enter into the physical domain part of the spectrum and which appears to be isolated.

In Table 8.8 we show the rates of convergence for $N = 3$ and $N = 5$. Although there appears to be no general theory to support this, there are partial results in [10] indicating a superior convergence rate of $\mathcal{O}(h^{2N+1})$, as we also see here for all modes not impacted by the machine precision. Note that the lowest mode is resolved to six decimal places with two third order elements per wavelength.

Penalty flux formulation

Let us now consider the use of a normal penalty flux. In Fig. 8.9 we show the eigenvalues for the discrete DG operator constructed with the penalty-based fluxes. As we expect from the L^2-stability result for the global discrete operator, all the eigenvalues are in the left half-plane. The first eight eigenvalues near the imaginary axis represent visually accurate eigenvalues. As we follow the eigenvalues to higher frequency, we see that their real parts become increasingly negative. This is caused by the corresponding eigenfunctions being successively further away from H_0 (div, Ω) which increases the normal jumps in the formula for the real part of the eigenvalue given in Eq. (8.29). As with the central and upwind flux results there is a spuriously large kernel of modes

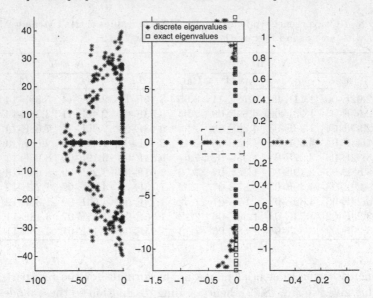

Fig. 8.9. Left: Full spectrum of a penalty type flux DG discretization of a $K = 16$ triangle hash mesh with a fourth order polynomial basis. Middle: Close up of region near the imaginary axis and the origin. Right: Closer view of the origin.

Table 8.9. h-Convergence study for the first 10 eigenvalues of the TM mode Maxwell operator with a penalty flux on the biunit square.

Mode	$N = 3$				$N = 5$			
	h	$h/2$	$h/4$	Rate	h	$h/2$	$h/4$	Rate
(1,1)	2.46E-05	5.74E-07	1.11E-08	5.55	9.32E-10	1.65E-12	3.29E-14	7.40
(1,2)	8.54E-04	1.69E-05	3.79E-07	5.57	3.78E-07	4.49E-10	7.07E-13	9.51
(2,1)	8.54E-04	1.69E-05	3.79E-07	5.57	3.78E-07	4.49E-10	7.82E-13	9.44
(2,2)	1.24E-03	6.37E-05	1.32E-06	4.94	5.87E-07	2.67E-09	3.77E-12	8.62
(1,3)	9.15E-03	1.25E-04	2.89E-06	5.82	1.65E-05	1.72E-08	2.47E-11	9.68
(3,1)	1.38E-02	2.15E-04	3.91E-06	5.89	4.58E-05	4.34E-08	4.22E-11	10.02
(2,3)	7.81E-03	2.54E-04	5.52E-06	5.23	1.62E-05	3.23E-08	4.69E-11	9.20
(3,2)	7.81E-03	2.54E-04	5.52E-06	5.23	1.62E-05	3.23E-08	4.69E-11	9.20
(1,4)	3.08E-02	8.74E-04	1.83E-05	5.36	1.40E-04	4.80E-07	5.56E-10	8.97
(4,1)	3.08E-02	8.74E-04	1.83E-05	5.36	1.40E-04	4.80E-07	5.56E-10	8.97

that have magnetic fields which do not actually belong to $H_0 (\mathrm{div}_0, \Omega)$. In addition to the overly large kernel, we observe eigenvalues on the real axis but close to the origin. These will correspond to spurious slowly decaying modes, which might be excited in a time-domain simulation and which are clearly not desirable.

The experimental rates of convergence are shown in Table 8.9, indicating rates close to $\mathcal{O}(h^{2N-1})$, although there is no known analysis to substantiate this.

Apart from the observations related to the kernel modes and the sub-optimal convergence rate, imposing weak normal continuity of the magnetic fields can cause problems in nonconvex domains where the magnetic field is not guaranteed to be continuous (i.e., there may be unbounded solutions that are excluded from the solution space). See [165] for further discussions and references directly related to such problems in electromagnetics.

Lax-Friedrichs flux formulation
Let us finally consider the case of a full Lax-Friedrichs flux. In Fig. 8.10 we show the eigenvalues for the discrete DG operator constructed with the Lax-Friedrichs-based fluxes on the "hash" mesh shown in Fig. 8.2. The results are quite similar to those shown for the upwind and the penalty fluxes. All the eigenvalues are in the left half-plane. The first eight eigenvalues represent visually accurate eigenvalues, and as we follow the eigenvalues to higher frequency, we see that their real parts become increasingly negative. This is caused by the corresponding eigenfunctions being successively further away from $H_0^1(\Omega)$ which increases all the jumps in the formula for the real part of the eigenvalue given in Eq. (8.29). Similar to the results based on the previous flux choices, we see a spuriously large kernel, corresponding to static modes

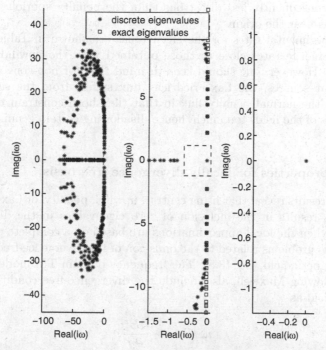

Fig. 8.10. Left: Full spectrum of a Lax-Friedrichs type flux DG discretization of a $K = 16$ hash mesh with a fourth order polynomial basis. Middle: Close up of region near the imaginary axis and the origin. Right: Closer view of the origin.

Table 8.10. h-Convergence study for the first 10 eigenvalues of the TM mode Maxwell operator with a Lax-Friedrichs flux on the biunit square.

Mode	$N = 3$				$N = 5$			
	h	$h/2$	$h/4$	Rate	h	$h/2$	$h/4$	Rate
(1,1)	3.31E-06	1.85E-08	7.49E-11	7.72	1.01E-10	4.89E-14	5.33E-15	7.11
(1,2)	3.07E-04	1.47E-06	7.66E-09	7.64	7.46E-08	2.30E-11	3.73E-14	10.47
(2,1)	3.07E-04	1.47E-06	7.66E-09	7.64	7.46E-08	2.30E-11	2.26E-14	10.83
(2,2)	1.17E-03	7.01E-06	3.81E-08	7.46	2.42E-07	1.85E-10	5.41E-14	11.05
(1,3)	5.01E-03	2.90E-05	1.88E-07	7.35	7.95E-06	2.82E-09	1.06E-12	11.42
(3,1)	6.08E-03	3.16E-05	1.89E-07	7.49	8.74E-06	2.84E-09	1.10E-12	11.46
(2,3)	1.07E-02	5.76E-05	2.98E-07	7.57	1.74E-05	3.73E-09	1.55E-12	11.71
(3,2)	1.07E-02	5.76E-05	2.98E-07	7.57	1.74E-05	3.73E-09	1.59E-12	11.69
(1,4)	3.32E-02	2.93E-04	2.00E-06	7.01	1.76E-04	9.19E-08	3.95E-11	11.04
(4,1)	3.32E-02	2.93E-04	2.00E-06	7.01	1.76E-04	9.19E-08	3.96E-11	11.04

with magnetic fields that do not actually belong to $H_0 (\text{div}_0, \Omega)$. In contrast to the penalty flux results, the overly large kernel shown here appears isolated from spurious real eigenvalues on the real axis. These will correspond to spurious modes that decay faster than the similar modes in the penalty case. These are consequently less disastrous than the penalty spurious modes on the real axis near the origin.

The experimental rates for the eigenvalues are shown in Table 8.10 and show convergence rates close to those obtained with the upwind flux (i.e., $\mathcal{O}(h^{2N+1})$). However, one should keep in mind that for non-convex domains and singular sources, the Lax-Friedrichs flux suffers from the same shortcomings as the normal penalty flux in that they both constrain the normal components of the fields too much, hence disallowing certain genuine physical modes.

Spectral properties for locally divergence free basis

The above results reveal that using central, upwind, penalty, or Lax-Friedrichs type fluxes results in the inclusion of zero eigenvalues in the discrete DG operators when the local approximations are based on a complete polynomial basis. These problems related to the omission of the divergence-free condition for the magnetic field, Eq. (8.8). The frequency domain TM mode equations should, following Maxwell, also include the divergence-free condition for the magnetic field as

$$i\omega \hat{H}^x = -\frac{\partial \hat{E}^z}{\partial y},$$

$$i\omega \hat{H}^y = \frac{\partial \hat{E}^z}{\partial x},$$

$$i\omega \hat{E}^z = -\frac{\partial \hat{H}^x}{\partial y} + \frac{\partial \hat{H}^y}{\partial x},$$

$$0 = \frac{\partial \hat{H}^x}{\partial x} + \frac{\partial \hat{H}^y}{\partial y}.$$

The latter condition guarantees that $\omega \neq 0$ except for the trivial null solution. One could naturally include this into the system above, although at increased cost. Furthermore, for the time-dependent problem, Eq. (8.7), this would result in a need to solve global problems, which is highly undesirable due to cost.

It is tempting to encode this condition into the scheme by restricting the local polynomial basis on each triangle to be locally divergence-free. This idea was first proposed in [70] in the context of Maxwell's equations and in [222] for inviscid magnetohydrodynamics.

We obtain such a basis of order N by considering the $(N + 1)$-th-order polynomial basis, $\{\psi_n(\boldsymbol{x})\}_{n=1}^{N_p}$ where

$$N_p = \frac{(N + 2)(N + 3)}{2},$$

and construct the vector basis, $\boldsymbol{\Psi}$, as

$$\boldsymbol{\Psi}_n = \left(\frac{\partial \psi_n}{\partial y}, \ -\frac{\partial \psi_n}{\partial x}\right),$$

which, by construction, obeys $\nabla \cdot \boldsymbol{\Psi}_n = 0$. We can then express the magnetic field as

$$\hat{H}^x = \sum_{n=2}^{N_p} \hat{H}_n \frac{\partial \psi_n}{\partial y}, \quad \hat{H}^y = \sum_{n=2}^{N_p} -\hat{H}_n \frac{\partial \psi_n}{\partial x},$$

where $\hat{H}_n$ are the coefficients for the vector function $\boldsymbol{H}$, represented by $(N+1)(N+4)/2$ 2-vector functions, each being locally divergence-free. This should be contrasted with the basic polynomial representation, which requires $(N+1)(N+2)$ degrees of freedom, suggesting a potential saving by as much as 50% in the number of degrees of freedom. Once the local basis is chosen, the variational equation is the same as Eq. (8.28) with the only modification being that the magnetic field test and trial functions is now given as the locally divergence-free basis $\boldsymbol{\Psi}$.

We repeat the $K = 16$ hash mesh experiment, replacing the basis for the magnetic field with the locally divergence-free basis. In this case, when coupled with an appropriate flux, we can guarantee a local empty kernel and some degree of separation for the spurious eigenvalues.

In Fig. 8.11 we show the block structure nature of the discrete operators using the locally divergence free basis. These should be contrasted with Fig. 8.6.

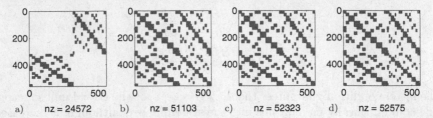

Fig. 8.11. DG-FEM operator based on a locally divergence-free basis for a $K = 16$ element hash grid with fourth order elements. (a) central fluxes; (b) upwind fluxes; (c) penalty fluxes; (d) Lax-Friedrichs fluxes.

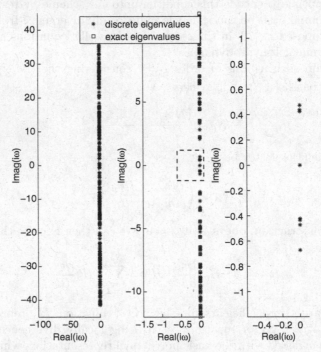

Fig. 8.12. Left: Full spectrum of a locally divergence- free, central flux DG discretization of a $K = 16$ triangle hash mesh with a fourth order polynomial basis. Middle: Close up of region near the imaginary axis and the origin. Right: Closer view of the origin.

Central flux formulation

In Fig. 8.12 we display the eigenvalues for the discrete DG operator constructed with central fluxes and locally divergence-free basis functions for the simple eigenvalue problem discussed previously. As we expect from the skew symmetric nature of the global discrete operator, all eigenvalues are on the imaginary axis up to machine precision. There should be no zero eigenvalues for this problem, but the discrete kernel is not empty despite the additional local divergence-free constraint for the magnetic field. In contrast to the results obtained with the simpler polynomial basis in the last section, we see that the first eight well resolved eigenvalues are interlaced with spurious eigenvalues, much in the way discussed in Example 8.1. There is no meaningful way we can separate the physical from the spurious eigenvalues, making this combination unusable.

Upwind Flux Formulation

In Figure 8.13 we show the eigenvalues for the discrete DG operator constructed with upwind fluxes on the hash mesh shown in Fig. 8.2. As this change of basis does not change the form of the L^2-stability result for the global discrete operator, all the eigenvalues are again expected to be in the left half-plane. The first eight eigenvalues, located near the imaginary axis, represent visually accurate eigenvalues. As we follow the eigenvalues to higher frequencies, their real parts become increasingly negative. There is a spuriously large kernel, corresponding to static modes with magnetic fields that do not actually belong to $H_0(\text{div}_0, \Omega)$ (i.e. the space of divergence free magnetic fields which satisfy the PEC boundary condition). In contrast to the original

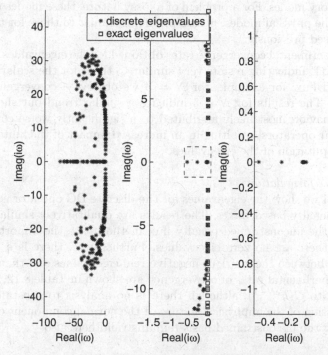

Fig. 8.13. Left: Full spectrum of a locally divergence-free, upwind flux DG discretization of a $K = 16$ triangle hash mesh with a fourth order polynomial basis. Middle: Close up of region near the imaginary axis and the origin. Right: Closer view of the origin.

Table 8.11. h-Convergence study for the first 10 eigenvalues of the TM mode Maxwell operator with a upwind flux and a locally divergence-free basis on the biunit square.

Mode	$N = 3$				$N = 5$			
	h	$h/2$	$h/4$	Rate	h	$h/2$	$h/4$	Rate
(1,1)	3.53E-06	1.54E-08	6.22E-11	7.90	9.98E-11	1.63E-13	4.87E-12	2.18
(1,2)	2.97E-04	1.18E-06	4.58E-09	7.99	9.52E-08	2.23E-11	2.61E-13	9.24
(2,1)	2.97E-04	1.18E-06	4.58E-09	7.99	9.52E-08	2.25E-11	1.70E-12	7.89
(2,2)	7.40E-05	5.48E-06	2.55E-08	5.75	7.13E-08	1.68E-10	5.45E-12	6.84
(1,3)	5.53E-03	2.77E-05	9.54E-08	7.91	1.13E-05	3.18E-09	4.18E-12	10.69
(3,1)	7.27E-03	3.29E-05	1.01E-07	8.06	1.32E-05	3.45E-09	2.97E-11	9.38
(2,3)	6.56E-03	4.91E-05	2.33E-07	7.39	1.32E-05	5.46E-09	5.08E-12	10.65
(3,2)	6.56E-03	4.91E-05	2.33E-07	7.39	1.32E-05	5.46E-09	2.03E-11	9.65
(1,4)	5.05E-02	3.96E-04	1.18E-06	7.68	3.62E-04	1.39E-07	1.61E-11	12.21
(4,1)	5.05E-02	3.96E-04	1.18E-06	7.68	3.62E-04	1.39E-07	3.28E-11	11.70

upwind method in Fig. 8.8, however, there is a cluster of unphysical eigenvalues on the real axis very close to the origin, representing slowly decaying nonoscillatory modes. For a problem of a lossy nature, these modes may interlace with the physical modes, posing problems similar to those for the central flux discussed previously.

The experimental convergence rates obtained for the eigenvalues are shown in Table 8.11, indicating results very similar to those for the scalar basis and the upwind flux; for example, for $N = 3$ we observe a convergence rate of $\mathcal{O}(h^{2N+1})$. The results for $N = 5$ indicate a similar trend but show a more uneven behavior, most likely attributed to a significantly worse conditioning of the linear operators, resulting in an increased impact of the finite precision in the computation of the eigenvalues.

Penalty flux formulation
In Fig. 8.14 we show the eigenvalues for the discrete DG operator constructed with the penalty-based fluxes. The results are qualitatively similar to those shown for the unconstrained penalty flux method with the important difference that there are no zero eigenvalues. Furthermore, there is a noticeable separation between the spurious negative real eigenvalues and the origin.

The experimental rates of convergence are shown in Table 8.12, indicating rates close to $\mathcal{O}(h^{2N-1})$, although there is no analysis to substantiate this. The discussion of the unphysical nature of the normal component constraints when using the unconstrained basis remains valid here also.

Lax-Friedrichs flux formulation
In Fig. 8.15 we finally show the eigenvalues for the discrete DG operator constructed with the locally divergence-free magnetic fields and the Lax-Friedrichs-based fluxes on the hash mesh. The results are again qualitatively

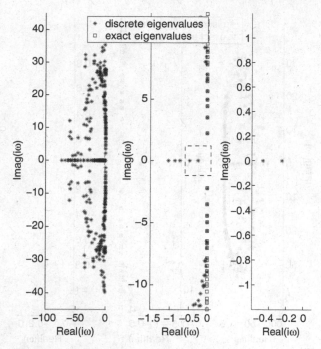

Fig. 8.14. Left: Full spectrum of a divergence-free, penalty-type flux DG discretization of a 16 triangle hash mesh with a fourth order polynomial basis. Middle: Close up of region near the imaginary axis and the origin. Right: Closer view of the origin.

Table 8.12. h-Convergence study for the first 10 eigenvalues of the TM mode Maxwell operator with a penalty flux and a locally divergence-free basis on the biunit square.

Mode	$N = 3$				$N = 5$			
	h	$h/2$	$h/4$	Rate	h	$h/2$	$h/4$	Rate
(1,1)	2.40E-05	4.77E-07	8.78E-09	5.71	9.28E-10	8.58E-13	7.20E-12	3.50
(1,2)	8.86E-04	1.61E-05	3.53E-07	5.65	3.26E-07	3.64E-10	4.91E-12	8.01
(2,1)	8.86E-04	1.61E-05	3.53E-07	5.65	3.26E-07	3.65E-10	1.51E-12	8.86
(2,2)	1.23E-03	6.18E-05	1.14E-06	5.04	5.87E-07	2.35E-09	1.66E-12	9.21
(1,3)	1.15E-02	1.14E-04	2.62E-06	6.05	2.75E-05	1.23E-08	1.96E-11	10.21
(3,1)	1.35E-02	1.97E-04	3.37E-06	5.98	3.43E-05	3.55E-08	3.75E-11	9.90
(2,3)	8.42E-03	2.43E-04	4.54E-06	5.43	1.30E-05	2.66E-08	1.14E-11	10.06
(3,2)	8.42E-03	2.43E-04	4.54E-06	5.43	1.30E-05	2.66E-08	4.46E-11	9.0
(1,4)	3.98E-02	8.38E-04	1.61E-05	5.64	2.87E-04	3.68E-07	3.38E-10	9.85
(4,1)	3.98E-02	8.38E-04	1.61E-05	5.64	2.87E-04	3.68E-07	3.54E-10	9.81

similar to those shown for the unconstrained Lax-Friedrichs method. As with the locally divergence-free penalty method, there are no zero eigenvalues and there is a clear separation between the spurious negative real eigenvalues and the origin.

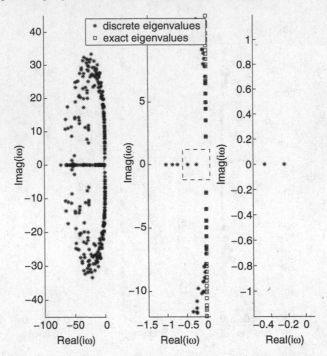

Fig. 8.15. Left: Full spectrum of a locally divergence-free, Lax-Friedrichs type flux DG discretization of a 16 triangle hash mesh with a fourth order polynomial basis. Middle: Close up of region near the imaginary axis and the origin. Right: Closer view of the origin.

As for the unconstrained Lax-Friedrichs case, we observe in Table 8.13 a convergence rate close to $\mathcal{O}(h^{2N+1})$ with a less clear picture for $N = 5$, quite possibly caused by poor conditioning of the operator.

Let us revisit the question of the size of the discrete kernel in a little more detail. First, assume that there exists a discrete eigenpair with $\omega = 0$ and $(\hat{H}^x, \hat{H}^y, \hat{E}^z)$ being non-zero. Then the local elementwise DG Lax-Friedrichs scheme is

$$0 = -\mathcal{D}_y \hat{E}_h^z + \frac{1}{2J}\mathcal{M}^{-1} \int_{\partial \mathsf{D}^k} \left(\hat{n}_y \left[\hat{E}_h^z\right] - \alpha \left[\hat{H}_h^n\right] \hat{n}_x - \beta \left[\hat{H}_h^\tau\right] \hat{n}_y\right) \boldsymbol{\ell}(\boldsymbol{x})\,d\boldsymbol{x},$$

$$0 = \mathcal{D}_x \hat{E}_h^z + \frac{1}{2J}\mathcal{M}^{-1} \int_{\partial \mathsf{D}^k} \left(-\hat{n}_x \left[\hat{E}_h^z\right] - \alpha \left[\hat{H}_h^n\right] \hat{n}_y + \beta \left[\hat{H}_h^\tau\right] \hat{n}_x\right) \boldsymbol{\ell}(\boldsymbol{x})\,d\boldsymbol{x},$$

$$0 = -\mathcal{D}_y \hat{H}_h^x + \mathcal{D}_x \hat{H}_h^y + \frac{1}{2J}\mathcal{M}^{-1} \int_{\partial \mathsf{D}^k} \left(-\gamma \left[\hat{E}_h^z\right] + \left[\hat{H}_h^x\right] \hat{n}_y - \left[\hat{H}_h^y\right] \hat{n}_x\right) \boldsymbol{\ell}(\boldsymbol{x})\,d\boldsymbol{x}.$$

The equation for the real part of the eigenvalue tells us that in this case, assuming $(\hat{\boldsymbol{H}}_h, \hat{E}_h^z)$ is L^2-normalized,

$$0 = -\sum_k \gamma \left(\left(\left[\hat{E}_h^z\right], \left[\hat{E}_h^z\right]\right)_{\partial \mathsf{D}^k} + \alpha \left(\left[\hat{H}_h^n\right], \left[\hat{H}_h^n\right]\right)_{\partial \mathsf{D}^k} + \beta \left(\left[\hat{H}_h^\tau\right], \left[\hat{H}_h^\tau\right]\right)_{\partial \mathsf{D}^k}\right).$$

Table 8.13. h-Convergence study for the first 10 eigenvalues of the TM mode Maxwell operator with a Lax-Friedrichs flux and locally divergence-free basis on the biunit square.

| Mode | | $N = 3$ | | | | $N = 5$ | | |
	h	$h/2$	$h/4$	Rate	h	$h/2$	$h/4$	Rate
(1,1)	3.61E-06	1.55E-08	6.33E-11	7.90	1.01E-10	3.46E-14	3.80E-13	4.03
(1,2)	3.59E-04	1.57E-06	6.65E-09	7.86	1.14E-07	3.21E-11	1.84E-12	7.96
(2,1)	3.59E-04	1.57E-06	6.65E-09	7.86	1.14E-07	3.25E-11	1.42E-12	8.14
(2,2)	1.31E-03	6.61E-06	2.62E-08	7.80	6.10E-07	1.63E-10	7.72E-13	9.80
(1,3)	5.71E-03	4.63E-05	2.11E-07	7.33	1.03E-05	5.32E-09	4.93E-12	10.49
(3,1)	1.05E-02	4.63E-05	2.11E-07	7.76	2.07E-05	5.39E-09	1.84E-13	13.37
(2,3)	1.12E-02	7.01E-05	3.05E-07	7.58	2.01E-05	6.78E-09	3.73E-11	9.52
(3,2)	1.12E-02	7.04E-05	3.05E-07	7.58	2.01E-05	6.78E-09	6.45E-12	10.79
(1,4)	6.90E-02	5.12E-04	2.64E-06	7.34	5.31E-04	1.96E-07	3.07E-12	13.68
(4,1)	6.90E-02	5.12E-04	2.64E-06	7.34	5.31E-04	1.96E-07	6.78E-11	11.45

If we assume that α, β, γ are all nonzero, as in the case of a Lax-Friedrichs flux, the jumps must all vanish and we are left with the local statements

$$0 = -\mathcal{D}_y \hat{E}_h^z,$$
$$0 = \mathcal{D}_x \hat{E}_h^z,$$
$$0 = -\mathcal{D}_y \hat{H}_h^x + \mathcal{D}_x \hat{H}_h^y.$$

Assuming that $\hat{E}_h^z = 0$ at the PEC boundary, this implies that $\hat{E}_h^z = 0$ everywhere. Furthermore, using a locally divergence-free basis, ψ_h, such that $\hat{H}_h^x = \frac{\partial \psi_h}{\partial y}, \hat{H}_h^y = -\frac{\partial \psi_h}{\partial x}$, we recover that the basis ψ_h is locally harmonic,

$$\frac{\partial^2 \psi_h}{\partial x^2} + \frac{\partial^2 \psi_h}{\partial y^2} = 0,$$

Furthermore, since ψ_h is a polynomial, we must thus be able to express the local magnetic field as

$$\hat{\boldsymbol{H}}_h = \boldsymbol{a}^k + \boldsymbol{b}^k x + \boldsymbol{b}_\perp^k y.$$

There are only four coefficients (since $\boldsymbol{b} \cdot \boldsymbol{b}_\perp = 0$) available per element to represent this magnetic field. We seek to further constrain these parameters by using the continuity condition $[\hat{\boldsymbol{H}}_h] = 0$. Taking the dot product of the jump in $\hat{\boldsymbol{H}}_h$ with the jump in the two $\boldsymbol{b}$-vectors for the elements sharing the edge, we obtain

$$0 = [\boldsymbol{b}] \cdot \left[\hat{\boldsymbol{H}}_h\right] = [\boldsymbol{b}] \cdot [\boldsymbol{a} + \boldsymbol{b}x + \boldsymbol{b}_\perp y]$$
$$= [\boldsymbol{b}] \cdot [\boldsymbol{a}] + [\boldsymbol{b}] \cdot [\boldsymbol{b}] \, x + [\boldsymbol{b}] \cdot [\boldsymbol{b}_\perp] \, y$$
$$= [\boldsymbol{b}] \cdot [\boldsymbol{a}] + [\boldsymbol{b}] \cdot [\boldsymbol{b}] \, x$$

and, similarly,

$$0 = [b_\perp] \cdot [a] + [b_\perp] \cdot [b_\perp] y$$

for all (x, y) on this edge. From these relationships we recover directly that b^k and a^k are both constant across all elements. Applying the $\hat{H}_h \cdot \hat{n} = 0$ boundary condition, we obtain that the only possible solution is the trivial $a = b = 0$ and, thus, $\hat{H}_h = 0$, confirming that the kernel of the Lax-Friedrichs-stabilized operator for divergence-free elements must be empty. Note also that we did not use any conditions on the tangential components, clarifying that it is the condition on the normal component that controls the spurious kernel. This also supports the discussion related to the penalty flux where we likewise recovered an empty kernel.

Unfortunately, as we have discussed several times, conditions on the normal components of the fields may lead to convergence problems for solutions with a strongly singular component.

Further examples

To further explore the ability of the discontinuous element formulation for solving two-dimensional Maxwell eigenproblems, we will present a number of additional examples for classical benchmark problems. These problems are taken from [88], where both details of the setup and results, analytic and computational as available, can be found. All subsequent results are obtained using unstabilized central fluxes and an unconstrained polynomial basis.

Metallic L-shaped domain

As a first test case, we consider the computation of the eigensolutions associated with a PEC L-shaped domain. The nonconvex nature of the outer boundary renders some of the eigenfunctions highly singular, making it a challenging test. We note that the first eigenfunction has a strongly unbounded singularity at the nonconvex corner, the second eigenfunction has a bounded derivative and the third and fourth eigenfunctions are analytic. The fifth eigenfunction is again strongly singular.

To compute the eigensolutions, we use both a sequence of cross-hatch grids, Fig. 8.16, as well as a fixed coarse finite element grid using elements of increasing order.

In Fig. 8.17 we show the error in the first five nonzero computed eigenvalues, computed using h- as well as order (N)-convergence. As expected we see a strong correlation between the convergence rate and the regularity of the corresponding eigenfunction. In particular, we observe exponential convergence with N of the eigenvalues corresponding to the analytic third and fourth eigenfunction. No spurious eigensolutions are encountered during the computation. The strong correlation between the smoothness of the eigenvector and the convergence rate for the associated eigenvalue is in agreement with the result in Theorem 8.2, although this result is not directly applicable

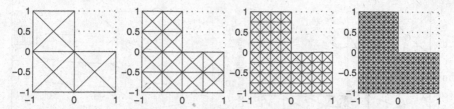

Fig. 8.16. Sequence of cross-hatch grids used to compute eigenvalues in PEC L-shaped domain, using h-refinement.

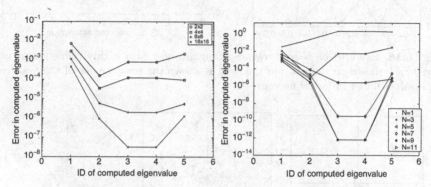

Fig. 8.17. Convergence of the first five eigenvalues in two-dimensional PEC L-shaped domain. On the left is shown h-convergence on the sequence of grids in Fig. 8.16 using a third order polynomial basis. On the right is illustrated N-convergence at a simple 12 element grid (leftmost grid in Fig. 8.16).

to this more complex case. As we shall see in Section 8.2.2, similar results are also known for a stabilized version of the scheme based on a central flux.

Given the strong singularities of some of the eigenfunctions, it seems natural that a combination of element, h, and order, N, refinement should result in improved accuracy. This is confirmed by the results in Fig. 8.18, where we show a highly refined grid and the spectrally converging eigenvalues computed using this grid.

Checkerboard material case
Consider also a case with a checkerboard pattern of materials inside a PEC cavity as illustrated in Fig. 8.19. The checkerboard pattern consists of two nonmagnetic materials with permittivities of $\varepsilon_r = 1.0$ and $\varepsilon_r = 0.1$. This yields singularities at the center cross with a worst-case behavior approximately like $r^{-0.6}$, where r measures the distance to the cross [88].

As in the previous test case we use a highly refined grid to capture local strongly singular eigenfunctions. A typical grid is shown in Fig. 8.19 and a close-up to illustrate the highly refined nature of the grid. In this case the smallest element has an edge length of $\mathcal{O}(10^{-6})$.

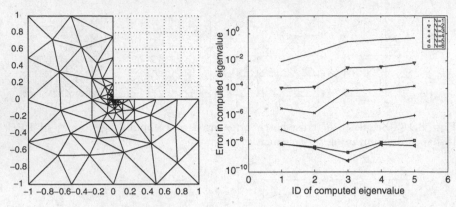

Fig. 8.18. Example of N-convergence of eigenvalues on a highly refined unstructured grid, shown on the left. On the right is shown the convergence of the first five eigenvalues under order refinement.

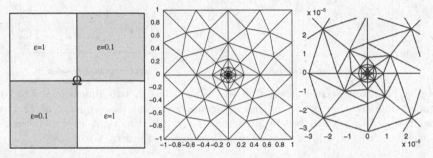

Fig. 8.19. On the left is shown the PEC square domain with a checkerboard pattern of dielectric materials. The middle figure illustrates a typical highly refined unstructured grid (312 elements) used to compute the eigenvalues, while the right shows a close up of the grid very close to the cross in the center. Note the scales.

In Fig. 8.20 we show the convergence of the first few eigenvalues, computed under order-convergence on the grid shown in Fig. 8.19. The worst eigenvector is apparently the second one, with the singularity behaving as $r^{-0.6}$. This limits the convergence order of this eigenvalue to less than linear. For the sixth eigenvalue, likewise associated with a strongly singular eigenfunction, the convergence rate is approximately linear. However, no spurious eigenvalues are found.

One should keep in mind that the checkerboard problem is not included in the equivalence result between the first-order system and the curl-curl equation discussed in Lemma 8.5. Thus, the connection between the curl-curl form and the Helmholtz problem is no longer rigid. However, as discussed previously, if the main problem is one of perturbations of the nullspace of the basis operator, of which we have some understanding, it appears unlikely that more

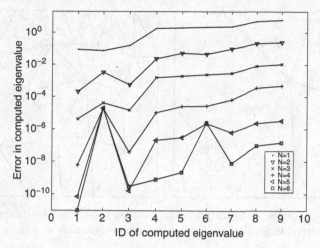

Fig. 8.20. Example of order convergence of the first nine eigenvalues for the checkerboard test case, computed on the highly refined grid shown in Fig. 8.19.

complex materials would make the situation worse for this particular concern. The results presented here support this argument.

A counter example

All results presented so far indicate that for the two-dimensional TM Maxwell eigenvalue problem, the problem with spurious modes is absent in the DG discretization.

Unfortunately, this is not true if the grid is sufficient far from quasi structured and a simple unstabilized central flux is used as in the previous examples. We revisit the TM vacuum problem but now with grids that are constructed to cause problems. Examples of such grids are shown in Fig. 8.21.

In Table 8.14 we show the first few eigenvalues for the two-dimensional TM Maxwell's operator, computed on the grids in Fig. 8.21. It is clear that small spurious eigenvalues are emerging, although the situation is not as critical as for the classic finite element case in Example 8.1. As we shall see in the next section, the situation is much more severe for the three-dimensional case. However, the solution to the problem developed in that case also applies directly to the two-dimensional case.

8.2.2 The three-dimensional eigenproblem

With some confidence in the performance of the discrete approximation to the two-dimensional eigenproblems and reasonable grids, let us consider the more complicated three-dimensional case.

The discussion on the connection between the eigensolutions to the first-order form and the curl-curl equations, stated in Lemma 8.5, still holds.

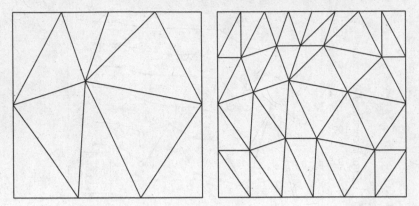

Fig. 8.21. Examples of distorted grids used to solve the two-dimensional TM Maxwell's eigenvalue problem with an unstabilized central flux.

Table 8.14. Emergence of spurious eigenvalues on distorted grids for the two-dimensional TM Maxwell problem using an unstabilized central flux. Spurious eigenvalues are highlighted in bold.

Exact ω^2	Mesh 1	Mesh 2	Mesh 3
2	2.00E+00	2.00E+00	2.00E+00
5	**3.03E+00**	**3.96E+00**	**4.33E+00**
5	**4.00E+00**	5.00E+00	5.00E+00
8	5.00E+00	5.00E+00	5.00E+00
10	5.00E+00	**5.04E+00**	**5.57E+00**
10	**6.33E+00**	8.00E+00	8.00E+00
13	**7.68E+00**	**8.81E+00**	1.00E+01
13	8.26E+00	1.00E+01	1.00E+01
17	9.89E+00	1.00E+01	1.00E+01
17	9.98E+00	**1.06E+01**	1.18E+01
18	1.24E+01	1.30E+01	1.30E+01
20	1.29E+01	1.30E+01	1.30E+01

However, the two-dimensional result, stated in Lemma 8.7, connecting the eigensolutions of the first-order problem and Helmholtz equation, does not seem to generalize to three dimensions. In other words, it is not clear that we can expect the same kind of stability of the large nullspace associated with the curl operator as we experienced in two dimensions.

To illustrate what may happen, we consider the simple eigenvalue problem associated with a vacuum-filled PEC cube, $[-1,1]^3$. The cube is filled with $K = 86$ elements, each of fourth order. In Fig. 8.22 we show the first 150 computed eigenvalues compared to the simple analytic ones. We immediately recognize the cloud of small spurious eigenvalues emerging in the spectrum, similar to what we observed in Example 8.1 for the continuous finite element method. The computations are done using unstabilized central fluxes.

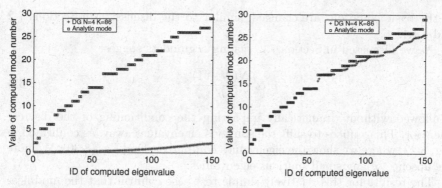

Fig. 8.22. On the left is shown the first 150 eigenvalues computed using a straightforward implementation. The right ($\sigma = N(N+1)/h$) shows results with the stabilization.

With our added understanding we can attribute this problem to a poor representation of the large nullspace of the operator, allowing a large number of spurious eigenvalues to appear. Restricting ourselves to the homogeneous case, we propose to consider the stabilized curl-curl formulation

$$\omega^2 \left(\varepsilon_r \hat{\boldsymbol{E}}_h, \boldsymbol{\phi}_h\right)_{\Omega,h} = \left(\tilde{\nabla}_D \times \boldsymbol{\phi}_h, \boldsymbol{p}_h\right)_{\Omega,h} - \sum_k \sigma \left(\hat{\boldsymbol{n}} \times [\boldsymbol{\phi}_h], \hat{\boldsymbol{n}} \times [\hat{\boldsymbol{E}}_h]\right)_{\partial \mathsf{D}^k},$$

$$\left(\boldsymbol{p}_h, \boldsymbol{\phi}_h\right)_{\Omega,h} = \left(\frac{1}{\mu_r}\tilde{\nabla}_D \times \hat{\boldsymbol{E}}_h, \boldsymbol{\phi}_h\right)_{\Omega,h}. \tag{8.30}$$

for all $\boldsymbol{\phi}_h \in \mathsf{V}_h^3$. The penalty parameter, σ, plays the role of a stabilization or conformity parameter. One should keep in mind, however, that the stabilization simply moves the spurious eigenvalues away from the center of attention by penalizing solutions that violate the tangential continuity condition. However, the stabilization only completely removes these modes for $\sigma \to \infty$.

One way to understand the impact of the above penalization is to consider the equivalent first order local problem

$$i\omega \left(\mu_r \hat{\boldsymbol{H}}_h, \boldsymbol{\phi}_h\right)_{\mathsf{D}^k} = -\left(\tilde{\nabla}_D \times \hat{\boldsymbol{E}}_h, \boldsymbol{\phi}_h\right)_{\mathsf{D}^k}^Y,$$

$$i\omega \left(\varepsilon_r \hat{\boldsymbol{E}}_h, \boldsymbol{\phi}_h\right)_{\mathsf{D}^k} = \left(\tilde{\nabla}_N \times \hat{\boldsymbol{H}}_h, \boldsymbol{\phi}_h\right)_{\mathsf{D}^k}^Z + \sqrt{\sigma}\left(\hat{\boldsymbol{n}} \times \hat{\boldsymbol{Q}}, \boldsymbol{\phi}_h\right)_{\partial \mathsf{D}^k},$$

$$i\omega \left(\hat{\boldsymbol{Q}}, \boldsymbol{\phi}_h\right)_{\partial \mathsf{D}^k} = -\sqrt{\sigma}\left(\hat{\boldsymbol{n}} \times [\hat{\boldsymbol{E}}_h], \boldsymbol{\phi}_h\right)_{\partial \mathsf{D}^k}^Y.$$

For $\hat{\boldsymbol{Q}}$ different from zero, $\hat{\boldsymbol{E}}_h$ must be strictly curl-conforming for ω vanishing, in agreement with the continuous equation. Choosing σ correctly is naturally at the heart of the problem; that is, increasing σ forces the solution toward tangential continuity, essentially approaching a curl-conforming discretization. Choosing σ too large, however, may impact the accuracy in an adverse manner as the basic equations are changed, similar to the discussion in Sections 7.2 and 8.1.1.

As we discussed in Section 7.2, scaling arguments suggest

$$\sigma \propto \frac{N(N+1)}{h}, \tag{8.31}$$

is allowed without dramatically impacting the conditioning of the discrete operator. This suffices to shift the spurious eigenvalues away, as confirmed in Fig. 8.22, where we show the eigenvalues computed using Eq. (8.30). We note the absence of any small spurious eigenvalues.

The results for this relatively simple test case confirm that the modification of the curl-curl operator is required to avoid the introduction of small spurious eigensolutions. In [34, 35] these observations are confirmed by extending the results in Theorem 8.2 to the second order curl-curl equation for Maxwell's equations; for example, one can expect an $\mathcal{O}(h^{2N})$ convergence rate of the eigenvalues for symmetric formulations. Furthermore, the stabilization relocates all spurious modes for h sufficiently small; that is, they accumulate toward ∞.

As successful as the above modification appears to be, a simple alternative is suggested by the discussion of the two-dimensional problem where we observed that the spurious modes supported by the scheme appear to be severely damped when an upwind flux is used. This suggests that using such a flux, while perhaps not removing the unphysical eigensolutions, yields a clear separation in the spectrum that would make the distinction between physical and spurious modes possible.

To illustrate this, we show in Fig. 8.23 the full eigenvalue spectrum for the first-order form solved in a $[-1, 1]^3$ PEC cube discretized using $K = 86$, $N = 3$ order elements and an upwind flux. As one would hope, the dissipation is indeed associated with the spurious part of the small eigenvalues.

While the initial results suggest that a straightforward discrete approximation of the three-dimensional Maxwell curl-curl problem may suffer from problems associated with a poor approximation of the large nullspace, the above results confirm that minor modifications suffice to overcome the problem. Which approach is to be preferred may well be problem dependent; that is, while the stabilization maintains the symmetry of the operator, large values of τ may adversely impact accuracy and problem stiffness. The use of upwind fluxes, however, is natural in the time domain and the above suggests a simple unified formulation.

8.2.3 Consequences in the time domain

Let us finally return to the original time-domain Maxwell's equations in an attempt to understand how the conclusions drawn from the analysis of the eigenvalue problems translates into the time domain.

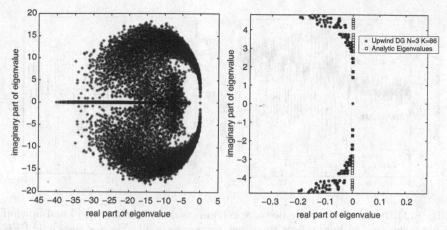

Fig. 8.23. On the left is plotted all computed eigenvalues for a $[-1,1]^3$ PEC cube discretized using $K = 86,\ N = 3$ order elements and an upwind flux, $\alpha = 1$. On the right is shown a close up of the region close to the origin, marking also the physical modes.

For the three-dimensional case, the existence of spurious modes for central fluxes suggest that there may be problems for which low-frequency spurious solutions need to be controlled and suppressed in time-domain computations. As we have already seen, the use of upwind fluxes appears to suffice for the control of these modes. However, inspired by the success of the stabilization of the curl-curl operator, one can likewise consider the locally modified semi-discrete time-domain scheme

$$\frac{d}{dt}\left(\varepsilon_r \boldsymbol{E}_h, \phi_h\right)_{\mathsf{D}^k} = \left(\tilde{\nabla}_N \times \boldsymbol{H}_h, \phi_h\right)_{\mathsf{D}^k}^{Z} + \sum_{k=1}^{K} \sqrt{\sigma_1}\left(\hat{\boldsymbol{n}} \times \boldsymbol{P}, \phi_h\right)_{\partial \mathsf{D}^k},$$

$$\frac{d}{dt}\left(\boldsymbol{P}, \phi_h\right)_{\partial \mathsf{D}^k} = -\frac{\sqrt{\sigma_1}}{2}\left(\hat{\boldsymbol{n}} \times [\boldsymbol{E}_h], \phi_h\right)_{\partial \mathsf{D}^k},$$

$$\frac{d}{dt}\left(\mu_r \boldsymbol{H}_h, \phi_h\right)_{\mathsf{D}^k} = -\left(\tilde{\nabla}_D \times \boldsymbol{E}_h, \phi_h\right)_{\mathsf{D}^k}^{Y} + \sqrt{\sigma_2}\left(\hat{\boldsymbol{n}} \times \boldsymbol{Q}, \phi_h\right)_{\partial \mathsf{D}^k},$$

$$\frac{d}{dt}\left(\phi_h, \boldsymbol{Q}\right)_{\partial \mathsf{D}^k} = -\frac{\sqrt{\sigma_2}}{2} - \left(\hat{\boldsymbol{n}} \times [\boldsymbol{H}_h], \phi_h\right)_{\partial \mathsf{D}^k}.$$

$$(8.32)$$

We have added two additional equations, although based directly on the previous discussion, only one should be needed; that is, in general, only one of σ_1 or σ_2 are different from zero. The additional equation is an ordinary differential equation over the face of the elements only; that is, apart from the additional storage, very little work is added. One can easily show that Eq. (8.32) reduces to Eq. (8.30) by taking $\sigma_1 = 0$ and eliminating $\boldsymbol{E}_h$ and $\boldsymbol{Q}$.

Choosing the σ's sufficiently large shifts the spurious modes up the spectrum and, as in the frequency domain, can be expected to remove the low-frequency spurious modes by weakly enforcing curl conformity. On the other

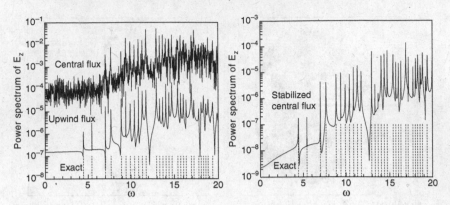

Fig. 8.24. In (a) we show the power spectrum computed using central and upwind fluxes. The vertical lines reflect the exact resonance frequencies for the cubic PEC resonator. In (b) the power spectrum computed using central flux with stabilization ($\sigma_1 = 0$, $\sigma_2 = 20$) is shown.

hand, a large value of σ is also likely to add additional stiffness to the problem, requiring a smaller explicit timestep. In this respect, one would like to choose σ as small as possible, although a scaling as in Eq. (8.31) is not expected to dramatically increase the stiffness.

To illustrate the impact of the stabilization in the time domain, we consider the simple PEC $[-1, 1]^3$ cavity with a very sharp initial pulse, placed off-center inside the cavity. By taking a time trace of the field components and computing the power spectrum through a Fourier transform, one can compute the eigenfrequencies (i.e., the eigenvalues) of the resonant modes. For this simple case, the exact ones are simply

$$\omega = \pi \sqrt{m^2 + n^2 + l^2}, \quad m + n + l \geq 2.$$

Using 286 elements at fourth order, we show in Fig. 8.24 the power spectrum of the E^z component based on integration until $T = 250$. Using an upwinded flux, we see an excellent agreement between the computed and the exact eigenfrequencies. However, a central flux without stabilization results in a very noisy spectrum where one cannot easily identify the resonance frequencies. This is in complete agreement with the results shown in Fig. 8.22, predicting a large number of low-frequency spurious modes. However, as also shown in Fig. 8.24, using the stabilized central-flux scheme with $\sigma_1 = 20$, $\sigma_2 = 0$ results in a clean lower spectrum, containing all the resonances as expected. Similar results can be obtained by taking $\sigma_1 = 0$ and $\sigma_2 = 20$. As long at the σ's scale as Eq. 8.31, the additional term introduces very little additional stiffness (i.e., the stabilization comes at very little cost).

It is worth noticing that the stabilized scheme in Eq. (8.32) is very similar to the internal penalty scheme for the wave equation recently introduced in [129, 130].

9

Curvilinear elements and nonconforming discretizations

All previous chapters have focused almost exclusively on the simplest cases of geometric discretizations where all boundaries are assumed to be piecewise linear and all elements share faces of equal size and order of approximation. However, one of the major advantages of discontinuous Galerkin (DG) methods lies in their flexibility to go beyond these cases and support more complex situation.

In this chapter we discuss the modifications required to extend the linear conforming elements to include the treatment of meshes containing curvilinear elements and/or non-conforming elements. As we will see, the required changes are limited, but the advantages of doing so can be dramatic in terms of improvements in accuracy and reductions in computational effort.

9.1 Isoparametric curvilinear elements

Let us begin with an example to illustrate the advantages of using curvilinear elements.

Example 9.1. We will seek the solution of the TM form of Maxwell's equations (see Section 6.5) in a unit radius, cylindrical, metallic cavity. One of the exact resonant modes is given by

$$H^x\left(x, y, t = 0\right) = 0, \quad H^y\left(x, y, t = 0\right) = 0,$$
$$E^z\left(x, y, t = 0\right) = J_6(\alpha_6 r) \cos(6\theta) \cos(\alpha_6 t),$$

where

$$\alpha_6 = 13.589290170541217,$$
$$r = \sqrt{x^2 + y^2}, \quad \theta = \text{atan2}\left(y, x\right),$$

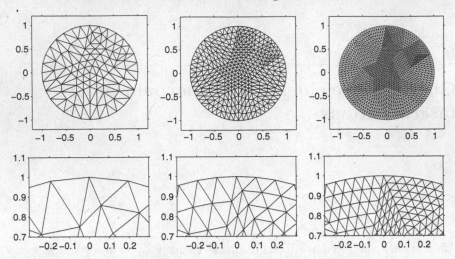

Fig. 9.1. Top: Sequence of meshes used in convergence test for time-domain Maxwell's equations solved using a DG method with straight-sided elements. Bottom: Zoom of top part of sequence of meshes.

Table 9.1. L^2-error in the electric field, E^z, and convergence rates for the sequence of meshes in Fig. 9.1 and a range of polynomial orders used to discretize a cylindrical cavity. Straight-sided elements are used to represent the geometry.

N	h	$h/2$	$h/4$	Rate
1	1.09E-01	3.78E-02	8.96E-03	1.80
2	2.37E-02	2.70E-03	3.58E-04	3.02
3	5.77E-03	1.09E-03	2.70E-04	2.21
4	4.39E-03	1.09E-03	2.72E-04	2.01
5	4.38E-03	1.09E-03	2.72E-04	2.00
6	4.40E-03	1.09E-03	2.72E-04	2.01
7	4.41E-03	1.09E-03	2.73E-04	2.01

and $J_6(z)$ is the sixth Bessel function of the first kind. This mode has six periods in the azimuthal direction. Homogeneous Dirichlet data are assumed for the electric field component.

In a simple numerical experiment we use meshes consisting of straight-sided elements as shown in Fig. 9.1, and integrated until final time $T = 0.5$, at which point we measure the L^2-error in the electric field. The estimated convergence rates are shown in Table 9.1 and it is evident that the solver from Section 6.5 achieves a sub optimal second-order rate of convergence.

This reduction of the convergence rate is caused by the physical boundary conditions, suitable for the actual cylinder, being applied on the straight-sided elements at the mesh boundary. Due to the improper representation of the curvilinear boundary, the convergence rate is significantly reduced (to about second order) even using high-order nodal approximation of the magnetic and electric fields. To remedy this, we must represent the geometry to the order of the scheme, resulting in the need to introduce a curvilinear representation of the cylinder boundary and to modify the underlying scheme accordingly.

The above problem is a prototypical example of the kind of issues one encounter when the geometry representation does not complement the accuracy of the solver. This results in situations where it is the geometry representation rather than the quality of the numerical method that determines the fidelity of the computed result. There are several examples in the literature of how this impacts different applications; for example, [21] illustrates the importance of using curved elements when solving a gas dynamics problems modeled by the Euler equations. Approximate techniques to address this problem are discussed in [210].

9.1.1 Forming the curvilinear element

In the following we will outline the steps required to modify elements near curvilinear boundaries so they conform accurately with the boundaries. Pragmatically, we cannot expect that a mesh generator will automatically place the nodes of these faces exactly on the desired boundary. In the first phase, we must therefore adjust the finite element mesh to ensure that the boundary mesh vertices lie on the boundary. For this we take the following approach:

1. Identify element faces that need to be curved.
2. Ensure that the vertices and face nodes forming these element faces are pushed onto the curvilinear boundary.
3. Update the volume node locations in the elements that have vertices on the curvilinear boundary.

In Fig. 9.2 we illustrate the process through an example of an eight order element, with one side touching the unit circle. The vertices in the element are ensured to touch the unit circle, by nudging them in the radial direction. The nodes on the interior of the edge are then redistributed by arc-length along the segment of the unit circle between two vertices. Finally, we apply the blending to create the mapping in the entire element.

The implementation of these steps are demonstrated in MakeCylinder2D.m, and in the following, we describe the individual components of this algorithm in parallel with the code itself. As input one provides a list of the element faces to be curved as well as information about the exact description of the

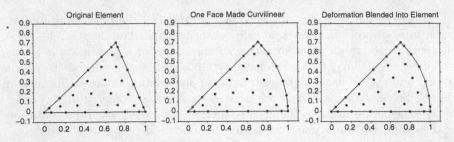

Fig. 9.2. Left: Original straight-sided element. Central: Nodes on the face are deformed to a circle. Right: The deformation is blended to the interior nodes.

curvilinear boundary, which, in this example, amounts to the radius and center of the cylinder.

```
                          ┌─────────────────┐
─────────────────────────│ MakeCylinder2D.m │─────────────────────
                          └─────────────────┘
1   function MakeCylinder2D(faces, ra,xo,yo)
2
3   % Function: MakeCylinder2D(faces, ra, xo, yo)
4   % Purpose:  Use Gordon-Hall blending with an isoparametric map to
5   %           modify a list of faces so they conform to a cylinder
6   %           of radius r centered at (xo,yo)
7   Globals2D;
8
9   NCurveFaces = size(faces,1);
10  vflag = zeros(size(VX));
11  for n=1:NCurveFaces
12
13    % move vertices of faces to be curved onto circle
14    k = faces(n,1); f = faces(n,2);
15    v1 = EToV(k, f); v2 = EToV(k, mod(f,Nfaces)+1);
16    theta1 = atan2(VY(v1),VX(v1)); theta2 = atan2(VY(v2),VX(v2));
17    newx1 = xo + ra*cos(theta1); newy1 = yo + ra*sin(theta1);
18    newx2 = xo + ra*cos(theta2); newy2 = yo + ra*sin(theta2);
19
20    % update mesh vertex locations
21    VX(v1) = newx1; VX(v2) = newx2; VY(v1) = newy1; VY(v2) = newy2;
22
23    % store modified vertex numbers
24    vflag(v1) = 1;  vflag(v2) = 1;
25  end
─────────────────────────────────────────────────────────────────
```

Once the finite element vertex nodes are correctly positioned, a list of elements containing these nodes is defined, and the volume node positions of the elements are adjusted to correctly match with the finite element mesh, as demonstrated in

---
MakeCylinder2D.m

```
27   % map modified vertex flag to each element
28   vflag = vflag(EToV);
29
30   % locate elements with at least one modified vertex
31   ks = find(sum(vflag,2)>0);
32
33   % build coordinates of all the corrected nodes
34   va = EToV(ks,1)'; vb = EToV(ks,2)'; vc = EToV(ks,3)';
35   x(:,ks) = 0.5*(-(r+s)*VX(va)+(1+r)*VX(vb)+(1+s)*VX(vc));
36   y(:,ks) = 0.5*(-(r+s)*VY(va)+(1+r)*VY(vb)+(1+s)*VY(vc));
```

This volume correction is important to ensure that the element modifications propagate to all elements sharing these vertices before adjustment.
After the straight-sided elements are correctly positioned, the next step is
to parameterize each segment of the curvilinear boundary by deforming the
straight-sided faces. Using the circular boundary as a specific example, we
find the polar angles of the vertices of the faces to be curved relative to the
polar coordinates centered at the cylinder axis by

MakeCylinder2D.m

```
38   for n=1:NCurveFaces   % deform specified faces
39     k = faces(n,1); f = faces(n,2);
40
41     % find vertex locations for this face and tangential coordinate
42     if(f==1) v1 = EToV(k,1); v2 = EToV(k,2); vr = r; end
43     if(f==2) v1 = EToV(k,2); v2 = EToV(k,3); vr = s; end
44     if(f==3) v1 = EToV(k,1); v2 = EToV(k,3); vr = s; end
45     fr = vr(Fmask(:,f));
46     x1 = VX(v1); y1 = VY(v1); x2 = VX(v2); y2 = VY(v2);
47
48     % move vertices at end points of this face to the cylinder
49     theta1 = atan2(y1-yo, x1-xo); theta2 = atan2(y2-yo, x2-xo);
```

To avoid severe distortion of the volume nodes and consequently destroy
the approximation property of the nodal basis, we deform the volume nodes
that happen to be on curved faces so that they maintain the same Legendre-
Gauss-Lobatto distribution in the parameter space of the curvilinear boundary
(i.e., the cylinder). A linear distribution of polar angles between the end-point
angles is created. A minor twist is introduced because of the multivalued
nature of the polar angle that is addressed by shifting these angles by multiples
of 2π until they have the same sign. For other types of curvilinear boundary
this may not be a problem. Once the end angles are computed correctly, we
recover the deformation required to shift the volume nodes on each curved face
to their new distribution (i.e., having a Legendre-Gauss-Lobatto distribution,
by arc-length, on the cylinder).

────────────────── MakeCylinder2D.m ──────────────────

```
51   % check to make sure they are in the same quadrant
52   if ((theta2 > 0) & (theta1 < 0)), theta1 = theta1 + 2*pi; end;
53   if ((theta1 > 0) & (theta2 < 0)), theta2 = theta2 + 2*pi; end;
54
55   % distribute N+1 nodes by arc-length along edge
56   theta = 0.5*theta1*(1-fr) + 0.5*theta2*(1+fr);
57
58   % evaluate deformation of coordinates
59   fdx = xo + ra*cos(theta)-x(Fmask(:,f),k);
60   fdy = yo + ra*sin(theta)-y(Fmask(:,f),k);
61
62   % build 1D Vandermonde matrix for face nodes and volume nodes
63   Vface = Vandermonde1D(N, fr);  Vvol  = Vandermonde1D(N, vr);
64   % compute unblended volume deformations
65   vdx = Vvol*(Vface\fdx); vdy = Vvol*(Vface\fdy);
```

Next, we blend [134] this boundary deformation of each curved face into
the element interiors so that the deformation is zero on the other two faces
of each triangle. This should be familiar to the diligent reader, as it is closely
related to the way that the nodes themselves are constructed in Section 6.1.
The geometry-specific deformation of the nodes is subsequently computed for
each such curved element and the element normals and geometric factors are
updated to account for the curvilinear boundary. This last sequence of tasks
is encoded in

────────────────── MakeCylinder2D.m ──────────────────

```
67   % blend deformation and increment node coordinates
68   ids = find(abs(1-vr)>1e-7); % warp and blend
69   if(f==1) blend = -(r(ids)+s(ids))./(1-vr(ids)); end;
70   if(f==2) blend =       +(r(ids)+1)./(1-vr(ids)); end;
71   if(f==3) blend = -(r(ids)+s(ids))./(1-vr(ids)); end;
72
73   x(ids,k) = x(ids,k)+blend.*vdx(ids);
74   y(ids,k) = y(ids,k)+blend.*vdy(ids);
75 end
76
77 % repair other coordinate dependent information
78 Fx = x(Fmask(:), :); Fy = y(Fmask(:), :);
79 [rx,sx,ry,sy,J] = GeometricFactors2D(x, y,Dr,Ds);
80 [nx, ny, sJ] = Normals2D(); Fscale = sJ./(J(Fmask,:));
```

9.1.2 Building operators on curvilinear elements

We recall from Section 3.2 that a generalized Vandermonde matrix, constructed with orthonormal polynomial basis functions, was used to compute

inner products needed in the implementation of the DG method. This approach is successful when the integrands are purely polynomial and the range of integration coincides with the range of the orthonormal polynomials. However, in this chapter we must consider integrals of nonpolynomial functions because of rational polynomial Jacobians induced by the coordinate mappings for curvilinear elements.

To handle the case where the integrands are non-polynomial (e.g., for curvilinear elements), it is recommended to carefully use cubature formulas to evaluate the elemental inner products, much as was discussed in Section 6.6.1 to deal with strongly nonlinear terms. A cubature is nothing more than a multidimensional version of the classic quadrature formulae; that is, a cubature is a set of N_c two-dimensional points $\{r_i^c, s_i^c\}_{i=1}^{N_c}$ with N_c associated weights $\{w_i^c\}_{i=1}^{N_c}$, where the number of points N_c depends on the desired maximum order of polynomial required to be integrated accurately. A survey of cubature formulas was performed in [84, 85, 294], where many different methods are reviewed. Here, we use the symmetric rules proposed in [315]. To compute an approximation of the inner product of two functions f and g on the reference element using an N-th-order cubature with $N_c(N)$ nodes, we evaluate

$$(f,g)_{\mathsf{D}^k} \approx \sum_{i=1}^{N_c} g\left(r_i^c, s_i^c\right) f\left(r_i^c, s_i^c\right) J_i^k w_i^c,$$

where J_i^k is the Jacobian of the polynomial map from the reference element to the physical element

$$J_i^k = \mathrm{Det}\left.\frac{\partial\left(x^k, y^k\right)}{\partial\left(r, s\right)}\right|_{(r_i^c, s_i^c)} = \left.\left(\frac{\partial x^k}{\partial r}\frac{\partial y^k}{\partial s} - \frac{\partial x^k}{\partial s}\frac{\partial y^k}{\partial r}\right)\right|_{(r_i^c, s_i^c)}.$$

Recall that (x, y) are both polynomial functions of (r, s) for the curvilinear elements and we have assumed that the coordinate transform relies on an isoparametric map of the form

$$x^k(r, s) = \sum_{i=1}^{N_p} \psi_i(r, s) x_i^k, \quad y^k(r, s) = \sum_{i=1}^{N_p} \psi_i(r, s) y_i^k,$$

for some set of coefficients $\{x_i^k, y_i^k\}_{i=1}^{N_p}$ on the k-th element using the polynomial basis $\{\psi_i\}_{i=1}^{N_p}$. A particularly convenient choice of basis functions is the multivariate Lagrange basis $\ell_i^k(x)$, discussed in Section 6.1, in which case the coefficients (x_i^k, y_i^k) correspond to the physical location of the i-th interpolation node on the k-th element. We can evaluate the geometric factors at the cubature nodes for use in the Jacobian by

$$\left.\frac{\partial x^k}{\partial r}\right|_{(r_i^c,s_i^c)} = \sum_{j=1}^{N_p} \left.\frac{\partial \ell_j^k}{\partial r}\right|_{(r_i^c,s_i^c)} x_j^k, \quad \left.\frac{\partial x^k}{\partial s}\right|_{(r_i^c,s_i^c)} = \sum_{j=1}^{N_p} \left.\frac{\partial \ell_j^k}{\partial s}\right|_{(r_i^c,s_i^c)} x_j^k \quad (9.1)$$

$$\left.\frac{\partial y^k}{\partial r}\right|_{(r_i^c,s_i^c)} = \sum_{j=1}^{N_p} \left.\frac{\partial \ell_j^k}{\partial r}\right|_{(r_i^c,s_i^c)} y_j^k, \quad \left.\frac{\partial y^k}{\partial s}\right|_{(r_i^c,s_i^c)} = \sum_{j=1}^{N_p} \left.\frac{\partial \ell_j^k}{\partial s}\right|_{(r_i^c,s_i^c)} y_j^k. \quad (9.2)$$

The right-hand-side derivatives of the Lagrange interpolants can be computed in a number of ways. Using previously introduced functionality, the most convenient approach is to view these as requiring the interpolation of the derivatives of Lagrange interpolant functions to cubature nodes (i.e., by introducing differentiation matrices as discussed in Section 3.2 and 6.2). We use Dmatrices2D.m to compute the derivatives of the Lagrange interpolant functions at the cubature nodes with

```
>> [cDr,cDs] = Dmatrices2D(N,cR,cS, V);
```

The partial derivatives of ℓ_j^k with respect to (r, s) are used to evaluate the geometric factors as in Section 6.2. These operations are all performed inside BuildCubatureMesh2D, which evaluates the Lagrange interpolants, their local derivatives, and the geometric factors at the cubature nodes.

──────────────── BuildCurvedOPS2D.m ────────────────

```
8   % 1. Create cubature information
9
10  % 1.1 Extract cubature nodes and weights
11  [cR,cS,cW,Ncub] = Cubature2D(intN);
12
13  % 1.1. Build interpolation matrix (nodes->cubature nodes)
14  cV = InterpMatrix2D(cR, cS);
15
16  % 1.2 Evaluate derivatives of Lagrange interpolants at cubature
17  % nodes
18  [cDr,cDs] = Dmatrices2D(N, cR, cS, V);
```

As well as changing the way that volume integrals are evaluated, we are also obliged to change the computation of the element boundary flux. To handle the surface integrals, we lay down a one-dimensional Gauss quadrature on each face of each element. To simplify the implementation process, the interpolation matrices are created on the reference element and then interpolated from the volume node data to the Gauss nodes on each face. Additional derivative matrices that allow us to evaluate the gradient of volume nodal data at the Gauss face nodes are also created. This is achieved in

─────────────── | BuildCurvedOPS2D.m | ───────────────

```
20  % 2. Create surface quadrature information
21
22  % 2.1 Compute Gauss nodes and weights for 1D integrals
23  [gz, gw] = JacobiGQ(0, 0, intN);
24
25  % 2.2 Build Gauss nodes running counter-clockwise on element faces
26  gR = [gz, -gz, -ones(size(gz))];
27  gS = [-ones(size(gz)), gz, -gz];
28
29  % 2.3 For each face
30  for f1=1:Nfaces
31    % 2.3.1 build nodes->Gauss quadrature interpolation and
32    % differentiation matrices
33    gV(:,:,f1) = InterpMatrix2D(gR(:,f1), gS(:,f1));
34    [gDr(:,:,f1),gDs(:,:,f1)] = Dmatrices2D(N, gR(:,f1), gS(:,f1), V);
35  end
```

To streamline the evaluation of derivatives, we consider each curved element and create customized physical space "x" and "y" nodal derivative matrices using an L^2-projection. The "x" derivative matrix, $\mathcal{D}_x$, is defined by

$$\sum_{l=1}^{N_p} \left(\ell_n, \mathcal{D}_{x,(l,m)}\ell_l \right)_{\mathsf{D}^k} = \left(\ell_n, \frac{\partial \ell_m}{\partial x} \right)_{\mathsf{D}^k}, \tag{9.3}$$

for $1 \leq n,m \leq N_p$, where ℓ_n is the n-th Lagrange interpolant basis function. For simplicity, we have dropped the direct reference to element k. We evaluate $\mathcal{D}_x$ as

$$\sum_{l=1}^{N_p} \left(\ell_n, \mathcal{D}_{x,(l,m)}\ell_l \right)_{\mathsf{D}^k} = \sum_{l=1}^{N_p} \left(\ell_n, \ell_l \right)_{\mathsf{D}^k} \mathcal{D}_{x,(l,m)},$$

$$\approx \sum_{l=1}^{N_p} \sum_{i=1}^{N_c} \ell_n \left(r_i^c, s_i^c \right) \ell_l \left(r_i^c, s_i^c \right) w_i^c J_i^c \mathcal{D}_{x,(l,m)},$$

and we can express this last term in matrix notation as

$$(\mathcal{V}^c)^T \mathcal{W}\mathcal{V}^c\mathcal{D}_x, \tag{9.4}$$

where $\mathcal{V}_{ij}^c = \ell_j \left(r_i^c, s_i^c \right)$ and the $\mathcal{W}$ matrix is diagonal with entries $\mathcal{W}_{ii} = w_i^c J_i^c$.

Using the chain rule and cubature approximation of the integrals the right-hand-side of Eq. (9.3) becomes

$$\left(\ell_n, \frac{\partial \ell_m}{\partial x} \right)_{\mathsf{D}^k} = \left(\ell_n, \frac{\partial r}{\partial x}\frac{\partial \ell_m}{\partial r} \right)_{\mathsf{D}^k} + \left(\ell_n, \frac{\partial s}{\partial x}\frac{\partial \ell_m}{\partial s} \right)_{\mathsf{D}^k}$$

$$= \sum_{i=1}^{N_c} \ell_n \left(r_i^c, s_i^c \right) w_i^c J_i^c \left(\frac{\partial r}{\partial x}\frac{\partial \ell_m}{\partial r} + \frac{\partial s}{\partial x}\frac{\partial \ell_m}{\partial s} \right)\Big|_{\left(r_i^c, s_i^c \right)}.$$

The presence of the nonconstant Jacobian and geometric factors requires individual treatment for each curvilinear element. Rewriting the above in matrix form, the implementation is straightforward. The following implementation demonstrates how custom differentiation and lift matrices are created for each curvilinear element and stored in a structure called cinfo.

```
──────────────────── BuildCurvedOPS2D.m ────────────────────
37 % 3. For each curved element, evaluate custom operator matrices
38 Ncurved = length(curved);
39
40 % 3.1 Store custom information in array of Matlab structs
41 cinfo = [];
42 for c=1:Ncurved
43   % find next curved element and the coordinates of its nodes
44   k1 = curved(c); x1 = x(:,k1); y1 = y(:,k1); cinfo(c).elmt = k1;
45
46   % compute geometric factors
47   [crx,csx,cry,csy,cJ] = GeometricFactors2D(x1,y1,cDr,cDs);
48
49   % build mass matrix
50   cMM = cV'*diag(cJ.*cW)*cV; cinfo(c).MM = cMM;
51
52   % build physical derivative matrices
53   cinfo(c).Dx = cMM\(cV'*diag(cW.*cJ)*(diag(crx)*cDr+diag(csx)*cDs));
54   cinfo(c).Dy = cMM\(cV'*diag(cW.*cJ)*(diag(cry)*cDr+diag(csy)*cDs));
```

Having computed the element-specific differentiation matrices for the curved element, we are left with just one more task. For the nodal solver, we must be able to extract the negative and positive traces of the solution at element boundary nodes and to lift the fluxes from these boundary nodes on each face into the interior of the elements using the lift matrix. The nodes at each face are no longer sufficient to extract data from the volume nodes directly, but we are obliged to interpolate the data to the Gauss nodes. Furthermore, the lift matrix is now specific to each face because of the nonconstant volume Jacobian involved in the inverse mass matrix and also the nonconstant surface Jacobian involved in the face mass matrix. Thus, we follow an approach as above to evaluate these element specific matrices for each curved face, but this time using a one-dimensional Gauss quadrature:

```
──────────────────── BuildCurvedOPS2D.m ────────────────────
56   % build individual lift matrices at each face
57   for f1=1:Nfaces
58     k2 = EToE(k1,f1); f2 = EToF(k1,f1);
59
60     % compute geometric factors
61     [grx,gsx,gry,gsy,gJ] = GeometricFactors2D(x1,y1,...
```

```
62        gDr(:,:,f1), gDs(:,:,f1));
63
64    % compute normals and surface Jacobian at Gauss points on face f1
65    if(f1==1) gnx = -gsx;        gny = -gsy;        end;
66    if(f1==2) gnx =  grx+gsx; gny =  gry+gsy; end;
67    if(f1==3) gnx = -grx;        gny = -gry;        end;
68
69    gsJ = sqrt(gnx.*gnx+gny.*gny);
70    gnx = gnx./gsJ;  gny = gny./gsJ;  gsJ = gsJ.*gJ;
71
72    % store normals and coordinates at Gauss nodes
73    cinfo(c).gnx(:,f1) = gnx;  cinfo(c).gx(:,f1)  = gV(:,:,f1)*x1;
74    cinfo(c).gny(:,f1) = gny;  cinfo(c).gy(:,f1)  = gV(:,:,f1)*y1;
75
76    % store Vandermondes for '-' and '+' traces
77    cinfo(c).gVM(:,:,f1) = gV(:,:,f1);
78    cinfo(c).gVP(:,:,f1) = gV(end:-1:1,:,f2);
79
80    % compute and store matrix to lift Gauss node data
81    cinfo(c).glift(:,:,f1) = cMM\(gV(:,:,f1)'*diag(gw.*gsJ));
82  end
```

9.1.3 Maxwell's equations on meshes with curvilinear elements

Having constructed the new curvilinear elements and all necessary integration and differentiation rules appropriate for these elements, we can revisit the solution of the time-domain Maxwell's equations, discussed in Example 9.1.

We recall the DG formulation to solve Maxwell's equations (see Section 6.5)

$$
\left(\phi_h, \frac{\partial \boldsymbol{H}_h}{\partial t} \right)_{\Omega,h} = (\phi_h, -\nabla \times \boldsymbol{E}_h)_{\Omega,h}
$$

$$
+ \frac{1}{2} \sum_{k=1}^{K} \left(\phi_h, (\hat{\boldsymbol{n}} \times [\boldsymbol{E}_h] + \hat{\boldsymbol{n}} \times \hat{\boldsymbol{n}} \times [\boldsymbol{H}_h]) \right)_{\partial \mathsf{D}^k},
$$

$$
\left(\phi_h, \frac{\partial \boldsymbol{E}_h}{\partial t} \right)_{\Omega,h} = (\phi_h, \nabla \times \boldsymbol{H}_h)_{\Omega,h}
$$

$$
+ \frac{1}{2} \sum_{k=1}^{K} \left(\phi_h, (-\hat{\boldsymbol{n}} \times [\boldsymbol{H}_h] + \hat{\boldsymbol{n}} \times \hat{\boldsymbol{n}} \times [\boldsymbol{E}_h]) \right)_{\partial \mathsf{D}^k},
$$

for all test functions, $\phi_h \in \mathsf{V}_h \times \mathsf{V}_h \times \mathsf{V}_h$ with $\mathsf{V}_h = \bigoplus_k P_N(\mathsf{D}^k)$ being the space of piecewise polynomial functions.

We will now upgrade the nodal DG scheme for solving the TM Maxwell's equations presented in Section 6.5 to accurately handle curvilinear domain boundaries. No substantial modifications are needed, since we can reuse the nodal DG right-hand-side function MaxwellRHS2D.m to evaluate the right-hand-side terms of Maxwell's equations and then replace the nodal DG treatment on the curvilinear elements with a hybrid cubature/Gauss DG treatment. This is demonstrated in the following piece of MaxwellCurvedRHS2D.m:

```
                           ┌─────────────────────┐
───────────────────────────│ MaxwellCurvedRHS2D.m │───────────────────────
                           └─────────────────────┘
1 function [rhsHx, rhsHy, rhsEz] = MaxwellCurvedRHS2D(cinfo,Hx,Hy,Ez)
2
3 % function [rhsHx, rhsHy, rhsEz] = MaxwellCurvedRHS2D(cinfo,Hx,Hy,Ez)
4 % Purpose   : Evaluate RHS flux in 2D Maxwell TM form
5
6 Globals2D;
7
8 [rhsHx,rhsHy,rhsEz] = MaxwellRHS2D(Hx, Hy, Ez);
```

Each curved element is then processed individually, taking into account the custom differentiation matrices and lift functions as discussed previously. First, we compute the local curls of the electric and magnetic fields by

```
                           ┌─────────────────────┐
───────────────────────────│ MaxwellCurvedRHS2D.m │───────────────────────
                           └─────────────────────┘
10 % correct residuals at each curved element
11 Ncinfo = length(cinfo);
12 for n=1:Ncinfo
13
14   % for each curved element computed L2 derivatives via cubature
15   cur = cinfo(n); k1 = cur.elmt; cDx = cur.Dx; cDy = cur.Dy;
16
17   rhsHx(:,k1) = -cDy*Ez(:,k1);
18   rhsHy(:,k1) =  cDx*Ez(:,k1);
19   rhsEz(:,k1) =  cDx*Hy(:,k1) - cDy*Hx(:,k1);
```

and, subsequently, the jump in the electric and magnetic fields at the Gauss nodes on each face of the element. These are then used to evaluate the flux functions, and the fluxes are lifted from the Gauss nodes to the element interior. Because the custom matrices are all set up in preprocessing, the additional implementation details to achieve this task are limited to the following script:

```
                           ┌─────────────────────┐
───────────────────────────│ MaxwellCurvedRHS2D.m │───────────────────────
                           └─────────────────────┘
21   % for each face of each curved element use Gauss quadrature based
22   % lifts
23   for f1=1:Nfaces
24     k2 = EToE(k1,f1);
25     gnx = cur.gnx(:,f1); gny = cur.gny(:,f1);
```

```
26       gVM = cur.gVM(:,:,f1); gVP = cur.gVP(:,:,f1);
27       glift = cur.glift(:,:,f1);
28
29       % compute difference of solution traces at Gauss nodes
30       gdHx = gVM*Hx(:,k1) - gVP*Hx(:,k2);
31       gdHy = gVM*Hy(:,k1) - gVP*Hy(:,k2);
32       gdEz = gVM*Ez(:,k1) - gVP*Ez(:,k2);
33
34       % correct jump at Gauss nodes on domain boundary faces
35       if(k1==k2)
36         gdHx = 0*gdHx; gdHy = 0*gdHy; gdEz = 2*gVM*Ez(:,k1);
37       end
38
39       % perform upwinding
40       gndotdH =  gnx.*gdHx+gny.*gdHy;
41       fluxHx =  gny.*gdEz + gndotdH.*gnx-gdHx;
42       fluxHy = -gnx.*gdEz + gndotdH.*gny-gdHy;
43       fluxEz = -gnx.*gdHy + gny.*gdHx   -gdEz;
44
45       % lift flux terms using Gauss based lift operator
46       rhsHx(:,k1) = rhsHx(:,k1) + glift*fluxHx/2;
47       rhsHy(:,k1) = rhsHy(:,k1) + glift*fluxHy/2;
48       rhsEz(:,k1) = rhsEz(:,k1) + glift*fluxEz/2;
49     end
50 end
```

With all the pieces in place, let us revisit Example 9.1.

Example 9.2. We seek the solution of the TM form of Maxwell's equations (see Section 6.5) in a unit radius, cylindrical, metallic cavity and choose the initial data to be

$$H^x (x,y,t=0) = 0, \quad H^y (x,y,t=0) = 0,$$
$$E^z (x,y,t=0) = J_6(\alpha_6 r) \cos(6\theta) \cos(\alpha_6 t),$$

where

$$\alpha_6 = 13.589290170541217,$$
$$r = \sqrt{x^2 + y^2}, \quad \theta = \text{atan2}\,(y,x),$$

and $J_6(z)$ is the sixth Bessel function of the first kind. This mode is an exact resonant mode with six periods in the azimuthal direction, and zero Dirichlet data are assumed for the electric field component.

Using the sequence of meshes in Fig. 9.3 we estimate orders of convergence obtainable with curvilinear elements shown in Table 9.2. We achieve very nearly optimal order of accuracy $\mathcal{O}(h^{N+1})$. Hence, we have eliminated the impact of the geometry on the accuracy through the introduction of a curvilinear truly body-conforming discretization.

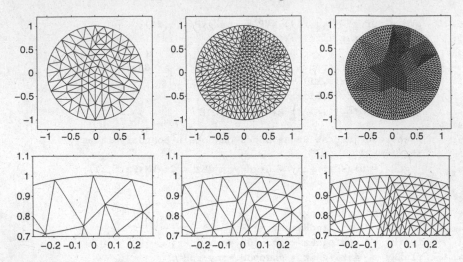

Fig. 9.3. Top: Sequence of meshes used in convergence test of Maxwell's equations with curvilinear elements. Bottom: Zoom of top part of meshes.

Table 9.2. L^2-error in the electric field, E^z, and convergence rates for a sequence of meshes in Fig. 9.3 and the range of polynomial orders used to discretize a cylindrical cavity. Curvilinear elements are used to represent the geometry. An $*$ marks that the accuracy is impacted by finite precision effects.

N	h	$h/2$	$h/4$	Rate
1	1.09E-01	3.78E-02	8.96E-03	1.80
2	2.21E-02	2.23E-03	2.05E-04	3.38
3	3.12E-03	1.92E-04	1.28E-05	3.97
4	6.01E-04	1.95E-05	5.88E-07	5.00
5	9.89E-05	1.69E-06	2.72E-08	5.92
6	1.74E-05	1.31E-07	9.81E-10	7.06
7	2.08E-06	8.97E-09	7.93E-11*	7.34*

9.2 Nonconforming discretizations

In many applications, it can be necessary to adjust the numerical resolution in a subset of the elements to control the global error of the computation. For the DG-FEM, this can be achieved by changing the local order of the approximation, known as order refinement, or by locally modifying the element size. In both cases, however, we have a situation where the order of approximation and/or the element sizes do not conform. This requires some additional attention to the construction of the numerical fluxes and the connectivity of the global grid.

Identifying which elements in a computational grid should be modified to increase the overall accuracy is the subject of error estimation. This is another very interesting area of research that is just now coming into maturity in the context of DG methods. A posteriori error norm estimates have recently been developed for some classes of linear problems [170, 171, 173] and for quasi-linear elliptic problems [177]. The use of adjoint methods for error estimation for conservation laws has been discussed extensively in [146, 147, 175, 176], highlighting the potential of such techniques for time-dependent conservation laws with shocks (e.g., the Euler equations of compressible gas dynamics). An alternative approach to the development of local error estimators is based on the phenomenon of superconvergence. This phenomenon was demonstrated for both elliptic [65] and hyperbolic problems [6, 120] and subsequently extended to multidimensional problems [7, 211] and nonlinear problems [8].

Robust and accurate local and global error estimation enables the adaptive, error-controlled solution of many different problems, possibly first demonstrated in [28] for a large-scale adaptive solution of conservation laws by using local ad-hoc error estimators. A rigorous approach is very complicated for the nonlinear problem and only few results are available [62]. Many of the different techniques in the above references are often problem-specific and a thorough discussion of these estimation techniques is beyond the scope of this text. We refer the interested reader to some of the many references for further exploration of this important topic.

Once an element has been tagged for a change of resolution, one needs to decide how to change the local resolution (i.e., by changing the order of approximation or by adjusting the element size). As we have already seen, the benefits of the high-order basis is most pronounced for locally smooth functions. Hence, a natural strategy is to increase the local order for smooth problems and decrease the local element size for nonsmooth solutions [15, 16, 17, 18, 19]. Such decisions rely on an indicator of the local smoothness, often called a smoothness indicator. For many problems – in particular, linear ones – this judgment is often easy to make, because the nonsmooth behavior is associated with geometric features, sources, or boundary conditions that are known a priori. The nonlinear problem is more complex and various ideas can be brought to bear on this problem; for example, one can estimate the decay rate of the spectral expansion coefficients as a measure of the local regularity [237, 238]. Another more rigorous approach is advocated in [93, 94], requiring computation and comparisons between several grids.

In the following, we will not address this, but simply assume that it has been decided, using some techniques, which elements require some level of refinement to appropriately control the computational error. We first focus attention on the case of element refinement. However, as we will discuss subsequently, very similar steps have be taken to enable order refinement.

9.2.1 Nonconforming element refinement

A major advantage of the ability to support nonconforming elements is that we do not need to propagate refinements beyond the elements selected for refinement. Consequently, very limited book keeping is required to perform the

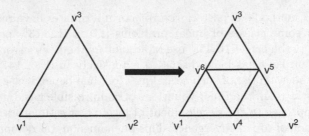

Fig. 9.4. Four-way element refinement used by Hrefine2D.m.

refinement in this way. We will consider the basic four-way element refinement procedure illustrated in Fig. 9.4.

Let us assume that a subset of elements are marked for element refinement. For these elements, we find the vertex triplets and use these to construct uniquely numbered vertices at the centers of the edges of those elements marked for refinement. These new vertices are renumbered to form a consecutive list following the existing numbering. Next, we move the vertices of the marked elements to their face centers and append three new elements for each marked element to complete the four-way split of the original marked triangles. Finally, we add coordinates for the face centers to the list of vertex coordinates.

A concise implementation of this sequence of operations is given in Hrefine2D.m. The first stage is to employ a unique numbering for each element face, determined for face f_1 of element k_1 with neighbor face f_2 of element k_2 by

$$\max\left(f_1 + 3\left(k_1 - 1\right), f_2 + 3\left(k_2 - 1\right)\right). \tag{9.5}$$

We have assumed the element/face pairs are indexed from 1. A vertex at the center of each face with the unique edge number is labeled as

---
Hrefine2D.m

```
1   function Hrefine2D(refineflag)
2
3   % function Hrefine2D(refineflag)
4   % purpose:  apply non-conforming refinement to elements labelled
5   %           in refineflag
6
7   Globals2D;
8
9   % 1.1 Count vertices
10  Nv   = length(VX(:));
11
12  % 1.2 Find and count elements to be refined
13  ref = sort(find(refineflag));
14  Nrefine = length(ref);
15
```

```
16   % 1.3 Extract vertex numbers of elements to refine
17   v1 = EToV(ref, 1); v2 = EToV(ref, 2); v3 = EToV(ref, 3);
18
19   % 1.4 Uniquely number all face centers
20   v4 = max( 1 + Nfaces*(0:K-1)', EToF(:,1) + Nfaces*(EToE(:,1)-1) );
21   v5 = max( 2 + Nfaces*(0:K-1)', EToF(:,2) + Nfaces*(EToE(:,2)-1) );
22   v6 = max( 3 + Nfaces*(0:K-1)', EToF(:,3) + Nfaces*(EToE(:,3)-1) );
```

Next, we extract the identifiers associated with the inserted face center vertices of the elements to be refined and renumber them consistently and consecutively with unique identifiers starting from $N_v + 1$. Recall, that N_v is the number of finite element vertices in the preexisting mesh.

───────────────────────── Hrefine2D.m ─────────────────────────
```
24   % 2.0 Extract face center vertices for elements to refine
25   v4 = v4(ref);       v5 = v5(ref);       v6 = v6(ref);
26
27   % 2.1 Renumber face centers contiguously from Nv+1
28   ids = unique([v4;v5;v6]);
29   newids(ids) = (1:length(ids))';
30   v4 = Nv+newids(v4)'; v5 = Nv+newids(v5)'; v6 = Nv+newids(v6)';
```

The three original vertices of each element are combined with their associated three new face center vertices to form the complex of four new elements in the refined mesh. The new set of four elements replaces the original element and appends three new elements to the end of the element-to-vertex list, EToV.

───────────────────────── Hrefine2D.m ─────────────────────────
```
32  % 2.2 Replace original triangle with triangle connecting edge centers
33  EToV(ref,:) = [v4,v5,v6];
34
35  % 3.0 Add extra triangles to EToV
36  EToV(K+1:K+3*Nrefine,1)=[v1;v2;v3]; % first  vertices of new elements
37  EToV(K+1:K+3*Nrefine,2)=[v4;v5;v6]; % second vertices of new elements
38  EToV(K+1:K+3*Nrefine,3)=[v6;v4;v5]; % third  vertices of new elements
```

In addition to updating EToV, we must also update the boundary condition-type array BCType to reflect the addition of boundary faces when an element adjacent to the boundary is chosen for refinement.

───────────────────────── Hrefine2D.m ─────────────────────────
```
40   % 3.1 Create boundary condition type for refined elements
41   bcsave = BCType(ref,:);
42   BCType(ref, :) = 0; % now internal faces
```

```
43
44    BCType(K+1:K+Nrefine, 1) = bcsave(:, 1);
45    BCType(K+1:K+Nrefine, 3) = bcsave(:, 3);
46
47    BCType(K+Nrefine+1:K+2*Nrefine, 1) = bcsave(:, 2);
48    BCType(K+Nrefine+1:K+2*Nrefine, 3) = bcsave(:, 1);
49
50    BCType(K+2*Nrefine+1:K+3*Nrefine, 1) = bcsave(:, 3);
51    BCType(K+2*Nrefine+1:K+3*Nrefine, 3) = bcsave(:, 2);
```

The coordinates of the face centers are computed and the vertex coordinate rows VX, and VY, are updated, along with the number of elements in the mesh, K. It should be noted that this routine only updates the finite element grid skeleton and further calls to StartUp2D.m are required to complete the refinement procedure.

─────────────────────────── Hrefine2D.m ───────────────────────────
```
53    % 3.2 Find vertex locations of elements to be refined
54    x1 = VX(v1);   x2 = VX(v2);   x3 = VX(v3);
55    y1 = VY(v1);   y2 = VY(v2);   y3 = VY(v3);
56
57    % 3.3 Add coordinates for refined edge centers
58    VX(v4) = 0.5*(x1+x2); VX(v5) = 0.5*(x2+x3); VX(v6) = 0.5*(x3+x1);
59    VY(v4) = 0.5*(y1+y2); VY(v5) = 0.5*(y2+y3); VY(v6) = 0.5*(y3+y1);
60
61    % 3.4 Increase element count
62    K = K+3*Nrefine;
```

Finally, we note that there may be faces in the refined mesh that have more than one physically adjacent (neighboring) face, but these connections are not reflected in the element-to-element EToE and element-to-face EToF connectivity arrays when they are updated by the routines discussed previously. In the following section, we will describe how to handle these type of non-conforming interfaces within the framework of DG methods.

Operators on nonconforming element meshes

Let us now extend the implementations in Chapter 6 to allow for the use of general non-conforming meshes. All previous implementations have assumed that elements sharing a face only share a full face. Thus, we need to develop some infrastructure to handle partial face fragments shared by elements.

The first step is to locate nonconforming element connections by examining all element faces that do not have conforming connections to other elements. We locate these nonconforming boundary faces and store them in a list including their element and face location.

```
                              ┌─────────────────┐
─────────────────────────────│ BuildHNonCon2D.m │─────────────────────
                              └─────────────────┘
 9    % 1. Build Gauss nodes
10    [gz, gw] = JacobiGQ(0, 0, NGauss-1);
11
12    % 1.1 Find location of vertices of boundary faces
13    vx1 = VX(EToV(:,[1,2,3])); vx2 = VX(EToV(:,[2,3,1]));
14    vy1 = VY(EToV(:,[1,2,3])); vy2 = VY(EToV(:,[2,3,1]));
15
16    idB = find(EToE==((1:K)'*ones(1,Nfaces)));
17    x1 = vx1(idB)'; y1 = vy1(idB)';
18    x2 = vx2(idB)'; y2 = vy2(idB)';
19
20    % 1.2 Find those element-faces that are on boundary faces
21    [elmtsB,facesB] = find(EToE==((1:K)'*ones(1,Nfaces)));
22    Nbc = length(elmtsB);
```

Next, we use the fact that only these boundary faces may participate in a partial face-face connection. For each of these nonconforming boundary faces, we perform an intersection test with every other member of the boundary face list.

The first check is to determine if the faces are collinear by computing the area of the quadrilateral formed by joining the two sets of two vertices bounding each element face being tested. If the area is greater than a specified tolerance, then the two faces do not share a face fragment. Otherwise, we must identify one of the three possible positive outcomes for the intersection test for two edges:

1. Face 1 contains all of face 2.
2. Face 1 contains part of face 2.
3. Face 1 is completely contained in face 2.

These situations are shown in Fig. 9.5.

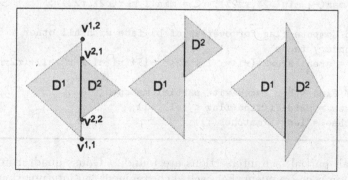

Fig. 9.5. Three possible positive face-face connections for the face intersect test.

The intersection result is prone to be confused by finite precision effects, so all intersection tests are preformed to a nonzero tolerance. A pseudo-code for the sequence of tests used to test if face 1 $\left(v^{1,1}\text{ to }v^{1,2}\right)$ and face 2 $\left(v^{2,1}\text{ to }v^{2,2}\right)$ intersect is as follows:

1. Compute the parameters r^1 and r^2 for the normal projection of the face 2 end vertices onto the line: $x = (1-r)v^{1,1} + rv^{1,2}$.
2. If $r^1, r^2 > 1$, then the test failed.
3. If $r^1, r^2 < 0$, then the test failed.
4. If $|r^1 - r^2| < \epsilon_2$ for some tolerance ϵ_2, the intersection is negligible.
5. If the code reaches this point, the edges intersect to the given tolerances.

BuildHNonCon2D.m

```
24  sk = 1;
25  % 2.1 For each boundary face
26  for b1=1:Nbc
27     % 2.2 Find element and face of this boundary face
28     k1 = elmtsB(b1); f1 = facesB(b1);
29
30     % 2.3 Find end coordinates of b1'th boundary face
31     x11 = x1(b1);   y11 = y1(b1);   x12 = x2(b1);   y12 = y2(b1);
32
33     % 2.4 Compute areas, lengths and face coordinates used in
34     % intersection tests comparing b1'th boundary face with
35     % all boundary faces
36     area1 = abs((x12-x11)*(y1-y11) - (y12-y11)*(x1-x11)); %scale
37     area2 = abs((x12-x11)*(y2-y11) - (y12-y11)*(x2-x11));
38     L    = (x12-x11)^2 + (y12-y11)^2 ;
39     r21 = ((2*x1-x11-x12)*(x12-x11) + (2*y1-y11-y12)*(y12-y11))/L;
40     r22 = ((2*x2-x11-x12)*(x12-x11) + (2*y2-y11-y12)*(y12-y11))/L;
41
42     % 2.5 Find range of local face coordinate
43     % (bracketed between -1 and 1)
44     r1 = max(-1,min(r21,r22)); r2 = min(1,max(r21,r22));
45
46     % 2.6 Compute flag for overlap of b1 face with all other
47     % boundary faces
48     flag = area1+area2+(r1<= -1 & r2<= -1)+(r1>=1 & r2>=1)+(r2-r1<tol);
49
50     % 2.7 Find other faces with partial matches
51     matches = setdiff(find(flag < tol),b1);
52     Nmatches = length(matches(:));
```

Once all partial face intersections are found, a Gauss quadrature grid is laid down on each fragment. For each such fragment, an interpolation matrix is computed for both the negative and positive traces of the neighbor elements at these Gauss nodes. Finally, a lift matrix is computed for each face fragment to lift the face data to the element on the negative trace. These steps are all collected in the following:

```
─────────────────────────  BuildHNonCon2D.m  ─────────────────────────
54   if(Nmatches>0)
55     % 3.1 Find matches
56     r1 = r1(matches); r2 = r2(matches);
57
58     % 3.2 Find end points of boundary-boundary intersections
59     xy11 = 0.5*[x11;y11]*(1-r1) +  0.5*[x12;y12]*(1+r1);
60     xy12 = 0.5*[x11;y11]*(1-r2) +  0.5*[x12;y12]*(1+r2);
61
62     % 3.3 For each face-face match
63     for n=1:Nmatches
64
65       % 3.4 Store which elements intersect
66       k2 = elmtsB(matches(n)); f2 = facesB(matches(n));
67       neighbors{sk}.elmtM = k1; neighbors{sk}.faceM = f1;
68       neighbors{sk}.elmtP = k2; neighbors{sk}.faceP = f2;
69
70       % 3.5 Build physical Gauss nodes on face fragment
71       xg = 0.5*(1-gz)*xy11(1,n) + 0.5*(1+gz)*xy12(1,n);
72       yg = 0.5*(1-gz)*xy11(2,n) + 0.5*(1+gz)*xy12(2,n);
73
74       % 3.6 Find local coordinates of Gauss nodes
75       [rg1,sg1] = FindLocalCoords2D(k1, xg, yg);
76       [rg2,sg2] = FindLocalCoords2D(k2, xg, yg);
77
78       % 3.7 Build interpolation matrices for volume nodes ->Gauss
79       % nodes
80       gVM = InterpMatrix2D(rg1,sg1); neighbors{sk}.gVM  = gVM;
81       gVP = InterpMatrix2D(rg2,sg2); neighbors{sk}.gVP  = gVP;
82
83       % 3.8 Find face normal
84       neighbors{sk}.nx = nx(1+(f1-1)*Nfp,k1);
85       neighbors{sk}.ny = ny(1+(f1-1)*Nfp,k1);
86
87       % 4.0 Build partial face data lift operator
88
89       % 4.1 Compute weights for lifting
90       partsJ = sqrt( (xy11(1,n)-xy12(1,n))^2 ...
91           + (xy11(2,n)-xy12(2,n))^2 )/2;
92       dgw = gw*partsJ/J(1,k1);
93
94       % 4.2 Build matrix to lift Gauss data to volume data
95       neighbors{sk}.lift = V*V'*(gVM')*diag(dgw);
96
```

```
97        sk = sk+1;
98     end
99  end
```

Maxwell's equations on meshes with nonconforming elements

As for the case of meshes with curvilinear elements, only minor algorithmic modifications are required to solve the time-domain Maxwell's equations on meshes with nonconforming elements. We again use the standard nodal DG-FEM on all elements, and then apply corrections to the right-hand-side terms for those elements with nonconforming connectivity. Since the list of boundary nodes now includes nodes on non-conforming interfaces, we replace these lists with the list of nodes on the "Wall" boundaries and then use the standard right-hand-side function for Maxwell's equations, MaxwellRHS2D.m, as follows:

MaxwellHNonConRHS2D.m

```
1  function [rhsHx, rhsHy, rhsEz] = ...
2       MaxwellHNonConRHS2D(Hx,Hy,Ez, neighbors)
3
4  % function [rhsHx, rhsHy, rhsEz] = ...
5  %      MaxwellHNonConRHS2D(Hx,Hy,Ez, neighbors)
6  % Purpose  : Evaluate RHS flux in 2D Maxwell TM form
7
8  Globals2D;
9
10 % 1.0 Only apply PEC boundary conditions at wall boundaries
11 savemapB = mapB; savevmapB = vmapB; mapB = mapW;  vmapB = vmapW;
12
13 % 1.1 Evaluate right hand side
14 [rhsHx, rhsHy, rhsEz] = MaxwellRHS2D(Hx, Hy, Ez);
15
16 % 1.2 Restore original boundary node lists
17 mapB = savemapB; vmapB = savevmapB;
```

In the curvilinear case, we replaced the right-hand-side terms completely for each curved element. In contrast, for the nonconforming case, we loop through the list of nonconforming face pairs and append the lift of the up-winded flux difference terms for each such face fragment to the right-hand-side residual of the appropriate element.

MaxwellHNonConRHS2D.m

```
19 % 2.0 Correct lifted fluxes at each non-conforming face fragment
20 Nnoncon = length(neighbors);
21 for n=1:Nnoncon
```

```
22   neigh = neighbors{n};
23
24   % 2.1 Extract information about this non-conforming face fragment
25   k1 = neigh.elmtM;  gVM = neigh.gVM;
26   k2 = neigh.elmtP;  gVP = neigh.gVP;
27   lnx = neigh.nx;    lny = neigh.ny;
28
29   % 2.2 Compute difference of traces at Gauss nodes on face fragment
30   ldHx = gVM*Hx(:,k1) - gVP*Hx(:,k2);
31   ldHy = gVM*Hy(:,k1) - gVP*Hy(:,k2);
32   ldEz = gVM*Ez(:,k1) - gVP*Ez(:,k2);
33
34   % 2.3 Compute flux terms at Gauss nodes on face fragment
35   lndotdH =  lnx.*ldHx+lny.*ldHy;
36   fluxHx =   lny.*ldEz + lndotdH.*lnx-ldHx;
37   fluxHy = -lnx.*ldEz + lndotdH.*lny-ldHy;
38   fluxEz = -lnx.*ldHy + lny.*ldHx   -ldEz;
39
40   % 2.4 Lift fluxes for non-conforming face fragments and update
41   % residuals
42   lift = neigh.lift;
43   rhsHx(:,k1) = rhsHx(:,k1) + lift*fluxHx/2;
44   rhsHy(:,k1) = rhsHy(:,k1) + lift*fluxHy/2;
45   rhsEz(:,k1) = rhsEz(:,k1) + lift*fluxEz/2;
46 end
```

This completes the modification required to handle nonconforming element interfaces in the DG-FEM used to solve the two-dimensional TM Maxwell's equations (see Section 6.5).

To show this in action, we repeat a convergence study for a single mode of a square perfectly electrically conducting (PEC) cavity but this time with a mesh that has some elements refined by Hrefine2D.m, leaving some nonconforming interfaces. The sequence of meshes are shown in Fig. 9.6 and the estimated order of convergence for $N = 1, ..., 8$ is shown in Table 9.3. The

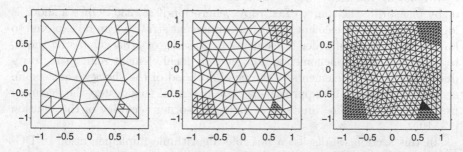

Fig. 9.6. Sequence of meshes used in convergence test for DG time-domain solution of the two-dimensional Maxwell's equations on non-conforming meshes.

Table 9.3. L^2-error in the electric field, E^z, and convergence rates for a sequence of meshes in Fig. 9.6 and the range of polynomial orders N used to discretize a square cavity. Nonconforming elements are used. An $*$ marks that the accuracy is impacted by finite precision effects.

N	h	$h/2$	$h/4$	Rate
1	1.10E-01	2.94E-02	7.48E-03	1.94
2	8.28E-03	9.94E-04	1.09E-04	3.12
3	9.44E-04	7.42E-05	4.28E-06	3.89
4	9.12E-05	2.92E-06	8.32E-08	5.05
5	9.80E-06	1.55E-07	2.30E-09	6.03
6	6.45E-07	4.93E-09	3.68E-11	7.05
7	5.93E-08	2.10E-10	7.86E-13*	8.10*
8	2.78E-09	5.59E-12*	3.96E-14*	8.05*

method achieves optimal order of accuracy, $\mathcal{O}(h^{N+1})$, and is not impacted by the nonconforming elements.

9.2.2 Nonconforming order refinement

In the previous subsection, we demonstrated how fairly arbitrary element refinement can be achieved. The flexibility in mesh refinement is made possible through the use of element nonconforming meshes. We are alternatively able to modify the resolution in desired locations in a mesh by assigning different polynomial order of approximation to each element. The DG framework is particularly flexible in this regard, and because the elements only communicate through fluxes, we are able to choose arbitrary orders of approximation in each element. As an example of how this might be useful, consider the example simulations from Section 5.9, that exhibit traveling shock waves. In such cases, we used a slope limiter to control the gradients in the vicinity of the shock. However, we could alternatively have applied an algorithm to locate the shock by examining inter element solution jumps [5, 6, 28, 120, 211], by evaluating the decay rate of the spectral expansion coefficients [237, 238], or by measuring the source of entropy generation [228, 230]. Once a shock is located, a combined element and order refinement strategy can be explored to lower the polynomial order in elements containing or adjacent to the shock, to avoid the Gibbs phenomenon, and employ local element refinement to reduce the size of the elements in the neighborhood of the shock to achieve an accurate solution. This approach has the potential to sidestep the accuracy problem with the slope limiter and yield high-order accuracy of the solution [15, 19].

In this section we discuss a possible algorithmic implementation of DG-FEM's on element-conforming meshes with variable polynomial order in each element. The basic premise behind an order nonconforming scheme is that the

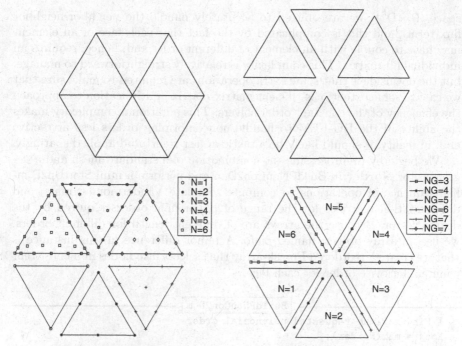

Fig. 9.7. Top left: Original six-element mesh. Top right: Exploded view. Bottom left: Basis nodes used on each element. Bottom right: Gauss nodes used for surface integrals on each element-element interface.

polynomial order is allowed to vary by element. We thus consider an element centric degree vector; that is, there is a vector of length K with the k-th entry holding the polynomial order to be used in that element. An example mesh of six elements with orders ranging from one to six is shown in Fig. 9.7.

We should point out to those readers familiar with the continuous finite element methods that there is no need in the DG-FEM to create a basis that has a degree vector that assigns polynomial orders to each edge, face, and element interior and furthermore require that the degree vector be continuous between these facets. Rather, because we are using broken spaces of polynomials, we merely require that the integrals used to evaluate the fluxes, transferring information between elements in the variational statement, are computed with sufficient accuracy. Thus the main considerations of the order nonconforming formulation are that we are able to handle the variable (elemental) degree vector and to compute the flux terms accurately at all interfaces.

In the following implementation examples, we outline the changes to the previously introduced implementations needed to handle nonconformity in the local order of approximation. The alert reader will notice that the implementations are somewhat more complicated and perhaps not suitable or relevant for the casual reader. Although each element uses a standard polynomial space $(P_N(D^k))$ we are obliged to accurately handle the neighbor-neighbor flux terms, and this is complicated by the fact that each face of an element may have to couple with an element of different order and, hence, requires an individual lift matrix. This complexity is relatively straightforward to manage, but the downside is that some extra operations are required to make sure that we can, to some extent, use efficient matrix-matrix multiplication to approach the efficiency of the constant order solvers. This additional complexity makes the ability of the DG-FEM to handle nonconforming orders less attractive and, in reality, it should likely be a tactic of last resort and applied sparingly.

We begin by preprocessing the combination of a standard mesh and a degree vector **Norder** in BuildPNonCon2D.m and perform a mini StartUp2D.m-like sequence of operations to compute template Vandermonde, mass, and differentiation matrices for the range of polynomial orders required by the degree vector. Because the arrays are of different length for different orders, we use an array object, named pinfo. A minor difference in this instance is that we create a reordered Fmask-array that selects face nodes in a consistent counter clockwise order for each face.

BuildPNonCon2D.m

```
 8 % Find maximum requested polynomial order
 9 Nmax = max(Norder);
10
11 % Mesh details
12 Nfaces = 3; NODETOL = 1e-12; VX = pVX; VY = pVY;
13
14 kmap = zeros(pK,2);
15 sk = 1;
16
17 % Perform a mini StartUp2D for the elements of each order
18 for N=1:Nmax
19
20    % load N'th order polynomial nodes
21    Np = (N+1)*(N+2)/2; Nfp = N+1;
22    [r,s] = Nodes2D(N); [r,s] = xytors(r,s);
23
24    % Find list of N'th order nodes on each face of the reference
25    % element
26    ids1 = find(abs(1+s)<NODETOL);
27    [foo, ids] = sort(r(ids1), 'ascend');   ids1 = ids1(ids);
28    ids2 = find(abs(r+s)<NODETOL);
29    [foo, ids] = sort(r(ids2), 'descend'); ids2 = ids2(ids);
30    ids3 = find(abs(1+r)<NODETOL);
31    [foo, ids] = sort(s(ids3), 'descend'); ids3 = ids3(ids);
32    Fmask = [ids1,ids2,ids3];
33
34    % Build reference element matrices
35    V = Vandermonde2D(N,r,s); invV = inv(V);
```

```
36    MassMatrix = invV'*invV;
37    [Dr,Ds] = Dmatrices2D(N, r, s, V);
38    LIFT = Lift2D();
39
40    % store information for N'th order elements
41    pinfo(N).Np = Np; pinfo(N).Nfp = Nfp; pinfo(N).Fmask = Fmask;
42    pinfo(N).r= r; pinfo(N).s=s;
43    pinfo(N).Dr=Dr; pinfo(N).Ds=Ds; pinfo(N).LIFT=LIFT; pinfo(N).V=V;
```

We next evaluate the geometric factors related to the subset of elements at each order. There is again significant repetition of the StartUp2D.m implementation, and we only need to mention the addition of an index array pinfo(N).ids that lists the location of the N-th-order interpolation nodes in a global vector of all interpolation nodes.

─────────────────── BuildPNonCon2D ───────────────────

```
45    % Find elements of polynomial order N
46    ksN = find(Norder==N); K = length(ksN);
47    if(K>0)
48
49        % Use the subset of elements of order N
50        EToV = pEToV(ksN,:); BCType = pBCType(ksN,:);
51        kmap(ksN,1) = 1:K; kmap(ksN,2) = N;
52
53        % Build coordinates of all the nodes
54        va = EToV(:,1)'; vb = EToV(:,2)'; vc = EToV(:,3)';
55        x = 0.5*(-(r+s)*VX(va)+(1+r)*VX(vb)+(1+s)*VX(vc));
56        y = 0.5*(-(r+s)*VY(va)+(1+r)*VY(vb)+(1+s)*VY(vc));
57
58        % Calculate geometric factors
59        [rx,sx,ry,sy,J] = GeometricFactors2D(x,y,Dr,Ds);
60        [nx, ny, sJ] = Normals2D();
61        Fscale = sJ./(J(Fmask,:));
62
63        % Calculate element connections on this mesh
64        [EToE, EToF] = tiConnect2D(EToV);
65        [mapM, mapP, vmapM, vmapP, vmapB, mapB] = BuildMaps2D();
66        BuildBCMaps2D();
67        pinfo(N).mapW = mapW;
68
69        % Compute triangulation of N'th order nodes on mesh
70        triN = delaunay(r, s);
71        alltri = [];
72        for k=1:K
73            alltri = [alltri; triN+(k-1)*Np];
74        end
75
76        % Store geometric inforatmion in pinfo struct
```

```
77      pinfo(N).rx = rx; pinfo(N).sx = sx; pinfo(N).ry = ry;
78      pinfo(N).sy = sy;
79      pinfo(N).nx = nx; pinfo(N).ny = ny; pinfo(N).sJ = sJ;
80      pinfo(N).J  = J;
81      pinfo(N).x  = x; pinfo(N).y = y; pinfo(N).Fscale = Fscale;
82
83      pinfo(N).K = K; pinfo(N).ks = ksN; pinfo(N).tri = alltri;
84
85      % Store location of the N'th order nodes in a global vector
86      pinfo(N).ids = reshape(sk:sk+K*Np-1, Np, K);
87      sk = sk+K*Np;
88    end
89 end
```

Thus far, we have only provided infrastructure to independently handle each of the subsets of nodes of the same order. We now need to determine how to evaluate the flux terms, taking into account the coupling between these different subsets of elements through the surface integrals.

To exemplify the steps needed, let us return to the solution of Maxwell's equations, given on variational form as

$$
\left(\phi_h, \frac{\partial \boldsymbol{H}_h}{\partial t}\right)_{\Omega,h} = (\phi_h, -\nabla \times \boldsymbol{E}_h)_{\Omega,h}
$$

$$
+ \frac{1}{2} \sum_{k=1}^{K} (\phi_h, (\hat{\boldsymbol{n}} \times [\boldsymbol{E}_h] + \hat{\boldsymbol{n}} \times \hat{\boldsymbol{n}} \times [\boldsymbol{H}_h]))_{\partial \mathsf{D}^k},
$$

$$
\left(\phi_h, \frac{\partial \boldsymbol{E}_h}{\partial t}\right)_{\Omega,h} = (\phi_h, \nabla \times \boldsymbol{H}_h)_{\Omega,h}
$$

$$
+ \frac{1}{2} \sum_{k=1}^{K} (\phi_h, (-\hat{\boldsymbol{n}} \times [\boldsymbol{H}_h] + \hat{\boldsymbol{n}} \times \hat{\boldsymbol{n}} \times [\boldsymbol{E}_h]))_{\partial \mathsf{D}^k},
$$

for all test functions, $\phi_h \in \mathsf{V}_h^3$. We observe that a naive implementation of the right-hand-side flux terms would inevitably require a loop through each face of each element to check the order of the neighboring face and then use an appropriate lift matrix depending on the orders of the neighboring elements. Unfortunately, such an approach will result in a very inefficient implementation due to frequent cache misses as these custom matrices are loaded and discarded repeatedly. Instead we notice that the following variational equation is equivalent to the above

$$
\left(\phi_h, \frac{\partial \boldsymbol{H}_h}{\partial t}\right)_{\Omega,h} = (\phi_h, -\nabla \times \boldsymbol{E}_h)_{\Omega,h}
$$

$$
+ \frac{1}{2} \sum_{k=1}^{K} \sum_{f=1}^{N\text{faces}} (\phi_h, \Pi^{k,f} (\hat{\boldsymbol{n}} \times [\boldsymbol{E}_h] + \hat{\boldsymbol{n}} \times \hat{\boldsymbol{n}} \times [\boldsymbol{H}_h]))_{\partial \mathsf{D}^k},
$$

$$\left(\phi_h, \frac{\partial \boldsymbol{E}_h}{\partial t}\right)_{\Omega,h} = (\phi_h, \nabla \times \boldsymbol{H}_h)_{\Omega,h}$$

$$+ \frac{1}{2} \sum_{k=1}^{K} \sum_{f=1}^{\text{Nfaces}} \left(\phi_h, \Pi^{k,f}\left(-\hat{n} \times [\boldsymbol{H}_h] + \hat{n} \times \hat{n} \times [\boldsymbol{E}_h]\right)\right)_{\partial \mathsf{D}^k}.$$

Here, $\Pi^{k,f}$ is a projection operator that performs an L^2-projection of the operand to the space of the functions supported on the negative trace of face f on element k. This is apparent because the inner product is zero for any function in the complement of this space of functions by L^2-orthogonality.

We follow Chapter 3 to perform such one dimensional projections, but this time, we evaluate the surface integrals by laying down a Gauss quadrature of max $\left(N^1, N^2\right) + 1$ points on each face of the reference element faces and proceed in the standard way. In Fig. 9.7 we show an example mesh where nodes of different orders are used in each element, then the elements are figuratively torn apart, and a Gauss quadrature is set up on each element interface in order to evaluate the above integrals.

BuildPNonCon2D

```
91  % For each possible order
92  for N1=1:Nmax
93    % generate face L2projection matrices
94    % (from order N2 to order N1 face space)
95    for N2=1:Nmax
96
97      % Set up sufficient Gauss quadrature to exactly perform surface
98      % integrals
99      [gz, gw] = JacobiGQ(0, 0, max(N1,N2));
100
101     % All edges have same distribution (note special Fmask)
102     rM = pinfo(N1).r(pinfo(N1).Fmask(:,1));
103     rP = pinfo(N2).r(pinfo(N2).Fmask(:,1));
104
105     % Build N2 to N1 projection matrices for '+' trace data
106     interpM = Vandermonde1D(N1, gz)/Vandermonde1D(N1, rM);
107     interpP = Vandermonde1D(N2, gz(end:-1:1))/Vandermonde1D(N2, rP);
108
109     % Face mass matrix used in projection
110     mmM = interpM'*diag(gw)*interpM;
111
112     pinfo(N1).interpP{N2} = mmM\(interpM'*diag(gw)*interpP);
113   end
114 end
```

To enable a vectorized approach for handling all the degrees of freedom; for examples, for the Runge-Kutta time integration operations or evaluation of initial conditions, we compute vectors storing the physical coordinates of all nodes in all elements, and a vector referencing the indices of these nodes for each element. Finally, to extract the correct values of the solution at each face to compute surface integrals we find, for each face of each element, the indices in the vector of nodes, for the nodes on the internal and external trace of the face.

BuildPNonCon2D

```
116  % Generate neighbor information for all faces
117  [EToE, EToF] = tiConnect2D(pEToV);
118
119  % For each possible polynomial order
120  for N1=1:Nmax
121
122    % Create a set of indexing arrays, one for each possible neighbor
123    % order
124    pinfo(N1).fmapM = cell(Nmax,1);
125    pinfo(N1).vmapP = cell(Nmax,1);
126
127    for N2=1:Nmax
128      pinfo(N1).fmapM{N2} = [];
129      pinfo(N1).vmapP{N2} = [];
130    end
131
132    % Loop through all elements of order N1
133    for k1=1:pinfo(N1).K
134
135      % Find element in original mesh
136      k1orig = pinfo(N1).ks(k1);
137
138      % Check all it's faces
139      for f1=1:Nfaces
140        % Find neighboring element
141        % (i.e. it's order and location in N2 mesh)
142        k2orig = EToE(k1orig,f1);
143        f2     = EToF(k1orig,f1);
144        k2     = kmap(k2orig,1);
145        N2     = kmap(k2orig,2);
146
147        % Compute location of face nodes of '-' trace
148        idsM = (k1-1)*pinfo(N1).Nfp*Nfaces + (f1-1)*pinfo(N1).Nfp;
149        idsM = idsM+1:idsM+pinfo(N1).Nfp;
150
151        % Find location of volume nodes on '+' trace of (k1orig,f1)
152        idsP = pinfo(N2).ids(pinfo(N2).Fmask(:,f2), k2);
153
154        % Store node locations in cell arrays
155        pinfo(N1).fmapM{N2} = [pinfo(N1).fmapM{N2}, idsM];
156        pinfo(N1).vmapP{N2} = [pinfo(N1).vmapP{N2}, idsP];
```

```
157
158     end
159   end
160 end
```

The code BuildPNonCon2D.m creates a large amount of the information that we would normally rely on StartUp2D.m to construct, complemented with the specific need for the surface mass matrices with different order test and trial spaces.

To evaluate the right-hand-side residuals with order-nonconforming discretizations, we loop in MaxwellPNonConRHS2D.m through each possible order and perform a similar sequence of operations as we did for the original approach using a single polynomial order. The most notable difference is that now we go through each possible polynomial order, N, and process the subset of elements of that order as before. A further difference is that the external trace of the solution at element boundaries is provided by sequentially looping through the possible polynomial orders of the neighboring elements and projecting the neighbor trace data onto the order N trace space. This is arranged in a vectorized fashion to avoid excessive computational overhead.

```
                        ┌─────────────────────┐
──────────────────────── │ MaxwellPNonConRHS2D.m │ ────────────────────────
                        └─────────────────────┘
1  function[rhsHx, rhsHy, rhsEz] = MaxwellPNonConRHS2D(pinfo,Hx,Hy,Ez)
2
3  % function[rhsHx, rhsHy, rhsEz] = MaxwellPNonConRHS2D(pinfo,Hx,Hy,Ez)
4  % Purpose  : Evaluate RHS flux in 2D Maxwell TM form
5
6  Globals2D;
7
8  % Initialize storage for right hand side residuals
9  rhsHx = zeros(size(Hx)); rhsHy = zeros(size(Hy));
10 rhsEz = zeros(size(Ez));
```

Next, on each element, we loop through each face and gather the negative and positive traces of the fields and use the projection matrices constructed by BuildPNonCon2D.m to evaluate the jump terms associated with each edge. In this specific implementation we rely heavily on our assumption that elements are straight-sided. This assumption was made to make this process as transparent as possible.

```
                        ┌─────────────────────┐
──────────────────────── │ MaxwellPNonConRHS2D.m │ ────────────────────────
                        └─────────────────────┘
12 % For each possible polynomial order
13 for N=1:length(pinfo)
14
15   % Extract information for this polynomial order
```

```
16   pinf = pinfo(N);
17   K = pinf.K;
18
19   % Check to see if any elements of this order exist
20   if(K>0)
21
22     % Find location of N'th order nodes
23     ids = pinf.ids; Fmask = pinf.Fmask;
24
25     % Extract N'th order nodes
26     HxN = Hx(ids); HyN = Hy(ids); EzN = Ez(ids);
27
28     % Extract '-' traces of N'th order nodal data
29     HxM = HxN(Fmask(:),:); HyM = HyN(Fmask(:),:);
30     EzM = EzN(Fmask(:),:);
31
32     % Storage for '+' traces
33     HxP = zeros(size(HxM)); HyP = HxP; EzP = HxP;
34
35     % For each possible order
36     for N2=1:length(pinfo)
37
38       % Check to see if any neighbor nodes of this order
39       % were located
40       if(length(pinf.fmapM{N2}(:))>0)
41
42         % L2 project N2'th order neighbor data onto N'th order
43         % trace space
44         interp = pinf.interpP{N2}; fmapM = pinf.fmapM{N2};
45         vmapP = pinf.vmapP{N2};
46
47         HxP(fmapM) = interp*Hx(vmapP); HyP(fmapM) = interp*Hy(vmapP);
48         EzP(fmapM) = interp*Ez(vmapP);
49       end
50     end
```

Finally, once the external traces have been projected, we proceed in almost exactly the same way as in the original scheme, but this time using the stored, order N-operators and geometric information retrieved from pinfo.

```
————————————————— | MaxwellPNonConRHS2D.m | —————————————————
52   % Compute jumps of trace data at faces
53   dHx = HxM-HxP;  dHy = HyM-HyP;  dEz = EzM-EzP;
54
55   % Apply PEC boundary condition at wall boundary faces
56   dHx(pinf.mapW) = 0; dHy(pinf.mapW) = 0;
57   dEz(pinf.mapW) = 2*EzM(pinf.mapW);
58
```

```
59      % evaluate jump in incoming characteristic variable
60      dR = -pinf.ny.*dHx + pinf.nx.*dHy + dEz ;
61
62      % Compute flux terms
63      fluxHx =  pinf.ny.*dR;
64      fluxHy = -pinf.nx.*dR;
65      fluxEz =          -dR;
66
67      % Evaluate local derivatives of fields
68      dHxdr = pinf.Dr*HxN; dHxds = pinf.Ds*HxN;
69      dHydr = pinf.Dr*HyN; dHyds = pinf.Ds*HyN;
70      dEzdr = pinf.Dr*EzN; dEzds = pinf.Ds*EzN;
71
72      % Compute physical derivatives of fields
73      dHxdy = pinf.ry.*dHxdr + pinf.sy.*dHxds;
74      dHydx = pinf.rx.*dHydr + pinf.sx.*dHyds;
75      dEzdx = pinf.rx.*dEzdr + pinf.sx.*dEzds;
76      dEzdy = pinf.ry.*dEzdr + pinf.sy.*dEzds;
77
78      % Compute right hand sides of the PDE's
79      rhsHx(ids) = -dEzdy          + pinf.LIFT*(pinf.Fscale.*fluxHx)/2.0;
80      rhsHy(ids) =  dEzdx          + pinf.LIFT*(pinf.Fscale.*fluxHy)/2.0;
81      rhsEz(ids) =  dHydx - dHxdy + pinf.LIFT*(pinf.Fscale.*fluxEz)/2.0;
82    end
83 end
```

The resulting implementation is a bit slower than the original simpler implementation on conforming meshes, partly due to the need to provide an element specific treatment. In fact, to achieve better efficiency, it may be preferable to use a uniform order approach at maximum order and then apply a filter to the solution to reduce the polynomial degree to the desired order in each element.

9.3 Exercises

1. An elliptical curve has the property that the sum of the straight-line distances from its two foci to any point on the ellipse is the same regardless of the choice of point on the ellipse. This property has been used in the construction of ellipsoidal "whispering galleries" that allow a speaker to whisper at one focus and be heard clearly at the other focus by a second listener even if the distance between the foci is large. This effect can be also observed in electromagnetic simulations.

   This exercise is designed to illustrate the importance of accurately representing the curvature of the ellipse to accurately simulate this effect.

a) Use Distmesh (or your favorite mesh generator) to construct a mesh for the domain

$$\left(\frac{x}{5}\right)^2 + \left(\frac{y}{4}\right)^2 \leq 1.$$

b) Use a basic mode for an electromagnetic pulse by imposing an initial condition of a Gaussian pressure pulse centered at the $x = 3$ focus of the ellipse with the following initial conditions:

$$H^x(x, y, t = 0) = 0,$$
$$H^y(x, y, t = 0) = 0,$$
$$E^z(x, y, t = 0) = \exp\left(-\delta\left((x - 3)^2 + y^2\right)\right).$$

The sharpness of the pulse is determined by the magnitude of δ. Increasing this parameter will make the focusing more pronounced, but also make the pulse harder to resolve. Set the FinalTime=10, as this is the travel time from one focus to boundary to the other focus.

c) Modify the MaxwellCurvedDriver2D.m to load a relatively coarse mesh for the ellipse refocusing problem and experiment with the size of δ to examine how well the coarse mesh can resolve the pulse. Next, try finer meshes and also higher polynomial orders, N. Do you observe the refocusing of the pulse at the second focus $x = -3$ and is there some polygonal deformation of the pulse as it refocuses? If so, please explain.

2. (Continued). In the above experiments, the boundary faces were treated with a default linear approximation of the ellipse boundary. To encode the elliptical boundary make the following adjustments:

a) Make sure the domain boundary faces of the mesh are labeled as Cyl type.

b) Modify the code in MakeCylinder2D.m to make the specified curved boundary faces conform to the ellipse described above rather than the originally specified circle.

c) You will need to add a second radius as input to the MakeCylinder2D.m function.

Repeat the original experiments but this time with the curvilinear elements. Do you observe better refocusing of the pulse and does the result improve with lesser polygonal artifacts when you increase the polynomial order compared with the original simulations?

10

Into the third dimension

Foreword to this chapter: We have included this chapter to provide a detailed description of how the discontinuous Galerkin methods, introduced for one-dimensional problems in Chapter 3 and two-dimensional problems in Chapter 6, can be extended to general three-dimensional problems. We will however caution against using the included Matlab scripts for anything other than experimentation and basic algorithm development for three-dimensional simulations. For more advanced developments, we refer to efficient C and C++ codes, which are available at http://www.nudg.org

In Chapter 6 we demonstrated the extension of the discontinuous Galerkin (DG) methods to problems in two spatial dimensions. This involved the introduction of nodal sets for the triangle and an orthonormal polynomial basis that we used as a reference basis for interpolation, differentiation, and the computation of inner products. In this chapter we will go further and consider the additional details required to extend this approach to three-dimensional domains.

As in previous chapters, we focus first on solving the conservation law

$$\frac{\partial u(\boldsymbol{x}, t)}{\partial t} + \nabla \cdot \boldsymbol{f}(u(\boldsymbol{x}, t), \boldsymbol{x}, t) = 0, \quad \boldsymbol{x} \in \Omega \in \mathsf{R}^3, \qquad (10.1)$$

$$u(\boldsymbol{x}, t) = g(\boldsymbol{x}, t), \quad \boldsymbol{x} \in \partial\Omega,$$

$$u(\boldsymbol{x}, 0) = f(\boldsymbol{x}).$$

We assume that boundary conditions are available at all inflow points (i.e., where the eigenvalues of the normal flux-Jacobian are negative).

To secure geometric flexibility and pave the way for the DG formulation, we assume that Ω can be tiled using K elements; that is,

$$\Omega \simeq \Omega_h = \bigcup_{k=1}^{K} \mathsf{D}^k,$$

where D^k is a straight-sided tetrahedron and the mesh is assumed to be geometrically conforming; that is, $\partial\Omega$ is approximated by a piecewise triangular polytrope with each triangle component being a face in a tetrahedron.

Let us follow the one-dimensional approach and assume that we can approximate $u(x, t)$ by elements in the piecewise polynomial space

$$\mathsf{V}_h = \bigoplus_{k=1}^{K} \mathsf{P}_N(\mathsf{D}^k),$$

where $\mathsf{P}_N(\mathsf{D}^k)$ represent an N-th-order polynomial basis defined on element D^k. We express this as

$$u_h^k(x, t) = \sum_{i=1}^{N_p} u_h^k(x_i, t)\ell_i^k(x),$$

where $\ell_i^k(x)$ is the multidimensional Lagrange polynomial based on the grid points, x_i, defined on element D^k. Note that in the multidimensional cases, N_p is very different from the order of the polynomial, N.

Following the previous discussion, we require the residual to be orthogonal to all test functions, $\phi_h \in \mathsf{V}_h$, yielding the local statements

$$\int_{\mathsf{D}^k} \left[\frac{\partial u_h^k}{\partial t} \ell_i^k(x) - f_h^k \cdot \nabla \ell_i^k(x) \right] dx = - \oint_{\partial \mathsf{D}^k} \hat{n} \cdot f^* \ell_i^k(x) \, dx$$

and

$$\int_{\mathsf{D}^k} \left[\frac{\partial u_h^k}{\partial t} + \nabla \cdot f_h^k \right] \ell_i^k(x) \, dx = \oint_{\partial \mathsf{D}^k} \hat{n} \cdot \left[f_h^k - f^* \right] \ell_i^k(x) \, dx,$$

as the weak and strong form, respectively, of the nodal DG method in three spatial dimensions. The missing piece is the specification of the numerical flux, f^*. We will primarily use the local Lax-Friedrichs flux

$$f^*(a, b) = \frac{f(a) + f(b)}{2} + \frac{C}{2}\hat{n}(a - b),$$

where C is the local absolute maximum of the normal flux-Jacobian

$$C = \max \left| \hat{n} \cdot \frac{\partial f}{\partial u} \right|,$$

although alternatives to this flux certainly are possible. In the most general case of a system, finding C requires one to bound the maximum eigenvalue of the normal flux-Jacobian matrix. The generalization to the multidimensional system is discussed in Chapter 2.

10.1 Modes and nodes in three dimensions

By leaning on the knowledge gained for the one dimensional approximation in Section 3.1, we now develop the tools needed for the polynomial interpolation on a tetrahedron.

We assume a local solution of the form

$$x \in \mathsf{D}^k: \quad u_h^k(\boldsymbol{x}, t) = \sum_{i=1}^{N_p} u_h^k(\boldsymbol{x}_i, t) \ell_i^k(\boldsymbol{x}) = \sum_{n=1}^{N_p} \hat{u}_n(t) \psi_n(\boldsymbol{x}),$$

where $\ell_i^k(\boldsymbol{x})$ is the multidimensional Lagrange polynomial based on the grid points, $\boldsymbol{x}_i$, and $\psi_n(\boldsymbol{x})$ is a three-dimensional polynomial basis.

Recall that N_p does not represent the order, N, of the polynomial, u_h^k, but, rather, the number of terms in the local expansion. These two are related as

$$N_p = \frac{(N+1)(N+2)(N+3)}{6},$$

for a polynomial of order N in three variables.

To simplify matters, we introduce a mapping, Ψ, which connects the general straight-sided tetrahedron, $\boldsymbol{x} \in \mathsf{D}^k$, with the standard tetrahedron, defined by

$$\mathsf{I} = \{ \boldsymbol{r} = (r, s, t) | (r, s, t) \geq -1; r + s + t \leq -1 \}.$$

To connect the two tetrahedra, we first assume that D^k is spanned by the four vertices, $(\boldsymbol{v}^1, \boldsymbol{v}^2, \boldsymbol{v}^3, \boldsymbol{v}^4)$, numbered counterclockwise and introduce the barycentric coordinates, $(\lambda^1, \lambda^2, \lambda^3, \lambda^4)$ with the properties that

$$0 \leq \lambda^i \leq 1, \quad \lambda^1 + \lambda^2 + \lambda^3 + \lambda^4 = 1. \tag{10.2}$$

Any point in the tetrahedron, spanned by the four vertices, can be expressed as

$$\boldsymbol{x} = \lambda^3 \boldsymbol{v}^1 + \lambda^4 \boldsymbol{v}^2 + \lambda^2 \boldsymbol{v}^3 + \lambda^1 \boldsymbol{v}^4.$$

In a similar fashion, we can naturally express

$$\begin{pmatrix} r \\ s \\ t \end{pmatrix} = \lambda^3 \begin{pmatrix} -1 \\ -1 \\ -1 \end{pmatrix} + \lambda^4 \begin{pmatrix} 1 \\ -1 \\ -1 \end{pmatrix} + \lambda^2 \begin{pmatrix} -1 \\ 1 \\ -1 \end{pmatrix} + \lambda^1 \begin{pmatrix} -1 \\ -1 \\ 1 \end{pmatrix}.$$

Combining this with Eq. (10.2), we recover

$$\lambda^1 = \frac{t+1}{2}, \quad \lambda^2 = -\frac{s+1}{2}, \quad \lambda^3 = -\frac{r+s+t+1}{2}, \quad \lambda^4 = \frac{r+1}{2},$$

and the direct mapping

$$\boldsymbol{x} = -\frac{r+s+t+1}{2}\boldsymbol{v}^1 + \frac{r+1}{2}\boldsymbol{v}^2 - \frac{s+1}{2}\boldsymbol{v}^3 + \frac{t+1}{2}\boldsymbol{v}^4 = \Psi(\boldsymbol{r}). \tag{10.3}$$

It is important to observe that the mapping is linear in r. This has the important consequence that any two straight-sided tetrahedra are connected through an affine mapping with a constant Jacobian. The metric for the mapping can be found directly since

$$\frac{\partial \boldsymbol{x}}{\partial \boldsymbol{r}} \frac{\partial \boldsymbol{r}}{\partial \boldsymbol{x}} = \begin{bmatrix} x_r & x_s & x_t \\ y_r & y_s & y_t \\ z_r & z_s & z_t \end{bmatrix} \begin{bmatrix} r_x & r_y & r_z \\ s_x & s_y & s_z \\ t_x & t_y & t_z \end{bmatrix} = \begin{bmatrix} 1 & 0 & 0 \\ 0 & 1 & 0 \\ 0 & 0 & 1 \end{bmatrix}.$$

Here, we use the standard notation of a_b to mean a differentiated with respect to b. From this, we recover

$$Jr_x = y_s z_t - z_s y_t, \quad Jr_y = -(x_s z_t - z_s x_t), \quad Jr_z = x_s y_t - y_s x_t,$$

$$Js_x = -(y_r z_t - z_r y_t), \quad Js_y = x_r z_t - z_r x_t, \quad Js_z = -(x_r y_t - y_r x_t), \quad (10.4)$$

$$Jt_x = y_r z_s - z_r y_s, \quad Jt_y = -(x_r z_s - z_r x_s), \quad Jt_z = x_r y_s - y_r x_s,$$

where the Jacobian J is given by

$$J = x_r(y_s z_t - z_s y_t) - y_r(x_s z_t - z_s x_t) + z_r(x_s y_t - y_s x_t); \quad (10.5)$$

and we can evaluate these expressions using Eq. (10.3), to obtain

$$(x_r, y_r, z_r) = \boldsymbol{x}_r = \frac{\boldsymbol{v}^2 - \boldsymbol{v}^1}{2}, \quad \boldsymbol{x}_s = \frac{\boldsymbol{v}^3 - \boldsymbol{v}^1}{2}, \quad \boldsymbol{x}_t = \frac{\boldsymbol{v}^4 - \boldsymbol{v}^1}{2}.$$

We are now back in the position where we can focus on the development of polynomials and operators defined on I only. Let us consider the local polynomial approximation

$$u(\boldsymbol{r}) = \sum_{n=1}^{N_p} \hat{u}_n \psi_n(\boldsymbol{r}) = \sum_{i=1}^{N_p} u(\boldsymbol{r}_i) \ell_i(\boldsymbol{r}),$$

where, as in the two-dimensional case, we define the expansion coefficients, $\hat{u}_n$, through an interpolation to avoid problems with the need to evaluate multidimensional integrals.

This yields the recognized expression

$$\mathcal{V}\hat{\boldsymbol{u}} = \boldsymbol{u},$$

where $\hat{\boldsymbol{u}} = [\hat{u}_1, \ldots, \hat{u}_{N_p}]^T$ are the N_p expansion coefficients and $\boldsymbol{u} = [u(\boldsymbol{r}_1), \ldots, u(\boldsymbol{r}_{N_p})]^T$ represents the N_p grid point values.

To ensure stable numerical behavior of the generalized Vandermonde matrix $\mathcal{V}$ with the entries

$$\mathcal{V}_{ij} = \psi_j(\boldsymbol{r}_i),$$

we need to identify a polynomial basis, $\psi_j(\boldsymbol{r})$, which is orthonormal on the reference tetrahedron, I. We must also identify families of points leading to good behavior of the interpolating polynomial defined on I. Such points can be viewed as a multidimensional generalization of the Legendre-Gauss-Lobatto points, although they do not have a quadrature formula associated with them.

The first issue is, at least in principle, easy to resolve, since we can begin by considering the canonical basis as

$$\psi_m(\boldsymbol{r}) = r^i s^j t^k, \quad (i,j,k) \geq 0; \quad i+j+k \leq N,$$

$$m = 1 + \frac{(11+12N+3N^2)i}{6} + \frac{(2N+3)j}{2} + k$$
$$- \frac{(2+N)i^2}{2} - ij - \frac{j^2}{2} + \frac{i^3}{6}, \quad 0 \leq i,j,k; i+j+k \leq N.$$

This spans the space of N-dimensional polynomials in three variables, (r,s,t). Based on the discussion in Chapter 3, we do not expect this to be a good basis. It is, however, complete and we can orthonormalize it through a Gram-Schmidt process to recover

$$\psi_m(\boldsymbol{r}) = \sqrt{8} P_i(a) P_j^{(2i+1,0)}(b)(1-b)^i P_k^{(2i+2j+2,0)}(c)(1-c)^{i+j}, \quad (10.6)$$

where

$$a = -2\frac{(1+r)}{s+t} - 1, \quad b = 2\frac{(1+s)}{1-t} - 1, \quad c = t.$$

Here, $P_n^{(\alpha,\beta)}(x)$ is the n-th-order Jacobi polynomial. Recall that $\alpha = \beta = 0$ is the Legendre polynomial, while the more general polynomials are discussed in Appendix A.

A library routine, Simplex3DP.m, to evaluate this basis is discussed in more detail in Appendix A. Evaluating the basis is done by

```
>> P = Simplex3DP(a,b,c,i,j,k);
```

with the meaning of (a,b,c) and (i,j,k) as in Eq. (10.6). The computation of (a,b,c) from (r,s,t) is accomplished through rsttoabc.m.

───────────────── rsttoabc.m ─────────────────

```
function [a,b,c] = rsttoabc(r,s,t)

% function [a,b,c] = rsttoabc(r,s,t)
% Purpose: Transfer from (r,s,t) -> (a,b,c) coordinates in triangle

Np = length(r);
a = zeros(Np,1); b = zeros(Np,1); c = zeros(Np,1);
for n=1:Np
  if(s(n)+t(n) ~= 0)
    a(n) = 2*(1+r(n))/(-s(n)-t(n))-1;
  else
```

```
   a(n) = -1;
 end
 if(t(n) ~= 1)
    b(n) = 2*(1+s(n))/(1-t(n))-1;
 else
    b(n) = -1;
 end
end
c = t;
return;
```

With the orthonormal polynomial basis we now need to identify a set of N_p points on I leading to well-behaved interpolations. One can approach this challenge of finding exactly N_p points for interpolation on the tetrahedron in several different ways [55, 163], leading to distributions and interpolations with very similar behavior.

We follow an extended version of the approach in Chapter 6 to compute nodal distributions for the tetrahedron. The construction of the coordinate transform can be generalized for the equilateral tetrahedron. For each face of the tetrahedron, we use the transform constructed for the triangle. These transforms are then blended into the interior of the tetrahedron. The vector warp functions for each of the four faces are

$$w^1 = g_1\left(\lambda^2, \lambda^3, \lambda^4\right) t^{1,1} + g_2\left(\lambda^2, \lambda^3, \lambda^4\right) t^{1,2},$$
$$w^2 = g_1\left(\lambda^1, \lambda^3, \lambda^4\right) t^{2,1} + g_2\left(\lambda^1, \lambda^3, \lambda^4\right) t^{2,2},$$
$$w^3 = g_1\left(\lambda^1, \lambda^4, \lambda^2\right) t^{3,1} + g_2\left(\lambda^1, \lambda^4, \lambda^2\right) t^{3,2},$$
$$w^4 = g_1\left(\lambda^1, \lambda^3, \lambda^2\right) t^{4,1} + g_2\left(\lambda^1, \lambda^3, \lambda^2\right) t^{4,2},$$

where the subscripts on g refer to the two components of these functions. These are computed in WarpShiftFace3D.m and evalshift.m

--------------------- WarpShiftFace3D.m ---------------------

```
function [warpx, warpy] = WarpShiftFace3D(p,pval, pval2, L1,L2,L3,L4)

% function [warpx, warpy] = WarpShiftFace3D(p,pval, pval2, L1,L2,L3,L4)
% Purpose: compute warp factor used in creating 3D Warp & Blend nodes

[dtan1,dtan2] = evalshift(p, pval, L2, L3, L4);
warpx = dtan1; warpy = dtan2;
return;
```

---
evalshift.m

```
function [dx, dy] = evalshift(p, pval, L1, L2, L3)

% function [dx, dy] = evalshift(p, pval, L1, L2, L3)
% Purpose: compute two-dimensional Warp & Blend transform

% 1) compute Gauss-Lobatto-Legendre node distribution
gaussX = -JacobiGL(0,0,p);

% 2) compute blending function at each node for each edge
blend1 = L2.*L3; blend2 = L1.*L3; blend3 = L1.*L2;

% 3) amount of warp for each node, for each edge
warpfactor1 = 4*evalwarp(p, gaussX, L3-L2);
warpfactor2 = 4*evalwarp(p, gaussX, L1-L3);
warpfactor3 = 4*evalwarp(p, gaussX, L2-L1);

% 4) combine blend & warp
warp1 = blend1.*warpfactor1.*(1 + (pval*L1).^2);
warp2 = blend2.*warpfactor2.*(1 + (pval*L2).^2);
warp3 = blend3.*warpfactor3.*(1 + (pval*L3).^2);

% 5) evaluate shift in equilateral triangle
dx = 1*warp1 + cos(2*pi/3)*warp2 + cos(4*pi/3)*warp3;
dy = 0*warp1 + sin(2*pi/3)*warp2 + sin(4*pi/3)*warp3;
return;
```

Here, $t^{f,1}$ and $t^{f,2}$ are two vectors forming orthogonal axes in the plane of face f specified for the origin-centered equilateral tetrahedron as

$$
\begin{aligned}
t^{1,1} &= (1,0,0), & t^{1,2} &= (0,1,0), \\
t^{2,1} &= (1,0,0), & t^{2,2} &= \left(0, \tfrac{1}{3}, \sqrt{\tfrac{8}{9}}\right), \\
t^{3,1} &= \left(-\tfrac{1}{2}, \sqrt{\tfrac{3}{4}}, 0\right), & t^{3,2} &= \left(-\sqrt{\tfrac{1}{12}}, \tfrac{1}{6}, \sqrt{\tfrac{8}{9}}\right), \\
t^{4,1} &= \left(\tfrac{1}{2}, \sqrt{\tfrac{3}{4}}, 0\right), & t^{4,2} &= \left(\sqrt{\tfrac{1}{12}}, -\tfrac{1}{6}, \sqrt{\tfrac{8}{9}}\right).
\end{aligned}
$$

The four scalar face blending functions are then given by

$$
b^1(\lambda^1, \lambda^2, \lambda^3, \lambda^4) = \left(\frac{2\lambda^2}{2\lambda^2 + \lambda^1}\right)\left(\frac{2\lambda^3}{2\lambda^3 + \lambda^1}\right)\left(\frac{2\lambda^4}{2\lambda^4 + \lambda^1}\right),
$$

$$
b^2(\lambda^1, \lambda^2, \lambda^3, \lambda^4) = \left(\frac{2\lambda^1}{2\lambda^1 + \lambda^2}\right)\left(\frac{2\lambda^3}{2\lambda^3 + \lambda^2}\right)\left(\frac{2\lambda^4}{2\lambda^4 + \lambda^2}\right),
$$

$$
b^3(\lambda^1, \lambda^2, \lambda^3, \lambda^4) = \left(\frac{2\lambda^1}{2\lambda^1 + \lambda^3}\right)\left(\frac{2\lambda^2}{2\lambda^2 + \lambda^3}\right)\left(\frac{2\lambda^4}{2\lambda^4 + \lambda^3}\right),
$$

$$
b^4(\lambda^1, \lambda^2, \lambda^3, \lambda^4) = \left(\frac{2\lambda^1}{2\lambda^1 + \lambda^4}\right)\left(\frac{2\lambda^2}{2\lambda^2 + \lambda^4}\right)\left(\frac{2\lambda^3}{2\lambda^3 + \lambda^4}\right).
$$

Table 10.1. Comparison of Lebesgue constants for a range of current nodal distributions on the tetrahedron, including the α-optimized set discussed in the text.

N	α	Λ Optimized	Λ [163]	Λ [55]	Λ Equidistant
1	0.0000	1.00	1.00	1.00	1.00
2	0.0000	2.00	2.00	2.00	2.00
3	0.0000	2.93	2.93	2.93	3.00
4	0.1002	4.07	4.08	4.11	4.88
5	1.1332	5.32	5.35	5.62	8.09
6	1.5608	7.01	7.34	7.36	13.66
7	1.3413	9.21	9.76	9.37	23.38
8	1.2577	12.54	13.63	12.31	40.55
9	1.1603	17.02	18.90	15.66	71.15
10	1.0153	24.36	27.19	-	126.20
11	0.6080	36.35	-	-	225.99
12	0.4523	54.18	-	-	409.15
13	0.8856	84.62	-	-	742.69
14	0.8717	135.75	-	-	1360.49
15	0.9655	217.70	-	-	2506.95

The combined deformation is then obtained as

$$g\left(\lambda^1, \lambda^2, \lambda^3, \lambda^4\right) = b^1 \boldsymbol{w}^1 + b^2 \boldsymbol{w}^2 + b^3 \boldsymbol{w}^3 + b^4 \boldsymbol{w}^4.$$

As with the triangle, we can increase the clustering of nodes near the boundary faces by modifying the blending functions

$$g\left(\lambda^1, \lambda^2, \lambda^3, \lambda^4\right) = \left(1 + \left(\alpha\lambda^1\right)^2\right) b^1 \boldsymbol{w}^1 + \left(1 + \left(\alpha\lambda^2\right)^2\right) b^2 \boldsymbol{w}^2$$
$$+ \left(1 + \left(\alpha\lambda^3\right)^2\right) b^3 \boldsymbol{w}^3 + \left(1 + \left(\alpha\lambda^4\right)^2\right) b^4 \boldsymbol{w}^4$$

for some parameter α.

The number of different node distribution types available for the tetrahedron is markedly smaller than for the triangle. In Table 10.1 we show Lebesgue constants for equally spaced nodes, our unoptimized and optimized nodes [55, 163]. We note that the new nodes are competitive with alternative nodal sets.

In Nodes3D.m, the α-optimized nodes are computed following the approach outlined above. If one needs elements of order higher than 15, one can either guess a value or take $\alpha = 0$, which will still lead to nodal sets with a reasonable Lagrange interpolant. In Fig. 10.1 we show α-optimized node sets for the tetrahedron for $N = 6, 8, 10, 12$.

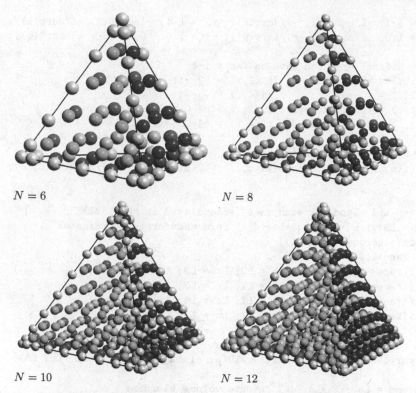

$N = 6$ $N = 8$

$N = 10$ $N = 12$

Fig. 10.1. α-Optimized tetrahedral node distribution for $N = 6, 8, 10, 12$.

Nodes3D.m

```
function [X,Y,Z] = Nodes3D(p)

% function [X,Y,Z] = Nodes3D(p)
% Purpose: compute Warp & Blend nodes
%  input:    p=polynomial order of interpolant
%  output: X,Y,Z vectors of node coordinates in equilateral tetrahedron

% choose optimized blending parameter
alphastore = [0;0;0;0.1002; 1.1332;1.5608;1.3413;1.2577;1.1603;...
              1.10153;0.6080;0.4523;0.8856;0.8717;0.9655];
if(p<=15); alpha = alphastore(p) ; else; alpha = 1. ; end

% total number of nodes and tolerance
N = (p+1)*(p+2)*(p+3)/6; tol = 1e-10;

[r,s,t] = EquiNodes3D(p); % create equidistributed nodes
L1 = (1+t)/2; L2 = (1+s)/2; L3 = -(1+r+s+t)/2; L4 = (1+r)/2;

% set vertices of tetrahedron
```

```
v1 = [-1, -1/sqrt(3), -1/sqrt(6)]; v2 = [ 1, -1/sqrt(3),-1/sqrt(6)];
v3 = [ 0,  2/sqrt(3), -1/sqrt(6)]; v4 = [ 0,  0,          3/sqrt(6)];

% orthogonal axis tangents on faces 1-4
t1(1,:) = v2-v1;          t1(2,:) = v2-v1;
t1(3,:) = v3-v2;          t1(4,:) = v3-v1;
t2(1,:) = v3-0.5*(v1+v2); t2(2,:) = v4-0.5*(v1+v2);
t2(3,:) = v4-0.5*(v2+v3); t2(4,:) = v4-0.5*(v1+v3);

for n=1:4 % normalize tangents
   t1(n,:) = t1(n,:)/norm(t1(n,:)); t2(n,:) = t2(n,:)/norm(t2(n,:));
end

% Warp and blend for each face (accumulated in shiftXYZ)
XYZ = L3*v1+L4*v2+L2*v3+L1*v4; % form undeformed coordinates
shift = zeros(size(XYZ));
for face=1:4
  if(face==1); La = L1; Lb = L2; Lc = L3; Ld = L4; end;
  if(face==2); La = L2; Lb = L1; Lc = L3; Ld = L4; end;
  if(face==3); La = L3; Lb = L1; Lc = L4; Ld = L2; end;
  if(face==4); La = L4; Lb = L1; Lc = L3; Ld = L2; end;

  % compute warp tangential to face
  [warp1 warp2] = WarpShiftFace3D(p, alpha, alpha, La, Lb, Lc, Ld);

  blend = Lb.*Lc.*Ld;   % compute volume blending

  denom = (Lb+.5*La).*(Lc+.5*La).*(Ld+.5*La);   % modify linear blend
  ids = find(denom>tol);
  blend(ids) = (1+(alpha.*La(ids)).^2).*blend(ids)./denom(ids);

  % compute warp & blend
  shift = shift+(blend.*warp1)*t1(face,:) + (blend.*warp2)*t2(face,:);

  % fix face warp
  ids = find(La<tol & ( (Lb>tol) + (Lc>tol) + (Ld>tol) < 3));
  shift(ids,:) = warp1(ids)*t1(face,:) + warp2(ids)*t2(face,:);
end
XYZ = XYZ + shift;
X = XYZ(:,1); Y = XYZ(:,2); Z = XYZ(:,3);
return;
```

One last issue is the fact that the nodes are computed in the equilateral tetrahedron while the orthonormal basis lives in I. Thus, we need to map the computed nodes into I using a mapping as discussed previously. This is done by xyztorst.m.

```
───────────────────────────── xyztorst.m ─────────────────────────────
function [r, s, t] = xyztorst(X, Y, Z)

% function [r,s,t] = xyztorst(x, y, z)
% Purpose : Transfer from (x,y,z) in equilateral tetrahedron
%           to (r,s,t) coordinates in standard tetrahedron

v1 = [-1,-1/sqrt(3), -1/sqrt(6)]; v2 = [ 1,-1/sqrt(3), -1/sqrt(6)];
v3 = [ 0, 2/sqrt(3), -1/sqrt(6)]; v4 = [ 0, 0/sqrt(3),  3/sqrt(6)];

% back out right tet nodes
rhs = [X';Y';Z'] - 0.5*(v2'+v3'+v4'-v1')*ones(1,length(X));
A = [0.5*(v2-v1)',0.5*(v3-v1)',0.5*(v4-v1)'];
RST = A\[rhs];
r = RST(1,:)'; s = RST(2,:)'; t = RST(3,:)';
return;
```

This completes the required developments; that is, we have identified both an orthonormal basis and a way to construct nodal sets, which are suitable for interpolation on the tetrahedron. With this, we can construct a well-behaved Vandermonde matrix for operations on the tetrahedron and generalize this to general tetrahedra by an affine mapping. Hence, as in the one-dimensional case, we can construct local approximations as

$$u(\boldsymbol{r}) \simeq u_h(\boldsymbol{r}) = \sum_{n=1}^{N_p} \hat{u}_n \psi_n(\boldsymbol{r}) = \sum_{i=1}^{N_p} u(\boldsymbol{r}_i)\ell_i(\boldsymbol{r}),$$

where $\boldsymbol{r}_i$ is the three-dimensional nodal set, and $\psi_n(\boldsymbol{r})$ is the orthonormal three-dimensional polynomial basis. We recall that the number of terms in the expansion is

$$N_p = \frac{(N+1)(N+2)(N+3)}{6},$$

for an N-th-order polynomial in three variables.

A result of this discussion is the stable construction of the Vandermonde matrix, $\mathcal{V}$, which establishes the connections

$$\boldsymbol{u} = \mathcal{V}\hat{\boldsymbol{u}}, \quad \mathcal{V}^T \boldsymbol{\ell}(\boldsymbol{r}) = \boldsymbol{\psi}(\boldsymbol{r}), \quad \mathcal{V}_{ij} = \psi_j(\boldsymbol{r}_i).$$

A script for initializing $\mathcal{V}$ is given in Vandermonde3D.m.

```
───────────────────────────── Vandermonde3D.m ─────────────────────────────
function [V3D] = Vandermonde3D(N, r, s, t);

% function [V3D] = Vandermonde3D(N, r, s, t);
% Purpose : Initialize the 3D Vandermonde Matrix,
%           V_{ij} = phi_j(r_i, s_i, t_i);
```

```
V3D = zeros(length(r),(N+1)*(N+2)*(N+3)/6);

% Transfer to (a,b) coordinates
[a, b, c] = rsttoabc(r, s, t);

% build the Vandermonde matrix
sk = 1;

for i=0:N % old ordering
  for j=0:N - i
    for k=0:N - i - j
      V3D(:,sk) = Simplex3DP(a,b,c,i,j,k);
      sk = sk+1;
    end
  end
end
return;
```

With $\mathcal{V}$, we can transform directly between a modal representation, using $\hat{u}_n$ as the unknowns, and a nodal form, using $u(\boldsymbol{r}_i)$. Furthermore, this supplies a way to evaluate the genuinely three-dimensional Lagrange polynomials, $\ell_i(\boldsymbol{x})$, for which no explicit expression is known.

10.2 Elementwise operations

With the fundamental tools in place, we continue the development of the key computational components. Following the two-dimensional approach, we need the mass-matrix

$$\mathcal{M}_{ij}^k = \int_{\mathsf{D}^k} \ell_i^k(\boldsymbol{x}^k)\ell_j^k(\boldsymbol{x}^k)\, d\boldsymbol{x}^k = J^k \int_{\mathsf{I}} \ell_i(\boldsymbol{r})\ell_j(\boldsymbol{r})\, d\boldsymbol{r},$$

where we have utilized that the transformation Jacobian, J, is a constant provided that D^k is a straight-sided tetrahedron, see Eq.(10.5).

Since $\mathcal{V}$ is constructed using an orthonormal basis, we recover

$$\mathcal{M}^k = J^k(\mathcal{V}\mathcal{V}^T)^{-1}.$$

The evaluation of the stiffness matrices follows the approach in Section 3.2 using the differentiation matrices. From the chain rule we immediately have

$$\frac{\partial}{\partial x} = \frac{\partial r}{\partial x}\mathcal{D}_r + \frac{\partial s}{\partial x}\mathcal{D}_s + \frac{\partial t}{\partial x}\mathcal{D}_t,$$

$$\frac{\partial}{\partial y} = \frac{\partial r}{\partial y}\mathcal{D}_r + \frac{\partial s}{\partial y}\mathcal{D}_s + \frac{\partial t}{\partial y}\mathcal{D}_t,$$

$$\frac{\partial}{\partial z} = \frac{\partial r}{\partial z}\mathcal{D}_r + \frac{\partial s}{\partial z}\mathcal{D}_s + \frac{\partial t}{\partial z}\mathcal{D}_t,$$

where the metric constants are given in Eq. (10.4).

To define the differentiation matrices, $\mathcal{D}_r$, $\mathcal{D}_s$ and $\mathcal{D}_t$, we use

$$\mathcal{V}_{r,(i,j)} = \left.\frac{\partial \psi_j}{\partial r}\right|_{r_i}, \quad \mathcal{V}_{s,(i,j)} = \left.\frac{\partial \psi_j}{\partial s}\right|_{r_i}, \quad \mathcal{V}_{t,(i,j)} = \left.\frac{\partial \psi_j}{\partial t}\right|_{r_i}.$$

If we recall the basis in Eq. (10.6) we can recover the spatial derivatives of the basis

$$\frac{\partial \psi_j}{\partial r} = \frac{\partial a}{\partial r}\frac{\partial \psi_j}{\partial a}, \quad \frac{\partial \psi_j}{\partial s} = \frac{\partial a}{\partial s}\frac{\partial \psi_j}{\partial a} + \frac{\partial b}{\partial s}\frac{\partial \psi_j}{\partial b}, \quad \frac{\partial \psi_j}{\partial t} = \frac{\partial a}{\partial t}\frac{\partial \psi_j}{\partial a} + \frac{\partial b}{\partial t}\frac{\partial \psi_j}{\partial b} + \frac{\partial \psi_j}{\partial c},$$

as implemented in GradSimplex3DP.m and this enables the computation of the entries of $\mathcal{V}_r$, $\mathcal{V}_s$, and $\mathcal{V}_t$, illustrated in GradVandermonde3D.m.

GradSimplex3DP.m

```
function [V3Dr, V3Ds, V3Dt] = GradSimplex3DP(a,b,c,id,jd,kd)

% function [V3Dr, V3Ds, V3Dt] = GradSimplex3DP(a,b,c,id,jd,kd)
% Purpose: Return the derivatives of the modal basis (id,jd,kd)
%          on the 3D simplex at (a,b,c)

fa = JacobiP(a,0,0,id);          dfa = GradJacobiP(a,0,0,id);
gb = JacobiP(b,2*id+1,0,jd);     dgb = GradJacobiP(b,2*id+1,0,jd);
hc = JacobiP(c,2*(id+jd)+2,0,kd); dhc = GradJacobiP(c,2*(id+jd)+2,0,kd);

% r-derivative
V3Dr = dfa.*(gb.*hc);
if(id>0);    V3Dr = V3Dr.*((0.5*(1-b)).^(id-1));    end
if(id+jd>0); V3Dr = V3Dr.*((0.5*(1-c)).^(id+jd-1)); end

% s-derivative
V3Ds = 0.5*(1+a).*V3Dr;
tmp = dgb.*((0.5*(1-b)).^id);
if(id>0);    tmp = tmp+(-0.5*id)*(gb.*(0.5*(1-b)).^(id-1));  end
if(id+jd>0) tmp = tmp.*((0.5*(1-c)).^(id+jd-1));            end
tmp = fa.*(tmp.*hc);
V3Ds = V3Ds+tmp;

% t-derivative
V3Dt = 0.5*(1+a).*V3Dr+0.5*(1+b).*tmp;
tmp = dhc.*((0.5*(1-c)).^(id+jd));
if(id+jd>0)
   tmp = tmp-0.5*(id+jd)*(hc.*((0.5*(1-c)).^(id+jd-1)));
end
tmp = fa.*(gb.*tmp); tmp = tmp.*((0.5*(1-b)).^id);
V3Dt = V3Dt+tmp;
```

```
% normalize
V3Dr = V3Dr*(2^(2*id+jd+1.5));
V3Ds = V3Ds*(2^(2*id+jd+1.5));
V3Dt = V3Dt*(2^(2*id+jd+1.5));
return
```

$\boxed{\text{GradVandermonde3D.m}}$

```
function [V3Dr,V3Ds,V3Dt] = GradVandermonde3D(N,r,s,t)

% function [V3Dr,V3Ds,V3Dt] = GradVandermonde3D(N,r,s,t)
% Purpose : Initialize the gradient of the modal basis (i,j,k)
%           at (r,s,t) at order N

V3Dr = zeros(length(r),(N+1)*(N+2)*(N+3)/6);
V3Ds = zeros(length(r),(N+1)*(N+2)*(N+3)/6);
V3Dt = zeros(length(r),(N+1)*(N+2)*(N+3)/6);

% find tensor-product coordinates
[a,b,c] = rsttoabc(r,s,t);

% Initialize matrices

sk = 1;
for i=0:N
  for j=0:N-i
    for k=0:N-i-j
      [V3Dr(:,sk),V3Ds(:,sk),V3Dt(:,sk)] = GradSimplex3DP(a,b,c,i,j,k);
      sk = sk+1;
    end
  end
end
return;
```

We can now finally define the required differentiation matrices

$$\mathcal{D}_r \mathcal{V} = \mathcal{V}_r, \ \mathcal{D}_s \mathcal{V} = \mathcal{V}_s, \ \mathcal{D}_t \mathcal{V} = \mathcal{V}_t,$$

with the corresponding stiffness matrices given by

$$\mathcal{S}_r = \mathcal{M}^{-1}\mathcal{D}_r, \ \mathcal{S}_s = \mathcal{M}^{-1}\mathcal{D}_s, \ \mathcal{S}_t = \mathcal{M}^{-1}\mathcal{D}_t.$$

The differentiation matrices are initialized using Dmatrices3D.m.

Dmatrices3D.m

```
function [Dr,Ds,Dt] = Dmatrices3D(N,r,s,t,V)

% function [Dr,Ds,Dt] = Dmatrices3D(N,r,s,t,V)
% Purpose : Initialize the (r,s,t) differentiation matrices
%           on the simplex, evaluated at (r,s,t) at order N

[Vr, Vs, Vt] = GradVandermonde3D(N, r, s, t);
Dr = Vr/V; Ds = Vs/V; Dt = Vt/V;
return;
```

The evaluation of the right-hand side of the DG schemes requires the computation of surface integrals like

$$\int_{\partial \mathsf{D}^k} \hat{\boldsymbol{n}} \cdot \boldsymbol{g}_h \ell_i^k(\boldsymbol{x}) \, d\boldsymbol{x},$$

where $\boldsymbol{g}_h$ is a trace polynomial, composed of the numerical flux or the jump in flux, depending on whether we use the weak or the strong form. We first split the integral into the four individual face components, each of the type

$$\int_{\text{face}} \hat{\boldsymbol{n}} \cdot \boldsymbol{g}_h \ell_i^k(\tilde{\boldsymbol{x}}) \, d\tilde{\boldsymbol{x}} = \sum_{j=1}^{Nf_p} \hat{\boldsymbol{n}} \cdot \boldsymbol{g}_j \int_{\text{face}} \ell_j^k(\tilde{\boldsymbol{x}}) \ell_i^k(\tilde{\boldsymbol{x}}) \, d\tilde{\boldsymbol{x}},$$

where Nf_p represents the number of nodes on a face, i.e.,

$$Nf_p = \frac{(N+1)(N+2)}{2}.$$

Also, $\tilde{\boldsymbol{x}}$ is the trace of $\boldsymbol{x}$ along the face. We have exploited that all tetrahedra are assumed to be straight-sided such that the outward pointing normal, $\hat{\boldsymbol{n}}$, is a constant along a face.

Thus, we need to compute face-mass matrices of the form

$$\mathcal{M}_{ij}^{k,f} = \int_{\text{face}} \ell_j^k(\tilde{\boldsymbol{x}}) \ell_i^k(\tilde{\boldsymbol{x}}) \, d\tilde{\boldsymbol{x}}.$$

However, exactly as in the two-dimensional case, all Lagrange polynomials where $\boldsymbol{r}_i$ does not reside on the face are exactly zero on the face over which integration is needed. This natural separation between inner and face modes implies that we only need to compute two-dimensional mass matrices. If we simply define the Vandermonde matrix, $\mathcal{V}^{2D}$, corresponding to the two-dimensional interpolation on the face, we have

$$\mathcal{M}^f = J^f (\mathcal{V}^{2D}(\mathcal{V}^{2D})^T)^{-1},$$

where J^f is the transformation Jacobian for the face – the ratio between the area of the face in D^k and in I, respectively.

To specify a simple way of implementing this surface integral, let us assume that we have defined a matrix, Fmask, of size $Nf_p \times 4$, which, in each column, contains the numbers for the nodes on the each of four faces of the tetrahedron. We can then use these to extract the face coordinates and, subsequently, form the local Vandermonde matrix and mass matrix along the edge.

As an example, along face 1 this can be done as

```
>> faceR = r(Fmask(:,1));
>> faceS = s(Fmask(:,1));
>> faceT = t(Fmask(:,1));
>> vFace = Vandermonde2D(N, faceR, faceS);
>> massFace1 = inv(vFace*vFace');
```

Let us then define the matrix, $\mathcal{E}$, of size $N_{\bar{p}} \times 4Nf_p$. The purpose of this is to compute the surface integral on l; that is,

$$(\mathcal{E}[g^1, g^2, g^3, g^4]^T) = \int_{\partial l} \hat{n} \cdot g_h \ell_i(r) \, dr.$$

We can form $\mathcal{E}$ directly by simply inserting the elements of four face-mass matrices into the positions corresponding to the face-nodes in the element.

This is all implemented in Lift3D.m, which returns the matrix LIFT of size $N_p \times 4Nf_p$ as

$$\text{LIFT} = \mathcal{M}^{-1}\mathcal{E}.$$

Lift3D.m

```
function [LIFT] = Lift3D(N,R,S,T)

% function [LIFT] = Lift3D(N, r, s, t)
% Purpose  : Compute 3D surface to volume lift operator used
%                in DG formulation

Globals3D;

Emat = zeros(Np, Nfaces*Nfp);

for face=1:Nfaces
  % process face
  if(face==1); faceR = R(Fmask(:,1)); faceS = S(Fmask(:,1)); end;
  if(face==2); faceR = R(Fmask(:,2)); faceS = T(Fmask(:,2)); end;
  if(face==3); faceR = S(Fmask(:,3)); faceS = T(Fmask(:,3)); end;
  if(face==4); faceR = S(Fmask(:,4)); faceS = T(Fmask(:,4)); end;

  VFace = Vandermonde2D(N, faceR, faceS);
  massFace = inv(VFace*VFace');

  idr = Fmask(:,face); idc = (face-1)*Nfp+1:face*Nfp;
```

```
    Emat(idr, idc) = Emat(idr, idc)+ massFace;
end

% inv(mass matrix)*\I_n (L_i,L_j)_{edge_n}
LIFT = V*(V'*Emat);
return
```

To simplify the notation and discretization of general partial differential equations, we define local operations following vector algebra. Thus, in Grad3D.m, we define the three-dimensional gradient operator as

$$v = (v_x, v_y, v_z) = \nabla u.$$

———————————————————— Grad3D.m ————————————————————
```
function [dUdx, dUdy, dUdz] = GradH3D(U)

% function [dUdx, dUdy, dUdz] = GradH3D(U)
% purpose: compute local elemental physical spatial derivatives of U

Globals3D;

% compute local derivatives on reference tetrahedron
dUdr = Dr*U;   dUds = Ds*U;   dUdt = Dt*U;

% compute physical spatial derivatives using the chain rule
dUdx = rx.*dUdr + sx.*dUds + tx.*dUdt;
dUdy = ry.*dUdr + sy.*dUds + ty.*dUdt;
dUdz = rz.*dUdr + sz.*dUds + tz.*dUdt;
return;
```
——

In Div3D.m, we obtain the local divergence of a three-dimensional vector field as

$$v = \nabla \cdot u.$$

———————————————————— Div3D.m ————————————————————
```
function [divU] = DivH3D(Ux, Uy, Uz)

% function [divU] = DivH3D(Ux, Uy, Uz)
% Purpose: compute local elemental physical spatial divergence
%          of (Ux,Uy,Uz)

Globals3D;

% compute local derivatives of Ux on reference tetrahedron
ddr = Dr*Ux;   dds = Ds*Ux;   ddt = Dt*Ux;
```

```
% dUx/dx
divU =  (rx.*ddr + sx.*dds + tx.*ddt);

% compute local derivatives of Uy on reference tetrahedron
ddr = Dr*Uy;  dds = Ds*Uy;  ddt = Dt*Uy;

% add dUy/dy to divergence
divU =  divU + (ry.*ddr + sy.*dds + ty.*ddt);

% compute local derivatives of Uz on reference tetrahedron
ddr = Dr*Uz;  dds = Ds*Uz;  ddt = Dt*Uz;

% add dUz/dz to divergence
divU =  divU + (rz.*ddr + sz.*dds + tz.*ddt);
return;
```

Finally, we define the three-dimensional curl-operator of a vector field as (Curl3D.m)

$$v = (v_x, v_y, v_z) = \nabla \times u.$$

Curl3D.m

```
function [curlx, curly, curlz] = CurlH3D(Ux, Uy, Uz)

% function [curlx, curly, curlz] = CurlH3D(Ux, Uy, Uz)
% purpose: compute local elemental physical spatial curl of (Ux,Uy,Uz)

Globals3D;

% compute local derivatives of Ux on reference tetrahedron
ddr = Dr*Ux;  dds = Ds*Ux;  ddt = Dt*Ux;

% increment curl components
curly =  (rz.*ddr + sz.*dds + tz.*ddt);
curlz = -(ry.*ddr + sy.*dds + ty.*ddt);

% compute local derivatives of Uy on reference tetrahedron
ddr = Dr*Uy;  dds = Ds*Uy;  ddt = Dt*Uy;

% increment curl components
curlx =          -(rz.*ddr + sz.*dds + tz.*ddt);
curlz =  curlz + (rx.*ddr + sx.*dds + tx.*ddt);

% compute local derivatives of Uz on reference tetrahedron
ddr = Dr*Uz;  dds = Ds*Uz;  ddt = Dt*Uz;
```

```
% increment curl components
curlx =  curlx + (ry.*ddr + sy.*dds + ty.*ddt);
curly =  curly - (rx.*ddr + sx.*dds + tx.*ddt);
return;
```

10.3 Assembling the grid

With all local operations in place, we can now address the question of the computation of the metric coefficients and the assembly of the global grid structure. Similar to what we did previously, we assume that we read in or construct the following information:

- Three row vectors (**VX,VY,VZ**) with the coordinates of the N_v vertices. These coordinate sets are numbered from 1 to N_v.
- An integer matrix, EToV, of size $K \times 4$ with the four vertex numbers in each row forming one element. It is always assumed that the 4 vertices are labeled in the same right-handed orientation. This is easily verified by checking that all Jacobians, Eq.(10.5), are positive.

Any grid generator will deliver this information to ensure a meaningful tessellation of the computational domain. With the commands

```
>> va = EToV(:,1)'; vb = EToV(:,2)'; vc = EToV(:,3)'; vd = EToV(:,4)'
>> x = 0.5*(-(1+r+s+t)*VX(va)+(1+r)*VX(vb)+(1+s)*VX(vc)+(1+t)*VX(vd));
>> y = 0.5*(-(1+r+s+t)*VY(va)+(1+r)*VY(vb)+(1+s)*VY(vc)+(1+t)*VY(vd));
>> z = 0.5*(-(1+r+s+t)*VZ(va)+(1+r)*VZ(vb)+(1+s)*VZ(vc)+(1+t)*VZ(vd));
```

we combine the mapping in Eq. (10.3) with the local (r,s,t) coordinates of the nodes in an element of order N and the vertex coordinates to compute the three arrays (x,y,z), each of size $N_p \times K$ with the spatial coordinates.

──────────────── | GeometricFactors3D.m | ────────────────

```
function [rx,sx,tx,ry,sy,ty,rz,sz,tz,J] = ...
        GeometricFactors3D(x,y,z,Dr,Ds,Dt)

% function [rx,sx,tx,ry,sy,ty,rz,sz,tz,J] = ...
%          GeometricFactors3D(x,y,z,Dr,Ds,Dt)
% Purpose  : Compute the metric elements for the local mappings
%            of the elements

% calculate geometric factors
xr = Dr*x; xs = Ds*x; xt = Dt*x;
yr = Dr*y; ys = Ds*y; yt = Dt*y;
zr = Dr*z; zs = Ds*z; zt = Dt*z;
```

```
J = xr.*(ys.*zt-zs.*yt) - yr.*(xs.*zt-zs.*xt) + zr.*(xs.*yt-ys.*xt);
rx =  (ys.*zt - zs.*yt)./J; ry = -(xs.*zt - zs.*xt)./J;
rz = (xs.*yt - ys.*xt)./J; sx = -(yr.*zt - zr.*yt)./J;
sy =  (xr.*zt - zr.*xt)./J; sz = -(xr.*yt - yr.*xt)./J;
tx =  (yr.*zs - zr.*ys)./J; ty = -(xr.*zs - zr.*xs)./J;
tz = (xr.*ys - yr.*xs)./J;
return;
```

This enables the computation of $(\nabla x, \nabla y, \nabla z)$ by using $\mathcal{D}_r$, $\mathcal{D}_s$, and $\mathcal{D}_t$ and, thus, the full metric for the mappings of the individual elements. This is implemented in GeometricFactors3D.m and we construct the $Nf_p \times 4$ array, Fmask, with the local node numbering as

```
>> fmask1 = find( abs(t+1) < NODETOL)';
>> fmask2 = find( abs(s+1) < NODETOL)';
>> fmask3 = find( abs(r+s+t+1) < NODETOL)';
>> fmask4 = find( abs(r+1) < NODETOL)';
>> Fmask = [fmask1;fmask2;fmask3;fmask4]';
>> Fx = x(Fmask(:), :); Fy = y(Fmask(:), :); Fz = z(Fmask(:),:);
```

where the latter extracts the (x, y, z)-coordinates of those nodes at the boundary of the elements.

The outward pointing vectors at the surfaces of the elements follow directly from the properties of the mapping; that is,

$$\hat{n}^1 = -\frac{\nabla t}{\|\nabla t\|}, \quad \hat{n}^2 = -\frac{\nabla s}{\|\nabla s\|}, \quad \hat{n}^3 = \frac{\nabla r + \nabla s + \nabla t}{\|\nabla r + \nabla s + \nabla t\|}, \quad \hat{n}^4 = -\frac{\nabla r}{\|\nabla r\|}.$$

These are computed using Eq. (10.4) based on (x, y, z). The Jacobian for the mapping along the face is obtained by the identity

$$J_s^1 = \|\nabla t\| J$$

and similarly for the three other faces. The computation of the normals and other metric information is all collected in Normals3D.m.

─────────────────────── Normals3D.m ───────────────────────

```
function [nx, ny, nz, sJ] = Normals3D()

% function [nx, ny, nz, sJ] = Normals3D()
% Purpose : Compute outward pointing normals at elements faces
%           as well as surface Jacobians

Globals3D;

[rx,sx,tx,ry,sy,ty,rz,sz,tz,J] = GeometricFactors3D(x,y,z,Dr,Ds,Dt);
```

```
% interpolate geometric factors to face nodes
frx = rx(Fmask(:), :); fsx = sx(Fmask(:), :); ftx = tx(Fmask(:), :);
fry = ry(Fmask(:), :); fsy = sy(Fmask(:), :); fty = ty(Fmask(:), :);
frz = rz(Fmask(:), :); fsz = sz(Fmask(:), :); ftz = tz(Fmask(:), :);

% build normals
nx = zeros(4*Nfp, K); ny = zeros(4*Nfp, K); nz = zeros(4*Nfp, K);
fid1 = (1:Nfp)'; fid2 = (Nfp+1:2*Nfp)';
fid3 = (2*Nfp+1:3*Nfp)'; fid4 = (3*Nfp+1:4*Nfp)';

% face 1
nx(fid1, :) = -ftx(fid1, :); ny(fid1, :) = -fty(fid1, :);
nz(fid1, :) = -ftz(fid1, :);

% face 2
nx(fid2, :) = -fsx(fid2, :); ny(fid2, :) = -fsy(fid2, :);
nz(fid2, :) = -fsz(fid2, :);

% face 3
nx(fid3, :) = frx(fid3, :)+fsx(fid3, :)+ftx(fid3, :);
ny(fid3, :) = fry(fid3, :)+fsy(fid3, :)+fty(fid3, :);
nz(fid3, :) = frz(fid3, :)+fsz(fid3, :)+ftz(fid3, :);

% face 4
nx(fid4, :) = -frx(fid4, :); ny(fid4, :) = -fry(fid4, :);
nz(fid4, :) = -frz(fid4, :);

% normalise
sJ = sqrt(nx.*nx+ny.*ny+nz.*nz);
nx = nx./sJ; ny = ny./sJ; nz = nz./sJ;
sJ = sJ.*J(Fmask(:), :);
return;
```

We are now left with having to combine the K elements into a continuous region of elements by computing the connectivity of the elements. In a departure from the connectivity algorithms we used for the one- and two-dimensional cases, we present a different algorithm [183]. For each face of each tetrahedron, the global vertex numbers for that face are used to compute a unique tag that any face sharing those same vertex numbers will possess. An array, consisting of $4K$ rows and two columns, is constructed with the unique labels forming the first column and the numbers 1 to $4K$, consecutively, in the second column. Face matches are found by sorting the rows of this array in tag order and then finding consecutive rows with the same tag number. These represent the face-to-face matches.

This is done in tiConnect3D.m, which provides two arrays of size $K \times 4$ with the information in terms of element-to-element (EToE) and element-to-face connectivity (EToF). Each of these two arrays holds, in position (i, j), the global element number (EToE) and the local face number (EToF), to which face j on element i connects.

---
tiConnect3D.m

```
function [EToE,EToF]= tiConnect3D(EToV)

% function [EToE,EToF]= tiConnect3D(EToV)
% Purpose: tetrahedral face connect algorithm due to Toby Isaac

Nfaces=4;
K = size(EToV,1);
Nnodes = max(max(EToV));

% create list of all faces 1, then 2, 3 & 4
fnodes = [EToV(:,[1,2,3]);EToV(:,[1,2,4]);...
          EToV(:,[2,3,4]);EToV(:,[1,3,4])];
fnodes = sort(fnodes,2)-1;

% set up default element to element and Element to faces connectivity
EToE = (1:K)'*ones(1,Nfaces); EToF = ones(K,1)*(1:Nfaces);

% uniquely number each set of three faces by their node numbers
id = fnodes(:,1)*Nnodes*Nnodes + fnodes(:,2)*Nnodes+fnodes(:,3)+1;
spNodeToNode=[id, (1:Nfaces*K)', EToE(:), EToF(:)];

% Now we sort by global face number.
sorted=sortrows(spNodeToNode,1);

% find matches in the sorted face list
[indices,dummy]=find( sorted(1:(end-1),1)==sorted(2:end,1) );

% make links reflexive
matchL = [sorted(indices,:)    ;sorted(indices+1,:)];
matchR = [sorted(indices+1,:) ;sorted(indices,:)];

% insert matches
EToE(matchL(:,2)) = matchR(:,3); EToF(matchL(:,2)) = matchR(:,4);
return;
```

We finally construct two vectors, **vmapM** and **vmapP**, each of length $4KNf_p$ (i.e., the total number of individual nodes along the faces of all tetrahedra). Assuming a global numbering of these nodes, the connectivity tables are used to connect the individual nodes along the edges directly with **vmapM** corresponding to the interior nodes, u^-, and **vmapP** to the exterior nodes, u^+. The vertices that do not connect to anything are assembled in the vector **vmapB** as shown in BuildMaps3D.m.

_____ | BuildMaps3D.m | _____

```
function [vmapM, vmapP, vmapB, mapB] = BuildMaps3D

% function [vmapM, vmapP, vmapB, mapB] = BuildMaps3D
% Purpose: Connectivity and boundary tables for nodes given
%          in the K # of elements, each with N+1 degrees of freedom.

Globals3D;

% number volume nodes consecutively
nodeids = reshape(1:K*Np, Np, K);
vmapM   = zeros(Nfp, Nfaces, K);
vmapP   = zeros(Nfp, Nfaces, K);

for k1=1:K
  for f1=1:Nfaces
    % find index of face nodes with respect to volume node ordering
    vmapM(:,f1,k1) = nodeids(Fmask(:,f1), k1);
  end
end

tmp = ones(1,Nfp);
for k1=1:K
  for f1=1:Nfaces
    % find neighbor
    k2 = EToE(k1,f1); f2 = EToF(k1,f1);

    % find find volume node numbers of left and right nodes
    vidM = vmapM(:,f1,k1); vidP = vmapM(:,f2,k2);

    xM = x(vidM)*tmp; yM = y(vidM)*tmp; zM = z(vidM)*tmp;
    xP = x(vidP)*tmp; yP = y(vidP)*tmp; zP = z(vidP)*tmp;

    % Compute distance matrix
    D = (xM -xP').^2 + (yM-yP').^2 + (zM-zP').^2;

    [idM, idP] = find(abs(D)<NODETOL);
    vmapP(idM, f1, k1) = vmapM(idP, f2, k2);
  end
end

vmapP = vmapP(:); vmapM = vmapM(:);

% Create list of boundary nodes
mapB = find(vmapP==vmapM); vmapB = vmapM(mapB);
return
```

Everything discussed in the above must be done for any problem being solved with a three-dimensional DG scheme. We have assembled these components in the startup scripts StartUp3D.m, also containing the definition of a few constants that will become useful later. We collect a number of constants and global variables in Globals3D.m.

────────────────────────── StartUp3D.m ──────────────────────────

```
% Purpose : Setup script, building operators, grid, metric,
%           and connectivity tables for 3D meshes of tetrahedra.

% Definition of constants
Np = (N+1)*(N+2)*(N+3)/6; Nfp = (N+1)*(N+2)/2; Nfaces=4;
    NODETOL = 1e-7;

% Compute nodal set
[x,y,z] = Nodes3D(N); [r,s,t] = xyztorst(x,y,z);

% Build reference element matrices
V = Vandermonde3D(N,r,s,t); invV = inv(V);
MassMatrix = invV'*invV;
[Dr,Ds,Dt] = Dmatrices3D(N, r, s, t, V);

% build coordinates of all the nodes
va = EToV(:,1)'; vb = EToV(:,2)'; vc = EToV(:,3)'; vd = EToV(:,4)';
x = 0.5*(-(1+r+s+t)*VX(va)+(1+r)*VX(vb)+(1+s)*VX(vc)+(1+t)*VX(vd));
y = 0.5*(-(1+r+s+t)*VY(va)+(1+r)*VY(vb)+(1+s)*VY(vc)+(1+t)*VY(vd));
z = 0.5*(-(1+r+s+t)*VZ(va)+(1+r)*VZ(vb)+(1+s)*VZ(vc)+(1+t)*VZ(vd));

% find all the nodes that lie on each edge
fmask1   = find( abs(1+t) < NODETOL)';
fmask2   = find( abs(1+s) < NODETOL)';
fmask3   = find( abs(1+r+s+t) < NODETOL)';
fmask4   = find( abs(1+r) < NODETOL)';
Fmask  = [fmask1;fmask2;fmask3;fmask4]';
Fx = x(Fmask(:), :); Fy = y(Fmask(:), :); Fz = z(Fmask(:), :);

% Create surface integral terms
LIFT = Lift3D(N, r, s, t);

% calculate geometric factors
[rx,sx,tx,ry,sy,ty,rz,sz,tz,J] = GeometricFactors3D(x,y,z,Dr,Ds,Dt);

% calculate geometric factors
[nx, ny, nz, sJ] = Normals3D();
Fscale = sJ./(J(Fmask,:));

% Build connectivity matrix
[EToE, EToF] = tiConnect3D(EToV);
```

```
% Build connectivity maps
[vmapM, vmapP, vmapB, mapB] = BuildMaps3D();

% Compute weak operators (could be done in preprocessing to save time)
[Vr, Vs, Vt] = GradVandermonde3D(N, r, s, t);
Drw = (V*Vr')/(V*V'); Dsw = (V*Vs')/(V*V'); Dtw = (V*Vt')/(V*V');
```

──────────────────────────────── Globals3D.m ────────────────────────────────

```
% Purpose: declare global variables

global Np Nfp N K
global VX VY VZ
global r s t
global Dr Ds Dt LIFT Drw Dsw Dtw
global Fx Fy Fz nx ny nz Fscale sJ
global vmapM vmapP vmapB mapB Fmask
global BCType mapI mapO mapW mapF mapC mapS mapM mapP mapD mapN
global vmapI vmapO vmapW vmapO vmapC vmapS vmapD vmapN
global rx ry rz sx sy sz tx ty tz J
global rk4a rk4b rk4c
global Nfaces EToE EToF EToV
global V invV
global x y z NODETOL

In = 1; Out = 2; Wall = 3; Far = 4; Cyl = 5; Dirichlet = 6;
Neuman = 7; Slip = 8;

% Low storage Runge-Kutta coefficients
rk4a = [            0.0 ...
        -567301805773.0/1357537059087.0 ...
        -2404267990393.0/2016746695238.0 ...
        -3550918686646.0/2091501179385.0 ...
        -1275806237668.0/842570457699.0];
rk4b = [ 1432997174477.0/9575080441755.0 ...
         5161836677717.0/13612068292357.0 ...
         1720146321549.0/2090206949498.0 ...
         3134564353537.0/4481467310338.0 ...
         2277821191437.0/14882151754819.0];
rk4c = [            0.0 ...
         1432997174477.0/9575080441755.0 ...
         2526269341429.0/6820363962896.0 ...
         2006345519317.0/3224310063776.0 ...
         2802321613138.0/2924317926251.0];
```

10.4 Briefly on timestepping

We have previously discussed at length how to choose the timestep to ensure a stable computation and there is nothing fundamentally new to add for the three-dimensional case.

The only concern that requires some attention is the computation of a characteristic length scale associated with a general element. If we continue to use the diameter of the inscribed sphere, we can compute this using the basic formula

$$d_{\text{inscribed}} = \frac{6V}{A_1 + A_2 + A_3 + A_4},$$

where V is the volume of the tetrahedron and A_i is the area of each of the four faces. If we use the simple approximations that $V = J$ (i.e., the mapping Jacobian), and $A_i = J_s^i$ (i.e., the surface mapping Jacobian), a reasonable estimate of the length scale is

$$h \simeq \min_{\Omega} \frac{J^k}{\min J_s^k},$$

and this is what we use. Once the length scale is identified, the rest follows directly from the discussion in Section 6.4 and we refer to the details there.

10.5 Maxwell's equations

In this example we extend the treatment in Section 6.5 of the transverse magnetic form of Maxwell's equations in two-dimensions to the full three-dimensional time-dependent Maxwell's equations as

$$\mu \frac{\partial \boldsymbol{H}}{\partial t} = -\nabla \times \boldsymbol{E}, \quad \varepsilon \frac{\partial \boldsymbol{E}}{\partial t} = \nabla \times \boldsymbol{H}.$$

In conservation form this becomes

$$\mathcal{Q} \frac{\partial \boldsymbol{q}}{\partial t} + \nabla \cdot \mathcal{F} = 0,$$

where

$$\boldsymbol{q} = \begin{bmatrix} \boldsymbol{H} \\ \boldsymbol{E} \end{bmatrix}, \quad \mathcal{Q} = \begin{bmatrix} \mu & 0 \\ 0 & \varepsilon \end{bmatrix}, \quad \mathcal{F} = \begin{bmatrix} -\hat{n} \times \boldsymbol{E} \\ \hat{n} \times \boldsymbol{H} \end{bmatrix} = \begin{bmatrix} \boldsymbol{F}_H \\ \boldsymbol{F}_E \end{bmatrix}.$$

Here, we have the magnetic vector field, $\boldsymbol{H} = (\tilde{H}^x, \tilde{H}^y, \tilde{H}^z)$, and the electric vector field, $\boldsymbol{E} = (\tilde{E}^x, \tilde{E}^y, \tilde{E}^z)$. These are all functions of $(\tilde{x}, \tilde{y}, \tilde{z}, \tilde{t})$. Furthermore, we have the magnetic permeability, $\mu(\boldsymbol{x})$, and the electric permittivity, $\varepsilon(\boldsymbol{x})$, reflecting the material coefficients.

In the following, we model a metallic air-filled cavity, $\Omega = [-1, 1]^3$. In this case, we can simplify the equations since $\mu = \mu_0$ and $\varepsilon = \varepsilon_0$ are the constant vacuum values. If we introduce the vacuum speed of light defined as

$$c_0 = \frac{1}{\sqrt{\varepsilon_0 \mu_0}} \simeq 3 \times 10^8 \text{ m/s},$$

we can consider the normalized system of equations in Cartesian coordinates on the form

$$\frac{\partial H^x}{\partial t} = -\frac{\partial E^z}{\partial y} + \frac{\partial E^y}{\partial z},$$

$$\frac{\partial H^y}{\partial t} = -\frac{\partial E^x}{\partial z} + \frac{\partial E^z}{\partial x},$$

$$\frac{\partial H^z}{\partial t} = -\frac{\partial E^y}{\partial x} + \frac{\partial E^x}{\partial y},$$

$$\frac{\partial E^x}{\partial t} = \frac{\partial H^z}{\partial y} - \frac{\partial H^y}{\partial z},$$

$$\frac{\partial E^y}{\partial t} = \frac{\partial H^x}{\partial z} - \frac{\partial H^z}{\partial x},$$

$$\frac{\partial E^z}{\partial t} = \frac{\partial H^y}{\partial x} - \frac{\partial H^x}{\partial y},$$

where the unit-free variables are obtained as

$$t = \frac{c_0 \tilde{t}}{L}, \quad x = \frac{\tilde{x}}{L}, \quad H = \frac{\tilde{H}}{H_0}, \quad E = (Z_0)^{-1} \frac{\tilde{E}}{H_0},$$

where H_0 is a unit magnetic field strength, $Z_0 = \sqrt{\mu_0/\varepsilon_0} \simeq 120\pi$ Ohm's is the vacuum impedance, and L is some reference length, often the wavelength of the phenomena of interest.

To specify the boundary conditions, we will assume that the walls of the cavity are purely metallic, in which case, the tangential components of the electric field vanishes at the wall (i.e., $\hat{n} \times E = 0$) and the magnetic field has vanishing normal components (i.e., $\hat{n} \cdot H = 0$).

To complete the formulation of the scheme we need only derive a numerical flux. As for the one-dimensional case, discussed in depth in Section 2.4, the numerical flux for the linear case can be obtained by the use of the Rankine-Hugoniot conditions along a normal, $\hat{n}$. In Section 6.5 we obtained the numerical fluxes

$$\hat{n} \cdot (F_H - F_H^*) = \frac{1}{2\{\{Y\}\}} \hat{n} \times (Y^+[E] + \alpha\hat{n} \times [H]]),$$

and

$$\hat{n} \cdot (F_E - F_E^*) = -\frac{1}{2\{\{Z\}\}} \hat{n} \times (Z^+[H] - \alpha\hat{n} \times [E]),$$

for the equations for the magnetic and electric vector fields, respectively. We recall that

$$[q] = q^- - q^+.$$

In both cases, we have the possibility of the piecewise constant material coefficients, represented by

$$Z^{\pm} = \frac{1}{Y^{\pm}} = \sqrt{\frac{\mu^{\pm}}{\varepsilon^{\pm}}},$$

as the local impedance and conductance, respectively.

The parameter, α, in the numerical flux can be used to control dissipation; for example, taking $\alpha = 0$ yields a non-dissipative central flux and $\alpha = 1$ results in the classic upwind flux.

In the simpler case considered here, we have $Z = Y = 1$ due to the normalization. This yields the numerical flux for

$$\hat{n} \cdot (\boldsymbol{F}_H - \boldsymbol{F}_H^*) = \frac{1}{2}\hat{n} \times ([\boldsymbol{E}] + \alpha\hat{n} \times [\boldsymbol{H}]),$$

and

$$\hat{n} \cdot (\boldsymbol{F}_E - \boldsymbol{F}_E^*) = -\frac{1}{2}\hat{n} \times ([\boldsymbol{H}] - \alpha\hat{n} \times [\boldsymbol{E}]).$$

We reorganize the fluxes for simplicity, using the vector identity $a \times (b \times c) = (a \cdot c)b - (a \cdot b)c$ to obtain

$$\hat{n} \cdot (\boldsymbol{F}_H - \boldsymbol{F}_H^*) = \frac{1}{2}\hat{n} \times [\boldsymbol{E}] - \frac{\alpha}{2}[\boldsymbol{H}] + \frac{\alpha}{2}(\hat{n} \cdot [\boldsymbol{H}])\hat{n},$$

and

$$\hat{n} \cdot (\boldsymbol{F}_E - \boldsymbol{F}_E^*) = -\frac{1}{2}\hat{n} \times [\boldsymbol{H}] - \frac{\alpha}{2}[\boldsymbol{E}] + \frac{\alpha}{2}(\hat{n} \cdot [\boldsymbol{E}])\hat{n}.$$

Following the standard DG approach, we recover the semidiscrete scheme to determine the approximation solutions $(\boldsymbol{E}_h, \boldsymbol{H}_h) \in \mathsf{V}_h^3$ as

$$\frac{dH_h^x}{dt} = -\mathcal{D}_y E_h^z + \mathcal{D}_z E_h^y + \frac{\mathcal{M}^{-1}}{2J} \int_{\partial \mathsf{D}^k} (-\hat{n}_y[E_h^z] + \hat{n}_z[E_h^y] + \alpha([H_h^x]$$
$$-(\hat{n} \cdot [\boldsymbol{H}_h])\hat{n}_x)\, \boldsymbol{\ell}(\boldsymbol{x})\, d\boldsymbol{x},$$

$$\frac{dH_h^y}{dt} = -\mathcal{D}_z E_h^x + \mathcal{D}_x E_h^z + \frac{\mathcal{M}^{-1}}{2J} \int_{\partial \mathsf{D}^k} (-\hat{n}_z[E_h^x] + \hat{n}_x[E_h^z] + \alpha([H_h^y]$$
$$-(\hat{n} \cdot [\boldsymbol{H}_h])\hat{n}_y)\, \boldsymbol{\ell}(\boldsymbol{x})\, d\boldsymbol{x},$$

$$\frac{dH_h^z}{dt} = -\mathcal{D}_x E_h^y + \mathcal{D}_y E_h^x + \frac{\mathcal{M}^{-1}}{2J} \int_{\partial \mathsf{D}^k} (-\hat{n}_x[E_h^y] + \hat{n}_y[E_h^x] + \alpha([H_h^z]$$
$$-(\hat{n} \cdot [\boldsymbol{H}_h])\hat{n}_z)\, \boldsymbol{\ell}(\boldsymbol{x})\, d\boldsymbol{x},$$

$$\frac{dE_h^x}{dt} = \mathcal{D}_y H_h^z - \mathcal{D}_z H_h^y + \frac{\mathcal{M}^{-1}}{2J} \int_{\partial \mathsf{D}^k} (\hat{n}_y[H_h^z] - \hat{n}_z[H_h^y] + \alpha([E_h^x]$$
$$-(\hat{n} \cdot [\boldsymbol{E}_h])\hat{n}_x)\, \boldsymbol{\ell}(\boldsymbol{x})\, d\boldsymbol{x},$$

$$\frac{dE_h^y}{dt} = \mathcal{D}_z H_h^x - \mathcal{D}_x H_h^z + \frac{\mathcal{M}^{-1}}{2J} \int_{\partial \mathsf{D}^k} (\hat{n}_z[H_h^x] - \hat{n}_x[H_h^z] + \alpha([E_h^y]$$
$$-(\hat{\boldsymbol{n}} \cdot [\boldsymbol{E}_h])\hat{n}_y) \,\boldsymbol{\ell}(\boldsymbol{x}) \, d\boldsymbol{x},$$

$$\frac{dE_h^z}{dt} = \mathcal{D}_x H_h^y - \mathcal{D}_y H_h^x + \frac{\mathcal{M}^{-1}}{2J} \int_{\partial \mathsf{D}^k} (\hat{n}_x[H_h^y] - \hat{n}_y[H_h^x] + \alpha([E_h^z]$$
$$-(\hat{\boldsymbol{n}} \cdot [\boldsymbol{E}_h])\hat{n}_z) \,\boldsymbol{\ell}(\boldsymbol{x}) \, d\boldsymbol{x},$$

where we have suppressed the element index, k, for simplicity. This algorithm is implemented in MaxwellRHS3D.m.

────────────────────────── | MaxwellRHS3D.m | ──────────────────────────

```
function [rhsHx, rhsHy, rhsHz, rhsEx, rhsEy, rhsEz] = ...
        MaxwellRHS3D(Hx,Hy,Hz,Ex,Ey,Ez)

% function [rhsHx, rhsHy, rhsHz, rhsEx, rhsEy, rhsEz] =
%          MaxwellRHS3D(Hx,Hy,Hz,Ex,Ey,Ez)
% Purpose  : Evaluate RHS flux in 3D Maxwell equations

Globals3D;

% storage for field differences at faces
dHx = zeros(Nfp*Nfaces,K); dHy = dHx; dHz = dHx;
dEx = zeros(Nfp*Nfaces,K); dEy = dEx; dEz = dEx;

% form field differences at faces
dHx(:)  = Hx(vmapP)-Hx(vmapM);   dEx(:)  = Ex(vmapP)-Ex(vmapM);
dHy(:)  = Hy(vmapP)-Hy(vmapM);   dEy(:)  = Ey(vmapP)-Ey(vmapM);
dHz(:)  = Hz(vmapP)-Hz(vmapM);   dEz(:)  = Ez(vmapP)-Ez(vmapM);

% make boundary conditions all reflective (Ez+ = -Ez-)
dHx(mapB) = 0;   dEx(mapB) = -2*Ex(vmapB);
dHy(mapB) = 0;   dEy(mapB) = -2*Ey(vmapB);
dHz(mapB) = 0;   dEz(mapB) = -2*Ez(vmapB);

alpha=1; % => full upwinding

ndotdH = nx.*dHx + ny.*dHy + nz.*dHz;
ndotdE = nx.*dEx + ny.*dEy + nz.*dEz;

fluxHx = -ny.*dEz + nz.*dEy + alpha*(dHx - ndotdH.*nx);
fluxHy = -nz.*dEx + nx.*dEz + alpha*(dHy - ndotdH.*ny);
fluxHz = -nx.*dEy + ny.*dEx + alpha*(dHz - ndotdH.*nz);

fluxEx =  ny.*dHz - nz.*dHy + alpha*(dEx - ndotdE.*nx);
fluxEy =  nz.*dHx - nx.*dHz + alpha*(dEy - ndotdE.*ny);
fluxEz =  nx.*dHy - ny.*dHx + alpha*(dEz - ndotdE.*nz);
```

```
% evaluate local spatial derivatives
[curlHx,curlHy,curlHz] = Curl3D(Hx,Hy,Hz);
[curlEx,curlEy,curlEz] = Curl3D(Ex,Ey,Ez);

% calculate Maxwell's right hand side
rhsHx = -curlEx + LIFT*(Fscale.*fluxHx/2);
rhsHy = -curlEy + LIFT*(Fscale.*fluxHy/2);
rhsHz = -curlEz + LIFT*(Fscale.*fluxHz/2);

rhsEx =  curlHx + LIFT*(Fscale.*fluxEx/2);
rhsEy =  curlHy + LIFT*(Fscale.*fluxEy/2);
rhsEz =  curlHz + LIFT*(Fscale.*fluxEz/2);
return;
```

The boundary condition for $\boldsymbol{E}$ is

$$\hat{\boldsymbol{n}} \times \boldsymbol{E} = 0,$$

and we need not specify any conditions on $\boldsymbol{H}$. This latter is implemented by simply setting $(\hat{\boldsymbol{n}} \times \boldsymbol{H})^* = \hat{\boldsymbol{n}} \times \boldsymbol{H}^-$ at the boundary. The homogeneous condition of $\hat{\boldsymbol{n}} \times \boldsymbol{E}$ can be implemented in different ways. We use a mirror principle, based on assigning $(\hat{\boldsymbol{n}} \times \boldsymbol{E})^* = -\hat{\boldsymbol{n}} \times \boldsymbol{E}^-$ such that $(\hat{\boldsymbol{n}} \times \boldsymbol{E})^* + \hat{\boldsymbol{n}} \times \boldsymbol{E}^+ = 0$. This is enforced as

$$\hat{\boldsymbol{n}} \times [\boldsymbol{E}] = 2\hat{\boldsymbol{n}} \times \boldsymbol{E}^-,$$

at all boundary points, identified by $\boldsymbol{vmapB}$.

With the semidiscrete formulation completed, we use an explicit low-storage Runge-Kutta method to integrate in time, exactly as in the one-dimensional cases discussed in Section 3.4.

The validation test used in Section 6.5 to validate the two-dimensional Maxwell solver is also a useful test for the full three-dimensional Maxwell solver, since it satisfies all the boundary conditions.

$$H^x(x,y,z,t) = -\frac{\pi n}{\omega} \sin(m\pi x) \cos(n\pi y) \sin(\omega t),$$

$$H^y(x,y,z,t) = \frac{\pi m}{\omega} \cos(m\pi x) \sin(n\pi y) \sin(\omega t),$$

$$E^z(x,y,z,t) = \sin(m\pi x) \sin(n\pi y) \cos(\omega t),$$

$$H^z = E^x = E^y = 0,$$

where the resonance frequencies, ω, are given as

$$\omega = \pi\sqrt{m^2 + n^2}, \quad (m,n) \geq 0.$$

For simplicity, we take $m = n = 1$.

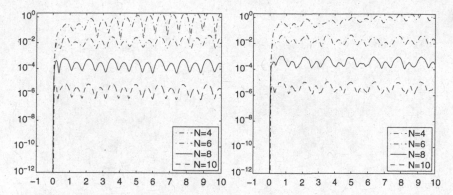

Fig. 10.2. Time trace of the discrete L^2-error for E^z for $T \in [0, 10]$, obtained using a DG method to solve Maxwell's equations with left) upwind fluxes, right) central fluxes for different values of N.

In Fig. 10.2 we show the convergence of the electric field, E^z, under order refinement. We show the results computed with both a central flux ($\alpha = 0$) and an upwind flux ($\alpha = 1$). Similar convergence behavior can be observed for the other field components.

When comparing the results, the scheme based on the upwind flux yields optimal convergence rates (i.e, $\mathcal{O}(h^{N+1})$, where $h = \sqrt[3]{K}$).

10.6 Three-dimensional Poisson equation

As a last example, we consider the solution of three-dimensional Poisson equation

$$\nabla \cdot \nabla u(\boldsymbol{x}) = f(\boldsymbol{x}), \ \ \boldsymbol{x} \in \Omega \subset \mathsf{R}^3,$$

with some specified Dirichlet boundary conditions

$$u(\boldsymbol{x}) = g(\boldsymbol{x}), \ \ \boldsymbol{x} \in \partial\Omega.$$

We consider a DG solver based on an internal penalty flux (IPDG) as discussed in Section 7.2, implemented in a direct way in PoissonIPDG3D.m. This illustrates the minor changes required to extend the one-dimensional formulation to this general three-dimensional case.

As the data for the test problem we take

$$f(\boldsymbol{x}) = -3\pi^2 \sin(\pi x) \sin(\pi y) \sin(\pi z), \ \ g(\boldsymbol{x}) = \sin(\pi x) \sin(\pi y) \sin(\pi z),$$

leading to an exact solution

$$u(\boldsymbol{x}) = \sin(\pi x) \sin(\pi y) \sin(\pi z).$$

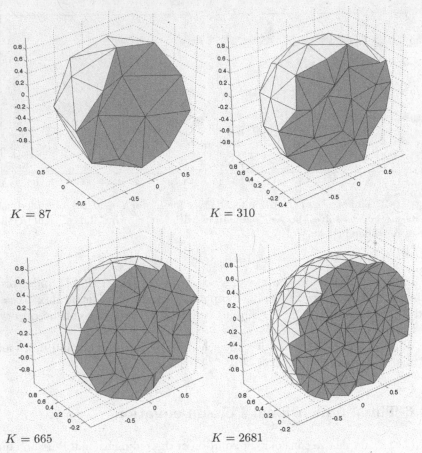

$K = 87$ $K = 310$

$K = 665$ $K = 2681$

Fig. 10.3. Cut-away view of the sequence of meshes used in the convergence test for the three-dimensional Poisson solver.

Table 10.2. Maximum pointwise error for solving the three-dimensional Poisson equation on the domains shown in Fig. 10.3.

N	$h = 0.6$	$h = 0.5$	$h = 0.3$	$h = 0.2$	Rate
2	1.90E-01	1.62E-01	6.29E-02	3.74E-02	2.53
3	9.63E-02	3.61E-02	2.12E-02	6.86E-03	3.66
4	1.75E-02	1.59E-02	2.51E-03	9.14E-04	4.68
5	8.92E-03	2.04E-03	5.60E-04	1.16E-04	6.17

To make things a bit interesting, we take the computational domain to be a triangulated sphere of unit radius, on a sequence of grids as illustrated in Fig. 10.3.

The results of the convergence test are shown in Table 10.2, with the error measured in the maximum norm. The convergence rate, estimated by a linear fit through the computed results with $h = K^{1/3}$, is seen to be close to the expected $\mathcal{O}(h^{N+1})$. The slight suboptimality is due to the relatively large errors in the preasymptotic regime where the solution is only marginally resolved on the coarse grids.

───────────── PoissonIPDG3D.m ─────────────

```
function [OP,MM] = PoissonIPDG3D()

% function [OP,MM] = PoissonIPDG3D()
% Purpose: Set up the discrete Poisson matrix directly
%          using IP. The operator is set up in the weak form

Globals3D;

% build local face matrices
massEdge = zeros(Np,Np,Nfaces);

% face mass matrix 1
Fm = Fmask(:,1); faceR = r(Fm); faceS = s(Fm);
V2D = Vandermonde2D(N, faceR, faceS);
massEdge(Fm,Fm,1) = inv(V2D*V2D');

% face mass matrix 2
Fm = Fmask(:,2); faceR = r(Fm); faceT = t(Fm);
V2D = Vandermonde2D(N, faceR, faceT);
massEdge(Fm,Fm,2) = inv(V2D*V2D');

% face mass matrix 3
Fm = Fmask(:,3); faceS = s(Fm); faceT = t(Fm);
V2D = Vandermonde2D(N, faceS, faceT);
massEdge(Fm,Fm,3) = inv(V2D*V2D');

% face mass matrix 4
Fm = Fmask(:,4); faceS = s(Fm); faceT = t(Fm);
V2D = Vandermonde2D(N, faceS, faceT);
massEdge(Fm,Fm,4) = inv(V2D*V2D');

% build local volume mass matrix
MassMatrix = invV'*invV;

% build DG derivative matrices
MM  = zeros(K*Np*Np, 3);   OP = zeros(K*Np*Np*(1+Nfaces), 3);

% global node numbering
entries = (1:Np*Np)'; entriesMM = (1:Np*Np)';
for k1=1:K
  if(~mod(k1,1000)) k1, end;
  rows1 = ((k1-1)*Np+1:k1*Np)'*ones(1,Np); cols1 = rows1';
```

```
% Build local operators
Dx = rx(1,k1)*Dr + sx(1,k1)*Ds + tx(1,k1)*Dt;
Dy = ry(1,k1)*Dr + sy(1,k1)*Ds + ty(1,k1)*Dt;
Dz = rz(1,k1)*Dr + sz(1,k1)*Ds + tz(1,k1)*Dt;

OP11 = J(1,k1)*(Dx'*MassMatrix*Dx ...
     + Dy'*MassMatrix*Dy + Dz'*MassMatrix*Dz);

% Build element-to-element parts of operator
for f1=1:Nfaces
  k2 = EToE(k1,f1); f2 = EToF(k1,f1);

  rows2 = ((k2-1)*Np+1:k2*Np)'*ones(1,Np); cols2 = rows2';

  fidM = (k1-1)*Nfp*Nfaces + (f1-1)*Nfp + (1:Nfp);
  vidM = vmapM(fidM); Fm1 = mod(vidM-1,Np)+1;
  vidP = vmapP(fidM); Fm2 = mod(vidP-1,Np)+1;

  id = 1+(f1-1)*Nfp + (k1-1)*Nfp*Nfaces;
  lnx = nx(id);  lny = ny(id);  lnz = nz(id);  lsJ = sJ(id);
  hinv = max(Fscale(id), Fscale(1+(f2-1)*Nfp, k2));

  Dx2 = rx(1,k2)*Dr + sx(1,k2)*Ds + tx(1,k2)*Dt;
  Dy2 = ry(1,k2)*Dr + sy(1,k2)*Ds + ty(1,k2)*Dt;
  Dz2 = rz(1,k2)*Dr + sz(1,k2)*Ds + tz(1,k2)*Dt;

  Dn1 = lnx*Dx  + lny*Dy  + lnz*Dz;
  Dn2 = lnx*Dx2 + lny*Dy2 + lnz*Dz2;

  mmE = lsJ*massEdge(:,:,f1);

  gtau = 2*(N+1)*(N+1)*hinv; % set penalty scaling
  if(EToE(k1,f1)==k1)
    OP11 = OP11 + ( gtau*mmE - mmE*Dn1 - Dn1'*mmE ); % ok
  else
    % interior face variational terms
    OP11        = OP11 + 0.5*( gtau*mmE - mmE*Dn1 - Dn1'*mmE );

    OP12 = zeros(Np);
    OP12(:,Fm2) =                   - 0.5*( gtau*mmE(:,Fm1) );
    OP12(Fm1,:) = OP12(Fm1,:) - 0.5*(       mmE(Fm1,Fm1)*Dn2(Fm2,:) );
    OP12(:,Fm2) = OP12(:,Fm2) - 0.5*(-Dn1'*mmE(:, Fm1) );
    OP(entries(:), :) = [rows1(:), cols2(:), OP12(:)];
    entries = entries + Np*Np;
  end
end
OP(entries(:), :)   = [rows1(:), cols1(:), OP11(:)];
MM(entriesMM(:), :) = [rows1(:), cols1(:), J(1,k1)*MassMatrix(:)];
```

```
   entries = entries + Np*Np; entriesMM = entriesMM + Np*Np;
end

OP  =   OP(1:max(entries)  -Np*Np,:);
OP  = myspconvert(OP, Np*K, Np*K, 1e-15);
MM  =   MM(1:max(entriesMM)-Np*Np,:);
MM  = myspconvert(MM, Np*K, Np*K, 1e-15);
return
```

10.7 Exercises

1. In Section 6.6 we introduced Euler's nonlinear equations of gas dynamics. In this exercise we consider those equations in three-dimensions but for small acoustic perturbations about a quiescent mean state. Dropping quadratic and higher-order terms in the perturbation reveals the classic linearized Euler equations of acoustics given by

$$\frac{\partial \boldsymbol{u}}{\partial t} = -\nabla p,$$

$$\frac{\partial p}{\partial t} = -\nabla \cdot \boldsymbol{u},$$

for the acoustic velocity, $\boldsymbol{u} = (u, v, w)$, and pressure, p, field. Following Section 10.5 a semi-discrete DG method for this system is given

$$\frac{\partial u_h}{\partial t} = -\mathcal{D}_x p_h + \frac{\mathcal{M}^{-1}}{2J} \int_{\partial \mathsf{D}^k} -\hat{n}_x \left(\alpha \llbracket \boldsymbol{u}_h \rrbracket - [p_h] \right) \boldsymbol{\ell}(\mathbf{x}) \, d\boldsymbol{x},$$

$$\frac{\partial v_h}{\partial t} = -\mathcal{D}_y p_h + \frac{\mathcal{M}^{-1}}{2J} \int_{\partial \mathsf{D}^k} -\hat{n}_y \left(\alpha \llbracket \boldsymbol{u}_h \rrbracket - [p_h] \right) \boldsymbol{\ell}(\mathbf{x}) \, d\boldsymbol{x},$$

$$\frac{\partial w_h}{\partial t} = -\mathcal{D}_z p_h + \frac{\mathcal{M}^{-1}}{2J} \int_{\partial \mathsf{D}^k} -\hat{n}_z \left(\alpha \llbracket \boldsymbol{u}_h \rrbracket - [p_h] \right) \boldsymbol{\ell}(\mathbf{x}) \, d\boldsymbol{x},$$

$$\frac{\partial p_h}{\partial t} = -\mathcal{D}_x u_h - \mathcal{D}_y v_h - \mathcal{D}_z w_h + \frac{\mathcal{M}^{-1}}{2J} \int_{\partial \mathsf{D}^k} \left(\llbracket \boldsymbol{u}_h \rrbracket - \alpha \hat{n} \cdot \llbracket p \rrbracket \right) \boldsymbol{l}(\mathbf{x}) \, d\boldsymbol{x}.$$

a) Create a new set of Matlab scripts for this system LinEulerDriver3D.m, LinEuler3D.m, LinEulerRHS3D.m, perhaps based on the MaxwellDriver 3D.m, Maxwell3D.m, MaxwellRHS3D.m scripts, using a low-storage Runge-Kutta method for timestepping. Assume reflective boundary conditions at all domain boundary faces, with external traces given by

$$\boldsymbol{u}^+ = -\boldsymbol{u}^-, \quad p^+ = p^-.$$

b) Devise a verification test case for the upwind DG solver ($\alpha = 1$) [Hint: consider a Fourier resonant mode of a cubic cavity] and determine the order of accuracy of the solver for, say, $N = 4$.

c) The time-domain linearized Euler solver you created in the previous exercise can be used to estimate the resonant frequencies of reflecting cavities. Design a cavity and create a three-dimensional mesh of tetrahedra using the DistMesh package.

To determine a range of resonant frequencies of the cavity, it can be excited with a Gaussian pressure pulse, and a microphone placed inside the cavity can be used to record a time history of the pressure at a specific site.

In the simulation, use the weights and element number provided by the service routine CODES3D/SAMPLE3D.M. First, call this routine to find the element and interpolation weights required to interpolate the nodal data to the sample (i.e., microphone location). To find the weights and element corresponding to an example location $\mathbf{x} = (0.1, 0.2, 0.3)$, use
>> [sampleweights, sampletet] = Sample3D(0.1, 0.2, 0.3);
and to interpolate a field p to this location use
>> psample = sampleweights*p(:,sampletet);
This computation can be performed at the end of every timestep and the result stored in an array.

At the end of the computation, use a Fourier transform of the sampled pressure time history to compute the dominant frequencies activated by the initial Gaussian pressure pulse. You may wish to monitor the pressure at a number of different microphone locations and compare the resulting sets of frequency responses. Keep in mind that for the upwind DG method, there are also initial solutions that damp out due to unphysical but strongly damped or underresolved modes excited by the initial condition so you should integrate to a reasonable final time (experiment to determine the effect of the choice of final time) and discard early samples of the pressure field.

d) In Chapter 8 we saw that the central flux DG was prone to generating spurious resonances for this kind of computation in the case of Maxwell's equations. Compute resonances for the cavity using, first, $\alpha = 0$ for a central flux computation and then $\alpha = 1$ for the upwind flux variant. Does the choice of α change the computed resonant response significantly?

e) We will now modify the linearized Euler equations discussed above to the case of small acoustic perturbations around a constant mean flow, $\mathbf{M}(\mathbf{x}) = (M_x(\mathbf{x}), M_y(\mathbf{x}), M_z(\mathbf{x}))$, resulting in the equations

$$\frac{\partial \mathbf{u}}{\partial t} + \mathbf{M} \cdot \nabla \mathbf{u} + (\mathbf{u} \cdot \nabla)\mathbf{M} = -\nabla p,$$

$$\frac{\partial p}{\partial t} + \mathbf{M} \cdot \nabla p + p \nabla \cdot \mathbf{M} = -\nabla \cdot \mathbf{u}.$$

Modify the scheme outlined in the previous example and implement the scheme for general smooth mean flow M.

f) Carefully discuss the number of boundary conditions needed for different values of the mean flow and the construction of the numerical flux – both upwind and central fluxes.

g) Devise a verification test case for the DG solver and determine the order of accuracy of the solver for, say, $N = 4$.

A

Appendix A: Jacobi polynomials and beyond

In the following we review a few properties of classical orthogonal polynomials, Gauss quadratures, and the extension of these ideas to simplices. We make no attempt to be complete and refer to the many excellent texts on classical polynomials for more details (e.g., [89, 111, 197, 296]).

The classical Jacobi polynomial, $P_n^{(\alpha,\beta)}(x)$, of order n is a solution to the singular Sturm-Liouville eigenvalue problem

$$\frac{d}{dx}(1 - x^2)w(x)\frac{d}{dx}P_n^{(\alpha,\beta)}(x) + n(n + \alpha + \beta + 1)w(x)P_n^{(\alpha,\beta)}(x) = 0, \quad \text{(A.1)}$$

for $x \in [-1, 1]$, where the weight function, $w(x) = (1-x)^\alpha(1+x)^\beta$. w-Weighted orthogonality of the polynomials is a direct consequence of Eq. (A.1). The polynomials are normalized to be orthonormal:

$$\int_{-1}^{1} P_i^{(\alpha,\beta)}(x)P_j^{(\alpha,\beta)}(x)w(x)\,dx = \delta_{ij}.$$

An important property of the Jacobi polynomials is [296]

$$\frac{d}{dx}P_n^{(\alpha,\beta)}(x) = \sqrt{n(n + \alpha + \beta + 1)}P_{n-1}^{(\alpha+1,\beta+1)}(x). \quad \text{(A.2)}$$

Also, recall the special case of $P_n^{(0,0)}(x)$, known as the Legendre polynomials.

While there is no known simple expression to evaluate the Jacobi polynomials, it is conveniently done using the recurrence relation

$$xP_n^{(\alpha,\beta)}(x) = a_n P_{n-1}^{(\alpha,\beta)}(x) + b_n P_n^{(\alpha,\beta)}(x) + a_{n+1}P_{n+1}^{(\alpha,\beta)}(x), \quad \text{(A.3)}$$

where the coefficients are given as

$$a_n = \frac{2}{2n + \alpha + \beta}\sqrt{\frac{n(n + \alpha + \beta)(n + \alpha)(n + \beta)}{(2n + \alpha + \beta - 1)(2n + \alpha + \beta + 1)}},$$

$$b_n = -\frac{\alpha^2 - \beta^2}{(2n + \alpha + \beta)(2n + \alpha + \beta + 2)}.$$

To get the recurrence started, we need the initial values

$$P_0^{(\alpha,\beta)}(x) = \sqrt{2^{-\alpha-\beta-1}\frac{\Gamma(\alpha+\beta+2)}{\Gamma(\alpha+1)\Gamma(\beta+1)}},$$

$$P_1^{(\alpha,\beta)}(x) = \frac{1}{2}P_0^{(\alpha,\beta)}(x)\sqrt{\frac{\alpha+\beta+3}{(\alpha+1)(\beta+1)}}\left((\alpha+\beta+2)x + (\alpha-\beta)\right).$$

Here, $\Gamma(x)$ is the classic Gamma function [4]. A Matlab script for evaluating Jacobi polynomials using the above procedure is given in JacobiP.m.

JacobiP.m
```
function [P] = JacobiP(x,alpha,beta,N);

% function [P] = JacobiP(x,alpha,beta,N)
% Purpose: Evaluate Jacobi Polynomial of type (alpha,beta) > -1
%          (alpha+beta <> -1) at points x for order N and returns
%          P[1:length(xp))]
% Note   : They are normalized to be orthonormal.

% Turn points into row if needed.
xp = x; dims = size(xp);
if (dims(2)==1) xp = xp'; end;

PL = zeros(N+1,length(xp));

% Initial values P_0(x) and P_1(x)
gamma0 = 2^(alpha+beta+1)/(alpha+beta+1)*gamma(alpha+1)*...
         gamma(beta+1)/gamma(alpha+beta+1);
PL(1,:) = 1.0/sqrt(gamma0);
if (N==0) P=PL'; return; end;
gamma1 = (alpha+1)*(beta+1)/(alpha+beta+3)*gamma0;
PL(2,:) = ((alpha+beta+2)*xp/2 + (alpha-beta)/2)/sqrt(gamma1);
if (N==1) P=PL(N+1,:)'; return; end;

% Repeat value in recurrence.
aold = 2/(2+alpha+beta)*sqrt((alpha+1)*(beta+1)/(alpha+beta+3));

% Forward recurrence using the symmetry of the recurrence.
for i=1:N-1
  h1 = 2*i+alpha+beta;
  anew = 2/(h1+2)*sqrt( (i+1)*(i+1+alpha+beta)*(i+1+alpha)*...
         (i+1+beta)/(h1+1)/(h1+3));
```

```
  bnew = - (alpha^2-beta^2)/h1/(h1+2);
  PL(i+2,:) = 1/anew*( -aold*PL(i,:) + (xp-bnew).*PL(i+1,:));
  aold =anew;
end;


P = PL(N+1,:)';
return
```

As is well known (see, e.g., [89]), there is a close connection between Jacobi polynomials and Gaussian quadratures for the approximation of integrals as

$$\int_{-1}^{1} f(x)w(x)\, dx = \sum_{i=0}^{N} f(x_i)w_i.$$

Here, (x_i, w_i) are the quadrature nodes and weights. It can be shown that if one chooses x_i as the roots of $P_{N+1}^{(\alpha,\beta)}(x)$ and the weights, w_i, by requiring the integration to be exact for polynomials up to order N, the above summation is in fact exact for f being a polynomial of order $2N+1$ – this is the celebrated Gaussian quadrature.

Finding the nodes and weights can be done in several ways, with perhaps the most elegant and numerically stable one being based on the recurrence, Eq. (A.3). Inspection reveals that setting $P_{N+1}^{(\alpha,\beta)}(x_i) = 0$ truncates the recurrence and the nodes, x_i, are the eigenvalues of a symmetric tridiagonal eigenvalue problem. The weights can be recovered from the elements of the eigenvectors; for the details, we refer to [131]. In JacobiGQ.m, an implementation of this algorithm is offered.

─────────────── JacobiGQ.m ───────────────

```
function [x,w] = JacobiGQ(alpha,beta,N);

% function [x,w] = JacobiGQ(alpha,beta,N)
% Purpose: Compute the N'th order Gauss quadrature points, x,
%          and weights, w, associated with the Jacobi
%          polynomial, of type (alpha,beta) > -1 ( <> -0.5).

if (N==0) x(1)=(alpha-beta)/(alpha+beta+2); w(1) = 2; return; end;

% Form symmetric matrix from recurrence.
J = zeros(N+1);
h1 = 2*(0:N)+alpha+beta;
J = diag(-1/2*(alpha^2-beta^2)./(h1+2)./h1) + ...
    diag(2./(h1(1:N)+2).*sqrt((1:N).*((1:N)+alpha+beta).*...
    ((1:N)+alpha).*((1:N)+beta)./(h1(1:N)+1)./(h1(1:N)+3)),1);
if (alpha+beta<10*eps) J(1,1)=0.0;end;
J = J + J';
```

```
% Compute quadrature by eigenvalue solve
[V,D] = eig(J); x = diag(D);
w = (V(1,:)').^2*2^(alpha+beta+1)/(alpha+beta+1)*gamma(alpha+1)*...
    gamma(beta+1)/gamma(alpha+beta+1);
return;
```

When solving partial differential equations using high-order and spectral methods, one often uses nodes based on Gauss-like quadrature points, as they are known to allow for high-order accurate interpolation (see Chapter 3). However, the pure Gauss points, computable with JacobiGQ.m, are less favorable since they do not include grid points at the end of the intervals. While not a major obstacle, it is often convenient to include these end points to impose boundary conditions.

One often uses Gauss-Lobatto points, given as the roots of $(1 - x^2) \times \frac{d}{dx} P_N^{(\alpha,\beta)}(x)$. Using Eq. (A.2), it is easily realized that the interior Gauss-Lobatto points are the $(N-2)$-th-order Gauss points of $P_{N-2}^{(\alpha+1,\beta+1)}(x)$ with the end points added. A simple routine utilizing this is shown as JacobiGL.m.

────────────────────────── JacobiGL.m ──────────────────────────

```
function [x] = JacobiGL(alpha,beta,N);

% function [x] = JacobiGL(alpha,beta,N)
% Purpose: Compute the N'th order Gauss Lobatto quadrature
%          points, x, associated with the Jacobi polynomial,
%          of type (alpha,beta) > -1 ( <> -0.5).

x = zeros(N+1,1);
if (N==1) x(1)=-1.0; x(2)=1.0; return; end;

[xint,w] = JacobiGQ(alpha+1,beta+1,N-2);
x = [-1, xint', 1]';
return;
```

A.1 Orthonormal polynomials beyond one dimension

The extension of polynomial modal expansions to the multidimensional case is a bit more complicated, mainly due to the added geometric variation of the domain (e.g., quadrilaterals/hexahedrals or triangles/tetrahedra).

In the case of domains/elements that are logically cubic, a simple dimension-by-dimension approach suffices – this approach is known as tensor products and is used widely; for example, to represent a function $u(x, y)$ on $[-1, 1]^2$, one can use a Legendre expansion

$$u_h(x,y) = \sum_{i,j=0}^{N} \hat{u}_{ij} P_i(x) P_j(y).$$

Note that this basis has $(N+1)^2$ terms, while only $\frac{1}{2}(N+1)(N+2)$ terms are needed for completeness in two dimensions. Orthonormality is clearly maintained in a dimension-by-dimension fashion.

For more complex elements (e.g., simplices), the construction of an orthonormal basis of order N is a bit more complicated. To ensure good approximation properties in finite domains, the polynomials should be orthonormal eigensolutions to a multidimensional singular Sturm-Liouville problem.

The construction of such polynomials has been pursued by several authors [103, 201, 261]. For the two-dimensional simplex

$$\mathsf{T}^2 = \{(r,s)|r,s \geq -1; r+s \leq 0\},$$

the N-th-order orthonormal basis is given as

$$\forall (i,j) \geq 0; i+j \leq N : \ \psi_{ij}(r,s) = \sqrt{2} P_i^{(0,0)}(a) P_j^{(2i+1,0)}(b)(1-b)^i,$$

where the extended coordinates $(a,b) \in [-1,1]^2$ relates to $(r,s) \in \mathsf{T}^2$ as

$$a = 2\frac{1+r}{1-s} - 1, \ \ b = s.$$

Note that there are exactly $\frac{1}{2}(N+1)(N+2)$ terms in the polynomial basis of order N. A script to evaluate the basis in the (a,b) coordinates is shown in Simplex2DP.m.

Simplex2DP.m

```
function [P] = Simplex2DP(a,b,i,j);

% function [P] = Simplex2DP(a,b,i,j);
% Purpose : Evaluate 2D orthonormal polynomial
%           on simplex at (a,b) of order (i,j).

h1 = JacobiP(a,0,0,i); h2 = JacobiP(b,2*i+1,0,j);
P = sqrt(2.0)*h1.*h2.*(1-b).^i;
return;
```

Similarly, one can derive an orthonormal basis for the three-dimensional simplex,

$$\mathsf{T}^3 = \{(r,s,t)|r,s,t \geq -1; r+s+t \leq -1\},$$

with the N-th-order orthonormal basis being

$$\forall (i,j,k) \geq 0; i+j+k \leq N :$$
$$\psi_{ijk}(r,s,t) = 2\sqrt{2} P_i^{(0,0)}(a) P_j^{(2i+1,0)}(b) P_k^{(2i+2j+2,0)}(b)(1-b)^i(1-c)^{i+j},$$

where the extended coordinates $(a, b, c) \in [-1, 1]^3$ relates to $(r, s, t) \in \mathsf{T}^3$ as

$$a = -2\frac{1+r}{s+t} - 1, \quad b = 2\frac{1+r}{1-t} - 1, \quad c = t.$$

Note that there are exactly $\frac{1}{6}(N+1)(N+2)(N+3)$ terms in the polynomial basis of order N. A script to evaluate the basis in the (a, b, c) coordinates is shown in Simplex3DP.m.

Simplex3DP.m

```
function [P] = Simplex3DP(a,b,c,i,j,k);

% function [P] = Simplex3DP(a,b,c,i,j,k);
% Purpose : Evaluate 3D orthonormal polynomial
%           on simplex at (a,b,c) of order (i,j,k).

h1 = JacobiP(a,0,0,i); h2 = JacobiP(b,2*i+1,0,j);
h3 = JacobiP(c,2*(i+j)+2,0,k);
P = 2*sqrt(2)*h1.*h2.*((1-b).^i).*h3.*((1-c).^(i+j));
return;
```

Orthogonality of the simplex basis is shown in [197] and the relation to Sturm-Liouville problems is discussed in [201]. The construction of orthonormal basis sets for canonical shapes (e.g., spheres and prisms) is discussed in more detail in [197, 201].

B

Appendix B: Briefly on grid generation

by Allan Peter Engsig-Karup

The generation of grids beyond the simple one-dimensional cases is a topic with a very significant literature of its own and we will make no attempt to do justice to this.

As has become clear during the discussion of the basic components in Chapters 3, 6, and 10, all that is required are lists of the N_v vertices, (VX, VY, VZ), and an array, EToV, of size $K \times$ Nfaces of the vertices forming the elements. In case several types of boundary condition are needed, a list, BCType, is also required, as we discussed in Section 6.4. It is always assumed that all elements are numbered counterclockwise on local elements. This is easily checked by considering the entries of EToV and (VX, VY, VZ).

For the one-dimensional case, it is straightforward to generate a simple grid generator that returns the required information. An example is given in MeshGen1D.m, which simply generates a one-dimensional equidistant grid.

```
———————————————————————| MeshGen1D.m |———————————————————————
function [Nv, VX, K, EToV] = MeshGen1D(xmin,xmax,K)

% function [Nv, VX, K, EToV] = MeshGen1D(xmin,xmax,K)
% Purpose  : Generate simple equidistant grid with K elements

Nv = K+1;

% Generate node coordinates
VX = (1:Nv);
for i = 1:Nv
  VX(i) = (xmax-xmin)*(i-1)/(Nv-1) + xmin;
end

% read element to node connectivity
EToV = zeros(K, 2);
for k = 1:K
```

```
   EToV(k,1) = k; EToV(k,2) = k+1;
end
return
```

For more complex problems (e.g., a nonuniform one-dimensional grid), and for generating two- and three-dimensional grids, a more advanced approach is required. This is the task of automated grid generation and is a large research topic in itself. Nevertheless, any standard grid generator will provide the required information in some format. All that is needed is to write a reader that takes into account the format of the particular grid file. This is often specific to a particular grid generator. An example is given in MeshReader1DGambit.m, which is written to read the native format of the grid generator *Gambit*, which is distributed as part of the Fluent Inc software suite and widely used.

MeshReaderGambit1D.m

```
function [Nv, VX, K, EToV] = MeshReader1DGambit(FileName)

% function [Nv, VX, K, EToV] = MeshReader1DGambit(FileName)
% Purpose  : Read in basic grid information to build grid
%
% NOTE     : gambit(Fluent, Inc) *.neu format is assumed

% Open file and read intro
Fid = fopen(FileName, 'rt');
for i=1:6; line = fgetl(Fid); end

% Find number of nodes and number of elements
dims = fscanf(Fid, '%d');
Nv = dims(1); K = dims(2);
for i=1:2; line = fgetl(Fid); end

% read node coordinates
VX = (1:Nv);
for i = 1:Nv
  line = fgetl(Fid);
  tmpx = sscanf(line, '%lf');
  VX(i) = tmpx(2);
end
for i=1:2; line = fgetl(Fid); end

% read element to node connectivity
EToV = zeros(K, 2);
for k = 1:K
  line   = fgetl(Fid);
  tmpcon = sscanf(line, '%lf');
  EToV(k,1) = tmpcon(4); EToV(k,2) = tmpcon(5);
end
```

```
% Close file
st = fclose(Fid);
return
```

To have a standalone approach, we include here a simple interface and brief discussion of the freely available DistMesh software, discussed in detail in

> P.O. Persson, G. Strang, *A Simple Mesh Generator in MATLAB*,
> SIAM Review, **46**(2), 329-345, 2004
> http://www-math.mit.edu/~persson/mesh/

This also serves as a more comprehensive user guide. DistMesh is convenient due to the ease of use and therefore a good starting point before moving on to more complex and advanced applications, which undoubtedly will require more advanced mesh generation software.

In the following, we explain how DistMesh can be used to automate the process of generating simple uniform and nonuniform meshes in both one and two horizontal dimensions. We will provide a few useful scripts and examples on how to setup and define a mesh, how to define the needed boundary maps from BCType for specifying boundary conditions, using signed distance functions to describe the geometries, and how to set up a simple periodic mesh.

B.1 Fundamentals

We illustrate the use of DistMesh to obtain $(\boldsymbol{VX}, \boldsymbol{VY})$ and EToV, needed to compute connectivities and so forth. DistMesh is based on the use of signed distance functions to specify geometries. These distance functions can then be combined and manipulated to specify whether one is inside or outside a particular geometry, with negative being inside the geometry.

The general calling sequence for DistMesh is

```
>> function [p,t] = distmeshnd(fd,fh,h0,bbox,pfix,varargin);
```

The arguments are

Input The six input variables are

  fd   Function which returns the signed distance from each node to the boundary.
  fh   Function which returns the relative edge length for all input points.
  h0   Initial grid size, which will also be the approximate grid size for an equidistant grid.
  bbox Bounding box for the geometry.

pfix   Array of fixed nodes.
varargin   Optional array with additional parameters to fd and fh if needed.

Output The two output variables are

p   List of vertices, i.e., VX=p(:,1)', VY=p(:,2)', etc
t   List of elements, EToV=t.

Let us first consider an example in which we generate a simple one-dimensional mesh of nonuniform line segments clustered about $x = 0$ on $\Omega = [-1, 1]$. We define a signed distance function as

```
% distance function for a circle about xc=0
>> fd=inline('abs(p)-1','p');
```

where p can be thought of as an x-coordinate. In this case, the distance function can be specified through an inline function, but this is not required.

To specify a nonuniform grid size, we must also specify an element size function that reflects the relative distribution over the domain; that is, the numbers in this function do not reflect the actual grid size. As an example, consider

```
% distribution weight function
>> fh=inline('abs(p)*0.075+0.0125','p');
```

The mesh is now generated through the sequence

```
% generate non-uniform example mesh in 1D using DistMesh
>> h0 = 0.025; % chosen element spacing
>> [p,t]=distmeshnd(fd,fh,h0,[-1;1],[]);
>> K = size(t,1); Nv = size(p,1);
```

Two tables are returned that completely defines the mesh and we can find the number of elements, K, and vertices, N_v, directly from the size of these arrays.

A uniform mesh can be generated by using the intrinsic function, @huniform as

```
% generate uniform example mesh in 1D using DistMesh
>> [p,t]=distmeshnd(fd,@huniform,h0,[-1;1],[]);
```

The vertex nodes are sorted to be in ascending order as

```
>> [x,i]=sort(p(t));
>> t=t(i,:);
>> t=sort(t','ascend')';
>> EToV = t; VX = p(:,1)';
```

This can now be returned directly to the solver. These simple steps are all collected in the small script MeshGenDistMesh1D.m.

```
function [Nv, VX, K, EToV] = MeshGenDistMesh1D()

% function [VX, K, EToV] = MeshGenDistMesh1D()
% Purpose  : Generate 1D mesh using DistMesh;

% distance function for a circle about xc=0
fd=inline('abs(p)-1','p');

% distribution weight function
fh=inline('abs(p)*0.075+0.0125','p');

% generate non-uniform example mesh in 1D using DistMesh
h0 = 0.025;             % chosen element spacing
[p,t]=distmeshnd(fd,fh,h0,[-1;1],[]);
K = size(t,1); Nv = K+1;

% Sort elements in ascending order
[x,i]=sort(p(t));
t=t(i,:);
Met=sort(t','ascend')';
EToV = t; VX = p(:,1)';
return
```

Two-dimensional grids can be created in a similar way. Consider as an example the need to generate an equidistant grid in a circular domain, bounded by $[-1, 1]^2$. Following the outline above, we continue as

```
>> fd = inline('sqrt(sum(p.^2,2))-1','p');
>> [p,t] = distmeshnd(fd,@huniform,0.2,[-1 -1; 1 1], []);
>> VX = p(:,1)'; VY = p(:,2)';
>> K = size(t,1); Nv = size(p,1);
>> EToV = t;
```

Note, that DistMesh does not guarantee a counterclockwise ordering of the element vertices and we therefore check and correct all element orderings to ensure this. This can be done in various ways; for example,

```
>> ax = VX(EToV(:,1)); ay = VY(EToV(:,1));
>> bx = VX(EToV(:,2)); by = VY(EToV(:,2));
>> cx = VX(EToV(:,3)); cy = VY(EToV(:,3));
>> D = (ax-cx).*(by-cy)-(bx-cx).*(ay-cy);
>> i = find(D<0);
>> EToV(i,:) = EToV(i,[1 3 2]);
```

There are many more examples in the DistMesh user guide where one also can find a list of a variety of simple functions; for example, to create two-dimensional rectangular grids, one can use

```
>> fd = inline('drectangle(p,-1,1,-1,1)','p');
>>[p,t] = distmeshnd(fd,@huniform,0.2,[-1 -1; 1 1], []);
>> VX = p(:,1)'; VY = p(:,2)';
>> K = size(t,1); Nv = size(p,1);
>> EToV = t;
```

possibly followed by sorting as needed. The process is illustrated in Mesh-GenDistMesh2D.m.

──────────────────────── | MeshGenDistMesh2D.m | ────────────────────────

```
function [VX, VY, K, EToV] = MeshGenDistMesh2D()

% function [VX, VY, K, EToV] = MeshGenDistMesh2D()
% Purpose  : Generate 2D square mesh using DistMesh;
% By Allan P. Engsig-Karup

% Parameters to set/define
%     fd     Distance function for mesh boundary
%     fh     Weighting function for distributing elements
%     h0     Characteristic length of elements
%     Bbox   Bounding box for mesh
%     param  Parameters to be used in function call with DistMesh

fd = inline('drectangle(p,-1,1,-1,1)','p');
fh = @huniform;
h0 = 0.25;
Bbox = [-1 -1; 1 1];
param = [];

% Call distmesh
[Vert,EToV]=distmesh2d(fd,fh,h0,Bbox,param);
VX = Vert(:,1)'; VY = Vert(:,2)';
Nv = length(VX); K  = size(EToV,1);

% Reorder elements to ensure counter clockwise orientation
ax = VX(EToV(:,1)); ay = VY(EToV(:,1));
bx = VX(EToV(:,2)); by = VY(EToV(:,2));
cx = VX(EToV(:,3)); cy = VY(EToV(:,3));

D = (ax-cx).*(by-cy)-(bx-cx).*(ay-cy);
i = find(D<0);
EToV(i,:) = EToV(i,[1 3 2]);
return
```

B.2 Creating boundary maps

While many examples and general grids are shown in the DistMesh user guide, the creation of special boundary maps for imposing different boundary conditions requires a few more steps.

A boundary table for all element faces is easily obtained from the EToE array as

```
>> BCType = int8(not(EToE));
```

However, if we want to be able to distinguish the different boundaries, we need an efficient way to specify boundary properties. This can conveniently be done using a distance function with the properties that $d = 0$ on the boundary and $d <> 0$ outside the boundary.

Let us consider an example. We set up a unit circle with a hole in the middle as

```
>> fd=inline('-0.3+abs(0.7-sqrt(sum(p.∧2,2)))');
>> [p,t]=distmeshnd(fd,@huniform,0.1,[-1,-1;1,1],[]);
```

which is easily generated with two lines of code using DistMesh.

We wish to define two maps *mapI* and *mapO* that allow us to specify different boundary conditions on the inner and the outer boundary of the unit circle with a hole. We note that the following distance functions describe the boundaries completely (as sketched in Fig. B.1):

```
>> fd_inner=inline('sqrt(sum(p.∧2,2))-0.4','p');
>> fd_outer=inline('sqrt(sum(p.∧2,2))-1','p');
```

Using these functions, we can determine index lists of the vertex nodes on the boundaries by

```
>> tol = 1e-8; % tolerance level used for determining nodes
>> nodesInner = find( abs(fd_inner(p))<tol );
>> nodesOuter = find( abs(fd_outer(p))<tol );
```

Here, tol is some given tolerance which may have to be changed depending on the grid. It has to be less than the local minimum distance between the nodes considered. It is recommended to always check the defined boundary maps. An easy way to do this is to plot them as

```
% choose a map to plot
>> MAP = nodesInner;
% show all vertices and circle out map nodes in red
>> plot(p(:,1),p(:,2),'k.',p(MAP,1),p(MAP,2),'ro');
```

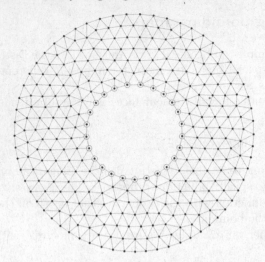

Fig. B.1. Mesh for a unit circle with a hole, generated using DistMesh. Vertex nodes are shown (black '·') as well as the zero-level set curve for fd_inner distance function ('-'), and the nodes in the map *nodesInner* ('o').

Now, we define the following conventions using constants

```
>> In = 1;
>> Out = 2;
```

and use them to correct the face information in the table, BCType, by

```
>> BCType = CorrectBCTable(EToV,BCType,nodesInner,In);
>> BCType = CorrectBCTable(EToV,BCType,nodesOuter,Out);
```

which makes use of BCType to insert the correct codes for boundaries into that array.

─────────────────────── CorrectBCTable.m ───────────────────────
```
function BCType = CorrectBCTable(EToV,BCType,mapnodes,BCcode)

% function BCType = CorrectBCTable(EToV,BCType,mapnodes,BCcode);
% Purpose: Setup BCType for boundary conditions in 2D
% By Allan P. Engsig-Karup

Globals2D;

VNUM = [1 2;2 3;3 1]; % face orientations

for k = 1:K
    % Test for each edge
    for l = 1:Nfaces
```

```
    m = EToV(k,VNUM(1,1)); n = EToV(k,VNUM(1,2));

    % if both points are on the boundary then it is a boundary
    % point!
    ok=sum(ismember([m n],mapnodes));
    if ok==2
        BCType(k,1)=BCcode;
    end;

    end
end
return
```

Once this is generated, one can follow the discussion in Section 6.4 to create the appropriate boundary maps. The creation of special boundary maps is illustrated using the script ConstructMap.m after having corrected the face information of BCType.

```
% Construct special face from the BCType table
>> mapI = ConstructMap(BCType, In);
>> mapO = ConstructMap(BCType, Out);
```

ConstructMap.m

```
function [map] = ConstructMap(BCType, BCcode)

% function map = ConstructMap(BCType, BCcode);
% Purpose: Construct boundary map from the BCType table
% By Allan P. Engsig-Karup, 07-12-2006.

Globals2D;

% Determine which faces in which elements which have the specified BCs:
%   fids = face id's, eids = element id's;
[eids,fids] = find(BCType==BCcode);

% initialize length of new map
map = [];

for n = 1 : length(fids) % go over each boundary face of BCcode type
    map = [map (eids(n)-1)*Nfaces*Nfp+[ (fids(n)-1)*Nfp + [1:Nfp] ]];
end
return
```

C

Appendix C: Software, variables, and helpful scripts

All codes discussed in this text can be downloaded freely at

http://www.nudg.org

The codes are distributed with the following disclaimer:

```
Permission to use this software for noncommercial
research and educational purposes is hereby granted
without fee. Redistribution, sale, or incorporation
of this software into a commercial product is prohibited.

THE AUTHORS OR PUBLISHER DISCLAIMS ANY AND ALL WARRANTIES
WITH REGARD TO THIS SOFTWARE,INCLUDING ALL IMPLIED
WARRANTIES OF MERCHANTABILITY AND FITNESS FOR ANY
PARTICULAR PURPOSE. IN NO EVENT SHALL THE  AUTHORS OR
THE PUBLISHER BE LIABLE FOR ANY SPECIAL, INDIRECT OR
CONSEQUENTIAL DAMAGES OR ANY DAMAGES WHATSOEVER
RESULTING FROM LOSS OF USE, DATA OR PROFITS.
```

C.1 List of important variables defined in the codes

In the following, we list the basic parameters and arrays used in the codes discussed throughout. This is done to help users understand the notation and enable them to quickly alter the codes to suit their needs. Note that the list is not exhaustive and only the essential parameters, arrays, and operators are included.

Table C.1. Basic variables, all of integer type.

Name	Function
N	Order of approximation in each element.
K	Number of elements.
Nv	Number of vertices in grid.
Nfp	Number of nodes on face.
Np	Number of nodes in each element.
Nfaces	Number of faces in each element.

Table C.2. Operators and local metric, all of real type.

Name	Size	Function
rx	$Np \times K$	Metric constant, r_x.
sx	$Np \times K$	Metric constant, s_x.
tx	$Np \times K$	Metric constant, t_x.
ry	$Np \times K$	Metric constant, r_y.
sy	$Np \times K$	Metric constant, s_y.
ty	$Np \times K$	Metric constant, t_y.
rz	$Np \times K$	Metric constant, r_z.
sz	$Np \times K$	Metric constant, s_z.
tz	$Np \times K$	Metric constant, t_z.
J	$Np \times K$	Volume Jacobian, J.
sJ	$(Nfaces{\cdot}Nfp) \times K$	Jacobian at surface nodes.
Fscale	$Nfaces \times K$	Fscale(i,k) is the ratio of surface to volume Jacobian. of face i on element k.
V	$Np \times Np$	Vandermonde matrix, $\mathcal{V}$.
invV	$Np \times Np$	Inverse Vandermonde matrix.
Mass	$Np \times Np$	Mass matrix $\mathcal{M}$ on I.
Dr	$Np \times Np$	Differentiation matrix $\mathcal{D}_r$ wrt r on I.
Ds	$Np \times Np$	Differentiation matrix $\mathcal{D}_s$ wrt s on I.
Dt	$Np \times Np$	Differentiation matrix $\mathcal{D}_t$ wrt t on I.
LIFT	$Np \times (Nfaces{\cdot}Nfp)$	Surface integral; $\mathcal{M}^{-1} \int_{\partial \mathrm{D}^k} \ell_i(\boldsymbol{x})\ell_j(\boldsymbol{x})d\boldsymbol{x}$.

Table C.3. Grid information, all of real type.

Name	Size	Function
VX	$1 \times Nv$	List of x-coordinates for vertices in grid.
VY	$1 \times Nv$	List of y-coordinates for vertices in grid.
VZ	$1 \times Nv$	List of z-coordinates for vertices in grid.
r	Np	Reference grid coordinate in I.
s	Np	Reference grid coordinate in I.
t	Np	Reference grid coordinate in I.
x	$Np \times K$	Physical grid coordinate in Ω
y	$Np \times K$	Physical grid coordinate in Ω
z	$Np \times K$	Physical grid coordinate in Ω
nx	$(Nfaces \cdot Nfp) \times K$	nx(i,k) is the $\hat{n}_x$ component of normal at face node i on element k.
ny	$(Nfaces \cdot Nfp) \times K$	ny(i,k) is the $\hat{n}_y$ component of normal at face node i on element k.
nz	$(Nfaces \cdot Nfp) \times K$	nz(i,k) is the $\hat{n}_z$ component of normal at face node i on element k.
Fx	$(Nfaces \cdot Nfp) \times K$	x-coordinates at faces of elements.
Fy	$(Nfaces \cdot Nfp) \times K$	y-coordinates at faces of elements.
Fz	$(Nfaces \cdot Nfp) \times K$	z-coordinates at faces of elements.

Table C.4. Connectivity information, all of integer type.

Name	Size	Function
EToV	$K \times$ Nfaces	EToV(k,i) is the vertex number in (VX,VY,VZ) for vertex i in element k.
EToE	$K \times$ Nfaces	EToE(k,i) is the global element number to which face i of element k connects.
EToF	$K \times$ Nfaces	EToF(k,i) is the local face number to which face i of element k connects.
BCType	$K \times$ Nfaces	BCType(k,i) is the boundary code for face i of element k.
vmapM	$(K \cdot \text{Nfaces} \cdot \text{Nfp}) \times 1$	Vector of global nodal numbers at faces for interior values, $u^- = u(\text{vmapL})$.
vmapP	$(K \cdot \text{Nfaces} \cdot \text{Nfp}) \times 1$	Vector of global nodal numbers at faces for exterior values, $u^+ = u(\text{vmapR})$.
vmapB	$(K \cdot \text{Nfaces} \cdot \text{Nfp}) \times 1$	Vector of global nodal numbers at faces for interior boundary values, $u^-(\partial\Omega) = u(\text{vmapB})$.
mapB		Vector of global face nodal numbers at interior boundary.
Fmask	Nfp $\times$ Nfaces	Local, elementwise array of face indices.
vmapI		Vector of global nodal numbers at faces with boundary condition "In".
vmapO		Vector of global nodal numbers at faces with boundary condition "Out".
vmapW		Vector of global nodal numbers at faces with boundary condition "Wall".
vmapF		Vector of global nodal numbers at faces with boundary condition "Far".
mapI		Vector of global face nodal numbers at faces with boundary condition "In".
mapO		Vector of global face nodal numbers at faces with boundary condition "Out".
mapW		Vector of global face nodal numbers at faces with boundary condition "Wall".
mapF		Vector of global face nodal numbers at faces with boundary condition "Far".

C.2 List of additional useful scripts

In the following, we provide a number of Matlab scripts that are not discussed as part of the text, but nevertheless prove useful when developing algorithms and visualizing the computed results.

PlotMesh2D.m

Description:

Function to plot the finite element mesh skeleton. Depends on global variables declared in Globals2D.m.

Example Usage:

PlotMesh2D()

Example Output:

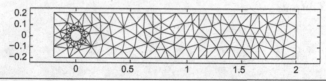

PlotDomain2D.m

Description:

Function to plot diagram of current domain, with boundary conditions annotated. Depends on global variables declared in Globals2D.m.

Example Usage:

PlotDomain2D()

Example Output:

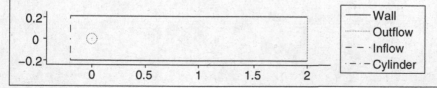

PlotField2D.m

Description:

Plot filled contour plot of nodal data in two dimensions, interpolated to equally spaced plotting points on each triangle (returns finite element type triangulation of plotting data). Depends on global variables declared in Globals2D.m. Warning: only works for fixed order elements.

Example Usage:

% Generate solution data
CurvedINSDriver2D;

% Render pressure field
[TRI,xout,yout,uout,interp] = PlotField2D(N, x, y, PR)

Example Output:

PlotContour2D.m

Description:

Plot user specified contours of a two-dimensional, triangular mesh, finite element data. Uses piecewise linear interpolation. Stand-alone routine not dependent on the global variables.

Example Usage:

```
% Generate solution data
CurvedINSDriver2D;

% Compute vorticity
[dummy,dummy,vort] = Curl2D(Ux, Uy);

% Compute triangulated data on finer grid
[TRI,xout,yout,uout,interp] PlotField2D(2*N, x, y, vort);

% Clear display and plot domain outline
clf;
PlotDomain2D();

% Plot contour lines, at 12 equally spaced levels
% between -6 and 6, on top of domain outline
hold on;
PlotContour2D(TRI, xout, yout, vortout, linspace(-6, 6, 12));
hold off;
```

Example Output:

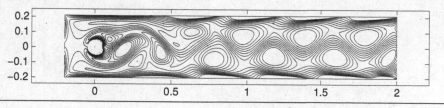

PlotAdaptiveContour2D.m

Description:

Plot line iso-contours of two-dimensional nodal solution data at user specified values. Uses an adaptive strategy to locate contours within specified tolerance.

Example Usage:

```
% Generate solution data
CurvedINSDriver2D;

% Compute vorticity
[dummy,dummy,vort] = Curl2D(Ux, Uy);

% Plot contours with adaptive contour finding algorithm
PlotAdaptiveContour2D(vort, linspace(-6,6,12), 1e-1)
```

Example Output:

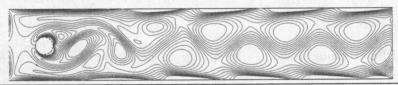

PlotContour3D.m

Description:

Plot user specified contours of three-dimensional data given at standard nodes on a mesh of tetrahedra represented in global variables. First interpolates to equispaced nodes on tetrahedron then uses piecewise linear interpolation to find contours.

Example Usage:

```
% Generate solution data
AdvecDriver3D;

% Plot contour lines, at 10 equally spaced levels in [.1,1]
% using order N equispaced interpolant of data
PlotContour3D(N, u, linspace(0.1, 1, 10));
```

Example Output:

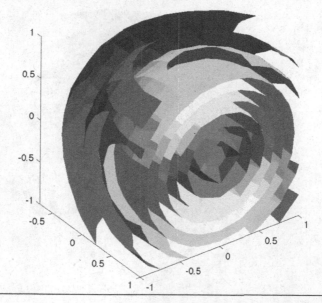

PlotAdaptiveContour3D.m

Description:

Plot user specified contours of three-dimensional data given at standard nodes on a mesh of tetrahedra represented in global variables. Uses adaptive refinement of elements until piecewise linear interpolation is sufficiently accurate to find contours within user specified tolerance.

Example Usage:

% Generate solution data
AdvecDriver3D;

% Plot contour lines, at 10 equally spaced levels in [.1,1]
% to an accuracy of 1e-2
PlotAdaptiveContour3D(u, linspace(0.1, 1, 10), 1e-2);

Example Output:

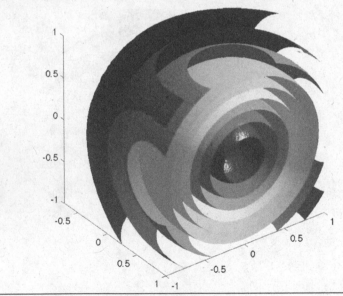

PlotSlice3D.m

Description:

Plot filled contour slice of three-dimensional nodal data in "x", "y", or "z" plane at user supplied offset. Uses globally defined variables in Globals3D.m.

Example Usage:

% Generate solution data
AdvecDriver3D;

% Plot contour lines, at five different slices, using 20 point
% interpolation in trianglulation of plane/element intersections.
clf;
PlotSlice3D(20, u, 'y', 0);
hold on
PlotSlice3D(20, u, 'x', 0)
PlotSlice3D(20, u, 'z', 0)
PlotSlice3D(20, u, 'y', 0.5)
PlotSlice3D(20, u, 'x', -0.5)
hold off

% Adjust plotting properties
axis equal; material shiny; camlight; camlight

Example Output:

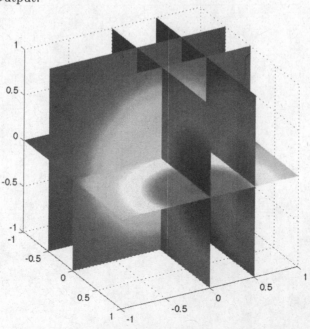

References

1. S. ABARBANEL AND A. DITKOWSKI, *Asymptotically stable fourth-order accurate schemes for the diffusion equation on complex shapes*, J. Comput. Phys. **133**(1997), pp. 279-288. Cited on page(s) 13
2. S. ABARBANEL AND D. GOTTLIEB, *On the construction and analysis of absorbing layers in CEM*, Appl. Numer. Math. **27**(1998), pp. 331-340. Cited on page(s) 240
3. R. ABGRALL AND H. DECONINCK (EDS.), Comput. Fluids **34**(4-5), 2005. Cited on page(s) 13
4. M. ABROMOWITZ AND I.A. STEGUN (EDS.), *Handbook of Mathematical Functions*, Dover Publications, New York, 1972. Cited on page(s) 446
5. S. ADJERID AND M. BACCOUCH, *The discontinuous Galerkin method for two-dimensional hyperbolic problems. Part I: Superconvergence error analysis*, J. Sci. Comput. **33**(2007), pp. 75-113. Cited on page(s) 12, 396
6. S. ADJERID, K. DEVINE, J. FLAHERTY, AND L. KRIVODONOVA, *A posteriori error estimation for discontinuous Galerkin solutions of hyperbolic problems*, Comput. Methods Appl. Mech. Eng. **191**(2002), pp. 1097-1112. Cited on page(s) 12, 387, 396
7. S. ADJERID AND T.C. MASSEY, *A posteriori discontinuous finite element error estimation for two-dimensional hyperbolic problems*, Comput. Methods Appl. Mech. Eng. **191**(2002), pp. 5877-5897. Cited on page(s) 12, 387
8. S. ADJERID AND T.C. MASSEY, *Superconvergence of discontinuous finite element solutions for nonlinear scalar hyperbolic problems* Comput. Methods Appl. Mech. Eng. **195**(2006), pp. 3331-3346. Cited on page(s) 12, 387
9. M. AINSWORTH, *Dispersive and dissipative behavior of high-order discontinuous Galerkin finite element methods*, J. Comput. Phys. **198**(2002), pp. 106-130. Cited on page(s) 91, 91, 92, 92
10. M. AINSWORTH, P. MONK, AND W. MUNIZ, *Dispersive and dissipative properties of discontinuous Galerkin finite element methods for the second-order wave equation*, J. Sci. Comput. **27**(2006), pp. 5-40. Cited on page(s) 93, 353
11. P.F. ANTONIETTI, A. BUFFA, AND I. PERUGIA, *Discontinuous Galerkin approximation of the Laplace eigenproblem*, Comput. Methods Appl. Mech. Eng. **195**(2006), pp. 3483-3503. Cited on page(s) 12, 337, 338, 338, 338

12. D.N. ARNOLD, *An interior penalty finite element method with discontinuous elements*, SIAM J. Numer. Anal. **19**(1982), pp. 724-760. Cited on page(s) 11, 11, 282, 296

13. D.N. ARNOLD, F. BREZZI, B. COCKBURN, AND L.D. MARINI, *Unified analysis of discontinuous Galerkin methods for elliptic problems*, SIAM J. Numer. Anal. **39**(2002), pp. 1749-1779. Cited on page(s) 11, 270, 282, 288, 292, 293, 296, 296

14. H. ATKINS AND C.-W. SHU, *Quadrature-free implementation of the discontinuous Galerkin method for hyperbolic equations*, AIAA J. **36**(1998), pp. 775-782. Cited on page(s) 11, 211

15. I. BABUSKA AND M.R. DORR, *Error estimates for the combined h- and p-versions of the finite element method*, Numer. Math. **37**(1981), pp. 257-277. Cited on page(s) 387, 396

16. I. BABUSKA, B.A. SZABO, AND I.N. KATZ, *The p-version of the finite element method*, SIAM J. Numer. Anal. **18**(1981), pp. 515-545. Cited on page(s) 387

17. I. BABUSKA AND B. GUO, *The h-p version of the finite element method - Part 1: The basic approximation results*, Comput. Mech. **1**(1986), pp. 21-41. Cited on page(s) 387

18. I. BABUSKA AND B. GUO, *The h-p version of the finite element method - Part 2: General results and applications*, Comput. Mech. **1**(1986), pp. 203-220. Cited on page(s) 387

19. I. BABUSKA AND T. STROUBOULIS, *The Finite Element Method and its Reliability*, Oxford University Press, Oxford, 2001. Cited on page(s) 293, 387, 396

20. F. BASSI, A. CRIVELLINI, S. REBAY, AND M. SAVINI, *Discontinuous Galerkin solution of the Reynolds-averaged NavierStokes and kω turbulence model equations*, Comput. Fluids **34**(2005), pp. 507-540. Cited on page(s) 316

21. F. BASSI AND S. REBAY, *A high-order discontinuous Galerkin finite element method solution of the 2D Euler equations*, J. Comput. Phys. **138**(1997), pp. 251-285. Cited on page(s) 11, 316, 320, 375

22. F. BASSI AND S. REBAY, *A high-order accurate discontinuous Galerkin finite element method for the numerical solution of the compressible Navier-Stokes equations*, J. Comput. Phys. **131**(1997), pp. 267-279. Cited on page(s) 11, 11, 11, 245, 315

23. F. BASSI AND S. REBAY, *Numerical evaluation of two discontinuous Galerkin methods for the compressible Navier-Stokes equations*, Int. J. Numer. Methods Fluids **40**(2002), pp. 197-207. Cited on page(s) 315

24. C.E. BAUMANN AND J.T. ODEN, *A discontinuous hp-Finite element method for convection-diffusion problems*, Comput. Methods Appl. Mech. Eng. **175**(1999), pp. 311-341. Cited on page(s) 11

25. J.P. BERENGER, *A perfectly matched layer for the absorption of electromagnetic waves*, J. Comput. Phys. **114**(1994), pp. 185-200. Cited on page(s) 240

26. J.P. BERENGER, *Three-dimensional perfectly matched layer for the absorption of electromagnetic waves*, J. Comput. Phys. **127**(1996), pp. 363-379. Cited on page(s) 240

27. C. BERNARDI AND Y. MADAY, *Polynomial interpolation results in Sobolev spaces*, J. Comput. Appl. Math. **43**(1992), pp. 53-80. Cited on page(s) 81, 82

28. R. BISWAS, K.D. DEVINE, AND J. FLAHERTY, *Parallel, adaptive finite element methods for conservation laws*, Appl. Numer. Math. **14**(1994), pp. 255-283. Cited on page(s) 12, 156, 387, 396

29. D. BOFFI, R.G. DURAN, AND L. GASTALDI, *A remark on spurious eigenvalues in a square*, Appl. Math. Lett. **12**(1999), pp. 107-114. Cited on page(s) 333

30. D. BOFFI. M. FARINA, AND L. GASTALDI, *On the approximation of Maxwell's eigenproblem in general 2D domains*, Comput. Struct. **79**(2001), pp. 1089-1096. Cited on page(s) 333

31. F. VAN DED BOS, J.J.W. VAN DER VEGT, AND B. J. GEURTS, *A multi-scale formulation for compressible turbulent flows suitable for general variational discretization techniques*, Comput. Methods Appl. Mech. Eng. **196**(2007), pp. 2863-2875. Cited on page(s) 316

32. F. BREZZI, B. COCKBURN, D.M. MARINI, AND E. SULI, *Stabilization mechanisms in discontinuous Galerkin finite element methods*, Comput. Methods Appl. Mech. Eng. **195**(2006), pp. 3293-3310. Cited on page(s) 11

33. F. BREZZI, L.D. MARINI, AND E. SULI, *Discontinuous Galerkin methods for first-order hyperbolic problems*, M3AS: Math. Models Methods Appl. Sci. **14**(2004), pp. 1893–1903, 2004. Cited on page(s) 10

34. A. BUFFA, P. HOUSTON, AND I. PERUGIA, *Discontinuous Galerkin computation of the Maxwell eigenvalues on simplicial meshes*, J. Comput. Appl. Math. **204**(2007), pp. 317-333. Cited on page(s) 12, 348, 370

35. A. BUFFA AND I. PERUGIA, *Discontinuous Galerkin approximation of the Maxwell eigenproblem*, SIAM J. Numer. Anal. **44**(2006), pp. 2198-2226. Cited on page(s) 12, 351, 370

36. A. BURBEAU, P. SAGAUT, AND C.-H. BRUNEAU, *A problem-independent limiter for high-order Runge-Kutta discontinuous Galerkin methods* J. Comput. Phys. **163**(2001), pp. 111-150. Cited on page(s) 224

37. A. BURBEAU AND P. SAGAUT, *Simulation of a viscous compressible flow past a circular cylinder with high-order discontinuous Galerkin methods*, Comput. Fluids **31**(2002), pp. 867-889. Cited on page(s) 315

38. A. BURBEAU AND P. SAGAUT, *A dynamic p-adaptive Discontinuous Galerkin method for viscous flow with shocks*, Comput. Fluids **34**(2005), pp. 401-417. Cited on page(s) 315

39. E. BURMAN AND B. STAMM, *Minimal stabilization for discontinuous Galerkin finite element methods for hyperbolic problems*, J. Sci. Comput. **33**(2007(), pp. 183-208. Cited on page(s) 10, 33

40. J.C. BUTCHER, *The Numerical Analysis of Ordinary Differential Equations: Runge-Kutta and General Linear Methods*, John Wiley & Sons, New York, 2003. Cited on page(s) 64, 93, 101, 157

41. N. CANOUET, L. FEZOUI, AND S. PIPERNO, *Discontinuous Galerkin time-domain solution of Maxwell's equations on locally-refined nonconforming Cartesian grids*, COMPEL **24**(2005), pp. 1381-1401. Cited on page(s) 110

42. C. CANUTO, M.Y. HUSSAINI, A. QUARTERONI, AND T.A. ZANG, *Spectral Methods in Fluid Dynamics*, Springer Series in Computational Physics, Springer-Verlag, New York, 1988. Cited on page(s) 186

43. C. CANUTO AND A. QUARTERONI, *Approximation results for orthogonal polynomials in Sobolev spaces*, Math. Comp. **38**(1982), pp. 67-86. Cited on page(s) 79, 79, 79

44. M.H. CARPENTER, D. GOTTLIEB, AND S. ABARBANEL, *Time-stable boundary conditions for finite-difference schemes solving hyperbolic systems: Methodology and application to high-order compact schemes*, J. Comput. Phys. **111**(1994), pp. 220-236. Cited on page(s) 13

45. M.H. CARPENTER, D. GOTTLIEB, AND C.-W. SHU, *On the conservation and convergence to weak solutions of global schemes*, J. Sci. Comput. **18**(2003), pp. 111-132. Cited on page(s) 120, 134

46. M.H. CARPENTER AND C. KENNEDY, *Fourth-order 2N-storage Runge-Kutta schemes*, NASA Report TM 109112, NASA Langley Research Center, 1994. Cited on page(s) 64

47. M.H. CARPENTER, J. NORDSTROM, AND D. GOTTLIEB *A stable and conservative interface treatment of arbitrary spatial accuracy*, J. Comput. Phys. **148**(1999), pp. 341-365. Cited on page(s) 13

48. P. CASTILLO, *Performance of discontinuous Galerkin methods for elliptic problems*, SIAM J. Sci. Comput. **24**(2002), pp. 524-547. Cited on page(s) 11, 11, 270

49. P. CASTILLO, *A review of the local discontinuous Galerkin (LDG) method applied to elliptic problems*, Appl. Numer. Math. **56**(2006), pp. 1307-1313. Cited on page(s) 270

50. P. CASTILLO, B. COCKBURN, I. PERUGIA, AND D. SCHÖTZAU, *An a priori error analysis of the local discontinuous Galerkin method for elliptic problems* SIAM J. Numer. Anal. **38**(2000), pp. 1676-1706. Cited on page(s) 11, 296

51. P. CASTILLO, B. COCKBURN, D. SCHÖTZAU, AND C. SCHWAB, *Optimal a priori error estimates for the hp-version of the local discontinuous Galerkin method for convection-diffusion problems*, Math. Comp. **71**(2002), pp. 455-478. Cited on page(s) 11

52. C. CHAUVIERE, J. S. HESTHAVEN, A. KANEVSKY, AND T. WARBURTON, *High-order localized time integration for grid-induced stiffness*, 2nd MIT Conference on Fluid Dynamics, Boston. Vol. II, pp. 1883-1886, 2003. Cited on page(s) 110

53. G. CHAVENT AND G. SALZANO, *A finite element method for the 1D flooding problem with gravity*, J. Comput. Phys. **45**(1982), pp. 307-344. Cited on page(s) 10

54. G. CHAVENT AND B. COCKBURN, *The local projection p^0-p^1-discontinuous Galerkin finite element for scalar conservation laws*, M^{2}AN **23**(1989), pp. 565-592. Cited on page(s) 10

55. Q. CHEN AND I. BABUSKA, *Approximate optimal points for polynomial interpolation of real functions in an interval and in a triangle*, Comput. Meth. in App. Mech. and Eng. **128**(1995), pp. 405–417. Cited on page(s) 175, 412, 414, 414

56. Y. CHENG AND C.-W. SHU, *A discontinuous Galerkin finite element method for directly solving the Hamilton-Jacobi equations*, J. Comput. Phys. **223**(2007), pp. 398-415. Cited on page(s) 11, 135

57. Y. CHENG AND C.-W. SHU, *A discontinuous Galerkin finite element method for time dependent partial differential equations with higher order derivatives*, Math. Comp 2007 – to appear. Cited on page(s) 260, 261, 261

58. E.T. CHUNG AND B. ENGQUIST, *Optimal discontinuous Galerkin methods for wave propagation*, SIAM J. Sci. Comput. **44**(2006), pp. 2131-2158. Cited on page(s) 11

59. P.G. CIARLET, *The Finite Element Method for Elliptic Problems*, SIAM Classic in Applied Mathematics **40**, SIAM Publications, Philadelphia, 2002. Cited on page(s) 87, 87, 261, 287, 292

60. B. COCKBURN, *Discontinuous Galerkin method for convection-dominated Problems*. In *High-Order Methods for Computational Physics*, T.J. Barth and H. Deconinck (Eds.), Lecture Notes in Computational Science and Engineering **9**, Springer-Verlag, Berlin, pp. 69-224, 1999. Cited on page(s) 88, 88, 145, 148, 152, 161

61. B. COCKBURN AND B. DONG, *An analysis of the minimal dissipation local discontinuous Galerkin method for convection diffusion problems*. J. Sci. Comput. **32**(2007), pp. 233-262. Cited on page(s) 10, 33

62. B. COCKBURN AND P.A. GREMAUD, *Error estimates for finite element methods for nonlinear conservation laws*, SIAM J. Numer. Anal. **33**(1996), pp. 522-554. Cited on page(s) 12, 387

63. B. COCKBURN AND J. GUZMAN, *Error estimates for the Runge-Kutta discontinuous Galerkin method for the transport equation with discontinuous initial data*, SIAM J. Numer. Anal. 2007 – submitted. Cited on page(s) 10

64. B. COCKBURN, S. HOU, AND C.-W. SHU, *TVB Runge-Kutta local projection discontinuous Galerkin finite element method for conservation laws IV: The multidimensional case*, Math. Comp. **54**(1990), pp. 545-581. Cited on page(s) 10, 224, 239

65. B. COCKBURN, G. KANSCHAT, I. PERUGIA, AND D. SCHÖTZAU, *Superconvergence of the local discontinuous Galerkin method for elliptic problems on Cartesian grids*, SIAM J. Numer. Anal. **39**(2001), pp. 264-285. Cited on page(s) 12, 387

66. B. COCKBURN, G. KANSCHAT, AND D. SCHTZAU, *A locally conservative LDG method for the incompressible Navier-Stokes equations*, Math. Comp. **74**(2005), pp. 1067-1095. Cited on page(s) 11, 12, 296

67. B. COCKBURN, G. KANSCHAT, D. SCHOTZAU, AND C. SCHWAB, *Local discontinuous Galerkin methods for the Stokes system*, SIAM J. Numer. Anal. **40**(2002), pp. 319-343. Cited on page(s) 12, 296

68. B. COCKBURN, G. KANSCHAT, AND D. SCHÖTZAU, *The local discontinuous Galerkin method for linearized incompressible fluid flow: A review*, Comput. Fluids **34**(2005), pp. 491-506. Cited on page(s) 11

69. B. COCKBURN, G.E. KARNIADAKIS, AND C.-W. SHU (EDS.), *Discontinuous Galerkin Methods. Theory, Computation and Application*, Lecture Notes in Computational Science and Engineering **11**, Springer-Verlag, Berlin, 1999. Cited on page(s) 13

70. B. COCKBURN, F. LI, AND C.-W. SHU, *Locally divergence-free discontinuous Galerkin methods for the Maxwell equations* , J. Comput. Phys. **194**(2004), pp. 588-610. Cited on page(s) 11, 357

71. B. COCKBURN, S.Y. LIN, AND C.-W. SHU, *TVB Runge-Kutta local projection discontinuous Galerkin finite element method for conservation laws III: One-dimensional systems*, J. Comput. Phys. **84**(1989), pp. 90-113. Cited on page(s) 10

72. B. COCKBURN, M. LUSKIN, C.-W. SHU, AND E. SULI, *Enhanced accuracy by post-processing for finite element methods for hyperbolic equations*, Math. Comp. **72**(2003), pp. 577-606. Cited on page(s) 10

478   References

73. B. COCKBURN, D. SCHÖTZAU, AND J. WANG, *Discontinuous Galerkin methods for incompressible elastic materials*, Comput. Methods Appl. Mech. Eng. **195**(2006), pp. 3184-3204.   Cited on page(s) 12, 296

74. B. COCKBURN AND C.-W. SHU (EDS.), J. Sci. Comput. **22**(1-3), 2005.   Cited on page(s) 13

75. B. COCKBURN AND C.-W. SHU, *TVB Runge-Kutta local projection discontinuous Galerkin finite element method for conservation laws II: General framework*, Math. Comp. **52**(1989), pp. 411-435.   Cited on page(s) 10, 11

76. B. COCKBURN AND C.-W. SHU, *The Runge-Kutta discontinuous Galerkin finite element method for conservation laws V: Multidimensional systems*, J. Comput. Phys. **141**(1998), pp. 199-224.   Cited on page(s) 11, 11, 224, 233, 234, 239

77. B. COCKBURN AND C.-W. SHU, *The Runge-Kutta local projection p^1-discontinuous Galerkin method for scalar conservation laws*, M^{2}AN **25**(1991), pp. 337-361.   Cited on page(s) 10

78. B. COCKBURN AND C.-W. SHU, *The local discontinuous Galerkin finite element method for convection-diffusion systems*, SIAM J. Numer. Anal. **35**(1998), pp. 2440-2463.   Cited on page(s) 11, 11, 251, 251, 253, 254, 257

79. B. COCKBURN AND C.-W. SHU, *Runge-Kutta discontinuous Galerkin methods for convection-dominated problems*, J. Sci. Comput. **16**(2001), pp. 173-261. Cited on page(s) 32, 135

80. G. COHEN, X. FERRIERES, AND S. PERNET, *A spatial high-order hexahedral discontinuous Galerkin method to solve Maxwell's equations in the time-domain*, J. Comput. Phys. **217**(2006), pp. 340-363.   Cited on page(s) 11

81. G. COHEN, X. FERRIERES, AND S. PERNET, *Discontinuous Galerkin methods for Maxwell's Equations in the time domain*, Comp. Ren. Phys. **7**(2006), pp. 494-500.   Cited on page(s) 11

82. G. COHEN AND M. DURUFLE, *Non spurious spectral-like element methods for Maxwell's equations* - preprint 2007.   Cited on page(s) 12

83. S. S. COLLIS, *Discontinuous Galerkin methods for turbulence simulation*. In *Proceedings of the 2002 Center for Turbulence Research Summer Program*, pp. 155-167.   Cited on page(s) 316

84. R. COOLS AND P. RABINOWITZ, *Monomial cubature rules since Stroud: A compilation*, J. Comput. Appl. Math. **48**(1993), pp. 309-326.   Cited on page(s) 211, 379

85. R. COOLS, *Monomial cubature rules since Stroud: A compilation - Part 2*, J. Comput. Appl. Math. **112**(1999), pp. 21–27.   Cited on page(s) 211, 379

86. S. CURTIN, R.M. KIRBY, J.K. RYAN, AND C.-W. SHU, *Post-processing for the discontinuous Galerkin method over non-uniform meshes*, SIAM J. Sci. Comput. 2007 – submitted.   Cited on page(s) 10

87. C.M. DAFERMOS, *Hyberbolic Conservation Laws in Continuum Physics*, Series: Grundlehren der mathematischen Wissenschaften **325**, Springer-Verlag, Berlin, 2005.   Cited on page(s) 118

88. M. DAUGE, *Benchmark computations for Maxwell equations for the approximation of highly singular solutions*, 2003. http://perso.univ-rennes1.fr/monique.dauge/benchmax.html.   Cited on page(s) 364, 365

89. P.J. DAVIS AND P. RABINOWITZ, *Methods of Numerical Integration*, Computer Science and Applied Mathematics, Academic Press, New York, 1975.   Cited on page(s) 212, 445, 447

90. P.J DAVIS, *Interpolation and Approximation.* Dover Publications, New York, 1975. Cited on page(s) 46, 49, 97

91. C. DAWSON (EDS.), Comput. Methods Appl. Mech. Eng. **195**(25-28), 2006. Cited on page(s) 13

92. J. DE LA PUENTE, M. KÄSER, M. DUMBSER, AND H. IGEL, *An arbitrary high order discontinuous Galerkin method for elastic waves on unstructured meshes IV: Anisotropy,* Geophys. J. Int. **169**(2007), pp. 1210-1228. Cited on page(s) 11

93. L. DEMKOWICZ, *Computing with hp-Adaptive Finite Elements: Volume 1, One and Two Dimensional Elliptic and Maxwell Problems,* Applied Mathematics & Nonlinear Science **7**, Chapman & Hall/CRC, London, 2006. Cited on page(s) 6, 387

94. L. DEMKOWICZ, *Computing with hp-Adaptive Finite Elements: Volume 2 Frontiers: Three Dimensional Elliptic and Maxwell Problems with Applications,* Applied Mathematics & Nonlinear Science **11**, Chapman & Hall/CRC, London, 2007. Cited on page(s) 6, 387

95. J. DESCLOUX, N. NASSIF, AND J. RAPPAZ, *On spectral approximation. Part 1: The problem of convergence,* RAIRO Anal. Numer. **12**(1978), pp. 97-112. Cited on page(s) 333

96. J. DESCLOUX, N. NASSIF, AND J. RAPPAZ, *On spectral approximation. Part 2: error estimates for the Galerkin method,* RAIRO Anal. Numer. **12**(1978), pp. 113-119. Cited on page(s) 333

97. M. DEVILLE, P. FISCHER, AND E.H. MUND, *High-Order Methods For Incompressible Fluid Flow,* Cambridge University Press, Cambridge, 2002. Cited on page(s) 301

98. W.S. DON, *Numerical study of pseudospectral methods in shock wave applications,* J. Comput. Phys. **110**(1994), pp. 103-111. Cited on page(s) 139, 145

99. W.S. DON AND D. GOTTLIEB, *Spectral simulations of supersonic reactive flows,* SIAM J. Numer. Anal. **35**(1998), pp. 2370-2384. Cited on page(s) 139, 145

100. W.S. DON AND C. QUILLEN, *Numerical simulation of reactive flow, Part I : resolution,* J. Comput. Phys. **122**(1995), pp. 244-265. Cited on page(s) 139, 145

101. W.S. DON AND A. SOLOMONOFF, *Accuracy enhancement for higher derivatives using Chebyshev collocation and a mapping technique,* SIAM J. Sci. Comput. **18**(1997), pp. 1040-1055. Cited on page(s) 101

102. J. DOUGLAS AND T. DUPONT, *Interior penalty procedures for elliptic and parabolic Galerkin methods,* Lecture Notes in Physics, Springer-Verlag, Berlin, 1976, **58**, pp. 207-216. Cited on page(s) 11, 11

103. M. DUBINER, *Spectral methods on triangles and other domains,* J. Sci. Comput. **6**(1991), pp. 345-390. Cited on page(s) 449

104. I.S. DUFF, A.M. ERSIMAN, AND J.K. REID, *Direct Methods for Sparse Matrices,* Oxford University Press, Oxford, 1986. Cited on page(s) 297, 300

105. M. DUMBSER, *Arbitrary High-Order Schemes for the Solution of Hyperbolic Conservation Laws in Complex Domains,* Springer-Verlag, New York, 2005. Cited on page(s) 112

106. M. DUMBSER AND M. KÄSER, *An arbitrary high order discontinuous Galerkin method for elastic waves on unstructured meshes II: The three-dimensional isotropic case,* Geophys. J. Int. **167**(2006), pp. 319-336. Cited on page(s) 11

107. M. DUMBSER, M. KÄSER, AND E.F. TORO, *An arbitrary high order discontinuous Galerkin method for elastic waves on unstructured meshes V: Local Time Stepping and p-Adaptivity*, Geophys. J. Int., 2007 – to appear. Cited on page(s) 11, 12, 109, 110

108. L. EMMEL, S.M. KABER, AND Y. MADAY, *Padé-Jacobi filtering for spectral approximations of discontinuous solutions*, Numer. Algorithms **33**(2003), pp. 251-264. Cited on page(s) 144

109. A. ENGSIG-KARUP, J. S. HESTHAVEN, H. BINGHAM, AND P. MADSEN, *Nodal DG-FEM solution of high order Boussinesq-type equations*, J. Eng. Math. **56**(2006), pp. 351-370. Cited on page(s) 11

110. A. ENGSIG-KARUP, J. S. HESTHAVEN, H. BINGHAM, AND T. WARBURTON, *DG-FEM solution for nonlinear wave-structure interaction using Boussinesq-type equations*, Costal Eng. 2007 - to appear. Cited on page(s) 11

111. A. ERDELYI (EDS.), *Higher Transcendental Functions, Vol. I-III*, Robert E. Krieger Publishing Company, Huntington, New York, 1953. Cited on page(s) 445

112. A. ERN AND J.-.L. GUERMOND, *Discontinuous Galerkin methods for Friedrichs' systems. I. General theory*, SIAM J. Numer. Anal. **44**(2006) pp. 753–778. Cited on page(s) 10

113. C. ESKILSSON AND S.J. SHERWIN, *A triangular spectral/hp discontinuous Galerkin method for modelling 2D shallow water equations*, Int. J. Numer. Meth. Fluids. **45**(2004), pp. 605-623. Cited on page(s) 11

114. C. ESKILSSON AND S.J. SHERWIN, *Discontinuous Galerkin spectral/hp element modelling of dispersive shallow water systems*, J. Sci. Comput. **22**(2005), pp. 269-288. Cited on page(s) 11

115. L.C. EVANS, *Partial Differential Equations*, American Mathematical Society, Providence, RI, 1998. Cited on page(s) 118

116. M. FEISTAUER AND V. KUERA, *On a robust discontinuous Galerkin technique for the solution of compressible flow*, J. Comput. Phys. **224**(2007), pp. 208-221. Cited on page(s) 315

117. X. FENG AND O. KARAKASHIAN, *Two-level additive Schwarz methods for a discontinuous Galerkin approximation of second order elliptic problems*, SIAM J. Numer. Anal. **39**(2001), pp. 1343-1365. Cited on page(s) 12, 297

118. L. FEZOUI, S. LANTERI, S. LOHRENGEL, AND S. PIPERNO, *Convergence and stability of a discontinuous Galerkin time-domain method for the 3D heterogeneous Maxwell equations on unstructured meshes*, Math. Model. Numer. Anal. **39**(2005), pp. 1149-1176. Cited on page(s) 348

119. K.J. FIDKOWSKI, T.A. OLIVER, J. LU, AND D.L. DARMOFAL, *p-Multigrid solution of high-order discontinuous Galerkin discretizations of the compressible Navier-Stokes equations*, J. Comput. Phys. **207**(2005), pp. 92-113. Cited on page(s) 12, 297

120. J. FLAHERTY, L. KRIVODONOVA, J.-F. REMACLE, AND M. SHEPHARD, *Aspects of discontinuous Galerkin methods for hyperbolic conservation laws*, Finite Elem. Anal. Design, **38**(2002), pp. 889 - 908. Cited on page(s) 12, 157, 387, 396

121. D. FUNARO, *A multidomain spectral approximation to elliptic equations*, Numer. Methods Partial Diff. Eq. **2**(1986), pp. 187-205. Cited on page(s) 13

122. D. FUNARO AND D. GOTTLIEB, *A new method of imposing boundary conditions in pseudospectral approximations of hyperbolic equations*, Math. Comp. **51**(1988), pp. 599-613. Cited on page(s) 13

123. D. FUNARO AND D. GOTTLIEB, *Convergence results for pseudospectral approximations of hyperbolic systems by a penalty-type boundary treatment*, Math. Comp. **57**(1991), pp. 585-596. Cited on page(s) 13

124. G. GASSNER, F.LÖRCHER, AND C.-D. MUNZ, *A Contribution to the Construction of Diffusion Fluxes for Finite Volume and Discontinuous Galerkin Schemes*, J. Comput. Phys. 2007 - submitted. Cited on page(s) 260

125. A GEORGE AND J.LIU, *Computer Solution of Large Sparse Positive Definite Systems*, Prentice-Hall, Englewood Cliffs, NJ, 1981. Cited on page(s) 298

126. E.H. GEORGOULIS AND E. SULI, *Optimal error estimates for the hp-version interior penalty discontinuous Galerkin finite element method*, IMA J. Numer. Anal. **25**(2005), pp. 205-220. Cited on page(s) 11

127. F.X. GIRALDO, J.S. HESTHAVEN, AND T. WARBURTON, *Nodal high-order discontinuous Galerkin method for the spherical shallow water equations*, J. Comput. Phys. **181**(2002), pp. 499-525. Cited on page(s) 11

128. G. GASSNER, F. LÖRCHER, AND C.-D, MUNZ, *A discontinuous Galerkin scheme based on a space-time expansion. II. Viscous flow equations in multi dimensions*, J. Comput. Phys. 2007 - submitted. Cited on page(s) 112

129. M. GROTE, A. SCHNEEBELI, AND D. SCHÖTZAU, *Interior penalty discontinuous Galerkin method for Maxwell's equations: Energy norm error estimates*, J. Comput. Appl. Math. **204**(2007), pp. 375-386. Cited on page(s) 11, 12, 372

130. M. GROTE, A. SCHNEEBELI, AND D. SCHÖTZAU, *Discontinuous Galerkin finite element method for the wave equation*, SIAM J. Num. Anal. **44**(2006), pp. 2408-2431. Cited on page(s) 11, 12, 372

131. G.H. GOLUB AND J.H. WELSCH, *Calculation of Gauss quadrature rules*, Math. Comp **23**(1969), 221-230. Cited on page(s) 447

132. J. GOODMAN AND R. LEVEQUE, *On the accuracy of stable schemes for 2D scalar conservation laws*, Math. Comp. **45**(1985), pp. 15-21. Cited on page(s) 239

133. J. GOPALAKRISHNAN AND G. KANSCHAT, *A multilevel discontinuous Galerkin method*, Numer. Math., **95**(2003), pp. 527–550. Cited on page(s) 12, 297

134. W.N. GORDON AND C.A. HALL, *Construction of curvilinear coordinate systems and application to mesh generation*, Int. J. Numer. Meth. Eng. **7**(1973), pp. 461-477. Cited on page(s) 175, 378

135. D. GOTTLIEB AND J.S. HESTHAVEN, *Spectral methods for hyperbolic problems*, J. Comput. Appl. Math. **128**(2001), pp. 83-131. Cited on page(s) 127, 137

136. D. GOTTLIEB AND S.A. ORSZAG, *Numerical Analysis of Spectral Analysis: Theory and Applications*, Society for Industrial and Applied Mathematics, Philadelphia, PA, 1977. Cited on page(s) 92

137. D. GOTTLIEB AND C.-W. SHU, *On the Gibbs phenomenon and its resolution*, SIAM Rev. **39**(1997), pp. 644-668. Cited on page(s) 137, 145

138. D. GOTTLIEB AND E. TADMOR, *Recovering pointwise values of discontinuous data with spectral accuracy*. In *Progress and Supercomputing in Computational Fluid Dynamics*, Birkhäuser, Boston, 1984. pp. 357-375. Cited on page(s) 143

139. S. GOTTLIEB AND C.-W. SHU, *Total variation diminishing Runge-Kutta schemes*, Math. Comp. **67**(1998), pp. 73-85. Cited on page(s) 12, 158, 158, 159

140. S. GOTTLIEB, C.-W. SHU, AND E. TADMOR, *Strong stability preserving high order time discretization methods*, SIAM Rev. **43**(2001), pp. 89-112. Cited on page(s) 12, 158, 159

141. A. GREENBAUM, *Iterative Methods for Solving Linear Systems* Frontiers in Applied Mathematics **17**, SIAM Publishing, Philadelphia, 1997. Cited on page(s) 297, 300

142. B. GUSTAFSSON, H.O. KREISS, AND J. OLIGER, *Partial Differential Equations and Difference Approximations*, John Wiley & Sons, New York, 2001. Cited on page(s) 2, 35

143. E. HAIRER, S.P. NØRSETT, AND G. WANNER, *Solving Ordinary Differential Equations I. Nonstiff Problems.*, Springer Series in Computational Mathematics **8**, Springer-Verlag, Berlin, 1987. Cited on page(s) 64, 93, 157, 301

144. E. HAIRER AND G. WANNER, *Solving Ordinary Differential Equations II. Stiff and Differential-Algebraic Problems.*, Springer Series in Computational Mathematics **14**, Springer-Verlag, Berlin, 1991. Cited on page(s) 64, 93, 157

145. A. HARTEN, P.D. LAX, AND B. VAN LEER, *On upstream differencing and Godunov-type schemes for hyperbolic conservation laws*, SIAM Rev. **25**(1983), pp. 35-61. Cited on page(s) 222, 222, 224

146. R. HARTMANN AND P. HOUSTON, *Adaptive discontinuous Galerkin finite element methods for nonlinear hyperbolic conservation laws*, SIAM J. Sci. Comput. **24**(2002), pp. 979-1004. Cited on page(s) 12, 387

147. R. HARTMANN AND P. HOUSTON, *Adaptive discontinuous Galerkin finite element methods for the compressible Euler equations*, J. Comput. Phys. **183**(2002), pp. 508-532. Cited on page(s) 12, 387

148. R. HARTMANN AND P. HOUSTON, *Symmetric interior penalty DG methods for the compressible Navier-Stokes equations I: Method formulation*, Int. J. Numer. Anal. Model. **3**(2006), pp. 1-20. Cited on page(s) 11

149. R. HARTMANN AND P. HOUSTON, *Symmetric interior penalty DG methods for the compressible Navier-Stokes equations II: Goal-oriented a posteriori error estimation*, Int. J. Numer. Anal. Model. **3**(2006), pp. 141-162. Cited on page(s) 11

150. B.T. HELENBROOK, D. MAVRIPLIS, AND H.L. ATKINS *Analysis of "p"-multigrid for continuous and discontinuous finite-element discretizations*, AIAA Paper 2003-3989, 2003. Cited on page(s) 12, 297

151. J.S. HESTHAVEN, *From electrostatics to almost optimal nodal sets for polynomial interpolation in a simplex*, SIAM J. Numer. Anal. **35**(1998), pp. 655-676. Cited on page(s) 47, 49, 175

152. J.S. HESTHAVEN, *A stable penalty method for the compressible Navier-Stokes equations. II. One-dimensional domain decomposition schemes*, SIAM J. Sci. Comput. **18**(1998), pp. 658-685. Cited on page(s) 13, 315

153. J.S. HESTHAVEN, *A stable penalty method for the compressible Navier-Stokes equations. III. Multi dimensional domain decomposition schemes*, SIAM J. Sci. Comput. **20**(1999), pp. 62-93. Cited on page(s) 13, 315

154. J.S. HESTHAVEN, *Spectral penalty methods*, Appl. Numer. Math. **33**(2000), pp. 23-41. Cited on page(s) 13

155. J.S. HESTHAVEN, *High-order accurate methods in time-domain computational electromagnetics. A review*, Adv. Imaging Elec. Phys. **127**(2003), pp. 59-123. Cited on page(s) 29

156. J.S. HESTHAVEN, P. DINESEN, AND J.P. LYNOV, *Spectral collocation time-domain modeling of diffractive optical elements*, J. Comput. Phys. **155**(1999), pp. 287-306. Cited on page(s) 100

157. J.S. HESTHAVEN AND D. GOTTLIEB, *A stable penalty method for the compressible Navier-Stokes equations. I. Open boundary conditions*, SIAM J. Sci. Comput. **17**(1996), pp. 579-612. Cited on page(s) 13, 315

158. J.S. HESTHAVEN AND D. GOTTLIEB, *Stable spectral methods for conservation laws on triangles with unstructured grids*, Comput. Methods Appl. Mech. Eng. **175**(1999), pp. 361-381. Cited on page(s) 13

159. J.S. HESTHAVEN, S. GOTTLIEB, AND D. GOTTLIEB, *Spectral Methods for Time-Dependent Problems*, Cambridge University Press, Cambridge, 2006. Cited on page(s) 23, 29, 45, 47, 54, 54, 79, 127, 186, 327

160. J.S. HESTHAVEN, S.M. KABER, AND L. LURATI, *Pade-Legendre interpolants for Gibbs reconstruction*, J. Sci. Comput. **28**(2006), pp. 337-359. Cited on page(s) 144

161. J.S. HESTHAVEN AND S.M. KABER, *Pade-Jacobi approximants*, J. Comput. Appl. Math. 2007 - submitted. Cited on page(s) 144

162. J.S. HESTHAVEN AND M. KIRBY, *Filering in Legendre spectral methods*, Math. Comput. 2007 - to appear. Cited on page(s) 143, 186

163. J.S. HESTHAVEN AND C. H. TENG, *Stable spectral methods on tetrahedral elements*, SIAM J. Sci. Comput. **21**(2000), pp. 2352-2380. Cited on page(s) 13, 412, 414, 414

164. J.S. HESTHAVEN AND T. WARBURTON, *High-order nodal methods on unstructured grids. I. Time-domain solution of Maxwell's equations*, J. Comput. Phys. **181**(2002), pp. 186-221. Cited on page(s) 11, 348

165. J.S. HESTHAVEN AND T. WARBURTON, *High order nodal discontinuous Galerkin methods for the Maxwell eigenvalue problem*, Royal Soc. London Ser A **362**(2004), pp. 493-524. Cited on page(s) 12, 290, 348, 355

166. P. HOUSTON, I. PERUGIA, A. SCHNEEBELI, AND D. SCHÖTZAU, *Mixed discontinuous Galerkin approximation of the Maxwell operator: The indefinite case*, Math. Model. Numer. Anal. **39**(2005), pp. 727-754. Cited on page(s) 12, 296

167. P. HOUSTON, I. PERUGIA, A. SCHNEEBELI, AND D. SCHÖTZAU, *Interior penalty method for the indefinite time-harmonic Maxwell equations*, Numer. Math. **100**(2005), pp. 485-518. Cited on page(s) 12, 296

168. P. HOUSTON, I. PERUGIA, AND D. SCHÖTZAU, *Mixed discontinuous Galerkin approximation of the Maxwell operator*, SIAM J. Numer. Anal. **42**(2004), pp. 434-459. Cited on page(s) 12, 296

169. P. HOUSTON, I. PERUGIA, AND D. SCHÖTZAU, *Mixed discontinuous Galerkin approximation of the Maxwell Operator: Non-stabilized formulation*, J. Sci. Comput. **22**(2005), pp. 325-356. Cited on page(s) 12, 296

170. P. HOUSTON, I. PERUGIA, AND D. SCHÖTZAU, *An a posteriori error indicator for discontinuous Galerkin discretizations of H(curl)-elliptic partial differential equations*, IMA J. Numer. Anal. **27**(2007), pp. 122-150. Cited on page(s) 12, 387

171. P. HOUSTON, I. PERUGIA, AND D. SCHÖTZAU, *Energy norm a posteriori error estimation for mixed discontinuous Galerkin approximations of the Maxwell operator*, Comput. Methods Appl. Mech. Eng. **194**(2005), pp. 499-510. Cited on page(s) 12, 387

172. P. HOUSTON, D. SCHÖTZAU, AND T. WIHLER, *An hp-adaptive mixed discontinuous Galerkin FEM for nearly incompressible linear elasticity*, Comput. Methods Appl. Mech. Eng. **195**(2006), pp. 3224-3246. Cited on page(s) 12, 296

173. P. HOUSTON, D. SCHÖTZAU, AND T. WIHLER, *Energy norm a posteriori error estimation for mixed discontinuous Galerkin approximations of the Stokes problem*, J. Sci. Comput. **22**(2005), pp. 347-370. Cited on page(s) 12, 387

174. P. HOUSTON, CH. SCHWAB, AND E. SULI, *Discontinuous hp-finite element methods for advection-diffusion-reaction problems*, SIAM J. Numer. Anal. **39**(2002), pp. 2133-2163. Cited on page(s) 11

175. P. HOUSTON, B. SENIOR, AND E. SULI, *hp-Discontinuous Galerkin finite element methods for hyperbolic problems: Error analysis and adaptivity*, Int. J. Numer. Methods Fluids **40**(2002), pp. 153-169. Cited on page(s) 12, 387

176. P. HOUSTON AND E. SULI, *hp-Adaptive discontinuous Galerkin finite element methods for first-order hyperbolic problems*, SIAM J. Sci. Comput. **23**(2002), pp. 1226-1252. Cited on page(s) 12, 387

177. P. HOUSTON, E. SULI, AND T.P. WIHLER, *A posteriori error analysis of hp-version discontinuous Galerkin finite element methods for second-order quasilinear elliptic problems*, IMA J. Numer. Anal. 2007 - to appear. Cited on page(s) 12, 387

178. C. HU AND C.-W. SHU, *A discontinuous Galerkin finite element method for Hamilton-Jacobi equations*, SIAM J. Sci. Comput. **21**(1999), pp.666-690. Cited on page(s) 11, 135

179. F.Q. HU AND H. ATKINS, *Eigensolution analysis of the discontinuous Galerkin method with non-uniform grids. Part I: One space dimension*, J. Comput. Phys. **182**(2002), pp. 516-545. Cited on page(s) 91, 92

180. F.Q. HU AND H. ATKINS, *Two-dimensional wave analysis of the discontinuous Galerkin method with non-uniform grids and boundary conditions*, AIAA paper 2002-2514, 11p Cited on page(s) 91

181. F.Q. HU, M.Y. HUSSAINI, AND P. RASETARINERA, *An analysis of the discontinuous Galerkin method for wave propagation problems*, J. Comput. Phys. **151**(1999), pp. 921-946. Cited on page(s) 91

182. T. HUGHES, *The Finite Element Method: Linear Static and Dynamic Finite Element Analysis*, Dover Publications, New York, 2000. Cited on page(s) 6, 6

183. T. ISAACS, Private communication, 2006. Cited on page(s) 427

184. G. JACOBS AND J.S. HESTHAVEN, *High-order nodal discontinuous Galerkin particle-in-cell methods on unstructured grids*, J. Comput. Phys. **214**(2006), pp. 96-121. Cited on page(s) 11

185. A. JAMESON, H. SCHMIDT, AND E. TURKEL, *Numerical solutions of the Euler equations by finite volume methods using Runge-Kutta time stepping schemes*, AIAA paper 1981-1259. Cited on page(s) 101, 103

186. P. JAWAHAR AND H. KAMATH, *A high-resolution procedure for Euler and Navier-Stokes computations on unstructured grids*, J. Comput. Phys. **164**(2000), pp. 165-203. Cited on page(s) 225

187. G. JIANG AND C.-W. SHU, *On a cell entropy inequality for discontinuous Galerkin methods*, Math. Comp. **62**(1994), pp. 531-538. Cited on page(s) 122

188. J. M. JIN, *The Finite Element Method in Electromagnetics*, John Wiley & Sons, New York, 1993. Cited on page(s) 332

189. V. JOHN, *Reference values for drag and lift of a two-dimensional time dependent flow around a cylinder*, Int. J. Numer. Methods Fluids **44**(2004), pp. 777-788. Cited on page(s) 308, 314, 314

190. C. JOHNSON AND J. PITKARANTA, *An Analysis of the discontinuous Galerkin method for a scalar hyperbolic equation*, Math. Comp. **46**(1986), pp. 1-26. Cited on page(s) 10, 88, 237, 237, 238

191. A. KANEVSKY, M.H. CARPENTER, D. GOTTLIEB, AND J.S. HESTHAVEN, *Application of implicit-explicit high-order Runge-Kutta methods to discontinuous Galerkin schemes*, J. Comput. Phys. 225(2007), pp. 1753-1781. Cited on page(s) 12, 315

192. A. KANEVSKY, M.H. CARPENTER, AND J.S. HESTHAVEN, *Idempotent filtering in spectral and spectral element methods*, J. Comput. Phys. 220(2006), pp. 41-58. Cited on page(s) 141

193. G. KANSCHAT, *Preconditioning methods for local discontinuous Galerkin discretizations*. SIAM J. Sci. Comput. **25**(2003), pp. 815-831. Cited on page(s) 12, 297

194. G. KANSCHAT AND R. RANNACHER, *Local error analysis of the interior penalty discontinuous Galerkin method for second order elliptic problems*, J. Numer. Math. **10**(2002), pp. 249-274. Cited on page(s) 11

195. G.E. KARNIADAKIS AND J.S. HESTHAVEN (EDS.), J. Eng. Math. 2006. Cited on page(s) 13

196. G.E. KARNIADAKIS, M. ISRAELI, AND S.A. ORSZAG, *High-order splitting methods for the incompressible Navier-Stokes equations*, J. Comput. Phys. **97**(1991), pp. 414-443. Cited on page(s) 301

197. G.E. KARNIADAKIS AND S.J. SHERWIN, *Spectral/hp Element Methods for CFD*, Numerical Mathematics and Scientific Computation. Clarendon Press, Oxford. 1999. Cited on page(s) 6, 212, 445, 450, 450

198. G.E. KARNIADAKIS AND S.J. SHERWIN, *Spectral/hp Element Methods in Computational Fluid Dynamics (2nd Ed.)*, Oxford University Press, Oxford, 2005. Cited on page(s) 301

199. M. KÄSER AND M. DUMBSER, *An arbitrary high order discontinuous Galerkin method for elastic waves on unstructured meshes I: The two-dimensional isotropic case with external source terms*, Geophys. J. Int. **166**(2006), pp. 855-877. Cited on page(s) 11

200. M. KÄSER, M. DUMBSER, J. DE LA PUENTE, AND H. IGEL, *An arbitrary high order discontinuous Galerkin method for elastic waves on unstructured meshes III: Viscoelastic attenuation*, Geophys. J. Int. **168**(2007), pp. 224-242. Cited on page(s) 11

201. T. KOORNWINDER, *Two-variable analogues of the classical orthogonal polynomials* in *Theory and Application of Special Functions*, R. A. Askey (Eds.), Academic Press, New York, 1975, pp. 435-495. Cited on page(s) 449, 450, 450

202. D.A. KOPRIVA, *A conservative staggered-grid Chebyshev multidomain method for compressible flows. II. A semi-structured method*, J. Comput. Phys. **128**(1996), pp. 475-488 Cited on page(s) 315

203. D.A. KOPRIVA, *A staggered-grid multidomain spectral method for the compressible Navier-Stokes equations*, J. Comput. Phys. **142**(1998) pp. 125-158. Cited on page(s) 315

204. D.A. KOPRIVA AND J.H. KOLIAS, *A conservative staggered-grid Chebyshev multidomain method for compressible Flows*, J. Comput. Phys. **125**(1996), pp. 244-261. Cited on page(s) 315

205. D. KOPRIVA, S.L. WOODRUFF, AND M.Y. HUSSAINI, *Discontinuous spectral element approximation of Maxwell's equations*. In *Discontinuous Galerkin Methods: Theory, Computation and Applications*, B. Cockburn, G.E. Karniadakis,

and C.W. Shu (Eds.), Lecture Notes in Computational Science and Engineering **11**, Springer-Verlag, Berlin, pp. 355-362. Cited on page(s) 11, 348

206. D. KOSLOFF AND H. TAL-EZER, *A modified Chebyshev pseudospectral method with an* $\mathcal{O}(N^{-1})$ *time step restriction*, J. Comput. Phys. **104**(1993), pp. 457-469. Cited on page(s) 98, 100, 100

207. L.S.G. KOVASZNAY, *Laminar flow behind a two-dimensional grid*, Proc. Camb. Philos. Soc. **44**(1948), pp. 58-62. Cited on page(s) 308

208. H.O. KREISS AND J. OLIGER, *Comparison of accurate methods for the integration of hyperbolic problems*, Tellus **24**(1972), pp. 199-215. Cited on page(s) 29

209. L. KRIVODONOVA, *Limiters for high-order discontinuous Galerkin methods*, J. Comput. Phys. **226**(2007), pp. 879-896. Cited on page(s) 157, 224

210. L. KRIVODONOVA AND M. BERGER, *High-order accurate implementation of solid wall boundary conditions in curved geometries*, J. Comput. Phys. **211**(2006), pp. 492-512. Cited on page(s) 375

211. L. KRIVODONOVA AND J. FLAHERTY, *Error estimation for discontinuous Galerkin solutions of multidimensional hyperbolic problems*, Adv. Comput. Math. **19**(2003), pp. 57-71. Cited on page(s) 12, 387, 396

212. L. KRIVODONOVA, J. XIN, J.-F. REMACLE, N. CHEVAUGEON, AND J.E. FLAHERTY, *Shock detection and limiting with discontinuous Galerkin methods for hyperbolic conservation laws*, Appl. Numer. Math. **48**(2004), pp. 323-338. Cited on page(s) 157, 224

213. C. LASSER AND A. TOSELLI, *An overlapping domain decomposition preconditioner for a class of discontinuous Galerkin approximations of advection-diffusion problems*, Math. Comp. **72**(2003), pp. 1215-1238. Cited on page(s) 12, 297

214. P.D. LAX, *Shock waves and entropy*. In *Proceeding of the Symposium at the University of Wisconsin*, 1971, pp. 603-634. Cited on page(s) 118

215. P.D. LAX AND R.D. RICHTMYER, *Survey of the stability of linear finite difference equations*, Comm. Pure Appl. Math. **9**(1956), pp. 267-293. Cited on page(s) 77

216. O. LEPSKY, C. HU, AND C.-W. SHU, *The analysis of the discontinuous Galerkin method for Hamilton-Jacobi equations*, Appl. Numer. Math. **33**(2000), pp.423-434. Cited on page(s) 11, 135

217. P. LESAINT AND P.A. RAVIART, *On a Finite Element Method for Solving the Neutron Transport Equation*. In *Mathematical Aspects of Finite Elements in Partial Differential Equations*, Academic Press, New York, 1974, pp. 89-145. Cited on page(s) 10, 88, 88, 236

218. R.J. LEVEQUE, *Finite Volume Methods for Hyperbolic Problems*, Cambridge University Press, Cambridge, 2002. Cited on page(s) 4, 32, 32, 33, 36, 116, 135, 150, 218, 224

219. D. LEVY, C.-W. SHU, AND J. YAN, *Local discontinuous Galerkin methods for nonlinear dispersive equations*, J. Comput. Phys. **196**(2004), pp. 751-772. Cited on page(s) 11, 260

220. B.Q. LI, *Discontinuous Finite Elements in Fluid Dynamics and Heat Transfer*, Springer Series in Computational Fluid and Solid Mechanics, Springer-Verlag, Berlin, 2006. Cited on page(s) 11

221. F. LI AND C.-W. SHU, *Reinterpretation and simplified implementation of a discontinuous Galerkin method for Hamilton-Jacobi equations*, Appl. Math. Lett. **18**(2005), pp. 1204-1209. Cited on page(s) 11

222. F. LI AND C.-W. SHU, *Locally divergence-free discontinuous Galerkin methods for MHD equations*, J. Sci. Comput. **22-23**(2005), pp. 413-442. Cited on page(s) 357

223. T.J. LINDE, *A practical, general-purpose, two-state HLL Riemann solver for hyperbolic conservation laws*, Int. J. Numer. Methods Fluids **40**(2002), pp. 391-402. Cited on page(s) 224

224. G. LIN AND G.E. KARNIADAKIS, *A discontinuous Galerkin method for two-temperature plasmas*, Comput. Methods Appl. Mech. Eng. **195**(2006), pp. 3504-3527. Cited on page(s) 11

225. Q. LIN AND A.H. ZHOU, *Convergence of the discontinuous Galerkin methods for a scalar hyperbolic equation*, Acta Math. Sci. **13**(1993), pp. 207-210. Cited on page(s) 10

226. J.G. LIU AND C.-W. SHU, *A high-order discontinuous Galerkin method for 2D incompressible flows*, J. Comput. Phys. **160**(2000), pp. 577-596. Cited on page(s) 11

227. Y. LIU AND C.-W. SHU, *Local discontinuous Galerkin methods for moment models in device simulations: formulation and one dimensional results*, J. Comput. Electr. **3**(2004), pp. 263-267. Cited on page(s) 11

228. I. LOMTEV AND G.E. KARNIADAKIS, *A discontinuous Galerkin method for the Navier-Stokes equations*, Int. J. Num. Methods Fluids **29**(1999), pp. 587-603. Cited on page(s) 315, 396

229. I. LOMTEV, *A discontinuous Galerkin method for the compressible Navier-Stokes equations in stationary and moving 3D domains*, PhD thesis. Brown University, 1999. Cited on page(s) 315

230. I. LOMTEV, C.B. QUILLEN, AND G.E. KARNIADAKIS, *Spectral/hp methods for viscous compressible flows on unstructured 2D meshes*, J. Comput. Phys. **144**(1998), pp. 325-357. Cited on page(s) 315, 396

231. F. LÖRCHER, G. GASSNER, AND C.-D, MUNZ, *A discontinuous Galerkin scheme based on a space-time expansion. I. Inviscid compressible flow in one-space dimension*, J. Sci. Comput. **32**(2007), pp. 175-199. Cited on page(s) 112

232. H. LUO, J.D. BAUM, AND R. LÖHNER, *A Hermite WENO-based limiter for discontinuous Galerkin method on unstructured grids*, J. Comput. Phys. **225**(2007), pp. 686-713. Cited on page(s) 11

233. J. LOVERICH AND U. SHUMLAK, *A discontinuous Galerkin method for the full two-fluid plasma model*, Comput. Phys. Commun. **169**(2005), pp. 251-255. Cited on page(s) 11

234. E. LUO AND H.O. KREISS, *Pseudospectral vs finite difference methods for initial value problems with discontinuous coefficients*, SIAM J. Sci. Comput. **20**(1998), pp. 148-163. Cited on page(s) 72

235. H. MA, *Chebyshev-Legendre super spectral viscosity method for nonlinear conservation laws*, SIAM J. Numer. Anal. **35**(1998), pp. 898-892. Cited on page(s) 134

236. Y. MADAY, S.M. OULD KABER, AND E. TADMOR, *Legendre pseudospetral viscosity method for nonlinear conservation laws*, SIAM J. Numer. Anal. **30**(1993), pp. 321-342. Cited on page(s) 134

237. C. MAVRIPLIS, *A posteriori error estimators for adaptive spectral element techniques*, Notes Numer. Methods Fluid Mech. **29**(1990), pp. 333-342. Cited on page(s) 387, 396

238. C. MAVRIPLIS, Adaptive mesh strategies for the spectral element method, Comput. Methods Appl Mech. Eng. **116**(1994), pp. 77-86. Cited on page(s) 387, 396

239. M. MOCK AND P. LAX, *The computation of discontinuous solutions of linear hyperbolic equations*, Comm. Pure Appl. Math. **31**(1978), pp. 423-430. Cited on page(s) 138

240. E. MONTSENY, S. PERNET, X. FERRIERES, AD G. COHEN, *Dissipative terms and local time-stepping improvements in a spatial high-order discontinuous Galerkin scheme for the time-domain Maxwell's equations*, J. Comput. Phys. 2007 – submitted. Cited on page(s) 93, 110

241. I. MOZOLEVSKI AND E. SULI, *A priori error analysis for the hp-version of the discontinuous Galerkin finite element method for the biharmonic equation*, Comput. Methods Appl. Math. **3**(2003), pp. 596-607. Cited on page(s) 11

242. I. MOZOLEVSKI, E. SULI, AND P. BOSING, *Discontinuous Galerkin finite element approximation of the two-dimensional Navier–Stokes equations in stream-function formulation*, Comm. Numer. Methods Eng. **23**(2006), pp. 447-459. Cited on page(s) 11

243. I. MOZOLEVSKI, E. SULI, AND P. BOSING, *hp-Version a priori error analysis of interior penalty discontinuous Galerkin finite element approximations to the biharmonic equation*, J. Sci. Comput. **30**(2006), pp. 465-491. Cited on page(s) 11

244. K. NAKAHASHI AND E. SAITOH, *Space-marching method on unstructured grid for supersonic flows with embedded subsonic regions*, AIAA J. **35**(1997), pp. 1280-1288. Cited on page(s) 236

245. C. NASTASE AND D. J. MAVRIPLIS, *High-order discontinuous Galerkin methods using a spectral multigrid approach* AIAA Paper 2005-1268, 2005. Cited on page(s) 12, 13, 297

246. C. NASTASE AND D. J. MAVRIPLIS, *Discontinuous Galerkin methods using an h-p multigrid solver for inviscid compressible flows on three-dimensional unstructured meshes*, AIAA Paper 2006-0107, 2006. Cited on page(s) 12

247. N.C. NGUYEN, P.-O. PERSSON, AND J. PERAIRE, *RANS solutions using high order discontinuous Galerkin methods*, Proc. of the *45th AIAA Aerospace Sciences Meeting and Exhibit*, January 2007. Cited on page(s) 316

248. J. NORDSTROM AND M.H. CARPENTER, *Boundary and interface conditions for high-order finite-difference methods applied to the Euler and NavierStokes equations* J. Comput. Phys. **148**(1999), pp. 621-645. Cited on page(s) 13

249. J.T. ODEN, I. BABUSKA, AND C.E. BAUMANN, *A discontinuous hp-finite element method for diffusion Problems*, J. Comput. Phys. **146**(1998). pp. 491-519. Cited on page(s) 11

250. CH. ORTNER AND E. SULI, *Discontinuous Galerkin finite element approximation of nonlinear second-order elliptic and hyperbolic systems*, SIAM J. Numer. Anal. **45**(2007), pp. 1370-1397. Cited on page(s) 12

251. S. OSHER, *Riemann solvers, the entropy condition, and difference approximations*, SIAM J. Numer. Anal. **21**(1984), pp. 217-235. Cited on page(s) 32

252. J. PERAIRE AND P.-O. PERSSON, *A Compact Discontinuous Galerkin (CDG) Method for Elliptic Problems*, SISC J. Sci. Comput. 2007 - to appear. Cited on page(s) 269

253. P.-O. PERSSON AND J. PERAIRE, *Sub-Cell Shock Capturing for Discontinuous Galerkin Methods*, AIAA Paper. 2007. Cited on page(s) 225

254. I. PERUGIA AND D. SCHÖTZAU, *An hp-analysis of the local discontinuous Galerkin method for diffusion problems*, J. Sci. Comput. **17**(2002), pp. 561-571. Cited on page(s) 11, 296, 296

255. I. PERUGIA AND D. SCHÖTZAU, *The hp-local discontinuous Galerkin method for low-frequency time-harmonic Maxwell equations*, Math. Comp. **72**(2003), pp. 1179-1214. Cited on page(s) 12, 296

256. I. PERUGIA, D. SCHÖTZAU, AND P. MONK, *Stabilized interior penalty methods for the time-harmonic Maxwell equations*, Comput. Methods Appl. Mech. Eng. **191**(2002), pp. 4675-4697. Cited on page(s) 12, 296

257. T.E. PETERSON, *A note on the convergence of the discontinuous Galerkin method for a scalar hyperbolic equation*, SIAM J. Numer. Anal. **28**(1991), pp. 133-140. Cited on page(s) 10, 237, 238, 238

258. S. PIPERNO, *L2-stability of the upwind first order finite volume scheme for the Maxwell equation in two and three dimensions on arbitrary unstructured meshes*, Math. Model. Numer. Anal. **34**(2000), pp. 139-158. Cited on page(s) 348

259. S. PIPERNO, *Symplectic local time-stepping in non-dissipative DGTD methods applied to wave propagation problems*, M2AN, **40**(2006), pp. 815-841. Cited on page(s) 110

260. S. PIPERNO, M. REMAKI, AND L. FEZOUI, *A non-diffusive finite volume scheme for the 3D Maxwell equations on unstructured meshes* SIAM J. Numer. Anal. **39**(2002), pp. 2089-2108. Cited on page(s) 348

261. J. PRORIOL, *Sur une Famille de polynomes à deux variables orthogonaux dans un triangle*, C. R. Acad. Sci. Paris **257**(1957), pp. 2459-2461. Cited on page(s) 449

262. J. QIU, M. DUMBSER, AND C.-W. SHU, *The discontinuous Galerkin method with Lax-Wendroff type time discretizations*, Comput. Methods Appl. Mech. Eng. **194**(2005), pp. 4528-4543. Cited on page(s) 12, 111

263. J. QIU, B.C. KHOO, AND C.-W. SHU, *A numerical study for the performance of the Runge-Kutta discontinuous Galerkin method based on different numerical fluxes*, J. Comput. Phys. **212**(2006), pp. 540-565. Cited on page(s) 33, 218, 224

264. J. QIU AND C.-W. SHU, *Hermite WENO schemes and their application as limiters for Runge-Kutta discontinuous Galerkin method: one dimensional case*, J. Comput. Phys. **193**(2003), pp. 115-135. Cited on page(s) 11

265. J. QIU AND C.-W. SHU, *Runge-Kutta discontinuous Galerkin method using WENO limiters*, SIAM J. Sci. Comput. **26**(2005), pp. 907-929. Cited on page(s) 11

266. J. QIU AND C.-W. SHU, *A comparison of troubled cell indicators for Runge-Kutta discontinuous Galerkin methods using WENO limiters*, SIAM J. Sci. Comput. **27**(2005), pp. 995-1013. Cited on page(s) 11

267. J. QIU AND C.-W. SHU, *Hermite WENO schemes and their application as limiters for Runge-Kutta discontinuous Galerkin method II: two dimensional case*, Comput. Fluids **34**(2005), pp. 642-663. Cited on page(s) 11

268. J. QIU AND C.-W. SHU, *Hermite WENO schemes for Hamilton-Jacobi equations*, J. Comput. Phys. **204**(2005), pp. 82-99. Cited on page(s) 11

269. W.H. REED AND T.R. HILL, *Triangular mesh methods for the neutron transport equation*, Los Alamos Scientific Laboratory Report, LA-UR-73-479, 1973. Cited on page(s) 10, 236

490   References

270. M. REMAKI AND L. FÉZOUI, *Une méthode de Galerkin discontinu pour la résolution des équations de Maxwell en milieu hétérogéne*, INRIA report **3501**, 1998. Cited on page(s) 11

271. J. I. RICHARDS AND H. K. YOUN, *Theory of Distributions. A Nontechnical Introduction*, Cambridge University Press, Cambridge, 1990. Cited on page(s) 30

272. G. R. RICHTER, *An optimal-order error estimate for the discontinuous Galerkin method*, Math. Comp. **50**(1988), pp. 75-88. Cited on page(s) 237

273. P.L. ROE, *Approximate Riemann solvers, parameter vectors, and difference Schemes*, J. Comput. Phys. **43**(1981), pp. 357-372. Cited on page(s) 219, 220

274. J. RYAN, C.-W. SHU AND H. ATKINS, *Extension of a post-processing technique for the discontinuous Galerkin method for hyperbolic equations with application to an aeroacoustic problem*, SIAM J. Sci. Comput. **26**(2005), pp. 821-843. Cited on page(s) 10

275. Y. SAAD, *Iterative Methods for Sparse Linear Systems (2nd edition)*, SIAM Publishing, Philadelphia, 2003. Cited on page(s) 282, 297, 300

276. D. SARMANY, M.A. BOTCHEV, AND J.J.W. VAN DER VEGT, *Dispersion and dissipation error in high-order Runge-Kutta discontinuous Galerkin discretisations of the Maxwell equations*, J. Sci. Comput. **33**(2007), pp. 47-74. Cited on page(s) 93

277. M. SCHÖFER AND S. TUREK, *The benchmark problem flow around a cylinder*. In *Flow Simulation with High-Performance Computers II*, E.H, Hirschel (Eds.). Notes on Numerical Fluid Mechanics **52**. Vieweg: Braunschweig, 1996; pp. 547566. Cited on page(s) 312

278. D. SCHÖTZAU, C. SCHWAB, AND A. TOSELLI, *Stabilized hp-DGFEM for incompressible flow*, Math. Models Meth. Appl. Sci. **13**(2003), pp. 1413-1436. Cited on page(s) 12, 296

279. D. SCHÖTZAU, C. SCHWAB, AND A. TOSELLI, *Mixed hp-DGFEM for incompressible flows*, SIAM J. Numer. Anal. **40**(2003), pp. 2171-2194. Cited on page(s) 12, 296

280. CH. SCHWAB, *p- and hp-Finite Element Methods: Theory and Applications in Solids and Fluid Dynamics*, Numerical Mathematics and Scientific Computing, Oxford University Press, Oxford, 1998. Cited on page(s) 79, 80, 81, 94

281. K. SHAHBAZI, *An explicit expression for the penalty parameter of the interior penalty method*, J. Comput. Phys. **205**(2005), pp. 401-407. Cited on page(s) 11, 270, 294

282. S.J. SHERWIN AND G.E. KARNIADAKIS, *A triangular spectral element method; Applications to the incompressible Navier-Stokes equations*, Comput. Meth. Appl. Mech. Eng. **123**(1995), pp. 189-229. Cited on page(s) 301

283. S.J. SHERWIN, R.M. KIRBY, J. PEIRO, R.L. TAYLOR, AND O.C. ZIENKIEWICZ, *On 2D elliptic discontinuous Galerkin methods*, Int. J. Num. Method **65**(2006), pp. 752-784. Cited on page(s) 11, 269, 290

284. S. SHERWIN, *Dispersion analysis of the continuous and discontinuous Galerkin formulations*. In *Discontinuous Galerkin Methods: Theory, Computation and Applications*, B. Cockburn, G.E. Karniadakis, and C.W. Shu (Eds.), Lecture Notes in Computational Science and Engineering **11**, Springer-Verlag, Berlin, 1999, pp. 425-431. Cited on page(s) 91

285. C.-W. SHU, *TVB Uniformly high-order schemes for conservation laws*, Math. Comp. **49**(1987), pp. 105-121. Cited on page(s) 10, 153

286. C.-W. SHU, *TVD Time-discretizations*, SIAM J. Sci. Stat. Comput. **9**(1988), pp. 1073-1084. Cited on page(s) 10, 12

287. C.-W. SHU AND P. WONG, *A note on the accuracy of spectral method applied to nonlinear conservation laws*, J. Sci. Comput. **10**(1995), pp. 357-369. Cited on page(s) 138

288. C.-W. SHU, *Different formulations of the discontinuous Galerkin methods for diffusion problems*. In *Advances in Scientific Computing*, Science Press, Beijing, 2001, pp. 144-155. Cited on page(s) 11, 243

289. P. SILVESTER, *Finite element solution of homogeneous waveguide problems*, Alta Frequenza **38**(1969), pp. 313-317. Cited on page(s) 331

290. J. SMOLLER, *Shock Waves and Reaction-Diffusion Equations*, Springer-Verlag, New York, 1983. Cited on page(s) 118

291. A. SOLONOMOFF, *A fast algorithm for spectral differentiation*, J. Comput. Phys. **98**(1992), pp. 174-177. Cited on page(s) 54

292. R. SPITERI AND S. RUUTH, *A new class of optimal high-order strong stability preserving time discretization methods*, SIAM J. Numer. Anal. **40**(2002), pp. 469-491. Cited on page(s) 158

293. G. STRANG AND G.J. FIX, *Analysis of the Finite Element Method* Prentice-Hall, Englewood Cliffs, NJ, 1973. Cited on page(s) 6

294. A. H. STROUD, *Approximate Calculation of Multiple Integrals*, Prentice-Hall, Englewood Cliffs, NJ, 1971. Cited on page(s) 211, 379

295. Y.Z. SUN AND Z.J. WANG, *Evaluation of discontinuous Galerkin and spectral volume methods for scalar and system conservation laws on unstructured grids*, Int. J. Numer. Meth. Fluids **45**(2004), pp. 819-838. Cited on page(s) 13

296. G. SZEGÖ, *Orthogonal Polynomials*, Colloquium Publications **23**, American Mathematical Society, Providence, RI, 1939. Cited on page(s) 45, 45, 79, 97, 445, 445

297. E. TADMOR, *The exponential accuracy of Fourier and Chebyshev differencing methods* SIAM J. Numer. Anal. **23**(1986), pp. 1-10. Cited on page(s) 79

298. J. TANNER AND E. TADMOR, *Adaptive mollifiers - high resolution recover of piecewise smooth data from its spectral information*, Found. Comput. Math. **2**(2002), pp. 155-189. Cited on page(s) 143

299. A. TAUBE, M. DUMBSER, D.S. BALSARA, AND C.-D. MUNZ, *Arbitrary high-order discontinuous Galerkin schemes for the magnetohydrodynamic equations*, J. Sci Comput. **30**(2007), pp. 441-464. Cited on page(s) 12, 112

300. M. TAYLOR, B. WINGATE, AND R.E. VINCENT, *An algorithm for computing Fekete points in the triangle*, SIAM J. Num. Anal. **38**(2000), pp. 1707–1720. Cited on page(s) 175

301. J.W. THOMAS, *Numerical Partial Differential Equations: Conservation Laws and Elliptic Equations*, Springer Texts in Applied Mathematics **33**, Springer-Verlag, New York, 1999. Cited on page(s) 116, 118, 150, 218, 224

302. V.A. TITAREV AND E.F. TORO, ADER SCHEMES FOR THREE-DIMENSIONAL NONLINEAR HYPERBOLIC SYSTEMS, J. Comput. Phys. **204**(2005), pp. 715-736. Cited on page(s) 12, 112

303. E.F. TORO, *Riemann Solvers and Numerical Methods for Fluid Dynamics. A Practical Introduction (2nd Ed.)*, Springer-Verlag, New York, 1999. Cited on page(s) 4, 36, 224

304. E.F. TORO, M. SPRUCE, AND W. SPEARES, *Restoration of the contact surface in the Harten-Lax-van Leer Riemann Solver*, J. Shock Waves **4**(1994), pp. 25-34. Cited on page(s) 223, 224

305. A. TOSELLI AND O. WIDLUND, *Domain Decomposition Methods - Algorithms and Theory*, Springer-Verlag, New York, 2005. Cited on page(s) 282

306. I. TOULOPOULOS AND J.A. EKATERINARIS, *High-order discontinuous Galerkin discretizations for computational aeroacoustics in complex domains*, AIAA J. **44**(2006), pp. 502-511. Cited on page(s) 11

307. L.N. TREFETHEN, *Spectral Methods in Matlab*, SIAM Publishing, Philadelphia, 2000. Cited on page(s) 29, 49

308. L. N. TREFETHEN AND D. BAU, III, *Numerical Linear Algebra*, SIAM Publishing, Philadelphia, 1997. Cited on page(s) 297, 300

309. L.N. TREFETHER AND M.R. TRUMMER, *An instability phenomenon in spectral methods*, SIAM J. Numer. Anal. **24**(1987) pp. 1008-1023. Cited on page(s) 96

310. S. TU AND S. ALIABADI, *A slope limiting procedure in discontinuous Galerkin finite element method for gasdynamic applications*, Int. J. Numer. Anal. Model. **2**(2005), pp. 163-178. Cited on page(s) 225, 230

311. B. VAN LEER AND S. NOMURA, *Discontinuous Galerkin for diffusion*, AIAA paper 2005-5108. Cited on page(s) 260

312. J.J. W. VAN DER VEGT AND H. VAN DER VEN, *Space-time discontinuous Galerkin finite element method with dynamic grid motion for inviscid compressible flows. I. General formulation*, J. Comput. Phys. **182**(2002), pp. 546-585. Cited on page(s) 112

313. H. VANDEVEN, *Family of spectral filters for discontinuous problems*, J. Sci. Comput. **8**(1991), pp. 159-192. Cited on page(s) 143

314. J.L. VOLAKIS, A. CHATTERJEE, AND L. KEMPEL, *Finite Element Methods for Electromagnetics: Antennas, Microwave Circuits and Scattering Applications*, IEEE Press, New York, 1998. Cited on page(s) 332

315. S. WANDZURA AND H. XIAO, *Symmetric quadrature rules on a triangle*, Comput. Math. Applica. **45**(2003), pp. 1829-1840. Cited on page(s) 379

316. Z.J. WANG, *Spectral (finite) volume method for conservation laws on unstructured grids: Basic formulation*, J. Comput. Phys. **178**(2002), pp. 210-251. Cited on page(s) 13

317. Z.J. WANG AND Y. LIU, *Spectral (finite) volume method for conservation laws on unstructured grids II: Extension to two-dimensional scalar equation*, J. Comput. Phys. **179**(2002), pp. 665-697. Cited on page(s) 13

318. T. WARBURTON, *Application of the discontinuous Galerkin method to Maxwell's equations using unstructured polymorphic hp-finite elements*. In *Discontinuous Galerkin Methods: Theory, Computation and Applications*, B. Cockburn, G.E. Karniadakis, and C.W. Shu (Eds), Lecture Notes in Computational Science and Engineering **11**, Springer-Verlag, Berlin, 1999, pp. 451-458. Cited on page(s) 11, 348

319. T. WARBURTON AND M. EMBREE, *The role of the penalty in the local discontinuous Galerkin method for Maxwell's eigenvalue problem*, Comput. Methods Appl. Mech. Eng. **195**(2006), pp. 3205-3223. Cited on page(s) 12, 341

320. T. WARBURTON AND J.S. HESTHAVEN, *On the constants in hp-finite element trace inverse inequalities*, Comput. Methods Appl. Mech. Eng. **192**(2003), pp. 2765-2773. Cited on page(s) 94, 292

321. T. WARBURTON AND T. HAGSTROM, *Taming the CFL number for discontinuous Galerkin methods on structured meshes*, J. Comput. Phys. 2006 – submitted. Cited on page(s) 105, 105

322. T. WARBURTON AND G. KARNIADAKIS, *A discontinuous Galerkin method for the viscous MHD equations*, J. Comput. Phys. **152**(1999), pp. 608-641. Cited on page(s) 11

323. T. WARBURTON, I. LOMTEV, R.M. KIRBY, AND G.E. KARNIADAKIS, *A discontinuous Galerkin method for the compressible Navier-Stokes equations on hybrid grids*, Proc. Tenth International Conference on Finite Elements in Fluids, M. Hafez and J.C. Heirich (Eds.), 1998. p. 604. Cited on page(s) 315

324. F.M. WHITE, *Viscous Fluid Flow (2nd Ed)*, McGraw-Hill, New York, 1991. Cited on page(s) 330

325. P. WOODWARD AND P. COLELLA, *The numerical simulation of two-dimensional fluid flows with strong shocks*, J. Comput. Phys. **54**(1984), pp. 115-173. Cited on page(s) 233, 234

326. L. WU AND H.O. KREISS, *On the stability definition of difference approximations for the initial boundary value problem*, Appl. Numer. Math. **12**(1993), pp. 213-227. Cited on page(s) 95

327. Y. XING AND C.-W. SHU, *High order well-balanced finite volume WENO schemes and discontinuous Galerkin methods for a class of hyperbolic systems with source terms*, J. Comput. Phys. **214**(2006), pp. 567-598. Cited on page(s) 134

328. Y. XING AND C.-W. SHU, *A new approach of high order well-balanced finite volume WENO schemes and discontinuous Galerkin methods for a class of hyperbolic systems with source terms*, Comm. Comput. Phys. **1**(2006), pp. 101-135. Cited on page(s) 134

329. Y. XU AND C.-W. SHU, *Local discontinuous Galerkin methods for three classes of nonlinear wave equations*, J. Comput. Math. **22**(2004), pp. 250-274. Cited on page(s) 260

330. Y. XU AND C.-W. SHU, *Local discontinuous Galerkin methods for nonlinear Schrödinger equations*, J. Comput. Phys. **205**(2005), pp. 72-97. Cited on page(s) 260

331. Y. XU AND C.-W. SHU, *Local discontinuous Galerkin methods for the Kuramoto-Sivashinsky equations and the Ito-type coupled KdV equations*, Comput. Methods Appl. Mech. Eng. **195**(2006), pp. 3430-3447. Cited on page(s) 260

332. J. YAN AND C.-W. SHU, *A local discontinuous Galerkin method for KdV type equations*, SIAM J. Numer. Anal. **40**(2002), pp. 769-791. Cited on page(s) 11, 260

333. J. YAN AND C.-W. SHU, *Local discontinuous Galerkin methods for partial differential equations with higher order derivatives*, J. Sci. Comput. **17**(2002), pp. 27-47. Cited on page(s) 11, 260

334. L. YUAN AND C.-W. SHU, *Discontinuous Galerkin method based on non-polynomial approximation spaces*, J. Comput. Phys. **218**(2006), pp. 295-323. Cited on page(s) 11

335. M. ZHANG AND C.-W. SHU, *An analysis of three different formulations of the discontinuous Galerkin method for diffusion equations*, Math. Models Methods Appl. Sci. **13**(2003), pp. 395-413. Cited on page(s) 11, 244

336. M. ZHANG AND C.-W. SHU, *An analysis of and a comparison between the discontinuous Galerkin and the spectral finite volume methods*, Comput. Fluids **34**(2005), pp. 581-592. Cited on page(s) 13

337. Q. ZHANG AND C.-W. SHU, *Error estimates to smooth solutions of Runge-Kutta discontinuous Galerkin methods for scalar conservation laws*, SIAM J. Numer. Anal. **42**(2004), pp. 641-666. Cited on page(s) 135, 161, 238

338. Q. ZHANG AND C.-W. SHU, *Error estimates to smooth solutions of Runge-Kutta discontinuous Galerkin methods for symmetrizable systems of conservation laws*, SIAM J. Numer. Anal. **44**(2006), pp. 1703-1720. Cited on page(s) 136, 161, 238

339. O.C. ZIENKIEWICZ AND R.L. TAYLOR, *Finite Element Method: Volume 1, The Basics (5th Ed.)*, Butterworth-Heinemann, Boston, 2000. Cited on page(s) 6

340. O.C. ZIENKIEWICZ AND R.L. TAYLOR, *Finite Element Method: Volume 2, Solid Mechanics (5th Ed.)*, Butterworth-Heinemann, Boston, 2000. Cited on page(s) 6

341. O.C. ZIENKIEWICZ AND R.L. TAYLOR, *Finite Element Method: Volume 3, Fluid Dynamics (5th Ed.)*, Butterworth-Heinemann, Boston, 2000. Cited on page(s) 6

342. O.C. ZIENKIEWICZ, R. TAYLOR, S.J. SHERWIN, AND J. PEIRO, *On discontinuous Galerkin methods*, Int. J. Num. Methods Eng. **58**(2003), pp. 1119-1148. Cited on page(s) 11

Index

Texts in Applied Mathematics

(*continued from page ii*)